ULTRASOUND:
ITS APPLICATIONS IN MEDICINE AND BIOLOGY

METHODS AND PHENOMENA
THEIR APPLICATIONS IN SCIENCE AND TECHNOLOGY

Series Editors

S.P. Wolsky
Director of Research, P.R. Mallory and Co., Inc.,
Burlington, Mass., U.S.A.

and

A.W. Czanderna
Senior Scientist, Solar Energy Research Institute,
Golden, Colorado, U.S.A.

Volume 1. A.W. Czanderna (Ed.), Methods of Surface Analysis
Volume 2. H. Kressel (Ed.), Characterization of Epitaxial Semiconductor Films
Volume 3. F.J. Fry (Ed.), Ultrasound: Its Applications in Medicine and Biology

In preparation

S.P. Wolsky and A.W. Czanderna (Eds.), Microweighing in Vacuum and Controlled Environments, Part I: Beam Microbalances

K. Moorjani and J.M.D. Coey, Disordered Solids

A.N. Dey, Lithium Batteries

ULTRASOUND: ITS APPLICATIONS IN MEDICINE AND BIOLOGY

Edited by

FRANCIS J. FRY

Ultrasound Research Laboratories of the Indianapolis Centre for Advanced Research
and
Indiana University School of Medicine, Indianapolis, Ind., U.S.A.

VOLUME 3 of

Methods and Phenomena: Their Applications in Science and Technology

PART I

ELSEVIER SCIENTIFIC PUBLISHING COMPANY
Amsterdam — Oxford — New York 1978

ELSEVIER SCIENTIFIC PUBLISHING COMPANY
335 Jan van Galenstraat
P.O. Box 211, Amsterdam, The Netherlands

Distributors for the United States and Canada:

ELSEVIER NORTH-HOLLAND INC.
52, Vanderbilt Avenue
New York, N.Y. 10017

Library of Congress Cataloging in Publication Data

Main entry under title:

Ultrasound : its applications in medicine and biology.

(Methods and phenomena, their applications in science and technology ; v. 3)
Includes bibliographies and index.
1. Ultrasonics in medicine. 2. Ultrasonics in biology. 3. Ultrasonics--Physiological effect. I. Fry, Francis J. II. Series. [DNLM: 1. Ultrasonics. 2. Ultrasonic therapy. WB515 U49]
R857.U48U47 616.07'54 77-17618
ISBN 0-444-41641-2

ISBN 0-444-41641-2 (Vol. 3)
ISBN 0-444-41642-0 (Part I)
ISBN 0-444-41643-9 (Part II)
ISBN 0-444-41640-4 (Series)

Printed in The Netherlands

CONTRIBUTORS TO PART I

D. Baker	Center for Bioengineering, University of Washington, Seattle, Washington, U.S.A.
R. C. Chivers	Physics Department, University of Surrey, Guildford, Surrey, United Kingdom.
T. Coakley	Department of Microbiology, University College, Cardiff, Wales, United Kingdom.
R. E. Daigle	Varian Associates, Palo Alto, California, U.S.A.
F. Dunn	Department of Electrical Engineering, Bioacoustics Research Laboratory, University of Illinois, Urbana Illinois, U.S.A.
F. K. Forster	Center for Bioengineering, University of Washington, Seattle, Washington, U.S.A.
F. J. Fry	Ultrasound Research Laboratories of the Indianapolis Center for Advanced Research and Indiana University School of Medicine, Indianapolis, Indiana, U.S.A.
C. R. Hill	Physics Division Institute of Cancer Research, Royal Marsden Hospital, Sutton, Surrey, United Kingdom.
R. W. Huggins	Physics Division Institute of Cancer Research, Royal Marsden Hospital, Sutton, Surrey, United Kingdom.
J. P. Jones	School of Medicine, Case Western Reserve University, Cleveland, Ohio, U.S.A.
A. C. Kak	Department of Electrical Engineering, Purdue University, West Lafayette, Indiana, U.S.A.
Y. Kikuchi	Tohoku Institute of Technology, Nagamachi - Koeji 19, Sendai 982, Japan.
D. Nicholas	Physics Division Institute of Cancer Research, Royal Marsden Hospital, Sutton, Surrey, United Kingdom.
W. L. Nyborg	Department of Physics, The University of Vermont, Burlington, Vermont, U.S.A.
W. D. O'Brien, Jr.	Department of Electrical Engineering, Bioacoustics Research Laboratory, University of Illinois, Urbana, Illinois, U.S.A.

METHODS AND PHENOMENA

PREFACE

Study of the interaction of ultrasound with biological systems involving ultrasound waves generated by man made devices began in the 1920's. Since that time the field has expanded and now includes areas of biological and medical research and applications to both diagnostic and therapeutic medicine.

Contributions to this two part volume are well balanced between first and second generation investigators each contributing in a pioneering sense to a particular portion of the field.

Because of the time period involved in preparation, all of the material is not of exactly the same chronological age, but this does not alter its relevance. Omissions from an originally anticipated scope include non-pulse echo imaging methods and two-dimensional real time imaging systems.

The material is of sufficient depth to be used as a text book and reference work by bioscience and bioengineering students and professionals. A considerable portion of Part II is readily understandable by the interested medical and science investigator having essentially no mathematics and physics background.

Chapter I provides an analytical framework for investigators interested in the application of physical science principles of ultrasound to biology and medicine. The approach used is that characteristic of linearized acoustics using series expressions for the dependent variables. Solution of the linearized equations for first and second order quantities yields appropriate approximate results for the selected world of reality in a not too limiting sense.

The interaction of ultrasonic waves with biological media may well include the phenomena of cavitation which is given an excellently balanced treatment in Chapter II.

Doppler methods using ultrasound were early conceived for measuring dynamic biological systems in a quantitative manner. The vigor of this activity through its evolution and present state of development is captivated in the presentation of Chapter III.

All ultrasonic systems employ a transducer. Chapter IV presents material in this field by an acknowledged pioneer who has influenced both the analytical analysis of transducer systems and the practical implementation of these analytical concepts.

The degree of maturity of a field in regard to its medical applications can, in large measure, be judged by the attention given to dosimetry. Although this aspect of ultrasound in medicine has not moved forward as rapidly as one would have liked, it is apparent that the material in Chapter V will aid considerably in setting a perspective for this endeavor.

The properties of absorption and dispersion in biological fluids and tissue are presented in Chapter VI, adding a discussion on these ultrasonic properties which is crucial to the understanding of interactions in all biological and medical areas using ultrasound.

Chapter VII was one of the earliest chapters to be received and its contents and subsequent work by this group of authors has been a dominant factor in the utilization of scattering in tissue to aid in quantitatively defining various tissue types.

Another method for aiding in the identification of tissue is a quantitatively derived acoustic impedance profile. Chapter VIII presents an analytical basis for an approach to this type of profiling and experimental model studies are presented to test the validity of the methodology. Additional work is an obvious next step if this type of methodology is to be tested in the world of live tissue *in vivo*.

Part II begins with an exciting Chapter IX which introduces a selected group of interactions and discusses mechanisms which are non-thermal in nature.

Chapter X introduces the subject of ultrasound as a therapeutic modality covering medical application areas in which the intensity levels are below 5 watts per cm^2. The primary thrust of this material is in relation to heat produced by ultrasound in medically important body sites aiding in their healing. It may well be that ultrasound in the intensity ranges discussed in this chapter will find use in tissues previously considered either inaccessible or inappropriate for application (brain and heart).

Ultrasound has had its most extensive impact on medicine in its application to visualization and identification of soft tissue structures. Because of its relatively hazard-free nature and its unique soft tissue visualization capability, it has found wide application in the manner discussed in Chapter XI. This chapter provides a concise synthesis of the instrumentation requirements and systems considerations which have made the work of these authors an outstanding success.

The human female breast offers unique possibilities for the application of ultrasound visualization and tissue identification in the medical management of a variety of disorders associated with this structure. Chapter XII develops the rationale for ultrasound scanning in breast tissue and delineates the difficulties in their solutions as this modality adds its contribution to solving a difficult diagnostic problem.

Chapter XIII introduces a whole new world to observation through the use of acoustic waves operating at 100 MHz. This frequency is embodied in an arrangement which produces images with 20 micron resolution and is appropriately called acoustic microscopy. The methodology permits the quantitative determination of sonic velocity, acoustic attenuation and impedance distribution in the material under observation.

Intense focused ultrasound remains a technique which has utility in specialized areas of biological and medical research. It has potential for wide scale usage but the appropriate environment and interested investigators must be found to carry the methods forward. In studies involving tissue selective lesions matching the precise size and shape of neural structures in brain, it is a method which is unequaled. Use of this methodology has permitted the uncovering of a quantitatively represented organizational pattern in the mammalian brain which has not been previously revealed. Time alone will decide its significance in the world of structural organization in brain.

The technical content of the work is completed by two appendices. These appendices summarize aspects of cavitation effects on organized mammalian tissue and thermal mechanisms in ultrasound tissue interactions.

The high frequency mechanical wave world considered in this two part volume is in its infancy by almost any standard of measurement. Its growth will be determined by the imagination and ingenuity of the hopefully well funded individuals who care to pursue its beckoning call. This world does not have a natural counterpart in many respects, so that many avenues will lead to interesting interactions with living things essentially untouched by this high frequency siren's call.

I would like to thank my co-authors for their forebearance in this undertaking.

I want to thank Dr. Julia F. Herrick, Dr. Elizabeth Kelly Fry, Kenneth Johnston and Dr. Jerome I. Kaplan for comments on some of the technical material, and Mrs. Wyonetta Hoffman who performed the final typing of all material and provided other assistance in bringing this material to completion.

Indianapolis, Indiana

Francis J. Fry

CONTENTS

CONTENTS of Part II

CHAPTER I

PHYSICAL PRINCIPLES OF ULTRASOUND

Wesley L. Nyborg

1. INTRODUCTION

In this chapter brief treatments are presented of physical topics which are important to medical and biological ultrasound. It is assumed that the reader has some background in mathematics and physics (to about the intermediate or advanced undergraduate level) but may have had little or no formal training in acoustics.

Ultrasound is typically defined as sound whose frequency is too high for perception by the human ear. For definiteness the lower frequency limit for ultrasound is often taken rather arbitrarily as 20 kHz. Whatever the lower limit it is clear that the fundamentals of ultrasound are the same as the principles of physical acoustics which apply to sound generally. A vast literature exists on general physical acoustics, much of which is relevant here; in this chapter it has been necessary to be selective. Numerous volumes exist to which the reader may turn for comprehensive treatments.[1-7] In the more specific subject of ultrasound in medicine, valuable presentations of physical topics have been made by Fry and Dunn[8] and by Wells.[9]

Since this volume deals with applications of ultrasound, the emphasis in this chapter is on defining useful concepts and presenting "working equations," rather than on detailed justifications of them. Nevertheless, some background theory is included in order to indicate assumptions and approximations on which working equations are based. Hopefully, with such explanatory background the reader will see how results cited here are to be applied.

Most of the useful equations of acoustics are arrived at by obtaining solutions of a few governing differential equations, which are taken up in section 2. These basic differential equations are nonlinear and are seldom solved exactly. Instead they are usually simplified by introducing approximations which are valid at small acoustic amplitudes. A method of successive approximations is described in Section 3. In carrying out this method the first-order approximation (taken up in Section 4) gives the usual results of linearized acoustics, which are indispensable for an understanding of wave propagation. The second-order approximation (treated in Section 5) yields equations for radiation pressure, acoustic streaming, and other aspects of a sound field which apply to methods for measuring power and intensity, and which are involved in production of sonic bioeffects.

2. BASIC DIFFERENTIAL EQUATIONS

We shall not take the space here to "derive" the fundamental equations from elementary principles. For derivations and related discussions the reader may consult a variety of textbooks and monographs on general acoustics,[1-7] hydrodynamics[10-12] or continuum mechanics.[13,14] In this brief treatment we present the equations, indicate briefly their physical significance, and define the symbols in terms of which they are expressed.

2.1 The Continuity Equation

When a statement that "mass is conserved" is applied to an arbitrary element of the medium we arrive at the first of our basic differential equations. In vector notation this is

$$\nabla\cdot\rho\underline{u} = -\frac{\partial\rho}{\partial t} \tag{1a}$$

where ρ is the density of the medium (mass/volume) and $\underline{u}$ is the velocity. Considering $\underline{u}$ to have components (u, v, w) along the x, y, z directions, resp., Eq. (1a) can be written

$$\frac{\partial(\rho u)}{\partial x} + \frac{\partial(\rho v)}{\partial y} + \frac{\partial(\rho w)}{\partial z} = -\frac{\partial\rho}{\partial t} \tag{1b}$$

Either of Eqs. (1) is frequently called a continuity equation.

2.2 The Dynamical Equation

Any small portion, or element, of a medium is, in general, acted on by forces which can be classified as either (i) contact forces or (ii) forces acting at a distance. In the first category are forces arising from stresses exerted on the element by the medium immediately external to it. These stresses arise ultimately from interactions between the molecules just inside the surface which envelopes the element, and the molecules just outside this surface. For an ideal solid these stresses will be elastic, corresponding to tensile, compressional or shearing deformations. For a Newtonian liquid, and for viscoelastic media, the stresses will be partly elastic and partly viscous.

The forces acting at a distance will usually arise from gravitational fields in situations considered here, but under some conditions they may arise from applied electric and magnetic fields. We let the vector $\underline{f}$ represent the net force of the first kind (contact force) acting on an element per unit volume, and $\underline{f}^*$ similarly the net force of the second kind (force at a distance). Then from the laws of dynamics it is found that

$$\underline{f} + \underline{f}^* = \rho\, D\underline{u}/Dt; \tag{2}$$

here the quantity $D\underline{u}/Dt$ gives the acceleration of the element, and is variously called the <u>hydrodynamical derivative</u>, the <u>material derivative</u> or the <u>total derivative</u>. The operator D/Dt can be written in x, y, z coordinates as

$$\frac{D}{Dt} = \frac{\partial}{\partial t} + u\frac{\partial}{\partial x} + v\frac{\partial}{\partial y} + w\frac{\partial}{\partial z} = \frac{\partial}{\partial t} + \underline{u}\cdot\nabla\ . \tag{3}$$

If the gravitational acceleration is $\underline{g}$ and if $\underline{f}^*$ arises entirely from gravity we realize that $\underline{f}^*$ is just $\rho\underline{g}$. Let the stress components relative to (x,y,z) axes be S_{xy}, etc.; for example, S_{xy} represents the component of the force in the x direction on unit area of a plane whose normal is along y. The sense of S_{xy} is positive if the medium in the +y direction from the plane exerts stress (across the plane) in the +x direction. Assume $\underline{g}$ is along the +z direction. Taking x-components of all terms in Eq. (2) we then obtain

$$\frac{\partial S_{xx}}{\partial x} + \frac{\partial S_{xy}}{\partial y} + \frac{\partial S_{xz}}{\partial z} = \rho\,\frac{Du}{Dt} = \rho\left[\frac{\partial u}{\partial t} + (\underline{u}\cdot\nabla)u\right]\ . \tag{4}$$

Corresponding equations for other axes are obtained by taking components along y, or components along z; the "z" equation will include a term ρg on the left hand side. It is sometimes convenient to use numbered axes (x_1, x_2, x_3) to represent (x,y,z) and the velocity components (u_1, u_2, u_3) for (u,v,w). Correspondingly the stress component S_{12} then represents S_{xy}, etc. In this notation equations such as Eq. (4) are written as

$$\sum_{j=1}^{3} \frac{\partial S_{ij}}{\partial x_j} = \rho \left[\frac{\partial u_i}{\partial t} + \sum_{j=1}^{3} u_j \frac{\partial u_1}{\partial x_j} \right] \tag{5}$$

where i may be either 1 or 2; when i = 1 this equation is identical to Eq. (4). When i = 3 the term ρg should be added on the left.

2.3 The Constitutive Equations

To set up a sound field we require one or more sources, and a material medium with elastic properties. For given sound sources the nature of the field these generate depends on the elastic and viscous characteristics of the medium, and also on its density and other properties. We assume that coefficients of elasticity and viscosity can be defined, in terms of which a constitutive equation can be written: this equation describes changes in size or shape of a volume element of the medium when stresses are applied. Before citing specific equations we first define displacement and strain quantities. Let ξ_i be the displacement in the x_i direction at a given point in a given state of strain. Then the strain component ε_{ij} is defined by

$$\varepsilon_{ij} = \tfrac{1}{2} \left(\frac{\partial \xi_i}{\partial x_j} + \frac{\partial \xi_j}{\partial x_i} \right) \tag{6}$$

Here a (diagonal) component with i = j represents longitudinal strain while a component for which $i \neq j$ represents shearing strain. The corresponding rate-of-strain component $\dot{\varepsilon}_{ij}$ is just $\partial \varepsilon_{ij}/\partial t$. It can be shown that in a rotation of coordinates the stress quantities S_{ij}, and also the strain quantities ε_{ij}, transform as components of Cartesian tensors of second rank. In terms of the stress and strain tensors we now write examples of constitutive equations for several media. For a linear isotropic solid the equation can be written

$$S_{ij} = \lambda\theta\delta_{ij} + 2\mu\varepsilon_{ij} \quad \text{(Linear isotropic solid)} \tag{7}$$

Here λ and μ are elastic constants and θ is the dilatation or "fractional increase in volume", given by the sum of diagonal strain components $(\varepsilon_{11} + \varepsilon_{22} + \varepsilon_{33})$; the Kronecker delta δ_{ij} is equal to 1 for i = j and zero otherwise. Since i and j can take on the values (1,2,3) Eq. (7) would seem to represent 9 separate equations; however, since the tensors S_{ij} and ε_{ij} are each symmetric the number reduces to six:

$$\begin{aligned} S_{11} &= \lambda\theta + 2\mu\varepsilon_{11} \\ S_{22} &= \lambda\theta + 2\mu\varepsilon_{22} \\ S_{33} &= \lambda\theta + 2\mu\varepsilon_{33} \\ S_{12} &= \mu\varepsilon_{12}; \quad S_{13} = \mu\varepsilon_{13}; \quad S_{23} = \mu\varepsilon_{23} . \end{aligned} \tag{8}$$

The constant μ is known as either the <u>modulus of shear elasticity</u> or the <u>rigidity modulus</u>. For the quantity λ it can be shown that

$$\lambda = B - (2/3)\mu , \tag{9}$$

where B is the bulk modulus of elasticity. The <u>compressibility</u> β is defined as the reciprocal of B. Expressions can be derived for the Young's modulus E and the Poisson's ratio ν in terms of B and μ;

$$E = \frac{9B\mu}{3B + \mu} ; \quad 2\nu = \frac{3B - 2\mu}{3B + \mu} . \tag{10}$$

For water and many soft tissues at ordinary temperatures B is approximately 2.2 in units of 10^{10} dyn/cm^2. In the same units E is about 1.2 for collagen, and for bone (bovine phalanx and femur) E and μ are, resp., about (22, 5.4) along the axis and (11.5, 4.4) in the transverse direction.[15] When $\mu \ll B$, as is true for soft tissues, we find from Eqs. (9) and (10) that $\lambda \simeq B$, $E \simeq 3\mu$ and $\nu \simeq 0.5$. More detailed information on properties of tissues may be found in other chapters of this volume.

For a motionless or inviscid fluid the constitutive equation is the same as in Eq. (7) except that μ is equal to zero; hence $\lambda = B$. For a viscous Newtonian fluid the constitutive equation is

$$S_{ij} = (B\theta + \lambda'\dot{\theta})\delta_{ij} + 2\eta\dot{\varepsilon}_{ij}. \quad \text{(Newtonian fluid)} \tag{11}$$

Here, as before, the dot indicates the time derivative $\partial/\partial t$; the constant η is the coefficient of shear viscosity and is tabulated for various fluids in handbooks. The second viscosity coefficient λ' is related to the coefficient of bulk viscosity B' by an equation analogous to Eq. (9); i.e.,

$$\lambda' = B' - (2\eta/3). \tag{12}$$

For sound waves of various frequencies it is usually found that λ' and B' are not constant but vary with the frequency. As we shall see later, the bulk-viscosity coefficient B' appears in expressions for the absorption coefficient α of a wave passing through Newtonian fluid; because of this, measurements of α are used to determine B'. In water B' is about equal to $2.8\ \eta$ at 20°C.

For a linear isotropic medium which exhibits shear elasticity as well as viscous behaviour a possible constitutive equation is

$$S_{ij} = (\lambda\theta + \lambda'\dot{\theta})\delta_{ij} + 2\mu\varepsilon_{ij} + 2\eta\dot{\varepsilon}_{ij}, \tag{13}$$

where λ, λ' are related to other constants by Eqs. (9) and (12). A substance to which Eq. (13) applies is sometimes referred to as a "Voigt medium". We see that Eq. (13) reduces to the constitutive equation for an isotropic solid (Eq. (7)) when λ' and η' are zero, and to that for a Newtonian liquid (Eq. (11)) when $\mu = 0$. Oestreicher[16] used Eq. (13) for a constitutive equation in his analysis of sound fields in tissue.

One sometimes defines a "mean pressure" $\bar{p}$ such that $-\bar{p}$ is the average of the diagonal stress components; thus

$$\bar{p} = -(S_{11} + S_{22} + S_{33})/3.$$

Substituting from Eq. (13) into this equation we find that $\bar{p}$ is simply related to the dilatation θ:

$$\bar{p} = -B\theta - B'\dot{\theta} . \tag{14}$$

In water-like media at ordinary frequencies the viscous term $B'\dot{\theta}$ in Eq. (14) is small compared to the elastic term $B\theta$. Specifically, if the sound is of angular frequency ω the viscous term is negligible when $\omega B' \ll B$. If also $\omega\eta \ll B$ the diagonal stress components S_{11}, S_{22}, S_{33} are essentially equal; there is then little distinction between $\bar{p}$ and the ordinary pressure p, the latter being equal to the negative of (for example) S_{11}.

All constitutive equations presented as examples so far consist of linear relations between components of stress, strain and rate-of-strain. Other relations have been given which include rate-of-stress components, and also higher-order time derivatives of both stress and strain.

Linear relations are valid for real media only at small values of the stress and strain quantities. At higher values the constitutive equations are nonlinear and the theory becomes difficult. We shall not attempt a systematic treatment of nonlinear constitutive equations, but instead discuss an approximation which is useful for viscous fluids. In this approximation we recognize that the stress components S_{ij} are nonlinearly related to the strain quantities ε_{ij}, but assume linear relations are adequate between the S_{ij} and the rate-of-strain quantities $\dot{\varepsilon}_{ij}$. With this in mind we have, instead of Eq. (11),

$$S_{ij} = [-p_e(\theta) + \lambda'\dot{\theta}]\ \delta_{ij} + 2\eta\dot{\varepsilon}_{ij} . \tag{15}$$

Here the viscosity coefficients λ' and η are assumed constant during a sonic period, while p_e is a nonlinear function of the dilatation θ. Magnitudes of the viscous contributions to S_{ij} (given by $|\lambda'\dot{\theta}|$ and $|2\eta\dot{\varepsilon}_{ij}|$) are typically much smaller than $|p_e|$; it is for this reason that linear forms are assumed for the viscous terms but not for the elastic term p_e. For the average of the diagonal stress components ($-\bar{p}$) we have, instead of Eq. (14),

$$\bar{p} = p_e - B'\dot{\theta} \tag{16}$$

We interpret p_e as the elastic contribution to the mean pressure $\bar{p}$ while $-B'\dot{\theta}$ is the viscous contribution; again, p_e has typically much the larger magnitude of the two. Information on the θ-dependence of p_e is available from pressure-volume or pressure-density data for various fluids. Acoustic data are usually presented in terms of coefficients in a power series defined by Fox and Wallace.[17] Regarding p_e as a function of the density difference $(\rho-\rho_o)$ this series is

$$p_e = P_o + A^*(\rho-\rho_o)/\rho_o + B^*\ (\rho-\rho_o)^2/2\rho_o^2 + \ldots\ , \tag{17}$$

where P_o is a reference pressure while ρ_o is the fluid density when $p = P_o$ and the constants A^*, B^* are given by

$$A^* = \rho_o\ (\partial p/\partial \rho)_{\rho=\rho_o} = \rho_o c_o^{\ 2} = B \tag{18}$$

$$B^* = \rho_o^2\ (\partial^2 p/\partial \rho^2)_{\rho=\rho_o}\ .$$

The ratio B^*/A^*, a measure of nonlinearity, has been determined for a number of fluids; results have been tabulated by Beyer.[18] When the ratio $(\rho-\rho_o)/\rho_o$ is small compared to unity it is approximately equal to $-\theta$ and Eq. (17) can be rewritten as a power series in θ. Neglecting terms of order θ^2 and higher, and substituting the resulting expression for p_e into Eq. (16), the latter reduces to Eq. (14) when $P_o = 0$.

2.4 Combined Equations

Using a suitable constitutive equation the stresses can be eliminated from the dynamical equation, Eq. (5). For a nonlinear viscous liquid to which Eq. (15) applies, the combined equation (neglecting gravity) can be written in vector form as

$$-\nabla p_e + [B' + (4/3)\eta]\ \nabla\dot{\theta} - \eta\nabla\times\nabla\times\underline{u} = \underline{f}\ , \tag{19a}$$

where

$$\underline{f} = \rho\ [\partial\underline{u}/\partial t + (\underline{u}\cdot\nabla)\underline{u}]\ . \tag{19b}$$

Here the velocity $\underline{u}$ has components u_i and is identical to $\dot{\underline{\xi}}$, where $\underline{\xi}$

is the displacement vector whose components ξ_i are referred to in Eq. (6); we realize then that $\dot{\theta}$ is identical to $\nabla \cdot \underline{u}$. The vector $\underline{f}$ is sometimes called the volume force and gives the force per unit volume on a fluid element exerted by stresses. A somewhat different form of Eq. (19a) can be obtained, when desired, by using the vector identity.

$$\nabla x \nabla x \underline{u} = \nabla \nabla \cdot \underline{u} - \nabla^2 \underline{u} \ . \tag{20}$$

It will be convenient for later use to combine the continuity equation, (Eq. (1)), with the expression for $\underline{f}$ in Eq. (19b) and obtain

$$\underline{f} = \partial(\rho \underline{u})/\partial t - \underline{F}' \ , \tag{21}$$

where

$$-\underline{F}' = \rho(\underline{u} \cdot \nabla)\underline{u} + \underline{u}\ (\nabla \cdot \rho \underline{u}) \ . \tag{22}$$

3. SUCCESSIVE APPROXIMATIONS TO THE BASIC EQUATIONS

3.1 Series Expressions for the Dependent Variables

For the nonlinear differential equations presented in Section 2 exact solutions are found only rarely. As a practical matter approximations are nearly always introduced, the results usually being valid when acoustic displacements are small. Systematic methods for obtaining approximate solutions have been introduced by Rayleigh,[1] Eckart[19] and others. In this section, Section 3, we describe an approach which can be called a "method of successive approximations." For definiteness we suppose the sound field is generated by one or more sources, all of the same angular frequency ω, with phase characterized by an angle α_r, and with displacement amplitude characterized by a nondimensional quantity ξ_r. Here ξ_r is the ratio of a characteristic source amplitude to a length characteristic of the sound field (such as the compressional wavelength λ). Then (for a given choice of the phase α_r) the pressure p at any position $\underline{r}$ at time t is a function of the source amplitude, and can be expressed as a power series in ξ_r:

$$p = P_o + a_1\xi_r + a_2\xi_r^2 + \ldots\ldots \ . \tag{23}$$

Here P_o is the pressure (assumed constant) when $\xi_r = 0$ while a_1 and a_2 are functions of $\underline{r}$ and t, but independent of ξ_r. It is convenient to rewrite Eq. (23) in the form

$$p = P_o + p_1 + p_2 + \dots \quad , \tag{24}$$

where p_1 is $a_1\xi_r$ and p_2 is $a_2\xi_r^2$, etc. In Eq. (24) the zero-order contribution P_o gives the pressure in the absence of sound; the first-order contribution p_1 is found to vary sinusoidally in time with the frequency ω of the source(s), its amplitude being proportional to ξ_r; the second-order contribution p_2 proves to have both a time-independent part, whose magnitude is proportional to ξ_r^2, and a second-harmonic component (with angular frequency 2ω) whose amplitude is proportional to ξ_r^2. When ξ_r is small it is clear that typically $|p_1| << P_o$ and $|p_2| << |p_1|$. Similar expressions can be written for the density ρ, the dilatation θ and the velocity $\underline{u}$, namely,

$$\begin{aligned} \rho &= \rho_o + \rho_1 + \rho_2 + \dots \\ \theta &= \theta_o + \theta_1 + \theta_2 + \dots \\ \underline{u} &= \underline{u}_1 + \underline{u}_2 + \dots \quad , \end{aligned} \tag{25}$$

where the subscripts identify contributions of various order, as for the pressure in Eq. (24). The zero-order contribution to $\underline{u}$ does not appear since we assume the fluid is quiescent in the absence of sound.

3.2 First-order Approximation

We now proceed by substituting series expressions from Eqs. (24) and (25) into some of the equations presented in Section 2. For the equation of continuity, Eq. (1), we obtain

$$\nabla\cdot[(\rho_o+\rho_1 + \rho_2 + \dots)\,(\underline{u}_1 + \underline{u}_2 + \dots)] = -\frac{\partial}{\partial t}(\rho_o + \rho_1 + \rho_2 + \dots) \tag{26}$$

when multiplications have been carried out on the left hand side of this equation we have terms of the type $\nabla\cdot(\rho_i\underline{u}_j)$, where i and j take on the values 0, 1, 2, etc. It follows from the discussion in Section 3.1 that such a term is proportional to ξ_r^{i+j}; stated otherwise, the term is "of order (i+j) in ξ_r." We see that (because u_o is assumed zero) there are no terms of order zero on the left hand side of Eq. (26); on the right hand side the sole zero-order term is $-\partial\rho_o/\partial t$ which must then be zero. Hence ρ_o is time-independent under the conditions assumed. Keeping first-order terms on both sides of Eq. (26) we obtain, assuming that ρ_o is constant in space as well as time,

$$\rho_o\dot{\theta}_1 = \rho_o\nabla\cdot\underline{u}_1 = -\partial\rho_1/\partial t \; ; \tag{27a}$$

this is the first-order approximation to the continuity equation. Integration with respect to time yields

$$\rho_1 = -\rho_o\theta_1 \; , \tag{27b}$$

since the integration constant must be zero. Considering now the constitutive equations taken up in Section 2.3, we recall that most of these are linear equations; they are valid only as relations between first-order contributions to the stress and strain quantities. Exceptions are Eqs. (15) through (18) which apply to viscous fluids. We substitute into Eq. (17) a series expression analogous to Eq. (24) for p_e and one from Eqs. (25) for ρ. Then, equating first-order terms, we obtain

$$p_{e1} = A^*\rho_1/\rho_o = B\rho_1/\rho_o = -\,B\theta_1, \tag{28}$$

where, as before, B is the bulk modulus of elasticity. With Eqs. (28) in mind it is readily seen that in the first-order approximation Eq. (16) becomes

$$\bar{p}_1 = -\,B\theta_1 - B'\dot{\theta}_1 \; ; \tag{29}$$

this is similar in form to Eq. (14), emphasizing the fact that the latter equation is valid only to first order. In some sound fields where viscous effects are important the diagonal stress components S_{11}, S_{22} and S_{33} are nevertheless essentially equal; this is especially likely if $B' \gg \eta$. The concept of a "mean pressure" is then unnecessary and we can replace $\bar{p}_1$ by p_1, the latter being equal to the negative of <u>any</u> diagonal stress component.

We now turn to Eqs. (19), which give the dynamical equation for a

Newtonian fluid; we assume $\underline{f}$ is given by Eqs. (21) and (22). Suppose series expressions are substituted for p_e, θ, $\underline{u}$ and ρ. The only zero-order term (since u_o is assumed zero) is on the left hand side of Eq. (19a); it is ∇P_o, which must then be zero. Thus P_o is constant in space, under the assumed conditions. Keeping first-order terms in Eqs. (19) we obtain

$$B\nabla\theta_1 + [B' + (4/3)\eta]\ \nabla\dot{\theta}_1 - \eta\nabla x\nabla x\underline{u}_1 = \rho_o\ \partial\underline{u}_1/\partial t, \tag{30}$$

where $\dot{\theta}_1 = \nabla\cdot\underline{u}_1$.

3.3 Second-order Approximation

Keeping terms up to second-order in Eq. (1), and noting cancellations of first-order terms because of Eq. (27), we obtain the second-order approximation to the continuity equation:

$$\rho_o\ \nabla\cdot\underline{u}_2 + \nabla\cdot(\rho_1\underline{u}_1) = -\partial\rho_2/\partial t\ . \tag{31}$$

From Eq. (17) we obtain a second-order approximation to the elastic part of the pressure:

$$p_{e2} = A^*\rho_2/\rho_o + B^*\ \rho_1^{\ 2}/2\rho_o^{\ 2}\ . \tag{32}$$

There is special interest in the second-order approximation to Eq. (19), with $\underline{f}$ given by Eqs. (21) and (22); letting $\nabla\cdot\underline{u}_2 = \dot{\theta}_2$ we obtain

$$-\nabla p_{e2} + [B' + (4/3)\eta]\ \nabla\dot{\theta}_2 - \eta\nabla x\nabla x\underline{u}_2 = \partial/\partial t\ (\rho_o\underline{u}_2 + \rho_1\underline{u}_1) - \underline{F}'_2\ , \tag{33}$$

where $\underline{F}'_2$ is the second-order approximation to $\underline{F}'$. For problems involving radiation pressure, acoustic streaming and related topics we are concerned with time-independent contributions to the dependent variables. Time-averaging all terms in Eq. (33) over a single cycle (or over many cycles) and letting $\langle X\rangle$ be the result of time-averaging for any quantity X we obtain

$$\nabla\ \langle p_{e2}\rangle + \eta\nabla x\nabla x\langle\underline{u}_2\rangle = \underline{F}\ , \tag{34a}$$

where $\underline{F}$ is the time-averaged value of $\underline{F}'_2$ and is given by

$$\underline{F} = -\rho_o\ \langle(\underline{u}_1\ \cdot\nabla)\ \underline{u}_1 + \underline{u}_1\ (\nabla\cdot\underline{u}_1)\rangle\ . \tag{34b}$$

In obtaining Eqs. (34) use has been made of the fact that in a steady state sound field, as is assumed here, it must be true for an observable quantity Y that $\langle \partial Y/\partial t \rangle$ is equal to zero. Tbis result is applied twice: once for $Y = \theta_2$, and again for Y equal to the quantity $(\rho_o \underline{u}_2 + \rho_1 \underline{u}_1)$. The quantity $\underline{F}$ has the units of force/volume; it is noted in Section 5.5 that Eq. (34a) is analogous to an equation used in hydrodynamics for viscous flow generated by a force field $\underline{F}$.

4. RESULTS: FIRST-ORDER SOLUTIONS

4.1 Introduction

In this section, Section 4, we present expressions which are descriptive of sound fields in a variety of situations of interest. They are obtained from solutions of the approximate equations discussed in Section 3.2, which are valid to first order in the nondimensional reference amplitude ξ_r. The expressions are the results of "linear acoustics" or "first-order acoustics," many of which are developed in standard textbooks. Our treatment of each topic will be relatively brief.

We now simplify our notation somewhat. To be consistent with Section 3 the first-order contributions to the pressure, velocity and other quantities should be designated as p_1, $\underline{u}_1$, etc. However, this is cumbersome and the practice in Section 4 will be to drop the use of subscript "1" to denote first-order quantities. Instead we follow a rather common convention of using the symbols p, $\underline{u}$ and θ to represent first-order contributions to the pressure, velocity and dilatation; however, for the corresponding contribution to the density we use $\delta\rho$. As before, the subscript "zero" denotes zero-order quantities, as in p_o , ρ_o , etc. We now rewrite Eqs. (27), (29) and (30) in the new notation:

$$\dot{\theta} = \nabla \cdot \underline{u} = - \frac{1}{\rho_o} \frac{\partial(\delta\rho)}{\partial t} \tag{35a}$$

$$\bar{p} = - B\theta - B'\dot{\theta} \; ; \tag{35b}$$

$$B \nabla\theta + [B' + (4/3)\eta] \nabla\dot{\theta} - \eta\nabla\times\nabla \times \underline{u} = \rho_o \partial\underline{u}/\partial t \; . \tag{35c}$$

Equations (35) are basic first-order equations for a Newtonian fluid. We obtain an equation in θ alone, a "wave equation", by operating on each term of Eq. (35c) with the divergence operator ($\nabla\cdot$) and replacing $\nabla\cdot\underline{u}$ by $\dot{\theta}$:

$$B\,\nabla^2\theta + [B' + (4/3)\eta]\,\nabla^2\dot{\theta} = \rho_o\ddot{\theta}\ . \tag{36}$$

It is by finding solutions of Eqs. (35) and (36) that most of the "standard" theoretical results of (linear) acoustics are obtained. One can proceed by first determining θ as a function of space and time, that is, by finding a suitable solution of Eq. (36). Then Eq. (35b) can be used to obtain the mean pressure $\bar{p}$, which often differs little from the pressure p. Equations (35a) and (35c) give relations between $\underline{u}$ and θ , of which solutions can be sought for the velocity $\underline{u}$.

4.2 Sinusoidal Time Dependence: Complex Notation

In some medical applications of ultrasound the generator is operated in a "continuous wave" or "CW" mode. In this mode of operation the first-order quantities p, θ, etc. which characterize the generated sound field at any given point all vary sinusoidally in time with frequency f, corresponding to an angular frequency $\omega = 2\pi f$. In other applications the ultrasound is generated in pulses, and the time-dependence of first-order field quantities is more complicated. However, it is a consequence of linear acoustics that results for pulses can be obtained by super position of results for CW fields, using theorems of Fourier and Laplace. It is therefore appropriate to give considerable emphasis to results for "single-frequency" or "monochromatic" fields in which the first-order quantities vary sinusoidally with time, being generated by one or more sources of angular frequency ω.

In describing such fields mathematically it is very convenient to use a method employing complex-number representations. Here each first-order dependent variable such as θ is represented by a complex number whose real part gives the physical quantity of interest. Validity of the method depends on the fact that operations carried out with these complex numbers yield the same results as if the real numbers had been used throughout.

The complex representation for θ, for example, is of the form $D\,e^{j(\omega t + d)}$, where D and d are time-independent but may be functions of space. A great convenience of the method comes from the fact that the complex representation of $\dot{\theta}$ is just $j\omega\theta$; thus the operation of differentiation with respect to time is replaced by the algebraic operation of multiplication by $j\omega$. For a single-frequency field Eqs. (35) become

$$j\omega\theta = \nabla\cdot\underline{u} = -\,j\omega\,\delta\rho/\rho_o\ ; \tag{37a}$$

$$\bar{p} = -\,(B + j\omega B')\,\theta\ ; \tag{37b}$$

$$\{B + j\omega[B' + (4/3)\eta]\}\,\nabla\theta - \eta\nabla x\nabla x\underline{u} = j\omega\rho_o\underline{u}\ . \tag{37c}$$

From Eq. (37b) we see that existence of bulk viscosity in a medium causes a phase shift between $\bar{p}$ and the negative dilatation $-\theta$. Since, by Eq. (37a), the quantity $-\theta$ is equal to $\delta\rho/\rho_o$ we see that bulk viscosity causes a phase difference between the first-order contributions to the mean pressure and the density, that is, between $\bar{p}$ and $\delta\rho$. In rewriting Eq. (36) it is convenient to cast it in the form

$$\nabla^2\theta + K^2\theta = 0, \tag{38a}$$

where K is a complex number such that

$$K^2 = \frac{\omega^2\rho_o}{B + j\omega[B' + (4/3)\eta]}\ . \tag{38b}$$

Here K is the complex propagation constant, which we rewrite in the form:

$$K = k - j\,\alpha\ , \tag{39}$$

where k and α are real quantities given by the following equations:

$$(k/k_o)^2 = \frac{(1+\varepsilon^2)^{1/2} + 1}{2(1+\varepsilon^2)} \tag{40a}$$

$$(\alpha/k_o)^2 = \frac{(1+\varepsilon^2)^{1/2} - 1}{2(1+\varepsilon^2)} \tag{40b}$$

$$k_o = \omega/c_o;\quad c_o = (B/\rho_o)^{1/2} \tag{40c}$$

$$\varepsilon = \omega[B' + (4/3)\eta]/B\ . \tag{40d}$$

For water at 20°C, η and B' have the values 1.01 cP and 2.82 cP, resp. At a frequency of 1 MHz the parameter ε is about 1.2×10^{-5}. Hence $\varepsilon \ll 1$ except at very high frequencies in media of very high viscosity; simplified expressions usually suffice for k and α:

$$k \cong k_o \; ; \; \alpha \cong \tfrac{1}{2} k_o \varepsilon \; . \tag{41}$$

4.3 Plane Longitudinal Travelling Waves (PTW's)

A plane wave of sound is characterized by two directions, which we define by the unit vectors $\underline{m}$ and $\underline{n}$. At any given instant the displacement varies only along the $\underline{m}$ direction, being the same everywhere in a plane whose normal is $\underline{m}$; the direction of the displacement is parallel (or antiparallel) to $\underline{n}$ everywhere, at all times. The vectors $\underline{m}$ and $\underline{n}$ are parallel for a longitudinal wave, and orthogonal for a shear wave. If $\underline{m}$ and $\underline{n}$ are each along the x direction Eq. (38a) becomes

$$\frac{\partial^2\theta}{\partial x^2} + K^2\theta = 0 \; . \tag{42}$$

A complex solution of this equation, remembering the time factor $e^{j\omega t}$, is

$$\theta = Ae^{j(\omega t - Kx + a)} \; , \tag{43}$$

where the amplitude A and phase factor $\underline{a}$ are real. By agreement the physical quantity, the dilatation in this instance, is given by the real part of the expression in Eq. (43); remembering that K is equal to $k - j\alpha$ we obtain

$$\mathrm{Re}(\theta) = Ae^{-\alpha x} \cos(\omega t - kx + a). \tag{44}$$

From Eq. (44) we see that the dilatation at any given distance x from the origin varies sinusoidally in time with angular frequency ω and with amplitude $Ae^{-\alpha x}$. Here α, whose negative is the imaginary part of the propagation constant, is the amplitude attenuation coefficient. The attenuation is, in general, the result of both absorption and scattering. Absorption is a process whereby mechanical energy is converted into heat, while scattering entails redirection of sonic energy to regions outside the primary field (the latter being usually an ultrasonic beam in medical applications). For a Newtonian fluid, to which Eq. (11) applies, the absorption coefficient is given by α in Eq. (41), with k_o and ε defined in Eq. (40). We see that α is proportional to $(B' + 4\eta/3)$, a linear combination of viscosity coefficients. Since B' and α are thus closely related, a method is presented which is the usual one for determining B': for a medium of interest α is measured and B' calculated using the relation contained in Eqs. (40) and (41).

For water α is proportional to ω^2 in the range of available measurements and, correspondingly, B' is independent of frequency (as is η). In other media the frequency-dependence of α departs from the "ω^2 relationship"; assuming Eq. (41) holds, B' evidently varies with frequency for these media. The subject of absorption in biological media is taken up in Chapter VI.

Equations (43) and (44) represent a wave travelling in the positive x direction; specifically, the maxima, minima and zeroes of the expression on the right hand side of Eq. (44) travel with velocity c equal to ω/k. The distance between consecutive maxima (or twice the distance between consecutive zeroes) is the wavelength λ, which is equal to c/f and is related to k, the real part of the propagation constant, by

$$\lambda = c/f = 2\pi/k. \qquad (45)$$

When, as is commonly true, $\varepsilon \ll 1$ the quantities c and k are essentially equal to c_o and k_o, resp.

Starting from the expression for θ in Eq. (43), analogous results for other parameters of the sound field can be obtained by using the first-order relations in Eqs. (37). Thus from Eq. (37a) the density contribution $\delta\rho$ is just equal to $-\rho_o\theta$. The mean pressure $\bar{p}$ is obtained nearly as simply from Eq. (37b): letting B* be a "complex coefficient of bulk elasticity" equal to $(B + j\omega B')$, we see that $\bar{p}$ is just $-B^*\theta$. When $\omega B'$ is negligible compared to B we see, of course, that B* reduces to B, the ordinary coefficient of bulk elasticity. To obtain the velocity $\underline{u}$, use is made of Eq. (37c); for plane wave fields the term involving $\nabla \times \underline{u}$ is zero. When effects of viscosity are negligible it is found that u is equal to $\bar{p}/\rho_o c_o$, where c_o is the velocity of sound given in Eq. (40c).

In most acoustics it is assumed that the diagonal stress components are essentially equal. When this is true we replace the symbol $\bar{p}$ (defined just previous to Eq. (14)) by the simpler one p; then each of the components S_{11}, S_{22}, S_{33} is approximately equal to - p. To test the validity of this approximation we compare S_{11} and S_{22} for the plane wave of Eq. (43), using the constitutive equation in Eq. (11). Remembering that here $\dot{\theta} = \dot{\varepsilon}_{11}$ we obtain

$$\frac{S_{11}}{S_{22}} = 1 + \frac{2j\omega\eta}{B + j\omega\lambda'} \quad . \tag{46}$$

We see from this equation that the magnitude of S_{11} is somewhat greater than that of S_{22} (the latter being equal to S_{33}). However, this difference is small except at very high frequencies in liquids of high viscosity. In water at 20°C the magnitude of the fraction on the right of Eq. (46) is 5.6×10^{-12} f, where f is the frequency in hertz. At 10 MHz in water the fractional excess of S_{11} over S_{22} is less than 1 part in 10^4. In this chapter we henceforth replace $\bar{p}$ by p.

For convenience we now list the first-order equations which apply to a plane wave, travelling in the positive x direction, when effects of viscosity and attenuation are negligible:

$$\begin{aligned} &\theta = A \cos (\omega t - kx) \\ &\delta\rho = -\rho_o\theta; p = - B\theta; \quad u = -c_o\theta \\ &p = \rho_o c_o u; \quad p = c_o^2\, \delta\rho \end{aligned} \tag{47}$$

The product $\rho_o c_o$ is called the characteristic impedance of the medium to which it applies. It is a significant quantity for a number of applications in acoustics. For water at $20^{o}C$ it has the value 1.48×10^5 in cgs units; then for a plane travelling wave with pressure amplitude 10^6 dyn/cm^2 (about 1 atm) we obtain from Eqs. (47) that the velocity amplitude is 6.8 cm/sec and (since B is 2.19×10^{10} dyn/cm^2) the dilatational amplitude is 4.6×10^{-5}. Since the dilatation θ represents the "fractional increase in volume" we see that for the wave considered the volume of any given mass element of fluid increases by a maximum of 4.6 parts in 10^5, and decreases by the same amount, during a sonic cycle. At the same time the density decreases and increases by about the same fraction. It is interesting to note that the magnitude of θ is equal to the magnitude of the "Mach number" u/c_o. Another important quantity for characterizing a plane wave is its intensity, but we defer discussion of this concept until Section 5.2.2.

In medical ultrasound, as in other applications of ultrasound, it is very common to use the plane-travelling-wave (PTW) relations discussed in this section, for analysis of the field. These relations, especially those in Eqs. (47), are relatively simple and lead to considerable insight. However, their accuracy is limited in several ways. First, the ultrasound may be pulsed, as is often true in diagnostic applications; the field is then a wave train of finite length. Special conditions prevail at the front and rear edges of the pulse; these "end-conditions" dominate the situation if the pulse is very short, with length of the order of several periods. Some special characteristics of pulsed fields are treated in Section 4.10.

Secondly, real fields differ from the PTW model because the transducers which generate the fields are finite in extent and, in general, do not execute a uniform vibratory pattern. At large distances from the source the field often has characteristics similar to those of the PTW in any small region, but the field varies from one region to another. At points near the source the field is especially complicated, and often has little resemblance to the PTW model. Theory used for treating the fields of finite sources is discussed in Section 4.8. If the source is small compared to the sonic wavelength the field may be that of a simple monopole; this field is treated in Section 4.7 and, especially in Chapter VII.

A third way in which real sound fields differ from the PTW model arises when boundaries are present which give rise to reflections. Some of these boundaries may be the surfaces of small suspended particles (such as biological cells, or intracellular bodies); when this is true the reflection process is often called scattering, a topic taken up in Section 4.9. Other boundaries may be of large size as, for example, are presented by the walls of a container, or the body surfaces of a mammal, or the air-water interface at the upper surface of water in a tank. Standing waves may then be set up; these are discussed in the next section, Section 4.4.

4.4 Plane Longitudinal Standing Wave

Since Eq. (42) is linear the sum of any two of its solutions is itself a solution. Consider then the solution

$$\theta = A_1 e^{j(\omega t - Kz + a_1)} + A_2 e^{j(\omega t + Kz + a_2)} , \qquad (48a)$$

of which the real part is

$$\theta = A_1 e^{-\alpha z} \cos(\omega t - kz + a_1) + A_2 e^{\alpha z} \cos(\omega t + kz + a_2) . \qquad (48b)$$

The expression for θ in either of Eqs. (48) is a solution of Eq. (36) and represents the sum of two PTW's, one of which travels in the positive z direction and the other in the negative z direction. The first of these we shall call the "original" PTW, and the second the "reflected" PTW. The total expression for θ is an appropriate solution to represent the ultrasonic field when a plane wave is projected from a transducer whose vibrating surface lies in a plane of constant z, and is reflected from an interface at another plane of constant z. For example, suppose that the transducer, submerged in a water tank, has its vibrating surface at z = 0 while the upper surface of the water is at z = H. Then if the tank is in air at atmospheric pressure, a solution which is suitable for most purposes is obtained by choosing the constants A_1, a_1, etc., such that $\theta=0$ at z=H; from Eq. (48b) we can then obtain, in real numbers, for the region $0 \leq z \leq H$:

$$\theta/2A = \sin \omega t \cosh \alpha(z-H) \sin k(z-H) - \cos \omega t \sinh \alpha(z-H) \cos k(z-H), \quad (49)$$

where A is the dilatation amplitude ($A_1 e^{-\alpha H}$ or $A_2 e^{\alpha H}$) of either PTW at z=H. It can be verified by direct substitution (keeping in mind Eqs. (40)) that the expression for θ in Eq. (49) is a solution of Eq. (36). If the attenuation is small so that $\alpha H << 1$ the expression for θ in Eq. (49) simplifies considerably. If also $\alpha << k$ we can use the relations in Eqs. (35) or (37) for the other first-order quantities, setting η and B' equal to zero. We obtain

$$\begin{aligned} &\theta = 2A \sin \omega t \sin k(z-H); \\ &\delta\rho = -\rho_o \theta; \quad p = -B\theta; \qquad (50) \\ &u = -2A\, c_o \cos \omega t \cos k(z-H), \end{aligned}$$

since $c_o = \omega/k$. A field such as is represented by Eqs. (50) is called a standing wave since, unlike the travelling wave, its zeroes and other characteristic points are fixed in space. When Eqs. (50) apply the first-order quantities θ, $\delta\rho$ and p are each zero when k(z-H) is equal to 0, π, 2π, etc., i.e., they are zero at the air-water interface and at a succession of planes whose distances (H-z) below the interface are integral multiples of the half-wavelength $\lambda/2$. Maxima of the amplitudes of θ, $\delta\rho$ and p occur midway between the planes at which zeroes occur; for each of these quantities, its maximum amplitude in the standing wave is equal to twice the corresponding amplitude in either of the constituent PTW's. For the velocity u the planes of zero and maximum amplitude are interchanged relative to those for θ, etc.; thus on planes of zero dilatation θ the velocity u is maximum, and on planes of maximum θ the velocity is zero. The maximum velocity amplitude is c_o times the maximum dilatational amplitude and $(\rho_o c_o)^{-1}$ times the maximum pressure amplitude.

The standing wave field represented in Eqs. (50) is relatively simple in nature, and provides a useful approximation to real ultrasonic fields under some circumstances. For Eqs. (50) to apply throughout the region $0 < z < H$ the attenuation must be negligible, that is, αH must be small compared to unity; the original and reflected PTW's are then of equal amplitude and uniform throughout the region. For information on the field when αH is not small we return to the more complicated expressions in Eqs. (48) or (49). When αH is large it is clear from Eqs. (48) that near the source transducer the dilatation θ is essentially that for the "original" PTW: since A_2 is equal to $A_1 e^{-2\alpha H}$ the amplitude of the reflected PTW is much less than that of the original near $z = 0$. From Eq. (49) we see that even when $\alpha H >> 1$ (if $\alpha << k$) a standing wave field similar to that described in Eqs. (50) applies near the air-water interface; here $\alpha(H - z)$ is small and the amplitudes of the two waves are nearly equal, though reduced in magnitude by a factor of $e^{-\alpha H}$ relative to the original PTW at $z = 0$. Primary features of the field are illustrated in Fig. 1; here the magnitude of θ, obtained from Eq. (49), is plotted <u>vs</u> z for different values of αH.

4.5 Reflection and Transmission at a Boundary

In Section 4.4 we discussed the situation where a PTW in water travels toward and away from an air boundary. The approximation used there, of zero dilatation at the boundary, amounts to an assumption that air is equivalent to vacuum. We now extend the topic and treat PTW's interacting at a plane boundary between two fluids of arbitrary density and compressibility. Here the sound field is usually assumed to consist of three PTW's: (1) an "incident" wave which is generated directly by the source transducer, (2) a "reflected" wave which also exists in the same medium as the source and (3) a "transmitted" wave which propagates in the second medium, across the boundary from the source. We assume that the boundary lies in the plane $z = 0$, that the source as well as the incident and reflected PTW's lie in

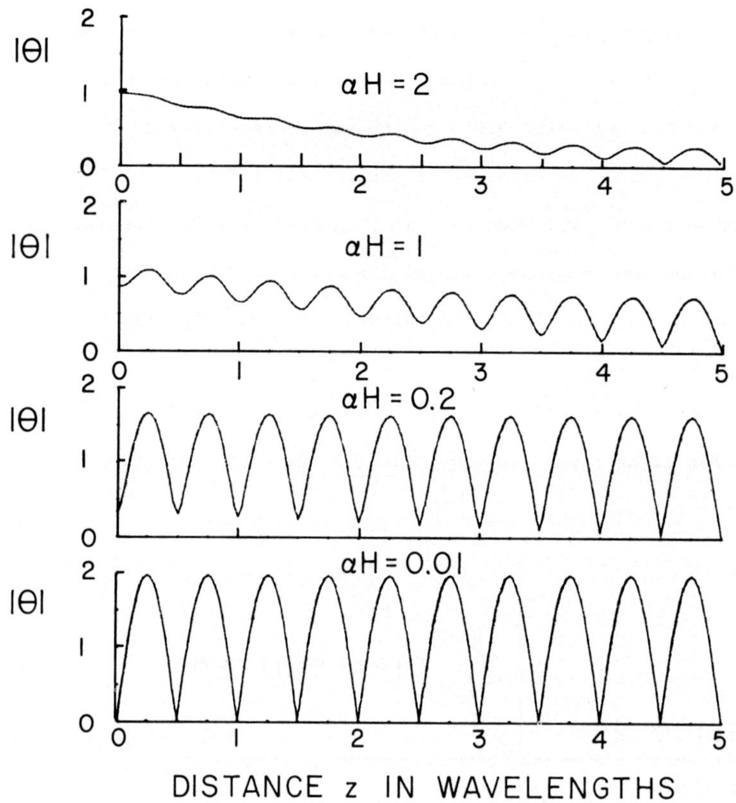

Fig. 1 Dilatation amplitude $|\theta|$ vs distance (in liquid) for reflected wave superimposed on original wave; plots based on Eqs. (49). Source is at $z = 0$; reflector (air-liquid interface) is at $z = 5$. Amplitude of original wave is unity at $z = 0$ in all plots; αH is attenuation (nepers) in source-reflector distance H.

the region of negative z, and that the transmitted wave propagates in the region of positive z. In the "source" medium the zero-order contribution to the density is ρ_o , the velocity of sound is c, and the elastic modulus is B; the corresponding quantities in the other medium are ρ_o', c' and B' .

Expressions for the field in either medium are required to satisfy Eqs. (35) and (36) with constants suitably chosen for that medium; however, effects of viscosity are neglected and thus the amplitude is uniform in each of the PTW's. In matching solutions at z=0 it is required that the first-order contribution to the pressure, and the normal component of velocity be continuous across this boundary; for example, the resultant pressure from the sum of the incident and reflected PTW's (which sum represents the solution for $z \leq 0$) must equal the pressure in the transmitted PTW (which represents the solution for $z \geq 0$), when each is evaluated at z=0. For a given amplitude and direction of propagation of the incident PTW, unique results are determined for the amplitude and direction of the reflected and transmitted PTW's.

For the directions the relationships are found to be according to rules enumciated by Snell for light rays (see Fig. 2): (i) Vectors along the

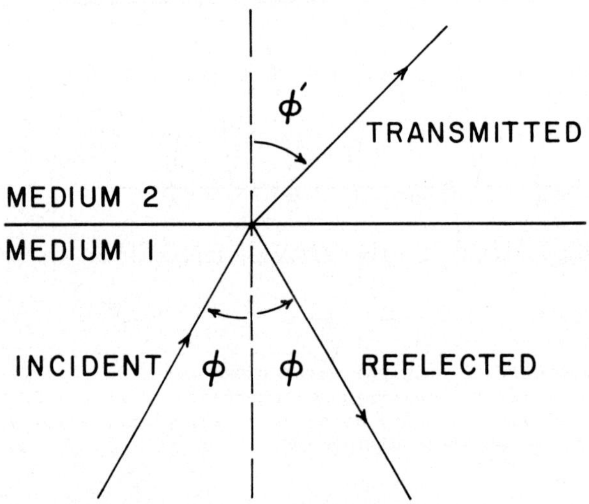

Fig. 2 Defining sketches for angles of incidence ϕ and of transmission ϕ'. The source is in Medium 1. The z direction is vertical here, and the x direction is to the right.

three propagation directions are coplanar with a vector normal to the boundary; (ii) The "angle of incidence" ϕ is equal to the "angle of reflection", as anticipated in Fig. 2; and (iii) The angle ϕ' between the normal and the propagation direction for the transmitted PTW is related to ϕ by the equation

$$\frac{\sin \phi'}{\sin \phi} = \frac{c'}{c} \quad . \tag{51}$$

We assume the dilatation in any of the PTW's is given by an expression of the form given in Eq. (43); however, K is replaced by the real constant k for Medium 1 and by k' for Medium 2; also x is replaced in Eq. (43) by a quantity which measures distance along the direction of propagation. Assuming the propagation directions lie in the xz plane the appropriate distance-quantity for the incident PTW is $x \sin \phi + z \cos \phi$; for the reflected wave ϕ is replaced by $(\pi - \phi)$, and for the transmitted wave ϕ is replaced by ϕ'. For example, the dilatation in the incident wave is given by

$$\theta = Ae^{j(\omega t - kx \sin \phi - kz \cos \phi + a)} . \tag{52}$$

For the reflected PTW we replace A and a, respectively, by A_r and a_r and for the transmitted PTW by A' and a' . It is found that the boundary conditions are satisfied if and only if $a = a_r$ and

$$\frac{A_r}{A} = \frac{m \cos \phi - (1 - C^2 \sin^2 \phi)^{\frac{1}{2}}}{m \cos \phi + (1 - C^2 \sin^2 \phi)^{\frac{1}{2}}} , \tag{53a}$$

where m and C are dimensionless constants defined by

$$m = \rho_o' c' / \rho_o c ; \quad C = c'/c . \tag{53b}$$

At normal incidence ϕ is zero and Eqs. (53) reduce to

$$\frac{A_r}{A} = \frac{m - 1}{m + 1} \tag{54}$$

From Eq. (54) we see that the amplitude ratio A_r/A is critically dependent on the impedance ratio m. If m is either very small or very large the reflected amplitude A_r is nearly equal in magnitude to the incident amplitude A. For example, if the source is in water and the second medium is air at standard conditions, the ratio m is equal to 0.00028 and A_r/A is equal to -0.9994, the minus sign indicating that the dilatation in the incident and reflected PTW's differ in phase by 180° at the boundary z=0. The resultant dilatation amplitude at z=0 is small compared to the incident amplitude, being equal to $A-A_r$ and hence only 0.0006 A; the approximation of "zero θ" at z=0, used in arriving at Eqs. (49) and (50) is thus justified in this situation.

A situation where m is very large is presented if the source is in air and the second medium is water. Then A_r/A would be 0.9994, a positive ratio, indicating that the incident and reflected dilatations are in phase at z=0. In this situation a standing wave is set up in the air. However, to be applicable here Eqs. (50) need to be somewhat modified: In this standing wave the dilatation amplitude $|\theta|$ is maximum at the boundary z=0, and at planes whose distance from the boundary is an integral number of half wavelengths; minima of $|\theta|$ occur at intermediate planes.

In general the ratio m is neither very large or very small and the magnitude $|A_r/A|$ is appreciably less than unity. In transmission of ultrasound across a boundary between tissues, or between tissue and an aqueous medium, the ratio m is often not very different from unity and the reflection ratio $|A_r/A|$ is relatively small. (See Chapter VIII). When A_r is less than A the field in the "source medium" is conveniently thought of as consisting of two superposed fields: (1) a PTW identical to the incident PTW but of reduced amplitude $(A - A_r)$ and (2) a standing wave field to which Eqs. (50) apply, except that A is replaced by A_r and ("phase-shifting") constants should be added to both the time t and the coordinate z.

It is common to refer to the concept of "intensity" (discussed in Section 5.2.2) in treating the reflection and transmission of PTW's at a boundary. Let I, I_r and I' be, respectively, the intensities of the incident, reflected and transmitted PTW's. Then the "intensity reflection coefficient" is defined as I_r/I; it proves to be equal to the square of the corresponding amplitude ratio, that is,

$$I_r/I = (A_r/A)^2 \,, \tag{55}$$

where A_r/A is given by either of Eqs. (53) or (54). The "intensity transmission coefficient" is defined similarly as I'/I. For normal incidence (i.e., when $\phi=0$) it is found that I' is equal to $I-I_r$ so that for this special condition we have

$$I'/I = 1 - (I_r/I). \tag{56}$$

However, at oblique incidence ($\phi\neq0$) this equation must be modified to take into account narrowing or widening of the transmitted PTW relative to the incident wave.

In Fig. 3 are plots of the intensity reflection coefficient I_r/I

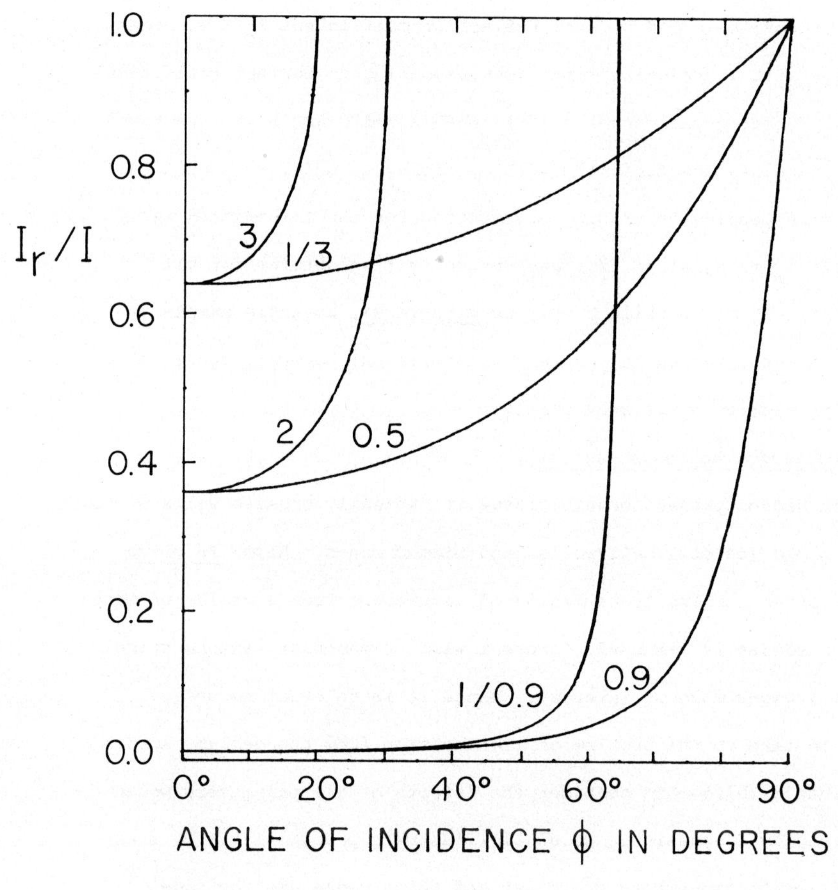

Fig. 3 Intensity reflection coefficient I_r/I vs angle of incidence ϕ, based on Eqs. (53) and (55). For each curve the impedance ratio m and velocity ratio c have the same value, the number indicated. E.g., for curve "3" both m and c are equal to 3.

based on Eqs. (53) and (55). These show I_r/I as a function of the angle of incidence ϕ for values of m and C in ranges relevant to biological media. Representative values from data given by Fry and Dunn[8] are: for transmission from water into fat, m = 0.93 and C = 0.96; from fat into muscle, m = 1.20 and C = 1.09; from brain tissue into skull bone m = 3.66 and C = 2.22; from skull bone into water m = 0.262 and C = 0.446. For transmission in reverse directions the reciprocals of these values for m and C would apply.

Several possibilities arise for the shape of a plot of I_r/I vs ϕ. (i) If m and C are both less than unity (i.e., when transmission is into the medium of smaller c and ρc) the reflection coefficient I_r/I increases monotonically with ϕ, reaching unity (corresponding to perfect reflection) when $\phi = \pi/2$ (grazing incidence). (ii) When m and C are both less than unity I_r/I increases monotonically with ϕ, rising to unity at a critical angle ϕ_c (corresponding to grazing orientation for the transmitted wave) such that $\sin \phi_c$ is equal to 1/C; perfect reflection obtains for all $\phi \geq \phi_c$. (iii) If m > 1 while C < 1, or *vice versa*, an angle exists at which I_r/I is zero, indicating perfect transmission; however, these latter possibilities seem to occur only rarely.

4.6 Transmission Across a Barrier

As ultrasound passes through tissue it typically crosses walls of body organs and blood vessels, cell walls, and biomembranes. Also, *in vitro* experiments often involve transmission of ultrasound from a medium in which a source transducer is immersed, across a wall or membrane, into a space occupied by a bio-sample of interest. Hence it is relevant to medical ultrasound to take up the problem of transmission from one medium, across a second, into a third. To consider the effects of all parameters would require a lengthy discussion. We shall illustrate possibilities by treating a special situation where the first and third media are the same and the boundaries are plane-parallel. Also we assume the fields consist of superposed PTW's which propagate normally to the boundaries.

We are thus led to consider transmission of ultrasound across a wall or barrier separating two spaces containing the same medium. Let the characteristic impedance of this medium be $\rho_o c$ and that of the wall material $\rho_o' c'$. Let the wall be of thickness ℓ, with boundaries at z=0 and z=ℓ. Assume the source is in the space of negative $\underline{z}$; in this same space will be incident and reflected PTW's, travelling in the positive and negative z direction, respectively, with amplitudes A and A^r, respectively. In the wall there will also be two PTW's, one directed along positive z and the other along negative z. Finally, in the space $z \geq \ell$ there will be a "transmitted" PTW of amplitude A', directed along positive z. Letting the intensity (see Section 5.2.2) of the incident and transmitted waves be I and I', respectively, the intensity transmission coefficient is defined as I'/I and is equal to $(A'/A)^2$. It is found that

$$I'/I = 4/[4 \cos^2 a_w + (m + m^{-1})^2 \sin^2 a_w] \quad , \tag{57}$$

where m is the ratio of the characteristic impedance of the wall and the external medium, while a_w gives the "phase path length" across the wall;

$$m = \frac{\rho_o' c'}{\rho_o c} \; ; \quad a_w = \frac{\omega \ell}{c'} = \frac{2\pi \ell}{\lambda'} \quad , \tag{58}$$

where λ' is the wavelength $(2\pi c'/\omega)$ in the wall. Plots of (I'/I) vs ℓ/λ are shown in Fig. 4, for several values of m. For any given value of m the plot is the same as for its reciprocal m^{-1}. For any m the ratio I'/I is unity, signifying "perfect transmission", when a_w is an integral multiple of π, that is, when the length ℓ is equal to a multiple of the half-wavelength $\lambda'/2$. Minima of I'/I, equal to $4(m+m^{-1})^{-2}$, occur when ℓ is equal to $\lambda'/4$ plus an integral multiple of $\lambda'/2$.

In biomedical applications the wall thickness ℓ is often so small compared to the wavelength λ' that $a_w \ll 1$. When this is true Eq. (57) can be written approximately as

$$I'/I = (1 + q^2)^{-1} \; ; \; q = (m + m^{-1})a_w/2 \; . \; (a_w \ll 1) \; . \tag{59}$$

The reflection coefficient I_r/I is just 1 - (I'/I). Equations (59) allow us to demonstrate the influence of even a small amount of air on the propa-

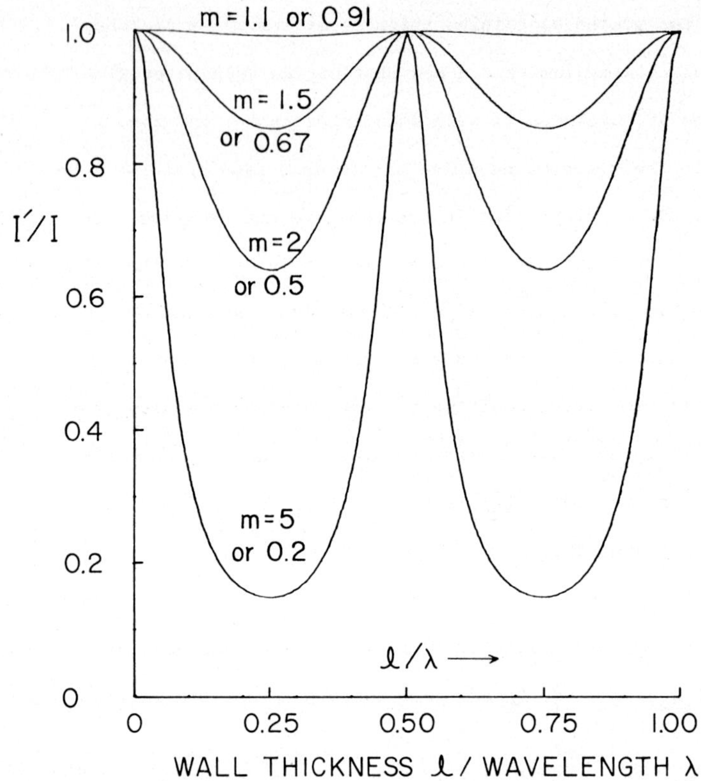

Fig. 4 Transmission coefficient I'/I for ultrasound passing through a wall of thickness ℓ, the wavelength being λ (in the wall). Curves are for different values of m, the ratio of characteristic impedance ("ρc") for the wall material and the medium outside. See Eqs. (57) and (58).

gation of ultrasound through water. Suppose the "wall" consists of a sheet of air whose thickness ℓ is only 1.0 μm. Here m is 0.00028 and at 1 MHz we find q = -33 and I'/I = 0.00092. That such a thin layer of air should reduce the transmitted intensity by more than a factor of 1000 is a result of the small value of m. For walls or membranes of biological organs, vessels or cells the ratio m is usually greater than unity, but not by a large factor. Hence for this application we can make the assumption that $q \ll 1$ and obtain, approximately

$$I'/I = 1 - q^2 \; ; \quad I_r/I = q^2. \quad (q \ll 1). \tag{60}$$

This equation also applies for thin sheets of plastic immersed in water as are often used in experiments. Consider such a sheet 10 μm thick, composed of a material whose density is 1.2 times that of water and in which the velocity of sound is 1.5 times that of water. Then m = 1.8 and, at 1 MHz, λ' = 2250 μm so q^2 becomes 0.00108 which is then equal to the intensity reflection coefficient.

If m is significantly greater than m^{-1} the quantity q becomes approximately equal to $\omega\rho_o{}'\ell/2\rho_o c$; here $\rho_o{}'\ell$ is the mass per unit area of the sheet.

4.7 Spherical Waves

In previous sections we have emphasized plane wave solutions. These give considerable insight on some aspects of ultrasonic fields, but are inadequate for others. Any real source-transducer is finite in size, of course, and its radiated field is complicated by interference and diffraction phenomena. In dealing with these and with other situations the concept of a "point source" is very useful. Such a source generates an outgoing spherical travelling wave (OSW) with spherical symmetry. An OSW is a reasonable approximation to the sound field of a very small pulsating source. In Chapter II it is applied to the field scattered by a gas bubble. However, its greatest importance arises from the use of superposed OSW's via Huyghen's principle (see Section 4.8) in mathematical description of diffraction fields.

In considering the OSW we find a solution of the wave equation, Eqs. (38); effects of viscosity are neglected and K is assumed equal to the real number k. In terms of a radial coordinate r, assuming spherical symmetry, the wave equation becomes

$$\partial^2\theta/\partial r^2 + 2r^{-1}\,\partial\theta/\partial r + k^2\theta = 0. \tag{61}$$

An appropriate solution for θ in complex notation is given below, together with expressions for other first-order quantities obtained by use of the relations in Eqs. (35) or (37) (with viscosities assumed zero):

$$\theta = \frac{A}{r} e^{j(\omega t - kr)}$$

$$\delta\rho = -\rho_o \theta \; ; \; p = - B\theta \; ; \tag{62}$$

$$u = -c_o \theta[1 - (j/kr)] \; .$$

Here u is the first-order velocity in the outward radial direction. From Eqs. (62) we see that in an OSW the functions for θ, $\delta\rho$ and p are similar to each other, but differ from that for u. For each of θ, $\delta\rho$ and p the phase varies with distance in the same way as for a PTW (see Eqs. (47)); for each the amplitude falls off with distance, being proportional to r^{-1}. For the velocity u the r-dependence is more complicated. When kr is very large u is proportional to θ, in the same way as for a PTW; however, when kr is small the velocity follows an inverse-square law, that is, $|u|$ is proportional to r^{-2}. This inverse square law reflects the fact that at points near a small source (specifically, such that $r << \lambda$) the flow is essentially that of an incompressible liquid flowing inward and outward from the source during its pulsations. In this "near field" the velocity differs in phase by 90° from any of the quantities θ, $\delta\rho$ or p.

4.8 Fields of Arrays and of Extended Sources

4.8.1 Small disk-source in an infinite rigid baffle

Transducers or transducer-arrays for generating ultrasound are often designed so that the effective vibrating surfaces lie in a single plane, which we shall call the "source plane". The ultrasonic field can then be represented, in principle, as a superposition of the spherical waves (OSW's) discussed in the previous section. Suppose the source plane is at z = 0 and O is a point on this plane. A basic situation is one where there is oscillatory movement near O, but none elsewhere on the source plane. Suppose a disk of small area dS vibrates with normal velocity given by the real part of $w_o e^{j\omega t}$, the disk being set flush in a rigid baffle of infinite extent (see Fig. 5). We consider the sound field generated in a medium occupying the half-space z > 0. When the dimensions of dS are small compared to the

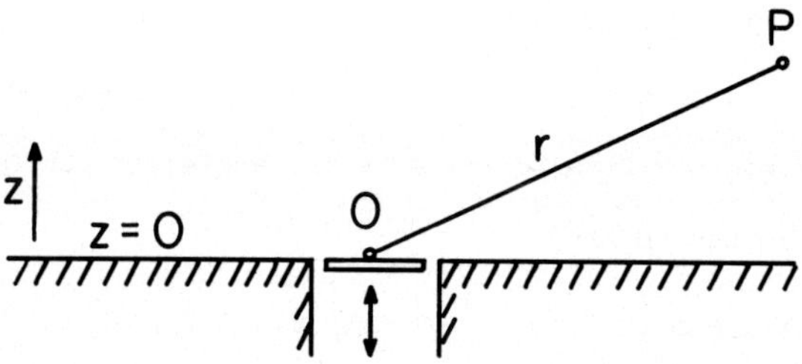

Fig. 5 Small disk-source in an infinite rigid baffle.

wave length λ and effects of viscosity are negligible, it is found from general theory that the field in this medium is that of an OSW. Specifically, the dilatation at any point P, at a distance r from 0, is given by (the real part of)

$$\theta = \frac{-j\omega w_o dS}{2\pi c_o^2 r} e^{j(\omega t - kr)} , \tag{63}$$

with the other first-order quantities $\delta\rho$, p and u given in terms of θ by Eqs. (62).

4.8.2 Linear array of sources

A promising development in medical ultrasound is the generation of beams using arrays which can be steered electronically. The principle can be illustrated rather simply. Consider an array of small disk-sources of the type just considered, set in an otherwise-rigid baffle in the plane z = 0. The sources are equally spaced along the x axis at the points $x = n\ell$, where n takes on all integral values from zero to N-1. The array thus extends from x=0 to x=L, where L is equal to $(N-1)\ell$. We assume the phase varies along the array, the normal velocity of the nth disk-source being given by $w_o e^{j(\omega t - \chi n)}$, where χ is a constant. At a point in the half-space z>0 the dilatation is obtained by summing contributions from all of the N elementary sources, using the expression in Eq. (63). This sum has the form

$$\theta(P) = A_1 e^{j\omega t} \sum_{n=0}^{n=N-1} r_n^{-1} e^{-j(kr_n + \chi n)} , \tag{64}$$

where A_1 is the factor $(-j\omega w_o dS/2\pi c_o^2)$, and r_n measures the distance from the nth source to the field point P. In general the sum is difficult to evaluate; we simplify it by choosing P in the xz plane at a large distance from the array. Let 0 be the origin of (x,y,z) coordinates and also the position of the source n = 0; also let ϕ be the angle between OP and the z axis. (Fig. 6). Then we have

$$r_n \simeq r_o - n\ell \sin\phi \ , \tag{65}$$

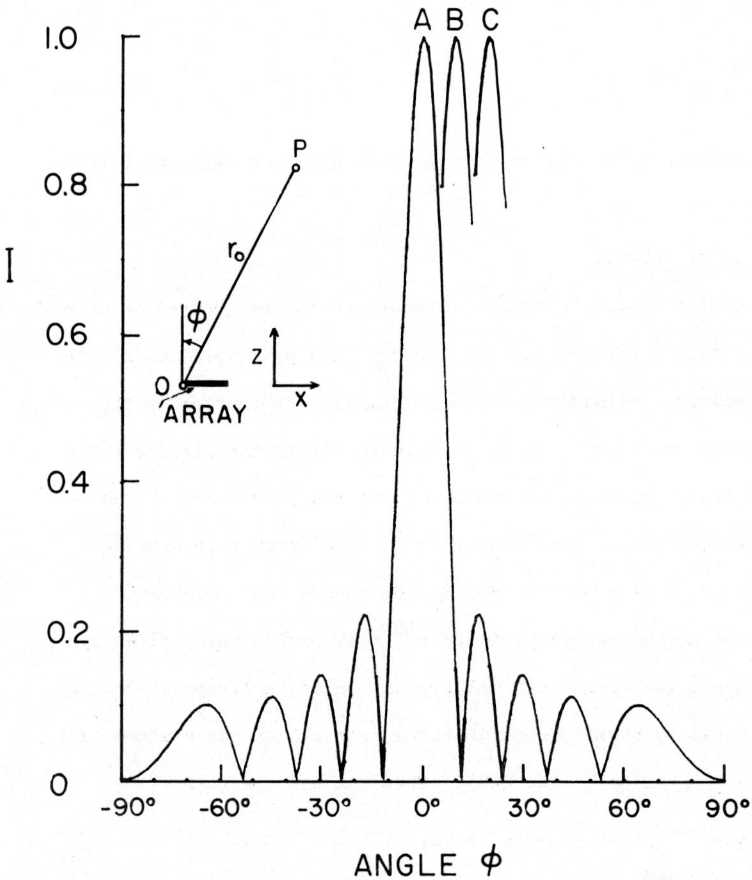

Fig. 6 Intensity vs angle in the far field of an array of 10 sources spaced one-half wavelength apart. For curve A all sources are in phase (χ=0). Steering is accomplished by shifting phases by amount χ from one source to the next; χ = 30° and 60°, resp., for (partial) curves B and C.

where r_o is the distance OP; the approximate expression in Eq. (65) is valid if $r_o >> N\ell$ and $r_o >> \lambda$. The sum on the right hand side of Eq. (64) is now approximately equal to $r_o^{-1}e^{-jkr_o}$ times the sum $\Sigma e^{j(I-\chi)n}$, where n takes on values from zero to N-1 and I is equal to $k\ell \sin \phi$. This sum is easily evaluated. We are especially interested in the square of the magnitude $|\theta(P)|$ since the intensity I(P) (see Section 5.2.2) at the point P is proportional to $|\theta(P)|^2$ under the assumed conditions. We find that I(P) has the following angular dependence:

$$I(P) = A_2 \frac{\sin[\frac{1}{2}N(k\ell \sin \phi - \chi)]}{\sin[\frac{1}{2} (k\ell \sin \phi - \chi)]} , \qquad (66)$$

where A_2 is independent of ϕ and $N\ell$ is the strength L of the array. Plots of I(P) vs ϕ are shown in Fig. 6 for several values of χ and for the choices $N = 10$ and $k\ell = \pi$, corresponding to an array of 10 sources spaced a half-wavelength apart. In the full curve A, for $\chi = 0$ we see a dominant peak at $\phi = 0^o$, as well as four minor peaks on each side. The device is steered by adjusting χ; setting χ equal to 30° causes the main peak to shift about 9.5^o, while $\chi = 60^o$ causes a total shift of 19.5^o.

Concepts for the linear array can be extended to analyze the performance of a two-dimensional array of sources in the plane z = 0. Such an array has a primary peak which, in principle, can be steered in any desired direction.

4.8.3 Piston-type source

Probably the most common type of transducer used for generation of medical ultrasound is the piezoelectric disk set into thickness oscillations. Typically the radius a of the disk will be significantly greater than the wavelength λ and the complications of diffraction must be taken into account. Considerable theoretical attention has been given to the difficult problem of analyzing the ultrasonic field generated by a piezoelectric disk. This problem is usually approached by considering the model situation of an oscillating piston set in an infinitely rigid baffle. Let us assume that, at rest, the baffle and piston surfaces of interest lie in the "source plane" z = 0. In this model the transducer is assumed to vibrate uniformly so that its plane surface, represented by the piston surface, vibrates normal to itself with uniform velocity $w_o e^{j\omega t}$. The baffle assumed in the model has the function of representing a condition of zero velocity in the z direction, everywhere in the source plane except at the vibrating piston. In the real situation the transducer may not vibrate completely uniformly; also, if it is surrounded by any baffle at all, the latter is often neither very extensive nor particularly rigid. Nevertheless the piston-in-baffle model, introduced for mathematical convenience, represents a useful approximation to many real situations. As in treating the array (Section 4.8.2) we proceed by superposing solutions for the elementary disk-source, of the type exhibited in Eq. (63). Here an integral is involved, rather than a sum; the dilatation $\theta(P)$ at an arbitrary point P in the space $z > 0$ is given by

$$\theta(P) = A_3 e^{j\omega t} \int_{S_o} r^{-1} e^{-jkr} \, dS . \tag{67}$$

Here A_3 is a constant, independent of time and the coordinates of P. The integral is over the surface S_o of the piston, which is assumed circular and of radius a, with center at 0. Because the field is symmetrical about an axis through 0 along the z direction, it suffices for location of P to use the spherical coordinates (r,ϕ); here r is the radial distance OP while ϕ is the polar angle, between the axis of symmetry and the direction OP. Results of this integration are not available in simple analytical closed form for arbitrary P. However, relatively simple expressions have been arrived at for selected regions in the field. As with the array, important and tractable results are obtained in the "far field", that is, at large distances r such that $r \gg a$ and $r \gg \lambda$. For the dilatation $\theta(P)$ in the far field we have

$$\theta(P) = -\frac{j\omega a^2 w_o}{2c_o^2 r} e^{j(\omega t - kr)} D(x) ,$$
$$D(x) = \frac{2J_1(x)}{x} \; ; \quad x = ka \sin\phi . \tag{68}$$

Here the function $D(x)$ is a "directivity factor"; it is expressed in terms of a Bessel function $J_n(x)$ of the first kind and of order $n = 1$.

The factor D(x) has been tabulated[2]; it is a "damped" oscillating function of x, whose first few (alternating) maxima and minima are 1.00 at x = 0, -0.133 at x = 5.15, 0.64 at x = 8.5 and -0.040 at x = 11.5 The first three zeroes occur at x = 3.83, 7.02 and 10.15. Corresponding to the far-field solutions in Eq. (68), the expressions for other first-order quantities are obtained by use of Eqs. (37), assuming negligible viscosity; that is,

$$\delta\rho = -\rho_o\theta \qquad p = -B\theta$$
$$\underline{u} = -(jB/\rho_o\omega)\ \nabla\theta \ . \tag{69}$$

We note that the velocity $\underline{u}$ has, in general, components along both r and θ directions; the component along r, in the far-field approximation, is $-c_o\theta$ as for a plane travelling wave (Eqs. (47)). The intensity I in the far field is proportional to $|\theta|^2$, just as for travelling plane and spherical waves (Section 5.2.2); hence the angular dependence of the intensity is the same as for the square of D(x).

While results like those of Eqs. (68) are very useful they are, for an extended source, valid only at large distances. In general, it is difficult to obtain convenient expressions for the "near field" but, fortunately, there are exceptions. For any point P along the axis of symmetry it has been shown that Eq. (67) yields

$$\theta(P) = -(w_o/c)\ e^{j\omega t}\ [e^{-jkr'} - e^{-jkr_o}];\ r' = (r_o^{\ 2} + a^2)^{\frac{1}{2}}\ ;$$
$$\delta\rho = -\ \rho_o\theta;\ p = -B\theta\ ; \tag{70}$$

here r_o is the distance OP, to the observation point P from the center O of the piston. Because of symmetry the velocity $\underline{u}$ can be written as u $\underline{i}_z$ where $\underline{i}_z$ is the unit vector along z; using the last of Eqs. (69) we find for u :

$$u = w_o e^{j\omega t}\ [(r_o/r')\ e^{-jkr'} - e^{-jkr_o}];\quad r' = (r_o^{\ 2} + a^2)^{\frac{1}{2}}\ . \tag{71}$$

It can be shown that in the "far field" ($r_o \gg a$; $r_o \gg ka^2$) the r-dependence for each of θ, δρ, p and μs is as $r_o^{-1} e^{j(\omega t - kr_o)}$, in agreement with far-field results discussed earlier. In the near field, results are much less simple, especially when the source radius $\underline{a}$ is appreciably larger than the wavelength λ. For example, instead of falling off monotonically with r the amplitude of θ passes through maxima and zeroes. On the axis it follows from Eq. (70) that zeroes of θ occur when $(r'-r_o)$ is equal to $m\lambda$, where m is a positive integer. Maxima of $|\theta|$ alternate with the zeroes; when $a >> \lambda$ the maximum most distant from the source occurs at

$$r_o \cong a^2/\lambda \ . \tag{72}$$

The parameter a^2/λ which appears in Eq. (72) is often taken (somewhat arbitrarily) as the distance from the source at which transition occurs between near-field and far-field behaviour. In general, relationships between first-order quantities are much more complex near the source than in the far field. For illustration we consider the ratio between the pressure p and the velocity u at various points along the axis; the ratio p/u is often called the "specific acoustic impedance". From Eqs. (70) and (71) we find that

$$\frac{p}{u} = \rho_o c \, \frac{e^{-jkr'} - e^{-jkr_o}}{(r_o/r')e^{-jkr'} - e^{-jkr_o}} \ ; \ r' = (a^2 + r_o^{\,2})^{\frac{1}{2}} \ . \tag{73}$$

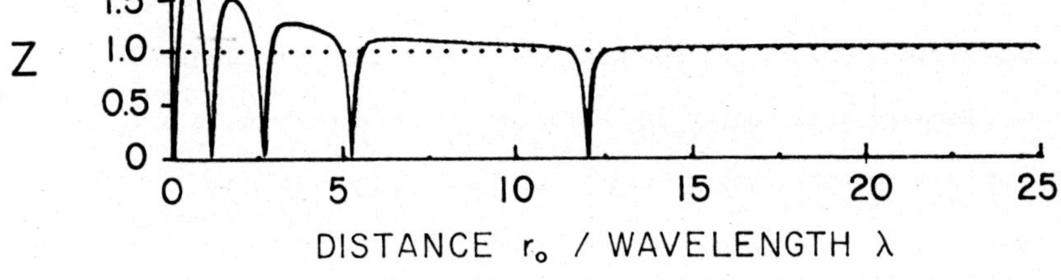

Fig. 7 Impedance p/u along the axis of a piston-type source, based on Eqs. (73). In lower curve Z is $|p/u|/\rho_o c$; in upper curve ζ is the phase angle by which the pressure p leads the velocity u. Both Z and ζ are plotted against the distance r_o in units of the wavelength λ. Radius $\underline{a}$ of piston equals 5λ. Last null of Z occurs at about $r_o = 12\lambda$, a little less than the distance $r_o = a^2/2\lambda$ (i.e., 12.5λ) at which the last null of $|p|$ occurs.

In Fig. 7 we see plots vs kr_o of the magnitude $|p/u|$ and of the phase angle ζ between p and u; these are based on Eq. (73). We see that both $|p/u|$ and ζ fluctuate considerably at points near the source. At large distances ζ approaches zero and $|p/u|$ approaches the constant $\rho_o c$, as for travelling plane and spherical waves.

4.9 Scattering; Relative Motion

Since this topic is taken up in Chapter VII the discussion here is brief. "Scattering" arises when sound passes through a medium containing bodies whose acoustical properties differ from those of the medium, and which therefore present inhomogeneities. The total field in the heterogeneous medium is often regarded as consisting of two parts: (i) the "incident", "original" or "unperturbed" field which would exist if the bodies were absent and the medium homogeneous, and (ii) the "scattered" field which exists because of the bodies. Theoretical determination of the first-order sound field in a medium with scatterers requires a suitable solution of a wave equation, such as Eq. (36). This solution must satisfy appropriate boundary conditions at the surface of each scattering body. To illustrate results we take up here the scattering from a single body, namely a small liquid sphere suspended in liquid.

Consider the incident wave to be a travelling wave as in Eq. (44), except that $\alpha = 0$; thus the dilatational amplitude is A, the angular frequency is ω, and the propagation constant k is real. Let the sphere be of radius a and let (r, ϕ) be spherical coordinates with origin at the center of the sphere; the polar angle ϕ is relative to the propagation direction for the plane wave. The densities ρ and ρ_o apply inside and outside the sphere, respectively, while χ_e is the ratio of the bulk elastic modulus B outside the sphere to that inside. Then, neglecting effects of viscosity, the scattered dilatation θ_{sc} at any point (r,θ) has been given by Rschevkin[20] as

$$\theta_{sc} = A\, e^{j(\omega t - kr)} \frac{k^2 a^3}{3r} \left[1 - \chi_e + \frac{3(\rho - \rho_o)}{2\rho + \rho_o} \cos\phi\right]; \; ka \ll 1 \qquad (74)$$

The condition $ka \ll 1$ means that the radius a is very small compared to the sonic wavelength. Since the (first-order) pressure p is proportional to θ we can, if we wish, replace θ_{sc} by the scattered pressure p_{sc} on the left-hand side of Eq. (74) and, at the same time, replace A by the pressure amplitude of the incident wave. If $\chi_e = 0$ and $\rho \gg \rho_o$ the result in Eq. (74) reduces to one given by Lamb[10] for a fixed rigid sphere.

We see that, especially because of the small factor k^2a^3/r, the scattered amplitude $|\theta_{sc}|$ is much less than the incident amplitude $|A|$. For example, if the radius a is 10μm and r is 10 cm, then for a frequency of 1 MHz in a waterlike medium ($k = 42\ cm^{-1}$) the factor k^2a^3/r is 1.8×10^{-7}. The weakness of scattering from a small body is seen also by considering its scattering cross section σ_{sc}. For an object on which a wave of intensity I is incident σ_{sc} is defined as an area such that $I\sigma_{sc}$ is equal to the total power scattered. Lamb[10] has shown that for a fixed rigid sphere ($\chi_e \ll 1$ and $\rho \gg \rho_o$) the quantity σ_{sc} is of the order of $(ka)^4$ times the geometric cross section πa^2. For the example cited $(ka)^4$ is equal to 3.1×10^{-6}. This small scattering cross section might be contrasted with results for the resonant gas bubble, discussed in Chapter II.

Examining the bracket [ ] on the right-hand side of Eqs. (74), we see that the scattered amplitude is composed of two parts: one, proportional to $1 - \chi_e$ and independent of angle, and the other proportional to $\cos \phi$. The first part disappears when $\chi_e = 1$, that is, when the internal and external liquids have equal elastic moduli; it arises from volume pulsation of the scatterer (specifically, from a difference between the pulsation of the sphere and that of an element of surrounding liquid). The second part arises from relative motion between the sphere and the surrounding liquid. Let the relative velocity be U_{rel}, equal to $u_s - u$, where u is the velocity of the liquid at a given instant as it would be in the absence of the sphere, while u_s is the velocity of the sphere at the same instant. Then when Eq. (74) applies we have (e.g., from Rschevkin[20]):

$$U_{rel}/u = 2(\rho - \rho_o)/(2\rho + \rho_o) \ . \tag{75}$$

Hence the bracket [ ] in Eq. (74) might be written

$$[\ ] = 1 - \chi_e + (3/2)(U_{rel}/u) \cos \phi \ . \tag{76}$$

Equation (75) applies when viscosity is neglected. A corresponding expression for U_{rel} has been given by Urick[21], applicable when the surrounding liquid is viscous:

$$U_{rel}/u = (\sigma - 1)\ [s^2 + (\sigma + \tau)^2]^{-\frac{1}{2}} e^{-i\psi} \ ;$$

$$\sigma = \rho/\rho_o \ ; \quad \tau = \tfrac{1}{2} + (9/4\beta a); \tag{77}$$

$$s = (9/4\beta a)\ [1 + (\beta a)^{-1}]; \qquad \tan \psi = s/(\sigma + \tau) \ .$$

The parameter β is defined in Section 4.11; its reciprocal β^{-1} is a characteristic thickness for the viscous boundary layers discussed there. When the viscosity is sufficiently small the boundary layer thickness is much smaller than the radius <u>a</u>, that is, $\beta a \gg 1$. Then the expression for U_{rel} in Eqs. (77) reduces to that in Eq. (75).

Besides its role in scattering, the relative motion discussed here has other implications. It has been shown by Carstensen and Schwan[22] that the ultrasonic absorption in a suspension of red cells is significantly affected by viscous losses generated by relative motion between the red cells and the surrounding liquid. Also the radiation forces on small particles is affected by the relative motion (Section 5.3.2), as is also the microstreaming near such particles (Section 5.5.3).

4.10 Pulsed Fields

In most of this chapter we have dealt with "continuous waves", that is, with sound fields generated by one or more sources maintained indefinitely at constant frequency and amplitude. This emphasis on continuous wave (CW) fields is usual in the pedagogy of physical acoustics since much insight is gained thereby. However, while CW fields are used in medical ultrasound, there are many applications, especially in diagnosis, where the ultrasound is generated in very short pulses. In this section we discuss some special features of pulsed fields. Following analysis presented by Coulson[23] we consider a wave generated by a single pulse; we assume it travels in the x direction and at t = 0 has the shape of a "Gaussian wave packet". Thus at t = 0 the dilatation θ_o is given by the real part of

$$\theta_o = Ae^{-x^2/2\Delta} e^{-ik_o x} . \tag{78}$$

The "standard deviation" $\sqrt{\Delta}$ is a measure of the spatial extent of the pulse; specifically we define the pulse length along x as δx, given by $2\sqrt{\Delta}$. From theory in Fourier analysis it is shown that, at t = 0, the pulse represented in Eq. (78) is equivalent to an infinite sum of CW waves, with a continuous range of values for k. For arbitrary time t the integral representing the infinite sum is

$$\theta(x,t) = A\left[\frac{\Delta}{2\pi}\right]^{\frac{1}{2}} \int_{k=-\infty}^{k=\infty} e^{-\frac{1}{2}\Delta(k-k_o)^2} e^{i(kx-\omega t)} dk. \tag{79}$$

From this equation the amplitude varies with k according to a Gaussian distribution with standard deviation $(1/\sqrt{\Delta})$; defining the k-width of the Fourier distribution as δk, given by $2/\sqrt{\Delta}$, we find that the product $\delta x \cdot \delta k$ is just equal to 4, a constant. It is reasonable to define the duration δt of the pulse as $\delta x/c_o$ and the angular-frequency band-width $\delta\omega$ of the Fourier distribution by $c_o\,\delta k$. We are led to the simple but important relations:

$$\delta t \cdot \delta\omega = \delta x \cdot \delta k = 4. \tag{80}$$

From Eq. (80) it follows that a short pulse necessarily corresponds to a wide band-width of frequency. For ultrasound of frequency 10 MHz a ten-cycle pulse corresponds to a duration of 10^{-6} sec and a band-width of $(\delta\omega/2\pi) = 0.64$ MHz.

All media are, at least, somewhat dispersive; that is, the <u>phase velocity</u> $c = \omega/k$ of a CW wave is frequency-dependent. Because of this it is found[23] that the envelope of the Gaussian wave pulse of Eq. (78) will travel with a <u>group velocity</u> c_g given by

$$c_g = (dk/d\omega)^{-1} . \tag{81}$$

Under the (common) conditions where Eqs. (41) apply, the ratio c_g/c is found to differ from unity by a quantity of the order of ε^2, where ε is defined in Eqs. (40); the group velocity is then not significantly different from the phase velocity.

More important than the dispersion is an effect on pulse propagation which arises because the absorption coefficient α of a CW wave depends on frequency. Because of this the amplitude distribution of the waves which comprise the packet changes as the pulse travels, thus changing its shape.

4.11 Boundary Layers

In this section we give brief attention to consequences of viscosity in fields where motion occurs tangentially to boundary surfaces. Consider a sound field established in a fluid near a rigid boundary which can be approximated locally as a plane at z = 0. Suppose that by ignoring viscosity a suitable "irrotational" solution of the dynamical equation, Eq. (30), has been obtained, and that from this the velocity component along x near the boundary is $u_{ao} \cos \omega t$. It is characteristic of the irrotational solution that u_{ao} may be nonzero as y approaches zero. Hence this "zero-viscosity" result does not satisfy the "nonslip" condition which is usually required at a rigid boundary. Fortunately, it is easy to obtain a more suitable approximate solution of Eq. (30), once the irrotational contribution is known. From this improved solution the velocity component along x is given near the boundary by

$$u_1 = u_{ao} \, [\cos \omega t - e^{-\beta z} \cos(\omega t - \beta z)]; \quad \beta = (\omega \rho_o / 2\eta)^{\frac{1}{2}} . \tag{82}$$

Since $u_1 = 0$ at z = 0 the nonslip condition is satisfied. At large values of z the velocity component u_1 reduces to the irrotational value $u_{ao} \cos \omega t$. Evidently u_1 in Eq. (82) represents a velocity component which increases from zero to the irrotational velocity u_{ao} in a distance of the order of β^{-1}. The fluid layer in which this velocity transition occurs is called the boundary layer. It is characterized by a thickness β^{-1} which is typically small; in water at a frequency of 1 MHz β^{-1} is about 0.56 µm.

Equation (82) is based on an assumption: the boundary layer thickness β^{-1} is small compared to a characteristic length ℓ of the irrotational field. Depending on the circumstances ℓ may be, for example, the smallest radius of curvature of the bounding surface, or the sonic wavelength. When the above assumption holds (i.e., when $\beta\ell >> 1$) the velocity gradient $\partial u_1/\partial z$ can be approximated by assuming u_{ao} nearly constant; at the boundary we obtain

$$(\partial u_1/\partial z)_{z=0} = \beta u_{ao} \cos \omega t \ . \tag{83}$$

Boundary layers are set up whenever sound fields exist in fluids such that motions occur tangentially to boundaries or interfaces. In tissues, fluid regions adjoin interfaces presented by the walls and membranes of body organs and cells. However, these interfaces do not, of course, represent rigid boundaries. Instead the interfaces themselves oscillate in a sound field and u_{ao} must be interpreted as a tangential component of velocity relative to the moving surfaces.

Boundary layers are discussed further in connection with heat generation (Section 5.2.4) and acoustic streaming (Section 5.5.3).

5. RESULTS FOR SECOND-ORDER QUANTITIES

5.1 Introduction

In Section 3 a general method is described for obtaining solutions of the basic nonlinear equations of acoustics. The basic situation is assumed to be that of a steady state sound field of angular frequency ω and typical nondimensional displacement amplitude ξ_r. In this method the first step is to find relevant solutions of simplified "first-order" equations, arrived at by linearizing the exact ones. These first-order equations are discussed in Section 3.2, and examples of solutions to them are taken up in Section 4. These solutions consist of expressions for first-order contributions to the dilatation, density, pressure and velocity. It is characteristic of the first-order contributions that they vary sinusoidally in time with angular frequency ω, the amplitude at any given point being proportional to ξ_r. In the next step (Section 3.3) improved approximations are obtained to the basic differential equations; these improved equations, valid to second-order in ξ_r, contain terms which are expressed as quadratic functions of the first-order solutions.

In the present section, Section 5, solutions are sought to the second-order equations; these give expressions for second-order contributions to the pressure, velocity or other acoustic variables. Such second-order solutions are useful in themselves, or as a source of intermediate information required, for example, in calculating the radiation force on an object in a sound field. We also take up a class of acoustic quantities which can be determined, to an approximation of second order in ξ_r, directly from quadratic combinations of first-order solutions. Included here are the energy densities, the intensity and the rate of energy dissipation.

We now revert back to the notation of Section 3; for example, the symbol p_i is used to denote a contribution to the pressure of order "i" in ξ_r. It is characteristic of second-order quantities in a steady state sound field that they consist of two parts: (1) a second-harmonic part which varies sinusoidally in time with frequency 2ω, twice that of the source(s), and (2) a part which is independent of time. Tbe time-independent part of p_2, for example, is represented by $< p_2 >$ and is obtained by averaging over a sonic period, or over a number of periods. In principle, real numbers are used for the first-order quantities when quadratic combinations are evaluated. However, when the product of a pair of these quantities is to be time-averaged, the following special rule holds and is convenient: Let Z_1 and Z_2 be two complex functions representing first-order quantities, their real parts being a_1 and b_1; each varies sinusoidally in time with angular frequency ω . Then

$$< a_1 b_1 > = \tfrac{1}{2} \,\mathrm{Re}\, (Z_1 Z_2^*) \; ; \tag{84}$$

here "Re ( )" denotes the "real part of" and "*" the complex conjugate.

5.2 Quantities Relating to Energy

5.2.1 Energy densities

In this and the next two subsections we take up acoustical quantities which are used in discussing the distribution, transport and dissipation of energy in sound fields. In the approximation usually accepted the energy density (energy per unit volume) is a quadratic function of first-order quantities. Thus in this approximation the kinetic energy density KE and potential energy density PE are given at any instant by $\frac{1}{2}\rho_o u_1^2$ and $\frac{1}{2}B\theta_1^2$, resp.; here u_1 and θ_1 are real quantities, and u_1 is the magnitude of the vector $\underline{u}_1$. The densities KE and PE are of second-order in the parameter ξ_r; as is characteristic of second-order quantities, each consists of two parts: a second-harmonic part and a time-independent part. Our interest is usually in the latter, that is, in the quantities:

$$< KE > = \tfrac{1}{2}\rho_o < u_1^2 > ; \quad < PE > = \tfrac{1}{2}B < \theta_1^2 > \quad . \tag{85}$$

We list here results for sound fields of several different types, discussed in Section 4. For the travelling wave represented by Eqs. (47) we find that the average kinetic and potential densities are equal and uniform throughout the wave. For these densities and their sum $< E >$ we have

$$\begin{aligned} &< KE > = < PE > = BA^2/4 ; \\ & \qquad\qquad\qquad\qquad \text{(Travelling wave)} \\ &< E > = \tfrac{1}{2}BA^2 . \end{aligned} \tag{86}$$

We recall that B is equal to $\rho_o c_o^2$ and A is the dilatational amplitude. Equations (86) can be expressed in terms of other quantities by recalling from Eqs. (47) that the amplitude of the density, pressure or velocity is equal to $\rho_o A$, BA or $c_o A$, resp.

For the standing wave represented by Eqs. (50) we find

$$\begin{aligned} &< KE > = BA^2 \cos^2 k(z - H) ; \\ &< PE > = BA^2 \sin^2 k(z - H) . \qquad \text{(Standing wave)} \\ &< E > = BA^2 . \end{aligned} \tag{87}$$

Here both the kinetic and potential energy densities vary periodically along z. The maxima of < KE > are at amplitude maxima of the first-order velocity ($u=u_1$), while those of < PE > are at amplitude maxima of the first-order dilatation (or density or pressure). However, the total energy density is uniform.

Some observations on the "additivity" of energy densities can be made by recalling from Section 4.4 that a standing wave is the sum of two oppositely directed travelling waves. From Eqs. (86) and (87) we find that while the total time-averaged energy density < E > for the standing wave is the sum of the < E >'s for the two travelling waves, a similar "superposition principle" does not hold for < KE > or < PE > .

For the spherical travelling wave of Eqs. (62) we find that

$$\langle KE \rangle = \frac{BA^2}{4r^2}\,[1 + (kr)^{-2}]\;;\qquad\qquad \text{(Spherical wave)} \qquad (88)$$

$$\langle PE \rangle = \frac{BA^2}{4r^2}\;.$$

In the near field where $kr \ll 1$ we see that the kinetic energy is dominant; however, at large distances < KE > and < PE > are equal, as for a plane travelling wave.

5.2.2 Intensity

To maintain the travelling plane wave represented by Eqs. (47), energy must be continually supplied by the source, even though effects of viscosity are neglected. A measure of the energy required is provided by the intensity I; this important acoustical quantity can be defined for a travelling plane wave as the time-averaged rate at which work is done per unit area in maintaining the wave. To calculate I for the travelling wave of Eqs. (47) consider a plane at $x = x_o$ in the sound field. The medium in the space $x < x_o$ does work across unit area of this plane (on the medium in the space $x > x_o$) at a rate equal to the product of pressure and velocity at $x = x_o$. It is customary to evaluate this rate, and hence the intensity I, by an approximation valid to order ξ_r^2; this is done by using first-order expressions for the pressure and velocity. Using the first-order relations in Eqs. (47) we then obtain several equivalent expressions for I:

$$I = < p_1 u_1 > = \tfrac{1}{2} B c_o A^2 = \tfrac{1}{2} B c_o |\theta_1|^2 ; \qquad (89a)$$

(Plane travelling wave)

also

$$I = \tfrac{1}{2} \rho_o c_o |u_1|^2 = |p_1|^2 / 2\rho_o c_o , \qquad (89b)$$

where $|\theta_1|$, $|u_1|$ and $|p_1|$ give the amplitudes of θ_1, u_1 and p_1, resp., and A is equal to $|\theta_1|$. From Eqs. (86) the time-averaged energy density $< E >$ is equal to $\tfrac{1}{2} BA^2$; hence an alternative expression for the intensity is

$$I = c_o < E > . \qquad (89c)$$

For the spherical wave represented by Eqs. (62) the intensity $< p_1 u_1 >$ is given by

$$I = B c_o A^2 / 2r^2 = \tfrac{1}{2} B c_o |\theta_1|^2 ; \qquad (90a)$$

also (Spherical wave)

$$I = |p_1|^2 / 2\rho_o c_o = \tfrac{1}{2} \rho_o c_o \, [1 + (kr)^{-2}]^{-1} \, |u_1|^2 . \qquad (90b)$$

Comparing Eqs. (89) and (90) we see that I is related to the amplitudes $|\theta_1|$ and $|p_1|$ in the same way for both spherical and plane travelling waves; the relation between I and the velocity amplitude $|u_1|$ is also the same for the two fields at large values of kr, but not at small values.

The intensity I is often interpreted as the "rate of energy flow across unit area per second" relative to a surface whose normal is along the direction of propagation. For a plane wave this interpretation is consistent with Eq. (89c), if we regard c_o as the velocity with which the energy (of density $< E >$) is transmitted. For a spherical wave the total time-averaged power P passing outward across a given spherical surface of constant r is by this interpretation equal to the product of the intensity and the area $4\pi r^2$. From Eqs. (90) we find that the power transmitted across the surface is, as it should be, independent of r:

$$P = 2\pi B c_o A^2 . \qquad (91)$$

For the standing wave of Eqs. (50) the intensity I is zero. Once this wave is established no energy is required to maintain it. It should be remembered that Eqs. (50) apply only when effects of viscosity and other sources of energy loss have been neglected. In a real field there are losses and, of course, energy is required to maintain the field.

In complex fields it is necessary to generalize the concept of intensity. We now define the intensity $\underline{I}$ at a given point P to be a vector whose direction is that of $\underline{u}_1$ at P, and whose magnitude is $< p_1 u_1 >$, where p_1 is the local value of the pressure and u_1 is the magnitude of the vector $\underline{u}_1$. Then, for example, in the near field of an extended transducer we find that $\underline{I}$ varies from point to point, both in magnitude and direction. Explicit results can be given for points on the axis of a piston-type source, to which Eqs. (70) and (71) apply. Here $\underline{I}$ is along the axis, its magnitude I being

$$I = \rho_o c_o w_o^2 \, [1 + (r/r')] \sin^2 \tfrac{1}{2} k(r' - r), \quad r' = (r^2 + a^2)^{\frac{1}{2}}. \qquad (92)$$

Since I is always positive in this equation, we see that the intensity is outward from the source at all points on the axis. However, especially if ka is large, I is highly nonuniform, passing through a series of maxima and zeroes. At large values of kr, where $r >> a$ and $r >> ka^2$, the intensity falls off proportionally to r^{-2}, as for a spherical wave.

The concept of intensity is not widely used in complex fields. This is especially because of experimental difficulties; measuring devices do not exist at present for direct measurement of the vector intensity from point to point in such fields as exist near an extended ultrasonic transducer.

5.2.3 Heat generation in a travelling plane wave

In a biological tissue or in any real medium the presence of a sound field is accompanied by irreversible processes; these cause continuous degradation of sonic energy into heat. Sonic heat production is an important phenomenon, since it is the basis of calorimetric methods for measuring the output of ultrasonic devices. (See Chapter V). It is also significant as a mechanism for bio-effects of ultrasound; the heating can easily bring about temperature elevations sufficient to alter biological structures and/or processes. (See Chapter IX). In this section we take up the process of heat generation as it occurs in a plane wave travelling through a homogeneous medium. In Section 5.2.4 we consider the heating in boundary layers which are established when sound fields interact with interfaces.

We must start with a satisfactory expression for the intensity in a wave travelling through a viscous medium. The expressions in Eqs. (89) were obtained on the basis of Eqs. (47), and hence on the assumption that effects of viscosity are negligible. They are nevertheless approximately valid even for viscous Newtonian liquids under typical conditions of frequency and viscosity. However, the expression $< p_1 u_1 >$, used for the intensity in obtaining Eqs. (89), is not adequate when viscosity is taken into account. For a plane wave travelling along the x direction in a viscous medium, the stress tensor is not isotropic and p_1 should be replaced by $-S_{11}$; here, consistent with the approximations used, S_{11} is understood to be the first-order contribution to the "11" stress component. In relating S_{11} to θ we have from Eqs. (11) and (12)

$$S_{11} = B\theta + [B' + (4\eta/3)]\,\dot{\theta}\ . \qquad (93)$$

Substituting S_{11} from this equation into the expression $< -S_{11}u_1 >$ for the intensity I we obtain, approximately,

$$I = I_o e^{-2\alpha x}\ , \qquad (94)$$

where I_o is constant, with α given by Eqs. (41) provided that $\alpha << k$. Under these conditions it is also found, as anticipated, that the intensity <u>at any given point in the wave</u> is related by Eqs. (89) to the amplitudes A, $|p_1|$, etc. at the same point.

The quantity 2α in Eqs. (94) is sometimes called the intensity coefficient of attenuation (or absorption) and α is then the corresponding amplitude coefficient. As used in Eq. (44) or Eq. (94) the coefficient α is expressed in nepers (Np) per unit length (e.g., Np/cm). For example, if α = 1 Np/cm then we see from Eqs. (44) that when the wave propagates a distance of one centimeter the dilatation amplitude is reduced by a factor of $e^{-1} = 0.37$, while from Eqs. (94) the intensity is reduced by a factor of $e^{-2} = 0.135$. The experimental values of the attenuation coefficient are often given in decibels (db) per unit length (e.g., db/cm). The conversion factor is such that 1 Np/cm is equal to 8.7 db/cm.

It is easily shown that when Eq. (94) applies, energy is lost from the sound field at the rate $-dI/dx$ or $2\alpha I$ per unit volume. This energy is evidently converted or "dissipated" irreversibly into heat. Letting q_v be the rate of heat production per unit volume we obtain

$$q_v = 2\alpha I \ . \tag{95}$$

As an example to illustrate units: If α is 0.1 Np/cm (as it is, approximately, for liver tissue at a frequency of 1 MHz) and I is 10 W/cm^2 we find that q_v is 2 joules or 0.48 calories per ml-sec. If this heat is not removed by conduction or radiation (as is true for a short time immediately after a sound field is applied) the temperature rises at a rate such that q_v is equal to $C_h\ \partial T/\partial t$; here C_h is the heat capacity per unit volume. For water-like media C_h is about 1 cal/ml$^{\circ}$C, and $\partial T/\partial t$ is then 0.48°C/sec in the example just cited. Temperature elevation as a mechanism for bioeffects of ultrasound is discussed further in Chapter IX.

5.2.4 Heat generation in a boundary layer

In considering mechanisms for bioeffects of ultrasound the question frequently arises whether heat is generated preferentially in regions of microscopic size, creating tiny "hot spots" whose presence would not be revealed in macroscopic temperature measurements. Possibilities present themselves in boundary layers of the type discussed in Section 4.11.

Here relatively high values exist of the velocity gradient, and the associated viscous shearing leads to heat generation. Suppose, as in Section 4.11, the velocity along x is u_1, and varies mainly with z. The velocity gradient is then $\partial u_1/\partial z$; let its magnitude be G. It can be shown that heat is generated by this motion at the time-averaged rate q_v per unit volume, given by

$$q_v = \eta < G^2 > ;$$

here < >, as before, represents the time average. Suppose the boundary layer is set up near a rigid surface by a plane wave which grazes the surface as it travels in the x direction. If the velocity in the plane wave is u_{ao} cos ωt we have from Eqs. (83) that on the surface G is βu_{ao} cos ωt and $< G^2 >$ is $1/2\beta^2 u_{ao}^2$. Substituting for β from Eqs. (82) and realizing (Cf Eqs. (89)) that the intensity I in the plane wave is $\rho_o c_o u_{ao}^{2}/2$, we find that q_v is given by the simple expression

$$q_v = kI/2; \quad k = \omega/c_o \ . \tag{96}$$

This is to be compared with the expression $q_v = 2\alpha I$ in Eqs. (95) for heat production by sound absorption in a travelling plane wave, away from boundaries. Since k is much larger than α in most biological media at megahertz frequencies we see that heat would indeed be produced preferentially in the boundary layer near a rigid surface under these conditions. However, as was pointed out in Section 4.11, the surfaces presented by biological membranes and walls are not <u>rigid</u> surfaces. As a plane wave passes, the velocity relative to a biological surface is less than the plane wave velocity u_{ao} cos ωt; hence G is reduced, and also q_v.

5.3 Radiation Pressure and Radiation Forces

5.3.1 Time-independent pressure as a field quantity

When the pressure in a liquid subject to sound is evaluated to second order in the reference parameter ξ_r, it is found that its time-average differs from the hydrostatic pressure in the absence of sound.

The average designated as $< p_2 >$ is a field quantity, representing at any given point the effective elevation of hydrostatic pressure at that point caused by the sound. Any object suspended in a sonicated liquid, or any portion of a material surface which bounds the liquid, is subjected to a time-independent force. This force is called a radiation force and the field quantity $< p_2 >$ is sometimes called a radiation pressure. (There is some variation in usage of the term "radiation pressure", however).

We first consider $< p_2 >$. Referring back to Eqs. (34) we identify $< p_2 >$ with $< p_{e2} >$ and neglect effects of viscosity. The viscosity-free fluid motion is irrotational: $\nabla x \underline{u}_1$ and $\nabla x \underline{u}_2$ are each zero. Using first-order relations it is then found that the vector function $\underline{F}$ can be written

$$\underline{F} = -\nabla L; \quad L = < KE > - < PE >; \qquad \text{(zero viscosity)} \qquad (97)$$

here L is the time-averaged Lagrangian energy density and is equal to the difference between the averages of the kinetic and potential energy densities; the latter are discussed in Section 5.2.1. From Eqs. (34a), when $\nabla x < u_2 >$ is zero, we now find that

$$< p_2 > + L = \text{constant.} \qquad \text{(zero viscosity)} \qquad (98)$$

If the conditions are such that $< p_2 >$ is obviously zero in some region where $L = 0$ (e.g., at any point outside an ultrasonic beam set up in liquid in an open vessel) we set the constant equal to zero and obtain for those conditions that $< p_2 >$ is equal to $-L$. Referring to results on energy densities from Section 5.2.1, we quickly obtain information on the pressure elevation $< p_2 >$ for several kinds of sound fields, neglecting effects of viscosity. For a plane travelling wave $< p_2 >$ is everywhere constant, and may be zero. For a plane standing wave $< p_2 >$ has its greatest values at pressure maxima, and its least values (less than the maxima by $2BA^2$) at the velocity maxima. In a spherical travelling wave $< p_2 >$ is constant (typically zero) at large values of r, but decreases as r decreases. The deficit in $< p_2 >$ is relatively large at points near the origin where $kr << 1$ and the kinetic energy dominates.

It is not easy to measure the field quantity $< p_2 >$ directly since a pressure-measuring device put into a liquid tends to alter $< p_2 >$ in its own vicinity. Instead this quantity usually manifests itself indirectly, especially thru its role in determining radiation forces.

5.3.2 Time-independent forces

At any point on the surface of a rigid fixed object immersed in a sonicated liquid the time-averaged pressure is just equal to P_o + $< p_2 >$; as before, P_o is the hydrostatic pressure. It must be realized that $< p_2 >$ is the quantity as it exists with the object present, which usually differs from $< p_2 >$ at the same point when the object is absent. To determine the radiation force on the object it is only necessary to calculate the integrated effect of the $< p_2 >$ surface distribution. For example, consider a plane wave of dilatation amplitude A and cross-sectional area S impinging normally on a plane x = 0 which is the front surface of a rigid fixed plate. Assuming no sound field exists behind the plate the radiation force on it is $< p_2 > S$, where $< p_2 >$ is evaluated at x = 0. In this situation $< p_2 >$ is equal to < PE >, since < KE > is zero at the motionless surface. Also, since the plate acts as a perfect reflector the dilatational amplitude at x = 0 is 2A, twice that in the incident wave. Hence < PE > at x = 0 is four times that in the incident wave; from Eqs. (86) we then obtain BA^2 for < PE > and BA^2s for the force. Letting the intensity in the incident wave be I, equal to $BcA^{2/2}$ (see Eqs. (89)), we have for the radiation force F_{rad}:

$$F_{rad} = 2IS/c. \qquad \text{(rigid plane)} \qquad (99)$$

Usually an immersed object is not fixed and rigid, but moves in response to the sound field. Calculation of the radiation force is then complicated by the fact that the time-averaged surface pressure $< p_2 > *$ must be calculated on the moving boundary. An adequate approximation, to second order in the reference parameter ξ_r, is given by

$$< p_2 >^* = < p_2 > + T; \quad T = < \underline{\xi}_n \cdot \nabla p_1 > . \qquad (100)$$

Here $< p_2 >$ is the time-averaged pressure at a fixed point near a given surface element and T gives the moving-boundary correction. In this correction term, $\underline{\xi}_n$ is the vector displacement (first-order approximation) of the element at any given time, while ∇p_1 is the gradient of first-order pressure in the liquid near the element.

A dramatic example of the importance of the term T is seen when we consider a travelling plane wave, just as in the previous example, but now incident on an air boundary at x = 0, and reflected from it. We calculate the radiation force on an imaginary film, of negligible thickness, at this boundary. Approximating the water-air interface as a free surface we take the dilatation to be zero at x = 0, and hence assume perfect reflection. Then $< PE >$ is zero there, so $< p_2 >$ is negative and equal to $- < KE >$ at x = 0. It is not hard to show that the correction term T is equal to just $2 < KE >$; that is, T has twice the magnitude of $< p_2 >$ and the opposite sign. We conclude that the radiation force F_{rad} is $S < KE >$. Because of the perfect reflection, $< KE >$ at the boundary is BA^2, four times the value of $< KE >$ in the incident wave. Hence we obtain

$$F_{rad} = BA^2 S = 2IS/c, \qquad \text{(Free surface)} \qquad (101)$$

where I is the intensity for the incident wave.

Another important simple example is that of a travelling plane wave incident on the plane surface (again, at x = 0) of a perfect absorber. Here it is assumed that there is no reflection; thus $< p_2 >$ is zero. Assuming the surface of the absorber moves with the undisturbed incident wave we find that T is equal to $BA^{2/2}$ and hence that

$$F_{rad} = IS/c. \qquad \text{(Perfect absorber)} \qquad (102)$$

A more detailed justification for Eq. (102) has been given by Rooney and Nyborg.[24]

Each of the results in Eqs. (99), (101), and (102) can be obtained by more simple and intuitive reasoning. To a first-order approximation in ξ_r, the momentum per unit volume in a medium traversed by a plane travelling wave is $\rho_o u_1$, and the rate at which this is transported along the propagation direction is $\rho_o u_1^2$ (to second-order in ξ_r). For a travelling wave of intensity I and cross-sectional area S the time-averaged rate of momentum transport proves to be IS/c, to second-order in ξ_r. The force on an object can be equated to the rate at which momentum is transported to it; assuming that the perfect absorber receives the momentum lost from the sound wave we immediately arrive at Eqs. (102). At the surface of a perfect reflector a sound wave is reversed in direction and its momentum, a vector quantity, is changed by 2IS/c per unit time; assuming the reflecting body experiences the same change leads to Eqs. (99) and (101).

Techniques involving measurements of the radiation force on a plane reflector or absorber are much used for determining the intensity or total power in ultrasonic beams. In other techniques the radiation force on a spherical object is measured. The theory for a sphere is similar in principle to that described above, but is more difficult to carry out. Since detailed results are discussed in Chapter V our discussion here is brief. For a sphere of radius a subject to a plane travelling wave of intensity I it is sometimes convenient to express the radiation force in the form

$$F_{rad} = KIS/c \, , \qquad \text{(Spherical object)} \qquad (103)$$

where S is the geometric cross-section πa^2; the constant K approaches unity at large values of ka but, in general, its value depends on properties of the sphere.[8,25-27]

In connection with bioeffects of ultrasound there is special interest in the radiation force on objects which are very small compared to the sonic wave length. A simple expression has been obtained for a small sphere in a region of a sound field where the phase is essentially constant. In this result, which applies in a standing wave, or near a small source or small scattering object, the vector radiation force $\underline{F}_{rad}$ is given[28-30] by

$$\underline{F}_{rad} = vD \, \nabla < KE > - \, v(1-\gamma) \, \nabla < PE > \, . \qquad (104a)$$

Here v is the volume of the sphere and γ is the ratio of its compressibility to that of the surrounding fluid; D is given by

$$D = 3(\rho-\rho_o)/(2\rho+\rho_o) \; , \qquad (104b)$$

where ρ is the density of the sphere and ρ_o that of the surrounding fluid. It is noteworthy that D is just 3/2 times the ratio U_{rel}/u quoted in Eqs. (75). Results for different fields can be obtained by using information on energy densities from Section 5.2.1. In a travelling plane wave < KE > and < PE > are uniform and Eqs. (104a) yields the erroneous result $\underline{F}_{rad}$ = 0; another (typically small) term should be added for this and comparable situations, involving spatial gradients of the phase.[29] However the expression for this term is lengthy and we omit it here.

In a standing wave, when expressions for < KE > and < PE > are inserted from Eqs. (87) the radiation force on a particle is found to vary periodically along the wave. The direction and magnitude of $\underline{F}_{rad}$ at any given point depends on the density and compressibility of the particle relative to corresponding properties of the fluid. When F_{rad} is sufficiently large compared to gravitational or other existing forces, in a suspension of identical particles, these latter will become distributed in layers one-half wavelength apart. The relevance of radiation forces to the stasis phenomenon reported by Pond, Woodward and Dyson[31] has been discussed by Gould and Coakley.[32]

From Eqs. (88) we see that in a spherical wave, near the origin, the kinetic energy density < KE > dominates over < PE > and the second term on the right of Eqs. (104a) is negligible. It is then found that small particles with density ρ greater than ρ_o migrate inward toward the origin while lighter particles for which $\rho < \rho_o$ move outward. In Chapter II this result is applied to the radiation force on a particle in suspension near a vibrating gas bubble. Related theory has been used to explain the migration of intracellular particles to a portion of cell wall contacted by a microvibrator.[33]

For a small gas bubble, smaller than resonant size, the potential energy term in Eqs. (104a) is dominant and $\underline{F}_{rad}$ becomes $v\gamma\, \nabla \langle PE \rangle$, an expression which agrees with results given in Chapter II.

Another special application of Eqs. (104) is to the interaction force between two small particles in a sound field.[34] Assume the spheres have radii a_1 and a_2, respectively, and density ρ; that they are in a liquid of density ρ_o with their centers separated a distance r; and that the line connecting their centers makes an angle ϕ with the direction in which the fluid would move in the sound field with the particles absent. The problem is approached by first determining the first-order field existing when only one of the spheres (say Sphere 1) is present; from this the energy density $\langle KE \rangle$ and $\langle PE \rangle$ are calculated for the field (near Sphere 1) in which the second sphere finds itself. Equation (104) is then applied to determine the force $\underline{F}_{rad}$ on this second sphere; this force is not, in general, along the line of centers but is equal and opposite to the force on the first sphere. We express the force $\underline{F}_{rad}$ as the vector sum of two components, F_r and F_ϕ, along the r and ϕ directions, respectively:

$$\begin{aligned} F_r &= F_o\,(3\cos^2\phi - 1) \\ F_\phi &= F_o\,\sin 2\phi \\ F_o &= 2\pi(\rho-\rho_o)^2 U_o^{\,2}\, a_1^3\, a_2^3/3\rho_o r^4 \,. \end{aligned} \qquad (105)$$

Here U_o is the velocity amplitude of the fluid in the absence of the particles. Provided that a_1, a_2 and r are all small compared to the sonic wavelength Eqs. (105) apply at any locality in a sound field of any kind; the fluid velocity, whose direction and amplitude U_o are referred to, is that existing in the selected locality. The expression for F_o in Eqs. (105) applies when the density ρ and ρ_o are not greatly different, so that $|\rho-\rho_o|$ is much less than ρ_o. Under these conditions the relative velocity U_r (see Eqs. (75)) between the particle and the fluid is $2(\rho-\rho_o)U_o/3\rho_o$. When F_o is expressed in terms of U_r Eqs. (105) are equivalent to results calculated by Konig[35] and cited by Rayleigh[1, Par. 253b]; however, the latter results are for solid spheres in air, for which $\rho >> \rho_o$ and $U_r \cong U_o$.

From Eqs. (105) it is seen that F_ϕ is zero at $\phi = 0$ and $\pi/2$; in these directions the force is radial, being outward for $\phi = 0$ and inward for $\phi = \pi/2$. The result for the $\phi = 0$ or "on-line" orientation is (for small separations) in agreement with more extended theory given by Embleton.[36] A consequence of the angular dependence of the radiation force is the formation of particle chains in suspensions subject to ultrasonic fields;[37] these are similar to the "pearl chains" reported in microwave fields.[38] In gaining an appreciation for the magnitude of the interaction force it is useful to compare F_o with the magnitude F_g of the gravitational force; for the particle of radius a_2 (volume v_2) we have that F_g is $v_2(\rho-\rho_o)g$. Note that v_2 cancels from the ratio F_o/F_g. If the sound field is a travelling plane wave of intensity I, the medium is water-like and $\rho = 1.1\ \rho_o$ we find that

$$F_o/F_g = 0.0067\ Ia_1^3/r^4 , \tag{106}$$

where a_1 and r are in cm while I is in W/cm^2. The radiation force drops off rapidly with separation distance, but is significant when r is not much greater than $a_1 + a_2$. Neighboring particles can be expected to come together and form a clump even in fields of modest intensities after sufficient time.[37]

5.4 Radiation Torque

5.4.1 Time-independent torque as a field quantity

When a suspended particle is observed through a microscope while subjected to ultrasound, it is often found to spin about its axis, either at a fixed position or while translating. Observations are nearly always made on inhomogeneities but it can be shown[39] that, in general, torque is exerted on any volume element of a sonicated medium, even when the latter is homogeneous. For a viscous fluid a specific expression can be cited by referring to the second-order time-averaged vector quantity $\underline{F}$ defined in Eqs. (34). The time-averaged torque $\underline{L}$ (to an approximation of order ξ_r^2) on a spherical element about its center is given by

$$\underline{L} = (I/2\rho_o)\ \nabla \times \underline{F}\ , \tag{107}$$

where I is the moment of inertia of the sphere about its center, and ρ_o is the fluid density; if $\underline{a}$ is the spherical radius $I/2\rho_o$ is equal to $4\pi a^5/15$. The torque quantity $\underline{L}$, proportional to $\nabla \times \underline{F}$, is a field quantity; it varies from point to point in a sonicated medium. As we see from Eqs. (34b) the force function $\underline{F}$, and hence also the torque function $\nabla \times \underline{F}$, depends only on the first-order velocity $\underline{u}_1$. It is found that $\nabla \times \underline{F}$ tends to be large in regions where $\underline{u}_1$ is nonuniform (e.g., near the edge of an ultrasonic beam, or an edge of an obstacle in the field). In homogeneous media direct determination of the quantity $\nabla \times \underline{F}$, or $\underline{L}$, would probably be difficult; instead its influence is felt indirectly, e.g., by its role in generating acoustic streaming. (Section 5.5). However, we shall see in Section 5.4.3 an example where $\underline{L}$ is applied directly in calculating the spinning velocity of a small absorbing body suspended in a sonicated liquid.

5.4.2 Orienting of asymmetric bodies

When a rod, disc or other asymmetric body is suspended in a sonicated liquid it tends to turn until it achieves a preferred orientation. The simplest kind of body to exhibit this behavior is probably a dumb-bell, formed by a rigid connection between two spheres. Assuming the connection itself is so thin that its contribution is negligible, the torque is easily obtained from analysis in Section 5.3.2. The moment L about Sphere 1 of the force $\underline{F}_{rad}$ on Sphere 2 is just

$$L = rF_\phi = rF_o \sin 2\phi\ , \qquad \text{(Dumb-bell)} \tag{108}$$

where F_o is given in Eqs. (105). Thus for $0 < \phi < \pi/2$ the torque is in the direction of increasing ϕ; the dumb-bell "prefers" the orientation $\phi = \pi/2$ where its axis is perpendicular to the direction of flow.

An expression was derived by Konig[40] and cited by Rayleigh[1, par. 253b] for the torque L on a disk of radius a small compared to the sonic wavelength λ, suspended in a medium whose density ρ_o is much less than its own. When the disk is suspended in a locality where the (undisturbed) first-order velocity has amplitude U_o and is directed at an angle ϕ relative to the axis of the disk, the time-averaged torque is

$$L = -(2/3)\rho_o a^3 U_o^2 \sin 2\phi . \qquad \text{(Disk in air)} \qquad (109)$$

The minus sign indicates that the torque tends to diminish ϕ; the disk "prefers" to be oriented with its axis along the flow. For disks suspended in liquids, Eqs. (109) should be modified to take into account first-order motion of the disk; U_o should be replaced by a relative velocity U_{rel} (as defined for a sphere in Section 4.9).

Orienting effects on small suspended particles might form the basis of useful techniques for assaying ultrasonic fields. In addition, particle orienting may be important in connection with effects of ultrasound on asymmetric cells, organelles or macromolecules.

5.4.3 Spinning of spherical bodies

A number of investigators[41-45,33,37] have observed steady rotation of intracellular bodies during sonation. A theoretical model for this situation is arrived at by using the results cited in Eqs. (107). While this equation was derived for an element of homogeneous fluid it is also applicable to a spherical inhomogeneity suspended in fluid provided that the specific acoustic impedance (e.g., $\rho_o c$) is similar for the fluid and the inhomogeneity. This condition is satisfied rather well for some biological situations. If the sonic torque on such a spherical body of radius a causes it to rotate with angular velocity $\dot{\phi}$ in a Newtonian liquid of shear viscosity coefficient η the drag torque will be $8\pi\eta a^3\dot{\phi}$.[10] In steady rotation we therefore have

$$\dot{\phi} = (a^2/30\eta)\ |\nabla \times \underline{F}| . \qquad (110)$$

One interesting situation involving sonic torque arises when an ultrasonic beam is reflected obliquely at a boundary. A torque distribution is then set up in a region near the boundary where the incident and reflected waves overlap. It is found that, at any given point P in this overlap region, $|\nabla \times \underline{F}|$ is proportional to the absorption coefficient α applicable to the medium at P. For an inhomogeneity in the form of a spherical body at P one then expects the body to spin with angular velocity given by Eqs. (110), with ϕ proportional to α/η; here α is the absorption coefficient applicable to the body-substance while η applies to the surrounding fluid. Especially high spin velocities are thus expected for bodies composed of material with a high absorption coefficient suspended in a medium of low viscosity. It is possible that situations of this kind can form the basis of methods for measuring absorption coefficients of small bodies.

5.5 Acoustic Streaming

5.5.1 General concepts

In this section, Section 5.5, we discuss briefly the phenomenon of time-independent flow which is commonly observed in sonated liquids. For this flow, called acoustic streaming, the basic differential equations are given in Eqs. (34). These are the same equations to which we referred in beginning the discussion of radiation pressure in Section 5.3. However, in treating acoustic streaming it is essential to include effects of viscosity whereas we neglected these in most of our discussion of radiation pressure. It has been shown that $\nabla \cdot < u_2 >$ is typically negligible; hence by a vector identity Eqs. (34a) may be re-written as

$$\nabla < p_{e2} > - \eta \nabla^2 < \underline{u}_2 > = \underline{F} . \tag{111}$$

The vector function $\underline{F}$ is again important. It is worth noting that Eq. (111) is identical with an equation used in hydrodynamics for analyzing fields of pressure and slow viscous flow resulting from externally applied force fields; in the latter application $\underline{F}$ represents the applied force per unit volume. In Eqs. (111) the second-order quantity $\underline{F}$ represents an "effective" or "pseudo" force field; however, the pressure and flow fields generated are the same as if $\underline{F}$ were a "real" force. (Thus $\underline{F}$ in Eqs. (34) and (111) is analogous to the centrifugal force referred to in connection with rotating systems). Equation (111) involves two dependent variables, $< p_{e2} >$ and $< \underline{u}_2 >$; a solution must therefore provide suitable functions for both the pressure and velocity. It is sometimes useful to obtain an equation in $< \underline{u}_2 >$ only; this can be done by taking the curl of both sides of Eqs. (111). The result can be written

$$\eta \nabla^2 \underline{R}_2 = - \nabla \times \underline{F} \; ; \; \underline{R}_2 = \nabla \times < \underline{u}_2 > \qquad (112)$$

here $\underline{R}_2$ is the vorticity associated with $< \underline{u}_2 >$.

In solving an acoustic streaming problem for a given situation one begins by determining the first-order solution; next $\underline{F}$ is calculated from Eqs. (34b); then a streaming equation (e.g., either of Eqs. (34a) or (111)) is solved subject to boundary conditions. In general, results are valid only when the magnitude of $< \underline{u}_2 >$ is small compared to that of $\underline{u}_1$. The quantity $< \underline{u}_2 >$ represents the time-averaged velocity at any point in space. Frequently it is desirable to know the time-averaged velocity $\underline{U}$ of an individual particle suspended in the fluid. Assuming the particle is isopycnic with the fluid and thus follows faithfully the fluid motion in its own vicinity, the velocity $\underline{U}$ is given by

$$\underline{U} = < \underline{u}_2 > + \underline{u}_T ; \quad \underline{u}_T = < \underline{\xi}_1 \cdot \nabla \underline{u}_1 > ; \qquad (113)$$

where $\underline{u}_T$ has been called the "velocity transform" and $\underline{\xi}_1$ is the instantaneous vector displacement of the observed particle, referred to its position when the averaging commences.

5.5.2 Quartz wind

The simplest situation is probably that of the "quartz wind" studied theoretically and experimentally by Eckart[19] and Liebermann.[46] Here a plane wave of ultrasound forms a beam which travels down the central region of a long tube of radius r_o, closed at each end, whose axis coincides with the z axis. The shape of the beam is idealized by assuming the intensity I is uniform in a cylinder of radius $\underline{a}$ and zero outside. Then

$$I = I_o e^{-2\alpha z} \quad \text{for } r \leq a \tag{114a}$$

$$I = 0 \quad \text{for } a < r < r_o \text{ .} \tag{114b}$$

Here I_o is a constant and r measures distance outward from the axis. When $\alpha << k$, Eq. (114a) applies to a wave in which the first-order velocity $\underline{u}_1$ is along z; its magnitude can be written as the product of $u_{10}e^{-\alpha z}$ and $\cos(\omega t - kz)$, with I_o equal to $\rho_o c\, u_{10}^{2/2}$. With $\underline{u}_1$ known, the "effective force" $\underline{F}$ can be evaluated; it lies along z and its magnitude F is found to be

$$F = 2\alpha I/c \text{ .} \tag{115}$$

(This nonzero value for F reflects the fact that Eq. (97) is not accurate for a viscous medium. If $\alpha << k$ it is approximately true, even in a viscous medium, that at any given point in a travelling plane wave $< KE > \cong < PE >$, as in Eqs. (86), and hence that L is nearly zero. However F is now given approximately by the sum of $- \nabla L$ and a term $2\alpha I/c$ representing momentum loss caused by absorption).

For the given field $\underline{F}$ a solution of Eq. (111) has been obtained. The pressure gradient $\nabla < p_{e2} >$ proves to be constant; the velocity field is such that on the axis $< \underline{u}_2 >$ is along z, and return flow occurs in the outer part of the tube. It is found that the transform $\underline{u}_T$ is usually negligible in this situation so that, from Eqs. (113) the velocities $\underline{U}$ and $< \underline{u}_2 >$ are equal. On the axis the velocity is

$$U = < u_2 > = C(y^2 - 1 - 2\ \ell n\ y) \ ; \qquad y = a/r_o \ ; \quad C = a^2 \alpha I_o/2c\eta \text{ .} \tag{116}$$

For $0 < y < 1$ the velocity U is always positive, increasing monotonically from zero as y decreases from 1 to 0. Since C is expressed in terms of the ratio α/η for the fluid, measurements of streaming velocity yield information on this ratio. Equations (116) are invalid if $y \simeq 1$ so that the beam nearly fills the tube. Under these conditions boundary layers developed at the tube walls, ignored in the above treatment, must be taken into account. The singularity at y = 0 reminds us of the idealization introduced by assuming that the beam has a "sharp edge" at r = a. The effective radius of a real beam is affected by diffraction and, of course, the radius a cannot really go to zero.

Acoustic streaming of the kind considered in this section is typically of rather large "scale." We define the latter quality in terms of a characteristic length ℓ^*. Let u_{ch} be a characteristic magnitude of $< \underline{u}_2 >$ or $\underline{U}$ in the field, and let G_{ch} be a characteristic magnitude of the velocity gradient. Then ℓ^* is a length such that G_{ch} is equal to u_{ch}/ℓ^*. For "quartz-wind" streaming, if y is not too small, the tube radius r_o is a reasonable choice for ℓ; the streaming fills the tube, velocity gradients being of the order u_{ch}/r_o where u_{ch} is, e.g., the axial velocity given in Eqs. (116). This kind of motion occurs in a beaker or tank in which an ultrasonic beam is generated by a piezoelectric transducer, for example, a quartz disk (hence "quartz wind"). This large-scale motion is sometimes a nuisance (e.g., in affecting measurements of radiation pressure) and sometimes useful (e.g., in aiding transport of heat or mass).

5.5.3 Near-boundary streaming; microstreaming

Steady flow patterns of much smaller scale are observed near surfaces where acoustic boundary layers are set up, of the kind discussed in Section 4.11. In the present context we refer to these layers, established in the first-order field, as "AC layers" to distinguish them from others to be discussed shortly. Typical of small scale flows are the eddies set up near the edges of transducers, near edges of obstacles in a sound field, near small vibrating objects and near vibrating membranes. The streaming near a vibrating gas bubble is taken up in Chapter II. At ultrasonic frequencies and in water-like media the flow is characterized by high velocity gradients in thin layers, called "DC boundary layers", associated with the "AC boundary layers" referred to earlier. The thickness of a DC layer tends to be comparable to that of an AC layer; it is thus given in order-of-magnitude by the parameter β^{-1} defined in Section 4.11. In the language of the previous section, Section 5.5.2, the characteristic length ℓ^* for the near-boundary streaming is β^{-1}. Since β^{-1} is in the sub-micron range at megahertz frequencies in water-like media the term "microstreaming" is quite appropriate for the near-boundary acoustic streaming under discussion.

Microstreaming is an important topic in connection with bio-effects of ultrasound, since the high velocity gradients and associated shearing stresses which exist in boundary layers can cause damage to biological cells, organelles and macromolecules. It is important to have adequate information on the microstreaming in situations of biological interest. Obtaining this information presents a challenge; the procedure for obtaining a detailed solution for a streaming problem is typically long and difficult. Fortunately a short cut exists when the boundary layer is thin as it is for megahertz frequencies in water-like media, and when the surfaces are not too highly curved. Let ℓ be a characteristic length for the irrotational approximation to the sound field in question. (See Section 4.11). Then if $\beta\ell \gg 1$ the acoustic streaming near a given point P on a boundary S is determined essentially by the local irrotational velocity field near P. This statement is the basis for a "thin boundary layer approximation."

Assuming the latter approximation holds, we return to Section 4.11; when $\beta\ell \gg 1$ the expression for u_1 in Eqs. (82) can be applied rather generally to any point P on a surface S where an acoustic boundary layer exists. In this expression the coordinates x and z belong to a local coordinate system, not necessarily Cartesian, appropriate to S at P. To represent a specific acoustic situation a suitable choice must be made for the irrotational velocity amplitude u_{ao}. With the latter known it is possible to find the first-order velocity component normal to the boundary and hence fully determine the vector $\underline{u}_1$. Then the force function $\underline{F}$ can be calculated from Eqs. (34b). Our interest here is primarily in the component

F_x along the x direction. We had noticed in Section 4.11 that β^{-1} is relatively small at megahertz frequencies in water-like media and hence that u_1 varies rapidly with distance z from the boundary. It is thus not surprising to find that in the boundary layer F_x is strongly dependent on z. Related to this is the finding that the magnitude $|\nabla \times \underline{F}|$ is relatively large in the boundary layer, compared to its values elsewhere. Since the torque function $\underline{L}$ discussed in Section 5.4.1 is proportional to $\nabla \times \underline{F}$ we are led to picture a "twisting" action exerted sonically on each element of medium in the boundary layer. In a fluid medium this action leads to generation of the microstreaming.

Results for the microstreaming in a number of situations have been obtained in the thin boundary-like approximation. We first state a result from Schlicting[12,54] which applies rather generally in "two-dimensional" situations, when the flow lines are in xz planes and when the irrotational first-order velocity $\underline{u}_a$ is divergence-free. Here the component of $\underline{u}_a$ along x is assumed to be $u_{ao} \cos \omega t$ where u_{ao} is a function of x; the streaming velocity U along x is given by

$$U = \omega^{-1} \partial(u_{ao})^2/\partial x \; [e^{-\beta z} \sin \beta z - (3/8)(1 - e^{-2\beta z})] \; . \qquad (117)$$

At large values of βz the velocity approaches a limiting value U_L of

$$U_L = -(3/8\omega) \; \partial(u_{ao})^2/\partial x \; . \qquad (118)$$

At the boundary z = 0 the velocity gradient becomes

$$(\partial U/\partial z)_{z=0} = -\;(2/3)\;\beta U_L \; . \qquad (119)$$

From these equations it is significant that the microstreaming velocity is proportional to the gradient $(\partial/\partial x)$ of $u_{ao}^{\;2}$, the squared velocity amplitude. This result exemplifies a consequence of acoustic streaming theory in general, that the motion depends on nonuniformity of $\underline{u}_a$, the first-order irrotational velocity field. In situations where Eq. (118) applies the direction of U_L is that in which $|u_{ao}|$ is decreasing.

We now consider two examples for application of the Schlicting result. The first is a problem first treated by Schlicting himself,[12] that of oscillatory flow relative to a fixed cylinder. When the flow at large distances outward from the cylinder is uniform and given by $V \cos \omega t$, the appropriate expression for u_{ao} at a given point P on the cylindrical surface is found to be $-2V \sin \phi$; here ϕ is the angle with which the direction of flow at infinity meets a line thru P normal to the cylindrical axis. As before, the amplitude u_{ao} is for the x component of irrotational velocity; in this example x is the arc length $a\phi$ measured along the surface on a circle coaxial with the cylinder. From Eqs. (118) and (119), using the above expression for u_{ao} (with x/a for ϕ), we obtain

$$U_L = -(3/2\omega a)\, V^2 \sin 2\phi \; ; \tag{120a}$$

$$(\partial U/\partial z)_{z=0} = (\beta/\omega a)\, V^2 \sin 2\phi \quad . \tag{120b}$$

From the angular dependence of these equations we see that the flow has symmetry such that the pattern is the same in each of four quadrants bounded by the zeroes at $\phi = 0$, $\pi/2$, π and $3\pi/2$. At distances from the surface large compared to the boundary layer thickness β^{-1} a set of four symmetrical circulations is set up (to which the boundary-layer result in Eqs. (117) is not applicable). It has been shown that for a given oscillatory amplitude the streaming relative to the cylinder is the same for (i) a fixed cylinder in an oscillating fluid (as treated here) as it is for (ii) an oscillating cylinder in a fixed fluid.[47]

In a technique reported by Williams, Hughes and Nyborg[48] the microstreaming near a transversely vibrating metal wire is used to study effects of sonically generated shear on biological cells. From Eqs. (120b) a typical magnitude of the velocity gradient at the wire surface is $\beta V^2/\omega a$. For a wire of radius 0.125 mm set into oscillation with 10 microns amplitude at a frequency of 20 kHz in a water-like medium we obtain 25,000 sec^{-1} for $\beta V^2/\omega a$. This "wire method" has subsequently been used in further studies by Rooney,[49] by Williams and O'Brien[50] and by Williams.[51]

Microstreaming qualitatively similar to that for the cylinder probably occurs near objects of various shapes suspended in liquid traversed by an ultrasonic beam. It has been reported by Connolly[52] that the streaming set up near a red cell in an ultrasonic beam generates shear sufficient to cause hemolysis at elevated temperatures. This flow is a consequence of relative motion between the cell and the surrounding fluid, a topic discussed in Section 4.9.

The second application of the Schlicting result is to a vibrating membrane.[53,54] If standing waves are set up in the membrane of amplitude ξ_o cos kx the first-order velocity amplitude u_{ao} will be of magnitude $\omega\xi_o$ sin kx and we obtain from Eqs. (118) that

$$U_L = -(3/8)\omega k\xi_o^2 \sin 2kx. \tag{121}$$

This result may be relevant to microstreaming set up within sonated biological cells. Techniques for setting walls of individual cells into vibration were investigated by Dyer and Nyborg.[43]

5.5.4 Piercy-Lamb flow

Another technique for sonic generation of steady flow was introduced by Piercy and Lamb,[55] as a means for measuring the attenuation coefficient α in a plane travelling wave. In this experiment the beam nearly filled the tube so that little quartz wind occurred, as expected from Eqs. (116) when $y \cong 1$. Instead the force $\underline{F}$, whose direction is along z and whose magnitude is given by $2\alpha I/c$ (Eqs. (115)), causes a gradient of time-averaged pressure $< p_{e2} >$ to be set up in the tube, with pressure increasing in the direction of propagation. In the Piercy-Lamb arrangement two small holes, H_1 and H_2, are provided in the wall of the sound-filled tube with H_1 at, say, z = 0 and H_2 at z = L. By means of these holes a fluid-filled tube of small bore, or capillary, is attached to the main tube as a side arm, and the difference in pressure $< p_{e2} >$ at H_1 and H_2 is communicated to its ends. This pressure difference causes flow to occur in the capillary (into which the sound field does not penetrate) which Piercy and Lamb measured for their purposes.

It was proposed by Gershoy and Nyborg[45] that fluid flow similar to Piercy-Lamb flow can occur in sonicated biological tissue. Specifically it was suggested that in a vacuolate plant cell vibration of the cell wall causes gradients of $< p_{e2} >$ in the water-like medium which fills the large central vacuole. This pressure is communicated across thin flexible membranes to narrow channels into which the sound field does not penetrate (because the absorption coefficient is relatively high there). In any such channel, flow occurs in the direction opposite to the communicated pressure gradient.

5.6 Generation of Second-Harmonic and Sum-Difference Frequencies

Since the basic equations of acoustics are nonlinear it is only at the lower amplitudes that a plane wave propagates without change in shape. At higher amplitudes distortion occurs, a subject which has been reviewed by Beyer.[18] When viscosity is ignored theory predicts that a wave initially sinusoidal in shape acquires a "sawtooth" form: the slope of amplitude vs distance becomes infinite at some point in the wave. For a plane wave propagating along the x axis, outward from a source at $x = 0$, the discontinuity theoretically occurs at $x = \ell^*$ such that

$$\ell^*/\lambda = (B/\pi p_{10}) \, [2 + (B^*/A^*]^{-1} \, . \qquad (122)$$

Here λ is the sonic wavelength; B^* and A^* are constants in the pressure-density relation defined in Eqs. (17) and (18); B is the bulk modulus of elasticity; and p_{10} is the first-order pressure amplitude at $x = 0$. For water Keck and Beyer[56] have suggested 5.5 for B^*/A^*; choosing B as 22,500 atm/cm^2 we find from Eqs. (122) that ℓ^*/λ is equal to $955/p_{10}$, if p_{10} is expressed in atmospheres. Hence if the travelling wave has an initial amplitude of 1 atm (corresponding to an intensity of 0.3 W/cm^2) the length ℓ^* is equal to 955 λ, a distance of 143 cm if the frequency is 1 MHz.

Distortion of the wave shape is, of course, evidence that the frequency spectrum is being altered; initially the spectrum has a single component of angular frequency ω but as the wave propagates harmonic components are generated of frequency 2ω, 3ω, etc. In the following brief discussion we consider primarily the second-harmonic component; for simplicity we consider only the initial stages of wave distortion, where $x << \ell^*$, and neglect effects of viscosity. For this purpose we return to Eqs. (19) and take the divergence of both sides; letting $B' = \eta = 0$ and using Eqs. (21) for $\underline{f}$ we obtain

$$-\nabla^2 p_e = \nabla \cdot \underline{f} = \partial/\partial t \, (\nabla \cdot \rho \underline{u}) - \nabla \cdot \underline{F}' \, , \tag{123}$$

where $\underline{F}'$ is given in Eqs. (22). From the continuity equation, Eqs. (1), we can replace $\nabla \cdot \rho \underline{u}$ by $-\partial\rho/\partial t$. Reorganizing the equation and subtracting the term $-c_o^2 \nabla^2 \rho$ from each side yields

$$\ddot{\rho} - c_o^2 \nabla^2 \rho = \nabla^2 (p_e - c_o^2 \rho) - \nabla \cdot \underline{F}' \, . \tag{124}$$

We now proceed to a second-order approximation to this equation. From the pressure-density relation in Eqs. (17) we find that the second-order contribution to $(p_e - c_o^2\rho)$ is $B^*\rho_1^2/2\rho_o^2$. Equation (124) then becomes

$$\ddot{\rho}_2 - c_o^2 \nabla^2 \rho_2 = (B^*/2\rho_o^2) \nabla^2 (\rho_1^2) - \nabla \cdot F'_2 \, , \tag{125}$$

$$-\underline{F}'_2 = \rho_o [(\underline{u}_1 \cdot \nabla) \underline{u}_1 + \underline{u}_1 \nabla \cdot \underline{u}_1]$$

This equation is of the kind given by Lighthill;[57] it was used by Westervelt in theory for scattering of sound by sound[58] and for the parametric acoustic array.[59] For our present application we apply the equation in a determination of second-harmonic developed in a plane wave. We assume the wave travels in the positive x direction, its first-order relations given as in Eqs. (47). Equations (125) then become

$$\ddot{\rho}_2 - c_o^2 \, \partial^2 \rho_2/\partial x^2 = [2 + (B^*/A^*)] \, \partial^2 (PE)/\partial x^2 \, , \tag{126}$$

where PE is the time-dependent potential energy density (also equal to the kinetic energy density), given at any instant by $c_o^2 \rho_1^2/2\rho_o$. This equation is equivalent to one used by Thuras, Jenkins and O'Neil[60] and also one by Keck and Beyer[56] (when the latter is reduced by neglecting viscosity). Assuming now that $\rho_1 = \rho_o\theta_1$, with the dilatation $\theta_1 = \theta$ given in Eqs. (47) the density PE can be determined. The right-hand side of Eqs. (126) is now evaluated, keeping only the second-harmonic part of PE. A solution of the equation is then obtained; this has a particularly simple form when it is simultaneously true that $kx \gg 1$ and $x \ll \ell^*$. Letting p_{20} be the pressure amplitude of the second harmonic at any distance z we obtain, approximately

$$p_{20}/p_{10} = 1/2\ x/\ell^* \ , \tag{127}$$

where ℓ^* is the characteristic length given in Eqs. (122). Thus the second-harmonic amplitude p_{20} increases linearly with the propagation distance x. We recall that when p_{10} is 1 atm and the frequency is 1 MHz the length ℓ^* is 143 cm; at a distance x of 14.3 cm (i.e., $\ell^*/10$) the predicted second-harmonic amplitude for a plane wave is thus 0.05 atm. Equation (127) is consistent with theory given by Gol'dberg.[61] It agrees with experimental results by Krasilnikov, et al.[62] for water at 1.5 MHz at pressure amplitudes p_{10} less than 5 atm and at distances x less than about 10-15 cm. At larger distances or higher amplitudes Eq. (127) is inadequate; as x increases the second-harmonic amplitude p_{20} passes through a maximum. The x-value for this maximum decreases with increasing p_{10}.

Near a real transducer the second-harmonic field is, understandably, more complex than in a single travelling plane wave. Studies of harmonic generation in the vicinity of a piston-type source have been carried out by Gould, Smith, Williams and Ryan.[63]

Equations (125) or (126), or their equivalents, have been used to treat situations where the first-order field is the superposition of two waves with angular frequencies ω_1 and ω_2, respectively. It is found that the sum and difference frequencies, $(\omega_1 + \omega_2)$ and $|\omega_1 - \omega_2|$ are generated. Of course, the second harmonic is just the sum frequency when

$\omega_1 = \omega_2$. See, for example, Thuras, *et al.*[60] In his proposal for a parametric acoustic array Westervelt[59] treated the field of two collinear plane travelling waves whose difference frequency $|\omega_1 - \omega_2|$ is much less than the separate frequencies ω_1 and ω_2. Here the high-frequency waves generate a low-frequency field with unique and useful directional characteristics.

In general, in a high amplitude sound field where the first-order spectrum consists of only one frequency, ω, the total spectrum contains not only the frequencies ω and 2ω, but also higher harmonics 3ω, 4ω, etc., as well as subharmonics $\omega/2$, $\omega/3$, etc. We shall not attempt to deal with this broad topic here, but instead refer the reader to more complete treatments of nonlinear acoustics; see, for example, a recent book on this subject by Beyer.[64]

REFERENCES

1. Lord Rayleigh, *Theory of Sound*, Dover Publications, Inc., New York (1945); includes historical introduction by R. B. Lindsay.

2. L. E. Kinsler and A. R. Frey, *Fundamentals of Acoustics*, John Wiley and Sons, Inc., New York (1962).

3. P. M. Morse and K. U. Ingard, *Theoretical Acoustics*, McGraw-Hill, New York (1968).

4. R. W. B. Stephens and A. E. Bate, *Acoustics and Vibrational Physics*, St. Martin's Press, New York (1966).

5. T. F. Hueter and R. H. Bolt, *Sonics*, John Wiley and Sons, Inc., New York (1955).

6. E. Skudrzyk, *The Foundations of Acoustics*, Springer-Verlag, New York (1971).

7. J. L. Hunter, *Acoustics*, Prentice-Hall, Englewood Cliffs, N.J. (1962).

8. W. J. Fry and F. Dunn, Ultrasound: Analysis and experimental methods in biological research. Chapter 6 in *Physical Techniques in Biological Research*, Vol. IV, ed. by W. L. Nastuk, Academic Press, New York (1962).

9. P. N. T. Wells, *Physical Principles of Ultrasonic Diagnosis*, Academic Press, New York (1969).

10. H. Lamb, *Hydrodynamics*, Dover Publications, New York (1945).

11. L. D. Landau and E. M. Lifshitz, *Fluid Mechanics*, Pergamon Press, London (1959).

12. H. Schlichting, *Boundary Layer Theory*, McGraw-Hill, New York (1955).

13. R. R. Long, Mechanics of Solids and Fluids, Prentice-Hall, Englewood Cliffs, N. J. (1961).

14. W. Prager, Introduction to Mechanics of Continua, Dover Publications, New York (1973).

15. W. L. Nyborg: Intermediate Biophysical Mechanics, Cummings Publishing Co., Menlo Park, California (1975), Chapt. 7.

16. H. L. Oestreicher, J. Acoust. Soc. Amer. 23 (1951) p. 707.

17. F. E. Fox and W. A. Wallace, Jr., J. Acoust. Soc. Amer. 26 (1954) p.994.

18. R. T. Beyer, Nonlinear Acoustics, in Physical Acoustics, ed. by W. P. Mason, Academic Press, Inc., New York (1965), Chapt. 10, Vol. 11B.

19. C. Eckart, Phys. Rev. 73 (1948) p. 68.

20. S. N. Rschevkin, The Theory of Sound, English translation by O. M. Blunn and P. E. Doak, Pergamon Press, New York (1963).

21. R. J. Urick, J. Acoust. Soc. Amer. 20 (1948) p. 283.

22. E. L. Carstensen and H. P. Schwan, J. Acoust. Soc. Amer. 31 (1959) p.185.

23. C. A. Coulson, Waves, Interscience Publishers, Inc., New York (1952).

24. J. A. Rooney and W. L. Nyborg, Amer. J. Phys. 40 (1972) p. 1825.

25. G. Maidanik, J. Acoust. Soc. Amer. 29 (1957) p. 738.

26. G. Maidanik and P. J. Westervelt, J. Acoust. Soc. Amer. 29 (1957) p. 936.

27. T. Hasegawa and K. Yosioka, J. Acoust. Soc. Amer. 58 (1975) p. 581.

28. L. P. Gor'kov, Soviet Phys.-Doklady 6 (1962) p. 773 [trans. from Dok. Aka. Nauk. SSSR 140 (1961) p. 88].

29. W. L. Nyborg, J. Acoust. Soc. Amer. 42 (1967) p. 947.

30. L. A. Crum, J. Acoust. Soc. Amer. 50 (1971) p. 157.

31. J. B. Pond, B. Woodward and M. Dyson, Phys. Med. Biol. 16 (1971) p.521.

32. R. K. Gould and W. T. Coakley, The effects of acoustic forces on small particles in suspension, in Finite-Amplitude Wave Effects in Fluids, ed. cy L. Bjørno, IPC Science and Technology Press, Ltd., (1974) pp. 252-257.

33. W. L. Wilson, F. J. Wiercinski, W. L. Nyborg, R. M. Schnitzler and F. J. Sichel, J. Acoust. Soc. Amer. 40 (1966) p. 1363.

34. A. Gershoy and W. L. Nyborg, Proceedings of the Second World Congress on Ultrasonics in Medicine, Rotterdam, 1973, Excerpta Medica, Amsterdam (1974) pp. 360-365.

35. W. König, Wied. Ann. t. 42 (1891) 353, p. 549.

36. T. F. W. Embleton, J. Acoust. Soc. Amer. 34 (1962) p. 1714.

37. W. L. Nyborg, D. L. Miller and A. Gershoy, Proceedings of Seventh Rochester International Conference on Environmental Toxicity, Plenum Publishing Co., New York (1974).

38. J. F. Herrick, Pearl Chain Formation, in Proceedings of Second Tri-Service Conference on Biological Effects of Microwave Energy, September (1958).

39. W. L. Nyborg, J. Acoust. Soc. Amer. 25 (1953) p. 938.

40. W. König, Wied. Ann. t. 43 (1891) p. 51.

41. E. N. Harvey, Biol. Bull. 59 (1930) p. 306.

42. F. O. Schmitt, Protoplasma 7 (1929) p. 332.

43. H. J. Dyer and W. L. Nyborg, IRE Trans. Med. Electron. ME-7 (1960) p. 163.

44. I. E. El'piner, I. M. Faikin and O. K. Basurmanova, Fed. Proc. 5 (1966) p. T716.

45. A. Gershoy and W. L. Nyborg, J. Acoust. Soc. Amer. 54 (1973) p. 1356.

46. L. N. Liebermann, Phys. Rev. 75 (1949) p. 1415.

47. P. J. Westervelt, J. Acoust. Soc. Amer. 27 (1955) p. 379.

48. A. R. Williams, D. E. Hughes and W. L. Nyborg, Science 169 (1970) p. 871.

49. J. A. Rooney, J. Acoust. Soc. Amer. 52 (1972) p. 1718.

50. A. R. Williams, W. D. O'Brien, Jr. and B. S. Coller, Ultrasound in Med. and Biol. 2 (1976) p. 113.

51. A. R. Williams, J. Acoust. Soc. Amer. 52 (1972) p. 688.

52. C. C. Connolly, Ph.D. Thesis, University of London (1969).

53. F. J. Jackson and W. L. Nyborg, J. Acoust. Soc. Amer. 30 (1958) p. 614.

54. W. L. Nyborg, Acoustic Streaming, in Physical Acoustics, ed. by W. P. Mason, Academic Press, New York (1965) Chapt. 11, Vol. 2B.

55. J. E. Piercy and J. Lamb, Proc. Roy. Soc. A226 (1954) p. 43.

56. W. Keck and R. T. Beyer, The Physics of Fluids, 3 (1960) p. 346.

57. M. J. Lighthill, Proc. Roy. Soc. (London) A222 (1952) p. 1.

58. P. J. Westervelt, J. Acoust. Soc. Amer. 29 (1957) p. 199.

59. P. J. Westervelt, J. Acoust. Soc. Amer. 35 (1963) p. 535.

60. A. L. Thuras, R. T. Jenkins and H. T. O'Neil, J. Acoust. Soc. Amer. 6 (1935) p. 173.

61. Z. A. Gol'dberg, Soviet Phys.-Acoust. (English Transl.) 3 (1957-8) 157, p. 340.

62. V. A. Krasilnikov, V. V. Shklovskaya-Kordy and L. K. Zarembo, J. Acoust. Soc. Amer. 29 (1957) p. 642.

63. R. K. Gould, C. W. Smith, A. O. Williams, Jr. and R. P. Ryan, J. Acoust. Soc. Amer. 40 (196-) p. 421.

64. R. T. Beyer, Nonlinear Acoustics, Naval Sea Systems Command, (1976). (Copies available from J. Neely, Code 06H1, Washington, D.C.).

CHAPTER II

CAVITATION; DYNAMICS OF GAS BUBBLES; APPLICATIONS

W. T. Coakley and Wesley L. Nyborg

1. PHYSICAL CONSIDERATIONS

1.1 Introduction

In a condensed medium the presence of cavities or gas-filled bubbles often has a profound effect on any sound field established in the medium. Such bodies of low density and high compressibility are effective in dissipating or redistributing acoustic energy. The absorption-scattering cross-section for a gas bubble, especially one of resonant size, is typically much greater than for a liquid droplet or solid particle of comparable size.

Perhaps even more important is the fact that when the ambient sound field is of moderate or high amplitude a gaseous body tends to be a site of vigorous activity. For such a body contracts and expands considerably during a sonic cycle, and the surrounding medium flows inward and outward with velocities much greater than those prevailing in the sound field when the body is absent. Also, at high amplitudes and as a consequence of nonlinearity, the pressure and temperature of the gas during contraction may achieve values much greater than those in the surrounding liquid. Except at very low amplitudes the dynamics of the gaseous body and the immediately surrounding medium is usually complex; the gas-liquid interface develops contortions which lead to the formation of liquid jets and gaseous micro-bubbles. As a consequence of the various kinds of activity set up in or near a small gaseous body of any of a number of physical, chemical and biological phenomena may occur. These include fragmentation of structures, acceleration of chemical reactions, and damage to biological cells and tissues.

The word "cavitation" is widely used as a general term for the activities simple or complex, of bubbles or cavities containing gas or vapor, in liquids or any media with liquid content. In this chapter we are, of course, chiefly concerned with sonically generated cavitation and its implications for medicine and biology. In this section (Section 1) we take up relevant physical principles and results of physical experimentation. Section 2 is concerned primarily with cavitation as it occurs in chemical and biological solutions and suspensions. Available information on sonically produced cavitation in plant and animal tissues is taken up elsewhere.

1.2 Cavitation Thresholds; Nuclei

It is rather commonly accepted that a sound field does not give rise to cavitation unless the amplitude exceeds a characteristic value, called the threshold value. For aerated water such as the "tap water" which flows from an ordinary water supply, the threshold pressure amplitude p_{oc} is often found to be about 1 atm. In general, however, reported values of the threshold vary considerably from one investigator (or laboratory) to another. This is partly because of variations in experimental conditions, and in methods used for detecting cavitation. In solutions which are optically transparent the method is usually visual inspection: an observer looks for gas bubbles which appear to be "active", or branching streaks called "streamers", or whitish clouds or "ghosts". Acoustic means are also much employed, and are especially useful for opaque media: here harmonics, subharmonics and/or general noise are taken as indices for cavitation. When visual, acoustic and other means are used to identify cavitation the threshold is often found to depend on the method used.

Much useful information has been obtained from investigations of cavitation thresholds in spite of typical uncertainties about their precise significance. The pressure amplitude threshold p_{oc} for cavitation is found to be very much a function of the experimental conditions, such as the ambient temperature and static pressure, and including the history of the medium. For tap water p_{oc} is increased by boiling the water, centrifuging it, or passing it through a fine filter; p_{oc} is also increased simply by allowing aerated water to stand, so that bubbles rise and disappear. The threshold is especially high for (degassed) liquids of high viscosity. Also p_{oc} appears to increase with increasing frequency, especially at high ultrasonic frequencies, well above 1 MHz.

From analysis of experimental findings on thresholds, it appears that under typical conditions in aqueous media and other liquids the occurrence of cavitation requires the pre-existence of small gaseous bodies; these are called cavitation nuclei, especially if they are too small to be visible. In freshly aerated aqueous solutions there are usually many gas bubbles, but the number decreases as they rise through buoyancy. A small bubble of radius R is found to rise at a rate v obtained by assuming its frictional drag is given approximately by the Stokes value for a rigid sphere of zero mass, namely,

$$v = 2\rho_o gR^2/9\eta; \tag{1}$$

here g is the acceleration due to gravity while ρ_o and η are, respectively, the density and shear viscosity coefficient for the liquid. For a bubble of radius R = 10 μm the rate of rise v in water is about 0.2 mm/sec. As time goes on an aqueous solution of moderate depth will become depleted of bubbles; since v is proportional to R^2 the large bubbles rise quickly and the small ones more slowly. Decreasing the ambient pressure will cause gas volumes to expand and will speed up the debubbling process. Heating the liquid also speeds the process, partly by decreasing the solubility of gases so that bubbles grow as gas comes out of solution, and partly by decreasing the viscosity coefficient η. Boiling provides vapor bubbles into which gas will diffuse; as these rise the liquid is quickly degassed. A liquid may thus be "degassed" by various means, such as by allowing it to stand for hours or days, by subjecting it to reduced pressure, or by boiling it. After this is done there will typically be no visible bubbles remaining. However, indirect evidence shows that unless unusual procedures are followed the liquid will still contain very small gas volumes or nuclei, with dimensions in the submicron range. These nuclei cannot be in the form of free spherical bubbles, for they would then be short-lived. To see this, consider an air-filled spherical nucleus of radius R in a liquid for which σ is the interfacial tension with air. The pressure p_i in the nucleus will exceed the pressure p_e which prevails in the surrounding liquid by the amount

$$p_i - p_e = 2\sigma/R . \tag{2}$$

We see that p_i is greater than p_e by an amount that increases with decreasing radius R and is very large for submicroscopic nuclei. If R is 0.5 μm and σ is

80 dyn/cm the pressure excess ($p_i - p_e$) is somewhat greater than 3 atm. Suppose the liquid contains dissolved air whose concentration is such as to be in equilibrium with gas-phase air at pressure p_a. Then if p_a is no greater than p_e (as is true if the liquid is not super-saturated) it follows that $p_i > p_a$. Under these conditions gas diffuses outward from the nucleus, at a rate which increases with ($p_i - p_a$). In water that is just saturated ($p_a = p_e$) with air it is calculated by Epstein and Plesset[1] that an air-filled nucleus of radius 10 μm will dissolve completely in a little less than 7 sec; if the radius is only 1 μm the time is less by more than a factor of 100.

To explain the long-time persistence of gas-filled nuclei in solution Harvey[2] suggested that the gas is trapped in cracks or corners in container walls or surfaces of suspended solid particles. In a crevice of suitable shape the gas-liquid interface may possess little outward convexity, if any; it may even be concave inwards. This means (Cf. Eq. (2)) that in a solution containing dissolved gas the excess of p_i over p_a may be very small (or zero) so that little or no gas is lost from the nucleus by diffusion. Another theory for explaining persistence of nuclei was advanced by Fox and Herzfeld,[3] and is especially appropriate for biological solutions or suspensions. By this theory organic molecules collect at the surface of a gas bubble and form a sturdy "skin" which prevents diffusion and gives mechanical strength to the nucleus. Still another mechanism for stabilization exists if the bubble has a free charge; this possibility has been discussed by Flynn.[4]

Since nuclei are critical for cavitation phenomena it is important to know whether they exist in living tissues. In experiments described by Harvey[2] no evidence was found for nuclei in normal quiescent blood. But there is considerable evidence from decompression studies (Fulton)[5] that nuclei are generated by muscular activity. The findings of Gramiak and Shah[6] are relevant here. Using an ultrasonic technique they observed echoes from blood in human cardiac chambers, certain of which they attribute to gas bubbles of microscopic size. They suggest that these microbubbles grow in zones of low pressure in the complex flow field. How long they persist as the blood proceeds through the normal circulation is as yet unknown, and their significance as cavitation nuclei is unknown.

In a different category is the gas which exists in the bronchi, tubes and alveoli of lung tissue, and the intercellular gas in plants. Here the gas-filled space is extensive, and losses by diffusion are continually restored from the atmosphere.

1.3 Radial Pulsation of a Gas Bubble; Linear Theory

In this chapter we shall use the word bubble to represent a small gas-filled space in a condensed medium (such as water, blood or other tissue). The most striking characteristic of the bubble is its compressibility; if it contains ideal gas at atmospheric pressure (P_{atm}) the adiabatic compressibility is given by $1/\gamma P_{atm}$, where γ is the ratio of specific heats. For air γ is 1.40 and the compressibility is therefore about 7×10^{-7} cm^2/dyn at atmospheric pressure, or 16,000 times greater than for water. When a sound field exists in the medium the bubble contracts and expands in response to the pressure variations. If the relative changes in volume are small, linear theory applies and yields much useful information. Suppose the sound field is such that the pressure p in the vicinity of the bubble is given by

$$p = P_o + p_o \cos \omega t; \quad (3)$$

here P_o is the prevailing static pressure in the absence of sound (usually approximately equal to P_{atm}); ω is the angular frequency and p_o a constant equal to the amplitude of the "driving pressure." In response to the driving pressure $p_o \cos \omega t$, the bubble pulsates. Let us assume its radius R varies with time according to the relation

$$R = R_o + \xi_b; \quad \xi_b = \xi_{bo} \cos (\omega t + \alpha - \pi); \quad (4)$$

here R_o is a constant, ξ_{bo} is the amplitude of the displacement ξ_b and $\alpha-\pi$ is chosen for later convenience as the phase angle between the displacement ξ_b and the driving pressure $p_o \cos \omega t$. As the bubble pulsates the surrounding liquid is set into motion. Assuming spherical symmetry the displacement ξ in the liquid at any given time is the same at all points at the same distance r from the center of the bubble, and is along the radial direction. If the frequency is low the liquid moves as if it were incompressible. This means that after any given time t the total volume which has been displaced outward through any (spherical) surface of constant r is independent of r. Thus the volume displacement $4\pi R^2 \xi_b$ at the bubble surface at a given time is equal to the volume displacement $4\pi r^2 \xi$ at any arbitrary value of r at that same time. Assuming ξ_b is small compared to the equilibrium radius R_o we set $R \simeq R_o$ and obtain

$$\xi/\xi_b = R_o^{\,2}/r^2; \tag{5}$$

thus the displacement in the fluid varies inversely with r^2. It follows that the velocity $\dot{\xi}$ also varies as r^{-2}. The assumption of incompressible liquid flow, on which these "inverse-square laws" are based, is valid for distances r small compared to the wavelength λ for sound waves in the liquid. An interesting result is obtained by determining the total kinetic energy (KE) of the liquid outside the bubble, assuming Eqs. (4) and (5) hold. For a liquid of density ρ_o this is easily shown to be just

$$KE = \tfrac{1}{2}\, m\dot{\xi}_b^{\,2}, \tag{6}$$

where the dot represents differentiation with respect to time and

$$m = 4\pi R_o^{\,3}\, \rho_o. \tag{7}$$

Thus the kinetic energy of the liquid is the same as that of a hypothetical small body, moving radially with the surface of the bubble, whose mass m is just three times that of the liquid displaced by the bubble.

Because the gas is of low density its kinetic energy is negligible but we must consider changes in its potential energy PE. For a gas whose ratio of specific heats is γ it is not hard to show that PE is given simply by

$$PE = \tfrac{1}{2}\, k_s \xi_b^{\,2} + P_o(V_o - V) , \tag{8}$$

where V is the volume of the bubble, V_o is its equilibrium volume at pressure P_o and

$$k_s = 12\pi\gamma\, R_o P_o. \tag{9}$$

It is assumed that $\xi_b \ll R_o$ and that changes in state occur adiabatically. Because of the low compressibility of the liquid, changes in its potential energy are negligible. Thinking of the gas in the interior of the bubble and the liquid on its exterior as a single system we can take the total mechanical energy (ME) of the system to be equal approximately to KE plus PE, where KE is given by Eq. (6) and PE by Eq. (8). The total ME varies with time since (i) ME is added to the system by sound sources and (ii) ME is lost through such irreversible processes as heat conduction and viscous dissipation. Since ME is characteristically dissipated at a rate proportional to the velocity squared, we set the dissipation rate equal to $b\dot{\xi}_b^{\,2}$, where b is a constant.

The energy supplied by sources is transmitted by the medium: The work done on the bubble by the external medium in unit time is just the rate of change $\dot{V}$ of the bubble volume multiplied by the negative of the ambient pressure. Hence we have

$$\frac{d}{dt}(PE + KE) + b\dot{\xi}_b^{\,2} = -\dot{V}\,(P_o + p_o \cos \omega t). \tag{10}$$

Letting $\dot{V}$ be given approximately by $4\pi R_o^{\,2}\,\dot{\xi}_b$ and substituting expressions for PE and KE from Eqs. (6) and (8) the following result is readily obtained:

$$m\ddot{\xi}_b + b\dot{\xi}_b + k_s\xi_b = -4\pi R_o^{\,2}\, p_o \cos \omega t. \tag{11}$$

This differential equation is a "forced harmonic oscillator" equation; it applies to a body of mass m attached to a spring of stiffness k_s and acted on by a force which varies sinusoidally with time. In the bubble problem under consideration the "effective mass" is given by m in Eq. (7), the "effective stiffness" by k_s in Eq. (9), and the "force" by the right hand side of Eq. (11). The quantity b is discussed later. Assuming b constant, a solution of Eq. (11) can be written (Eq. (4)) as

$$\xi_b = \xi_{bo} \cos(\omega t + \alpha - \pi), \tag{12}$$

where the amplitude ξ_{bo} is a positive number given by

$$\frac{\xi_{bo}}{R_o} = \frac{p_o}{3\gamma P_o}\,\chi(\Omega,\delta)\ ; \tag{13}$$

here the nondimensional quantity χ is defined by

$$\chi = [(1-\Omega^2)^2 + \Omega^2\delta^2]^{-\frac{1}{2}}; \tag{14}$$

in this expression Ω is the relative frequency ω/ω_o, where ω_o and δ are defined by

$$\omega_o^{\ 2} = k_s/m; \quad \delta = b/m\omega_o = \omega_o b/k_s. \tag{15}$$

The phase angle α satisfies the relation

$$\alpha = \arctan\,[\Omega\delta/(1 - \Omega^2)]\ ; \tag{16}$$

for definiteness the branch of the arctangent is chosen for which α increases from zero to π as Ω increases from zero to infinity. Since $\alpha = 0$ when $\Omega = 0$ we see from Eq. (12) that ξ_b then differs in phase by π from the driving pressure $p_o \cos \omega t$. This is expected since at low frequency ($\omega \ll \omega_o$) the bubble contracts when the pressure increases and expands when it decreases. On the other hand ξ_b is in phase with the driving pressure at high values of Ω. Letting ξ_{oo} be the value of ξ_{bo} at $\Omega = 0$ we see that

$$\xi_{oo}/R_o = p_o/3\gamma P_o = 0.24\ p_o/P_o\ ; \tag{17}$$

here the number 0.24 is for air or any other diatomic gas ($\gamma = 1.4$). From Eqs. (13) and (17) we see that χ is the ratio ξ_{bo}/ξ_{oo}, and thus can be regarded as a "nondimensional" or "relative" amplitude. In Fig. 1 plots are shown of χ vs the relative frequency Ω, for two values of the damping constant δ. We see that χ peaks at approximately $\Omega = 1$, i.e., at a driving frequency ω equal to ω_o, the resonance frequency. Equation (14) shows that χ rises to a peak value of δ^{-1} at resonance. Thus the resonant bubble vibrates with a displacement amplitude equal to δ^{-1} times its low frequency amplitude, the latter being given (since $\chi = 1$) by ξ_{oo} in Eq. (17). At frequencies above ω_o the relative amplitude χ decreases to zero as Ω increases. Since $\omega_o^{\ 2} = k_s/m$ we have from the expressions for k_s and m in Eqs. (7) and (9) that

$$\rho_o\omega_o^{\ 2}R_o^{\ 2} = 3\gamma P_o\ . \tag{18}$$

For an air bubble ($\gamma = 1.40$) in water ($\rho_o = 1.00$ cgs) at atmospheric pressure this leads to the formula

$$f_o R_o = 3.28 \; , \tag{19}$$

where R_o is in millimeters while f_o is the resonance frequency ($\omega_o/2\pi$) in kHz. For example, a bubble of radius 0.1 mm is resonant at a frequency of 32.8 kHz. Equations (18) and (19) for the resonance frequency have been tested experimentally and found valid at frequencies as high as 100 kHz. At frequencies of 1 MHz and higher these equations need to be modified by including

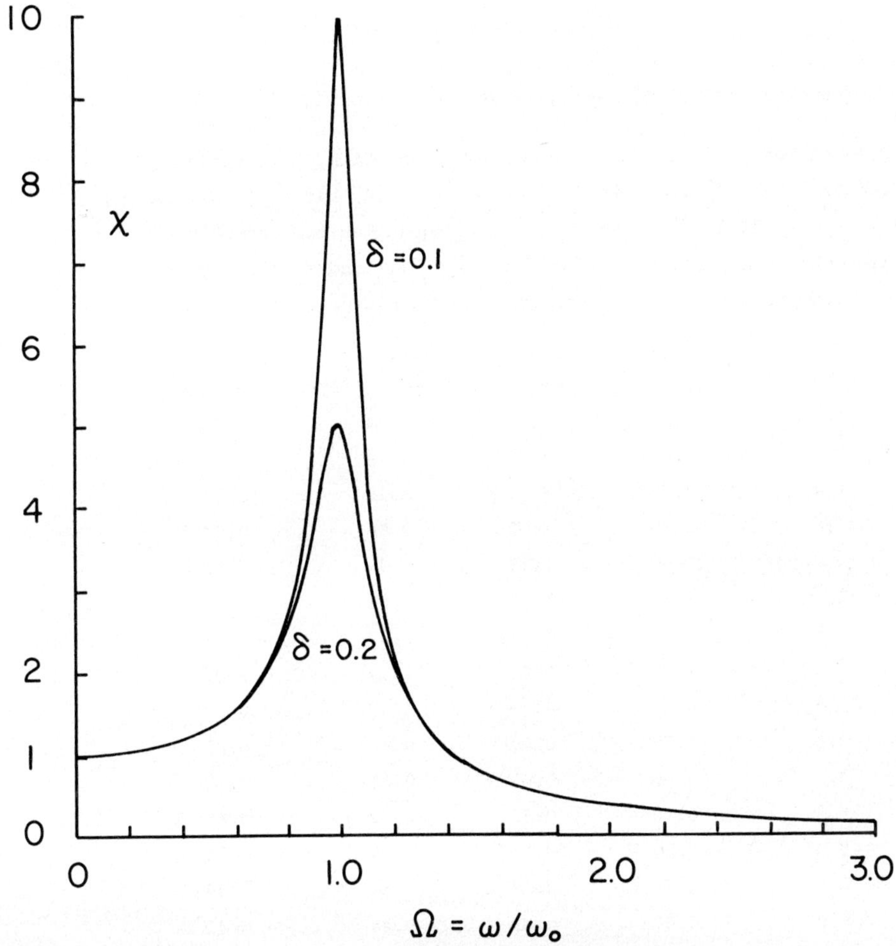

Fig. 1 Plots (for δ = 0.2) of nondimensional amplitude $\chi = \xi_{bo}/\xi_{oo}$ versus nondimensional frequency $\Omega = \omega/\omega_o$, based on Eq. (14).

effects of surface tension. This may be done in Eqs. (9), (13), (17) and (18) by substituting P_o' for P_o, where

$$P_o' = P_o + \frac{2\sigma(3\gamma - 1)}{3\gamma R_o} = P_o\left(1 + \frac{2\sigma(3\gamma - 1)}{3\gamma R_o P_o}\right) ; \qquad (20)$$

here σ is the gas-liquid surface tension. Calculations based on Eq. (18), modified in this way, are shown in Table 1. Miller[7a] has determined R_o experimentally to be about 3.5 μm at 1 MHz in water; this is in agreement with Eq. (20) if $\gamma = 1.2$.

Table 1

Frequencies for Volume Resonance of Air Bubbles in Water

Frequencies, in MHz, are from Eq. (18), in which P_o is replaced by P_o' from Eq. (20); $f_o(1.4)$ gives $\omega_o/2\pi$ for $\gamma = 1.4$, and $f_o(1)$ similarly for $\gamma = 1$; σ is 73 dyn/cm. At megahertz frequencies the effective values of γ lie between 1 and 1.4. Primed values $f_o'(1)$ and $f_o'(1.4)$ give corresponding frequencies for negligible surface tension ($\sigma = 0$).

R_o	$f_o(1)$	$f_o'(1)$	$f_o(1.4)$	$f_o'(1.4)$
1.0	3.87	2.76	4.74	3.26
1.2	3.09	2.30	3.78	2.72
1.4	2.56	1.97	3.12	2.33
1.6	2.19	1.72	2.66	2.04
1.8	1.90	1.53	2.31	1.81
2.0	1.68	1.38	2.04	1.63
2.5	1.30	1.10	1.57	1.31
3.0	1.06	0.92	1.27	1.09
3.5	0.89	0.79	1.07	0.93
4.0	0.77	0.69	0.92	0.82
4.5	0.68	0.61	0.81	0.73
5.0	0.60	0.55	0.72	0.65
6.0	0.50	0.46	0.59	0.54
7.0	0.42	0.39	0.50	0.47
8.0	0.37	0.35	0.44	0.41
9.0	0.32	0.31	0.38	0.36
10.0	0.29	0.28	0.34	0.33

1.4 Damping of Radial Vibrations

1.4.1 Radiation Damping

The damping of a pulsating bubble arises from several causes. Radiation damping arises from the fact that a pulsating bubble in a body of liquid generates a spherical wave which travels outward into the liquid. Energy is continually carried away by the wave; this is easily shown for a wave of single frequency ω from the fact that the pressure in the spherical wave at the bubble surface has a component in phase with the velocity. If the velocity at the surface is $\dot{\xi}_b$ the in-phase pressure component there is given by $\rho_o c(kR_o)^2 \dot{\xi}_b$, where c is the velocity of sound in the liquid and k is the propagation constant ω/c. The work done per unit time by the bubble on the medium per unit area of its surface is then $\rho_o c(kR_o\dot{\xi}_b)^2$. For the entire bubble this should be multiplied by $4\pi R_o^2$, yielding an expression of the form $b\dot{\xi}_b^2$; here $b = b_{rad}$ has the same significance as in Eqs. (10) and (11). The time-average of this rate-of-work is not zero; hence the bubble continually loses energy by generating the spherical wave. The damping constant δ_{rad} arising from radiation losses is equal (Eq. (15)) to $b_{rad}/m\omega_o$ and is found to be

$$\delta_{rad} = 2\pi R_o f_o/c \qquad (21)$$

for a resonant bubble. For air bubbles in water, in the frequency range where Eq. (18) or (19) applies the product $R_o f_o$ is independent of frequency for bubbles of resonant size and δ_{rad} is about 0.014. At a frequency of 1 MHz and higher the damping from radiation is small compared to other contributions.

1.4.2 Viscous Damping

Viscous damping arises from energy "dissipation", i.e., irreversible conversion of mechanical energy into heat caused by shearing action set up in the liquid as the spherical wave passes through. In this wave the velocity u is radial and, assuming incompressible flow near the bubble as in Eq. (5), is given by $R_o^2\dot{\xi}_b/r^2$, where $\dot{\xi}_b$ is the velocity at the bubble surface. Adapting an expression from Lamb[8] (Par. 329) for a "dissipation function", the rate of energy dissipation is $3\mu(\partial u/\partial r)^2$ per unit volume, where μ is the (dynamic) coefficient of shear viscosity for the liquid. In the wave this volume rate of dissipation becomes $12\mu R_o^4 \dot{\xi}_b^2/r^6$. To obtain the total rate of energy lost the latter expression is integrated over all space outside the bubble, though (since the integrand drops off rapidly with r) there are no effective contributions to the integral beyond a few bubble radii from the center. The result can be expressed in the form $b\dot{\xi}_b^2$, where $b = b_{vis}$ is equal to $16\pi\mu R_o$. Letting δ_{vis} be the contribution to damping from viscous dissipation we obtain

$$\delta_{vis} = \frac{b_{vis}}{\omega_o m} = \frac{4\mu}{\rho_o \omega_o R_o^2} \; ; \quad \omega_o = 2\pi f_o \; . \tag{22}$$

here f_o is the resonance frequency for the bubble. For an air bubble in water at room temperature, resonant at 1 MHz (radius about 3.3 μm) we obtain 0.06 for δ_{vis}.

1.4.3 Thermal Damping

Another contribution to the damping arises from heat conduction. The theory required for deriving the corresponding (thermal) damping coefficient δ_{th} is rather lengthy and we shall omit it here. However, we can understand in a simple way why heat conduction leads to net loss of vibratory energy by considering the flows of entropy. As the gas in a bubble contracts and expands during a sonic cycle the temperature in the surrounding liquid remains nearly constant while the temperature of the gas (except for relatively small bubbles) rises and falls during the cycle. There is therefore a temperature difference between the gas and liquid, which varies periodically with time. Hence heat is conducted alternately out from and into the gas. Suppose the temperature of the gas at time t is T(t), given by the sum of a constant part T_c and a part $T_s(t)$ that varies sinusoidally with time. Starting from an instant when T_s is zero the gas returns to the same temperature T_c and the same thermodynamic state at the end of each cycle. If the rate of heat flow out from the bubble at time t is $\dot{q}(t)$ the corresponding rate at which entropy flows outward is given by the momentary value of $\dot{q}/T$. Clearly T is relatively large when $\dot{q}$ is positive, and small when $\dot{q}$ is negative. Suppose $\dot{q}$ were to vary sinusoidally with time, its average value being zero; then the corresponding average value of $\dot{q}/T$ over a cycle would be nonzero negative. But this would be impossible: Entropy is a state function and must return to its initial value when the gas returns to its original state. The difficulty is removed by assuming that in order for a "vibratory steady state" to be achieved, i.e., one in which the temperature repeats itself every cycle, the time-averaged or constant temperature T_c must rise to a value higher than the liquid temperature T_o. For then there will be a steady heat flow $\dot{q}'$ outward from the bubble, superposed on the oscillatory part $\dot{q}$. Associated with the outward heat flow is an outward flow of entropy equal to $\dot{q}'/T_c$. In order for the entropy to return to the same value at the end of each cycle the average outward flow during a cycle must be zero; thus, when $T_s << T_c$,

$$\dot{q}'/T_c + < \dot{q}/T > \cong 0 \tag{23}$$

where the bracket < > indicates time-average over a cycle. The existence of a steady heat flow $\dot{q}'$ outward from the bubble means, of course, that vibratory energy is being continually degenerated into heat. The rate of energy loss can be expressed in the form $b\dot{\xi}_b^{\,2}$, as with other damping mechanisms. Devin[7] has reviewed and corrected a treatment by Pfriem of the heat conduction problem; in so doing he has determined the values of $b = b_{th}$ and the damping constant δ_{th}. The required expressions are too lengthy to be repeated here, but values of δ_{th} are plotted in Figure 2. The abscissae are expressed in terms of the nondimensional parameter $2\phi R_o$, where ϕ is the reciprocal of the thermal diffusion layer thickness in the gas and is given by

$$\phi = (\omega/2D_{th})^{\frac{1}{2}} \; ; \; D_{th} = K/\rho s_p \; . \tag{24}$$

Here D_{th} is the thermal diffusivity, K the thermal conductivity coefficient, ρ the density and s_p the specific heat at constant pressure, all for the gas phase. For air at room temperature D_{th} is equal to about 0.18 cm^2/sec and hence at a frequency of 1 MHz the quantity ϕ has the value 4200 cm^{-1}. For a bubble resonant at 1 MHz the approximate radius R_o is 3.3 μm, $2\phi R_o$ is 2.8,

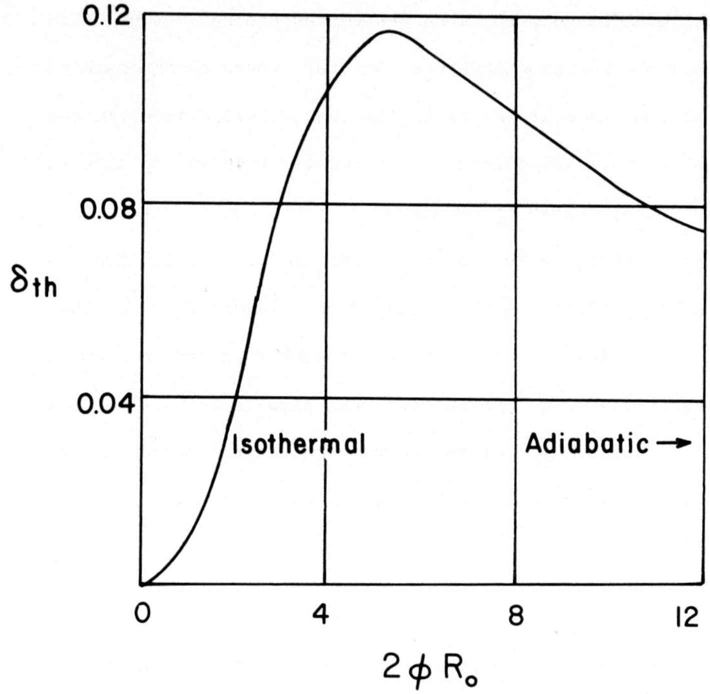

Fig. 2 Plot of thermal damping coefficient δ_{th} for a resonant gas bubble versus the nondimensional parameter $2\phi R_o$; ϕ is defined by Eq. (24). From Devin.[7]

and δ_{th} is about 0.07. At large values of $2\phi R_o$ the bubble responds nearly adiabatically during a vibration cycle, while at these rather small values the temperature remains more nearly constant. The total damping constant δ_{th} is calculated by adding the three contributions:

$$\delta_{tot} = \delta_{rad} + \delta_{vis} + \delta_{th} \ . \qquad (25)$$

For a resonant bubble at 1 MHz in water the respective constants are 0.014, 0.06 and 0.07, yielding a value for δ_{tot} of 0.14. As the frequency is raised above 1 MHz the contribution δ_{rad} increases (rather slowly), δ_{vis} increases and δ_{th} decreases. Of course, δ_{vis} becomes larger for liquids of higher viscosity.

Results given above for the damping constant are primarily for resonance conditions. More general expressions have been given by Eller.[8a] In Sections 1.3 and 1.5 we assume that b and δ are constant and equal to their resonance values, an approximation which is valid for frequencies near those for resonance, but not otherwise.

1.5 Scattering by a Gas Bubble

1.5.1 The Scattered Pressure Field

From results given previously in this section we know that when a sound field is set up in a liquid where a gas bubble exists, the latter will be set into volume oscillations. Hence it will act as a secondary source. This means that it will generate a sound field, the "scattered" field, which is superposed on the original, or "ambient" field. Let ρ_o be the liquid density and let the velocity of the bubble surface be $\dot{\xi}_b$, where $\dot{\xi}_b$ is given by Eqs. (12) through (16). Assuming spherical symmetry the scattered field will be an outgoing spherical wave in which the sound pressure p_s near the bubble is given approximately by the real part of

$$p_s = j\omega\rho_o R_o^2 \dot{\xi}_b / r \,. \tag{26}$$

Here R_o is the bubble radius and r measures distances from the bubble center; this approximation is valid for distances r small compared to the wavelength. Assuming $\dot{\xi}_b$ is given by Eqs. (12)-(16) we obtain p_s as the real part of

$$p_s = \frac{\Omega^2 p_o R_o / r}{[(1 - \Omega^2)^2 + \Omega^2\delta^2]^{\frac{1}{2}}} e^{j(\omega t + \alpha)} . \tag{27}$$

Here α is the phase angle between the scattered pressure p_s and the ambient pressure $p_o \cos \omega t$, and satisfies Eq. (16). As in that equation the multi-valued arctangent function is evaluated so that α increases monotonically from zero to π as Ω increases from zero to infinity; we note that $\alpha = \frac{1}{2}\pi$ when $\Omega = 1$. Hence in the scattered and ambient fields the pressures are in phase at low frequencies, but the phase difference between them increases with frequency ω, being $\frac{1}{2}\pi$ at the resonance frequency ω_o and approaching π for $\omega >> \omega_o$.

Relationships between the phase and magnitude of pressure in the scattered and ambient fields are shown by the plots in Figure 3. Here the vector $\underline{B}$ represents the complex amplitude of the scattered pressure, i.e., the expression obtained on the right hand side of Eq. (27) by letting t = 0; thus

$$\underline{B} = p_o \Omega^2 R_o r^{-1} [(1 - \Omega^2)^2 + \Omega^2 \delta^2]^{-\frac{1}{2}} e^{j\alpha} \tag{28}$$

For comparison the complex amplitude $\underline{A}$ of the ambient field, equal to the real number p_o, is shown on the same figure. The damping constant δ has the values 0.1 and 0.2 for the two curves; the pressure is evaluated at $r = R_o$, i.e., at the bubble surface. We see that the phase of $\underline{B}$ changes rapidly in the vicinity of the resonant condition $\Omega = 1$, especially between $\Omega = 0.9$ and $\Omega = 1.1$. For Ω greater than about 1.5 the scattered amplitude $\underline{B}$ is nearly oppositely directed to $\underline{A}$, and at large values of Ω the vectors $\underline{B}$ and $\underline{A}$ almost cancel, leading to essentially zero pressure amplitude at the bubble surface. At resonance the scattered pressure amplitude at the bubble surface is $1/\delta$ times the ambient amplitude.

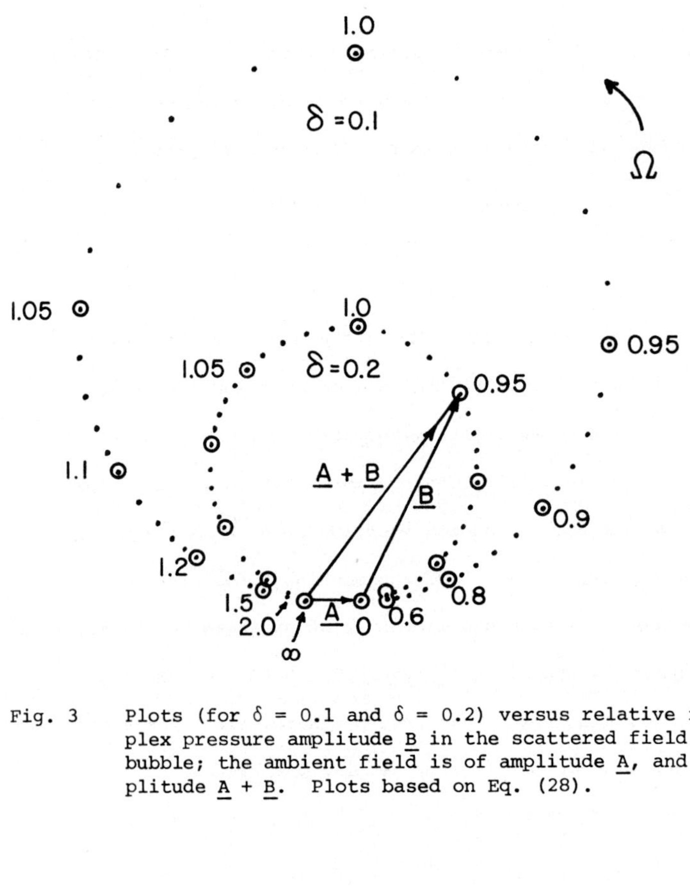

Fig. 3 Plots (for $\delta = 0.1$ and $\delta = 0.2$) versus relative frequency Ω of the complex pressure amplitude $\underline{B}$ in the scattered field at the surface of a gas bubble; the ambient field is of amplitude $\underline{A}$, and the total field has amplitude $\underline{A} + \underline{B}$. Plots based on Eq. (28).

Similar plots for $r > R_o$ would be similar in shape to those in Figure 3, but the loop traced by the vector $\underline{B}$ for any given value of δ would be reduced in size. For example, at a distance of one radius from the bubble surface, i.e., at $r = 2R_o$, the phase of $\underline{B}$ for given Ω and δ would be the same as in Figure 3, but the magnitude $|\underline{B}|$ would always be reduced by a factor of two.

In Figure 4 the total pressure amplitude in the field, i.e., the magni-

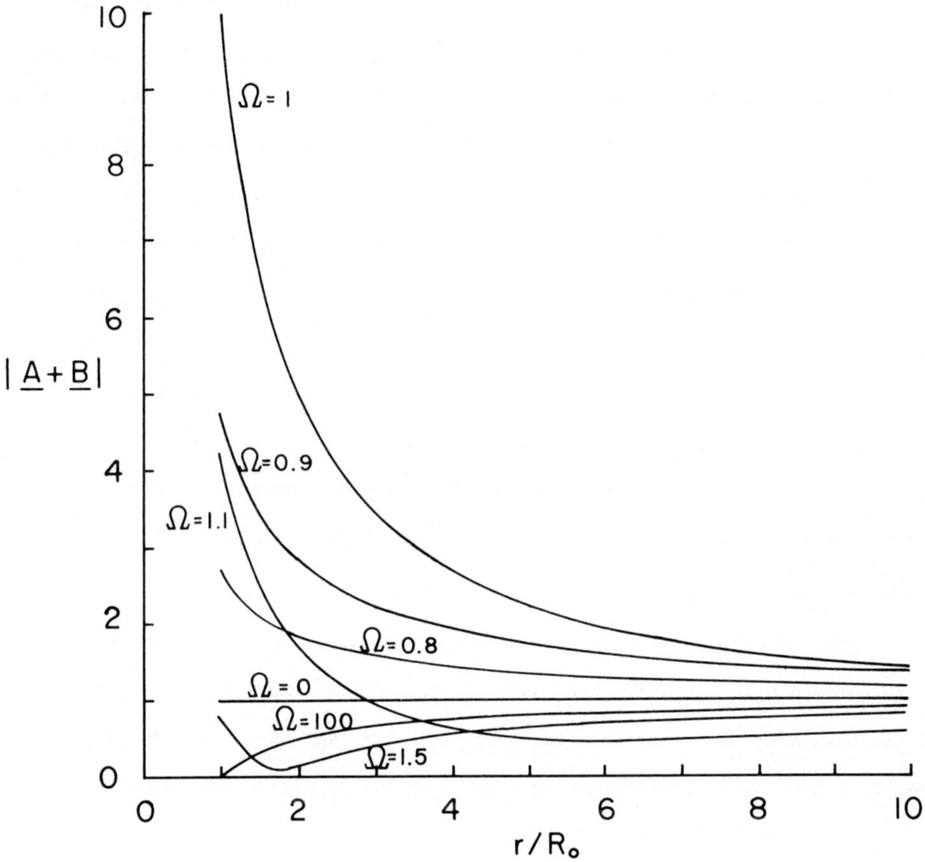

Fig. 4 Plot of total pressure amplitude $|\underline{A} + \underline{B}|$ in the field of a bubble versus nondimensional distance r/R_o for different values of Ω; $\underline{B}$ and $\underline{A}$ are defined in Eq. (28) and subsequent discussion. $\delta = 0.1$.

tude of $|\underline{A} + \underline{B}|$, is plotted against the nondimensional distance r/R_o for various values of Ω, all for $\delta = 0.1$. At $\Omega = 0$, i.e., at zero frequency, the pressure amplitude is everywhere equal to the ambient value, since the scattered pressure is then zero. For $\Omega < 1$, i.e., at driving frequencies ω equal to or less than the resonant value ω_o, the pressure amplitude has its greatest value at $r = R_o$, decreases monotonically with increasing r, and approaches the ambient value p_o at large r. At $\Omega = 1$ the greatest pressure amplitude is slightly more than 10 p_o, at $\Omega = 0.9$ it is 4.7 p_o, and at $\Omega = 0.8$ it is 2.7 p_o.

For $\Omega > 1$ the plots are more complex. For $\Omega = 1.1$ the greatest pressure amplitude is 4.2 p_o, and occurs at $r = R_o$; as r increases the pressure amplitude decreases to a broad minimum of about 0.4 p_o, occurring at about $r = 6\ R_o$, then slowly rises to approach p_o at large r. For $\Omega = 1.5$ the pressure amplitude at $r = R_o$ is only 0.8 p_o; with increasing r the pressure amplitude falls to a minimum of about 0.1 p_o at $r = 1.8\ R_o$, then rises to approach p_o. This behavior results from the fact that for $\Omega > 1$, the real part of the scattered amplitude is oppositely directed to the amplitude of the ambient pressure (the latter being real by definition). The minimum occurs approximately when the respective real parts are equal in magnitude, and thus cancel. For very high frequencies, $\Omega = 100$, the cancellation occurs at $r = R_o$; then the resultant pressure amplitude is zero at $r = R_o$ and increases monotonically with distance from the bubble.

1.5.2 Absorption-Scattering Cross-Section of a Gas Bubble

When an ultrasonic beam passes through a medium containing gas bubbles it sets these into vibration and, as a result, energy is lost from the beam. If the time-averaged power lost per bubble is $\langle\dot{W}\rangle$ and the intensity of the beam is I the absorption-scattering cross section σ_{as} is defined such that

$$\langle\dot{W}\rangle = \sigma_{as}\, I \ . \qquad (29)$$

In Eq. (10) the term $b\dot{\xi}_b^{\,2}$ gives the instantaneous rate at which mechanical energy is lost; its time-average gives the quantity $\langle\dot{W}\rangle$ we seek. The constant b is related to the damping constant δ by Eq. (15) so we may write, e.g.:

$$<\dot{W}> = m\omega_o\delta \ < \dot{\xi}_b^{\,2} > ; \tag{30}$$

here δ is the total damping coefficient, identical to the quantity δ_{tot} defined in Eq. (25).

Especially interesting and important is the condition of resonance ($\omega = \omega_o$). Under this condition we notice that the left hand side of Eq. (11) reduces to $b\dot{\xi}_b$, and hence we can readily solve for $\dot{\xi}_b$. Substituting this result into Eq. (30) and letting $\delta = \delta_{tot}$ we obtain

$$< \dot{W} > = 2\pi R_o p_o^{\,2}/\rho_o\omega_o\delta_{tot} \ . \tag{31}$$

For a plane wave travelling through a medium in which c is the velocity of sound, the intensity I is equal to $p_o^{\,2}/2\rho_o c$. Using this expression in Eq. (31) and comparing with Eq. (29), we find that the cross-section σ_{as} is given by

$$\sigma_{as} = (2\lambda_o/\pi\delta R_o)\ \pi R_o^{\,2}, \tag{32}$$

where λ_o is the ultrasonic wavelength in the medium at frequency ω_o, and $\pi R_o^{\,2}$ is the geometric cross-section. In water, when Eq. (19) holds, we find that λ_o/R_o is 457 and hence obtain

$$\sigma_{as} = (291/\delta)\ \pi R_o^{\,2} \ . \tag{33}$$

If $\delta = 0.1$ the absorption-scattering cross-section for a resonant gas bubble is nearly 3000 times its geometric cross-section. Clearly the resonant gas bubble is extremely effective in absorbing and scattering energy from a sound wave.

1.6 The Vibrating Bubble as a Heat Source

As noted in the previous section, a gas bubble in the path of an ultrasonic beam extracts energy from it. Part of the energy (a fraction equal to $\delta_{rad}/\delta_{tot}$) is used in generating the scattered waves described in Section 1.5. Another part, associated with δ_{vis}, is converted into heat by viscous dissipation in a shell of liquid surrounding the bubble. The remainder of the extracted energy, associated with δ_{th}, appears as heat in the gas phase, as a consequence of heat conduction. It is easy to obtain an expression for $< \dot{Q} >$, the total time-averaged rate of heat generation by a bubble. Returning to Eq. (31) we realize that $<\dot{Q}>$ is just the part of $<\dot{W}>$ associated with δ_{vis} and δ_{th}, and obtain

$$<\dot{Q}> = <\dot{W}> \cdot \frac{\delta_{vis} + \delta_{th}}{\delta_{tot}} . \tag{34}$$

We now consider the vibrating bubble as a steady heat source and solve the heat conduction problem; for the temperature T in the surrounding liquid we find that

$$T = T_o + \frac{<\dot{Q}>}{4\pi K'r} , \tag{35}$$

where T_o is the temperature in the absence of sound, K' is the thermal conductivity coefficient for the liquid, and r is distance from the bubble center. The temperature T_b in the bubble is obtained by letting $r = R_o$; substituting from Eqs. (34) and (35) we obtain

$$T_b - T_o = \frac{p_o^2}{2\rho_o\omega_o K'} \cdot \frac{\delta_{vis} + \delta_{th}}{\delta_{tot}^2} . \tag{36}$$

In water $K' = 6.0 \times 10^4$ ergs/sec cm °C; at 1 MHz for a resonant bubble we take $\delta_{tot} = 0.145$ and the sum of δ_{vis} and δ_{th} to be 0.13 (Section 1.4) at this frequency the temperature rise is then

$$T_b - T_o = 8.2 \times 10^{-12} p_o^2 , \tag{37}$$

where p_o is to be expressed in dyn/cm^2. For a pressure amplitude of 0.1 atm or about 10^5 dyn/cm^2 (corresponding to an intensity of 0.003 W/cm^2 in a travelling plane wave) the temperature rise $T_b - T_o$ is 0.08 °C.

1.7 Rectified Diffusion

It is easily demonstrated that a gas bubble immersed in a liquid containing dissolved gas will grow when set into vibration of sufficient amplitude. The process has been termed "rectified diffusion" by Harvey.[9] Since the theory is rather involved we shall not attempt to present it here, but only cite some of the more important results. Consider a gas bubble of radius R_o, set into vibration at angular frequency ω with radial surface amplitude ξ_{bo}.

Let D be the diffusion coefficient for the gas molecules in the surrounding liquid and define β by

$$\beta = (\omega/2D)^{\frac{1}{2}} \; ; \tag{38}$$

then β^{-1} is the <u>diffusion layer thickness</u>. For the O_2 molecule in water at 20°C the coefficient D is 2.4×10^{-5} cm^2/sec; at a frequency of 1 MHz the corresponding value of β^{-1} is 0.028 μm. Let $< J >$ be the time-averaged rate at which gas enters the bubble *via* sonically aided diffusion. Then, according to a result given by Hsieh and Plesset,[10] modified according to a suggestion of M. Strasberg (private communication) we have

$$< J > = 24\pi D R_o C_o \left[1 + \frac{2\sigma}{P_o R_o}\right] \left[\frac{\xi_{bo}}{R_o}\right]^2 \; ; \tag{39}$$

here σ is the surface tension while C_o is the concentration of the gas which would be dissolved in the liquid if it were saturated at the static pressure P_o. The Hsieh-Plesset equation was arrived at by a method of successive approximations to the nonlinear diffusion equations; it was assumed that changes in volume occur isothermally, that $\xi_{bo} << R_o$, and that $\beta R_o >> 1$. If C_o has the units of mass/volume the quantity $< J >$ gives the increase in mass of the bubble per unit time.

In the absence of sound gas diffuses across the bubble surface at a steady rate J* such that

$$J^* = -4\pi \, C_o R_o D \, [1 - \Gamma + (2\sigma/P_o R_o)]; \tag{40}$$

here Γ is the ratio C_∞/C_o, where C_∞ is the concentration of dissolved gas in the liquid in the absence of a bubble. From Eq. (40) we see that J* is negative, indicating outward flow, when $\Gamma < 1$. The net rate of flow is given by the sum of $< J >$ in Eq. (39) and J* in Eq. (40). At small values of the relative amplitude ξ_{bo}/R_o the bubble shrinks, while at large values it grows. For the critical condition where $< J >$ is equal to $-J^*$ the amplitude is given by

$$(\xi_{bo}/R_o)^2 = (1 - \Gamma + \Sigma)/6(1 + \Sigma), \; \Sigma = 2\sigma/P_o R_o \; . \tag{41}$$

This equation provides a critical value of the relative displacement amplitude

ξ_{bo}/R such that bubble shrinkage occurs for smaller values, and bubble growth for larger ones. Assuming Eqs. (13-15) apply, the quantity ξ_{bo}/R can be eliminated in terms of the relative pressure amplitude p_o/P_o; specifically (representing isothermal conditions by setting $\gamma = 1$) ξ_{bo}/R_o is then equal to $\chi p_o/3P_o'$, where P_o' is given in Eq. (20); then Eq. (41) becomes

$$\frac{p_o^2}{(P_o')^2} = \frac{3(1 - \Gamma + \Sigma)}{2\ (1 + \Sigma)}\ [(1 - \Omega^2)^2]\ . \tag{42}$$

This equation is similar to one given by Safar[10a]; it is a rather simple extension of one derived by Strasberg[11] from results of Hsieh and Plesset;[10] it reduces to the Hsieh-Plesset-Strasberg equation when $\Omega = 0$, except for the factor $1 + \Sigma$ in the denominator which was inadvertently omitted. From Eq. (42) we see that the pressure amplitude threshold for bubble growth by rectified diffusion may be very low for resonant bubbles ($\Omega = 1$). Thus for an air bubble in water resonant at 1 MHz the ratio Σ is about 0.5; setting $\delta = 0.14$ we obtain from Eq. (42) that $p_o/P_o' = 0.10$ and $p_o/P_o \cong 0.18$. Hence for this situation bubble growth occurs for p_o greater than 0.18 atm, corresponding to an intensity of 6 mW/cm^2 in a plane travelling wave.

Eller and Flynn[12] solved the nonlinear equations for bubble motion and diffusion by computer methods and, in this way, avoided small-amplitude restrictions. They also carried out analytical solutions for small-amplitude conditions, including effects of resonance and arbitrary values of γ, though assuming no damping. Their approximate analytical solution resembles Eq. (42) but is, in general, more complex and too lengthy to be cited here. Eller[13] has shown that when $\Sigma \ll 1$ and $\Gamma = 1$ the Eller-Flynn approximate result reduces to

$$\frac{p_o^2}{P_o^2} = \frac{3\gamma\Sigma}{2}\ \frac{(1 - \Omega^2)^2}{[1 - (\Omega^2/8)]}\ . \tag{43}$$

This result, given by Eller, may be compared with one obtained from the extended Hsieh-Plesset-Strasberg equation, Eq. (42). While the two equations are similar for $\Gamma = 1$, they differ in that Eq. (42) includes effects of damping, while Eq. (43) contains the factors γ and $[1 - (\Omega^2/8)]$. Eller and Flynn[12] found that their analytical solution gave an adequate approximation to the computer-generated results under some circumstances but, in general, they differ by an amount that can be considerable.

Experimental tests by Strasberg[11,14] at 24.5 kHz, by Eller[13,15] at 11 kHz and 26.6 kHz, and by Gould[16] at 20 kHz showed reasonable agreement with either of Eqs. (42) or (43). In these tests the bubbles were small compared to resonant size, so that effects of damping were negligible and $\Omega^2/8 \ll 1$. There is uncertainty about the value of γ to be used; for air it lies between 1 and 1.4.

Miller[7a] has carried out experiments at 1 MHz with resonant bubbles, in which he was able to obtain data on rectified diffusion. His measurements on the threshold for bubble growth by rectified diffusion are in reasonable agreement with Eq. (42).

Referring back to Eq. (39), this gives the rate at which gas flows into a bubble during rectified diffusion. This may be modified by using Eqs. (13-15) to eliminate the displacement amplitude ξ_{bo} in terms of the driving pressure amplitude p_o. An alternative, rather similar, approximate expression has been derived by Eller.[15] In the experimental tests by Eller[13,15] at frequencies of 11 kHz and 26.6 kHz the rate of bubble growth was found to be much greater than predicted. Other experiments were carried out by Gould;[16] he found that rate of growth to be given approximately by the Eller theory when conditions were such as to minimize acoustic microstreaming near the bubble. But when surface waves were established and vigorous microstreaming generated, the rate of gas transport was greatly increased. No adequate theory is available for this important phenomenon.

1.8 Radiation Forces

1.8.1 The Radiation Force on a Bubble in a Standing Wave

In a standing-wave sound field any gas bubbles which are smaller than resonant size tend to collect at maxima of pressure amplitude, while larger ones move to minima. This behavior exemplifies the action of radiation forces on highly compressible bodies. These forces are very important since they affect the movements and spatial distributions of bubbles in a sound field. It is not hard to understand the origin of the sonically generated force on a compressible body. In a field where a pressure gradient ∇p exists a body of volume V is acted on by a force given by $-V\,\nabla p$. If the pressure everywhere varies sinusoidally in time with frequency f the volume V and the gradient ∇p will also vary in time with the same frequency f. Unless the functions $V(t)$ and $\nabla p(t)$ differ in phase by $\pm 90^{\circ}$ (or $\pm 270^{\circ}$, etc.) the time-averaged product is nonzero, and hence the body is acted on by a nonzero time-averaged force. To obtain a specific result for a spherical gas bubble let $<\underline{F}>$ be the time-averaged (vector) force; then

$$<\underline{F}> = -<V\,\nabla p>\ , \tag{44}$$

where the bracket $<\ >$ signifies the average over a cycle. Assume the pressure p is given by Eq. (3), where it is understood that the pressure amplitude p_o may be a function of the vector position $\underline{r}$, i.e., $p_o = p_o(\underline{r})$; assume however, that $\nabla P_o = 0$. The volume of the bubble at any instant is $4\pi R^3/3$ where R is given by Eq. (4) as $R_o + \xi_b$, and the displacement ξ_b is given by Eqs. (12-16). Assuming that $\xi_{bo} << R_o$, the expression for V can be written approximately as

$$V = V_o\ (1 + 3\xi_b R_o^{-1}), \tag{45}$$

where V_o is the volume $4\pi R_o^3/3$ in the absence of vibration. Returning to Eq. (44), evaluating V and p as indicated above, and time-averaging, we obtain

$$<\underline{F}> = \frac{V_o p_o\ \nabla p_o}{2\gamma P_o'}\ F_o(\Omega,\delta)\ , \tag{46a}$$

where P_o' is defined in Eq. (20) and

$$F_o\ (\Omega,\delta) = \frac{1 - \Omega^2}{(1-\Omega^2)^2 + \Omega^2\delta^2}\ . \tag{46b}$$

We see from Eqs. (46) that the direction of the radiation force $<\underline{F}>$ depends on the relative frequency $\Omega = \omega/\omega_o$. This force is along the direction of ∇p_o if $\Omega < 1$, and is along the opposite direction if $\Omega > 1$. Hence a relatively small bubble whose resonance frequency ω_o is greater than the driving frequency ω will be driven in the direction of increasing pressure amplitude p_o. This agrees with observations, referred to earlier, that such small bubbles tend to collect at maxima of p_o. As might be expected, bubbles of subresonant size are important biologically; since they move to maxima of p_o they are set into vibration of high amplitude, and cause alterations in cells and other structures in these regions. Effects on plant cells at pressure antinodes of a standing wave field were reported by Goldman and Lepeschkin,[17] and the significance to their work of radiation forces on bubbles (sometimes called "Bjerknes forces") was pointed out by Blake.[18]

It should be realized that in the megahertz frequency range bubbles of subresonant size are very small and are unlikely to be detected except by special methods. Much more obvious are the larger bubbles, for which $\omega_o < \omega$. Since these drift to minima of p_o they are often relatively inactive; however, in a complicated field such as a quasi-stationary wave situation the sound pressure minima may not be greatly less than the pressure maxima and surface wave activity of high mode order may occur on the bubbles (see page 113).

More detailed features of the function F_o (Ω,δ) are shown in Figure 5; here F_o is plotted against Ω for $\delta = 0.1$ and $\delta = 0.2$. We see that F_o equals unity at low frequencies, arises slowly at first with increasing Ω, then more rapidly as resonance is neared. For $\delta = 0.1$ the peak value of F_o is 5.26 and occurs at $\Omega = 0.95$; for $\delta = 0.2$ the peak is 2.78 and occurs at $\Omega = 0.89$. As Ω increases slightly beyond the peak the value of F_o drops precipitously, falling through zero at $\Omega = 1$. For $\delta = 0.1$ the greatest negative value of F_o is -4.76 and occurs at $\Omega = 1.05$; for $\delta = 0.2$ the greatest negative value is -2.27 and is at $\Omega = 1.10$. (If δ is sufficiently small we find that the greatest positive and negative values of F_o are $\pm$ $(1/2\delta)$ and occur at $\Omega = 1 \mp \frac{1}{2}\delta$). When Ω increases further F_o remains negative but its magnitude decreases, to approach zero at high frequency.

Theory similar to that outlined above for the radiation force on a gas bubble has been tested experimentally by Eller.[19] For bubbles of radius varying

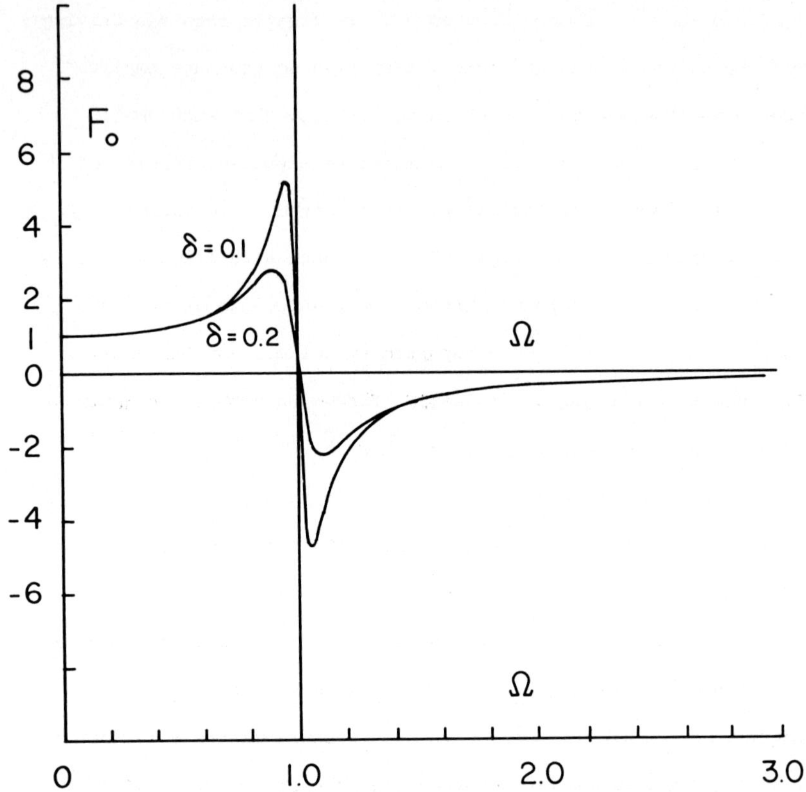

Fig. 5 Plot versus Ω of nondimensional force function F_o defined in Eq. (46b), for $\delta = 0.1$ and $\delta = 0.2$.

from 23 to 79 μm in a standing wave field at 27 kHz, he found reasonable agreement (discrepancies ranging from 1% to 34%) between theory and experiment. Gould[20] has used gas bubbles for determining the pressure amplitude in a standing wave field; the force is measured on a bubble of known size in the field, and from this the pressure amplitude is calculated. Based on this approach he has developed a novel method for calibrating small hydrophones, and suggests that it should be valid in the frequency range 3 to 90 kHz.

We can estimate the speed U_r at which a bubble moves relative to the fluid if U_r is not too great. According to Crum and Eller[21] the drag force on a bubble of radius R_o at low speeds is the Stokes value for a rigid sphere of equilibrium radius, namely $6\pi\eta R_o U_r$. (But Strasberg[14] quotes results for radii less than 60 μm which correspond to a drag force of $7.4\pi\eta R_o U_r$). For a force given by $< F >$ in Eqs. (46) we obtain

$$U_r = \frac{< F >}{6\pi\eta R_o} = \frac{R_o^2 p_o \nabla p_o}{9\eta\gamma P_o'} \quad F_o(\Omega,\delta) \ . \tag{47}$$

As an example consider a standing wave field in which p_o is given by p_m sin kx, where $k = \omega/c$. Then ∇p_o is along x, its magnitude being $kp_m \cos kx$. The velocity U_r is also along x, with magnitude U_r given by

$$U_r = \frac{\omega p_m^2 R_o^2 F_o(\Omega,\delta)}{18\eta\gamma c P_o'} \quad \sin 2kx. \tag{48}$$

Consider an air bubble with radius 1 μm in water at atmospheric pressure, subject to a sound field at frequency 1 MHz and pressure amplitude 0.1 atm. Since $\Omega << 1$ we have $F_o = 1$. We then obtain from Eq. (48)

$$U_r = 0.17 \sin 2kx \ \text{(mm/sec)}.$$

Since the wavelength at 1 MHz in water is 1.5 mm the time required for a bubble to travel to the nearest pressure maximum (no more than one-fourth wavelength away), is only a few seconds. Hence even at this very low amplitude (corresponding to an intensity of 3 mW/cm^2 in a plane travelling wave) small bubbles quickly travel to the pressure maxima.

Theory similar to the above has been tested experimentally by Crum and Eller,[21] who measured speeds of bubbles in a standing wave field. At 26.8 kHz, for bubble radii of about 60 μm, and at relatively low amplitudes, they found good agreement between theory and experiment.

1.8.2 The Radiation Force on a Bubble in a Travelling Wave

In the previous section the theory is primarily for a sound field in which there are spatial variations in the amplitude of the pressure, but not in its phase. In a plane travelling wave a reverse situation holds: the phase varies, along the propagation direction, but the amplitude is everywhere constant. To treat this situation we assume that the amplitude p_o is constant in Eq. (3), while ωt is replaced by $\omega t+kx$. Then the same replacement

($\omega t+kx$ for ωt) should be made in the expression for ξ_b in Eq. (4). Using Eqs. (44) and (45) we now obtain, instead of Eqs. (46), the result

$$< F > = \left[\frac{V_o k p_o^2}{2\gamma P_o'}\right] \cdot \left[\frac{\Omega\delta}{(1 - \Omega^2)^2 + \Omega^2\delta^2}\right] \quad (49)$$

It is interesting to compare Eq. (49) with Eqs. (46). When the latter applies to a standing wave so that $p_o = p_m \sin kx$ (as is assumed in obtaining Eq. (48)) the maximum value for the magnitude of $p_o \nabla p_o$ is just $\frac{1}{2}k\ p_m^2$. When $p_o \nabla p_o$ is replaced by $\frac{1}{2}kp_m^2$ the fraction multiplying F_o in Eq. (46a) is just half the first bracket [ ] on the right hand side of Eq. (49), provided that there the amplitude p_o is set equal to p_m. When δ is sufficiently small the greatest value of the second bracket on the right hand side of Eq. (49) is δ^{-1} and occurs when $\Omega = 1$; we recall that the greatest magnitude of F_o is $1/2\delta$ when δ is small, but occurs when Ω is $1 \pm \frac{1}{2}\delta$. Thus (if δ is small) the greatest radiation force < F > on a gas bubble in a travelling wave of pressure amplitude p_o is just four times the corresponding force in a standing wave when the maximum amplitude p_m is equal to p_o.

It is also worth noting that when $\Omega = 1$ the expression for < F > in Eq. (49) is just equal to $<\dot{W}>/c$, where $<\dot{W}>$ is given in Eq. (31). From Eq. (29) the power loss $<\dot{W}>$ is just equal to σ_{as} I; expressions for the absorption-scattering cross section σ_{as} are given in Eqs. (32) and (33). That < F > should be related in this way to $<\dot{W}>$ and σ_{as} was suggested by Westervelt[21a] and is discussed in Chapter I.

1.8.3 Radiation Forces Between Bubbles

It is well known that when two bubbles, say, A and B, are in the same sound field a radiation force acts on each along the line connecting them. Whether the force is attractive or repulsive depends on their resonant frequencies ω_A and ω_B, respectively, relative to the frequency ω of the ambient field. Suppose $\omega_A >> \omega$; then from results of Section 1.8.1 we know that a radiation force will act on A in the direction of increasing pressure amplitude. Suppose also that $\omega_B > \omega$; then we know from the plots in Figure

4 (for Ω = 0.8, 0.9 and 1) that the pressure amplitude in the field near B decreases monotonically with distance from B. Hence the radiation force on A will be toward B, and the bubbles will come together.

Now suppose that B is relatively large, so that $\omega_B < \omega$; in particular, suppose Ω = 1.5 for B. Then from the corresponding plot in Figure 4 we see that the pressure amplitude p_o decreases with distance from the bubble for distances $(r - R_o)$ less than about 0.8 R_o from the bubble surface, but increases with r for greater distances. Hence a small bubble A (for which we still assume $\omega_A >> \omega$) will be attracted to B at short distances, but repelled at larger distances. The "attractive region" becomes smaller as Ω increases, approaching zero for large Ω ; thus if B is very much larger than the resonant size, A will be repelled from it at all distances.

It should be realized that the above discussion of interactions is valid only when the bubble A is relatively small, so that $\omega_A >> \omega$. In this situation the presence of A has little effect on the vibration of bubble B. In general, however, the pulsation of each bubble is influenced by the presence of the other, a phenomenon which has been discussed by Nyborg and Storm.[22]

1.8.4 Radiation Force on a Particle Near a Bubble

A pulsating bubble can sometimes act as an effective collector of particles in its own vicinity. To see this we refer to an expression given in Chapter 1 for the radiation force on a small particle in terms of gradients of the time-averaged density of kinetic and potential energy. Near a gas bubble whose size is in the vicinity of that for resonance only the kinetic energy is of consequence. An expression for the radiation force F directed radially from the bubble on a neighboring particle of volume V can then be written

$$F = \frac{3V(\rho-\rho_o)}{2\rho+\rho_o} \frac{\partial \langle KE \rangle}{\partial r} \quad . \qquad (50)$$

Here <KE> is the time-averaged kinetic energy density, ρ is the density of the particle and ρ_o that of the medium. If the particle is spherical of radius a and Stokes law applies, the force F causes the particle to move with velocity $\dot{r}$ equal to $F/6\pi\eta a$, where η is the viscosity of the medium. Using Eq. (50), we can then write approximately

$$\dot{r} = \frac{3s}{2\rho + \rho_o} \frac{\partial \langle KE \rangle}{\partial r} , \tag{51}$$

where s is the "sedimentation coefficient" referred to in centrifugation theory, given for a sphere by

$$s = 2a^2 (\rho - \rho_o)/9\eta . \tag{52}$$

In the vicinity of a bubble of radius R_o vibrating with amplitude ξ_{bo} at angular frequency ω the time-averaged kinetic energy density $\langle KE \rangle$ is given by

$$\langle KE \rangle = \rho_o \omega^2 \xi_{bo}{}^2 R_o{}^4 / 4r^4 . \tag{53}$$

The gradient has its greatest value at the surface of the bubble, namely,

$$\partial \langle KE \rangle / \partial r = - \rho_o \omega^2 \xi_{bo}{}^2 / R_o , \tag{54}$$

but drops off rapidly with distance, according to the inverse fifth power of the radius r.

It is instructive to make an approximate calculation, using theory from Section 1.3 and the present section. At a frequency of 1 MHz in water a driving pressure amplitude of about 0.05 atm will set a resonant bubble ($R_o \simeq 3.3$ μm) into oscillation with an amplitude of $R_o/10$. For this condition $\partial\langle KE \rangle/\partial r$ at the bubble surface ($r = R_o$) is 1.3×10^8 in cgs units. For platelets in saline ($\rho - \rho_o = 0.05$ g/ml, a = 1 μm, $\eta = 0.01$ P) the sedimentation coefficient s is about 10^{-8} sec. From Eq. (51) the speed $|\dot{r}|$ is approximately equal to 1.3 cm/sec at $r = R_o$; at $r = 2R_o$ the speed is less by a factor of 2^5 and is 0.4 mm/sec. An easy calculation shows that under the assumed conditions all platelets in the region $R_o < r < 2R_o$ will arrive at the bubble surface within 0.003 sec. A vibrating bubble can thus be an effective collector of particles, as was stated earlier. This has been demonstrated experimentally in unpublished work by D. L. Miller at the University of Vermont.

1.9 Microstreaming near a Vibrating Bubble

It is shown in Chapter 1 that sound fields give rise to time-independent circulation, i.e., acoustic streaming, in fluid media. Near small obstacles in a sound field, near a vibrating membrane, or near a small sound source, the streaming may be of microscopic scale, and is hence sometimes called microstreaming. Of special importance to bio-effects of ultrasound is the microstreaming set up near a vibrating bubble.

Elder[23] made detailed studies of the motion near vibrating bubbles of near-resonant size at 10 kHz. In his experiments the bubble was situated near a solid surface. He found that the streaming pattern might take on any of four different forms, depending on the purity of the liquid, and on the amplitude of the vibration. Elder also made determinations of the streaming speeds and, under some conditions, found order-of-magnitude agreement with an approximate theory suggested by Nyborg.[24] In this theory the bubble is assumed to be a simple source, generating a spherical wave. As the wave travels, it grazes a solid surface S on which the bubble rests, and a time-dependent acoustic boundary layer is established. In this boundary layer a time-independent field of "effective force" is set up, which gives rise to the microstreaming. (See Chapter 1). Associated with the microstreaming is another boundary layer, this one time-independent, set up near the surface S. Both boundary layers are characterized by the thickness δ' ($=\delta$) defined in Chapter 1 as $(2\nu/\omega)^{\frac{1}{2}}$, where ω is the angular frequency of the sound and ν is the kinematic viscosity coefficient.

At the surface S the velocity is assumed to be zero, but its tangential component increases rapidly with distance z from the boundary, achieving a "characteristic" or "limiting" value U_L when z reaches a value somewhat in excess of δ'. According to the "simple source" theory the limiting velocity at a distance r from the center of a hemispherical bubble of radius a resting on a boundary is given by

$$U_L = \omega\xi_{bo}^{\;2}\, a^4\, r^{-5}, \tag{55}$$

where ξ_{bo} is, as before, the displacement amplitude at r = a, i.e., at the bubble surface. At r = a the limiting velocity becomes

$$U_L = \omega\xi_{bo}^2/a. \quad (56)$$

The velocity gradient G associated with the microstreaming has its greatest value in the boundary layer and can be approximated as U_L/δ'; also the associated shearing stress S is ηG and we obtain at r = a

$$G = \omega\xi_{bo}^2/a\delta' \; ; \; S = \eta\omega\xi_{bo}^2/a\delta' \; . \quad (57)$$

At a frequency of 1 MHz near a bubble of radius 5 μm vibrating with amplitude 0.5 μm we have from Eq. (56) that the limiting speed at r = a is 31 cm/sec. The boundary layer thickness δ' is $(2\nu/\omega)^{1/2}$ and at 1 MHz in water is very small, having the value 0.56 μm. High values of the velocity gradient and shearing stress exist in the boundary layer near the bubble; from Eq. (57) these are given approximately by

$$G = 5.5 \times 10^5 \text{ sec}^{-1}; \quad S = 5500 \text{ dyn/cm}^2 \; .$$

The microstreaming associated with bubble motion can be very significant biologically, just as was seen for the microstreaming discussed in Chapter I. Suspended objects, such as cells or macromolecules, are carried in streaming orbits and may be brought momentarily into the boundary layer near a bubble once during each traverse of an orbit. When in this boundary layer they may be distorted or fragmented by the high shearing stress. Rooney[25] carried out experiments with canine erythrocytes suspended in viscous dextran solutions at room temperature using a vibrating bubble at 20 kHz as his source of microstreaming. Under these circumstances leakage of hemoglobin from the cells was obtained when the calculated shearing stress exceeded about 4500 dyn/cm^2. These results are comparable to those obtained with microstreaming set up near small vibrating wires, discussed in Chapter I.

In Eqs. (55-57) and subsequent discussion we have considered microstreaming set up near a vibrating gas bubble which rests on a solid boundary. Detailed theory has been developed by Davidson and Riley[26] for streaming near a pulsating and translating bubble freely suspended in liquid, away from any boundary; for this situation experimental tests have not yet been carried out. When $R_o >> \delta'$ the streaming speeds near such a free bubble are much smaller than those near a bubble resting on a solid boundary. However, when R_o is comparable to δ' the streaming speed for the two situations are of the same order of magnitude.

When surface waves are set up on the bubble the microstreaming is greatly altered. As mentioned in Section 1.7, Gould[16] has found surface-wave associated microstreaming to have a profound effect on transport of gas to or from a vibrating bubble.

Still another aspect of microstreaming arises in plant tissue because of the presence of intercellular gas. When the gas is set into vibration by an ambient ultrasonic field, it sets adjacent cell walls into vibration; because of this, microstreaming tends to occur within the cells bounded by these walls (Nyborg, Miller and Gershoy).[27]

1.10 Surface Waves

Up to this point in the present chapter the gas bubble is usually assumed to vibrate with complete symmetry about its center. This assumption leads to simplifications in the theory, and useful results are obtained. But actually, the assumption of symmetry is valid only under restricted conditions. In water, waves can be established on the bubble surface even at rather low amplitudes. These surface waves are important for several reasons: (1) They give rise to increased transport of heat and dissolved gas via microstreaming; (2) At high amplitudes they become distorted, so that microbubbles and jets are formed; and (3) They contribute to the acoustic spectra scattered by bubbles. We now consider the nature of these waves, and how they are set up.

While surface waves are usually part of a complex and chaotic activity under high-amplitude conditions, they may be coherent and of relatively simple form at the lower amplitudes. In the simplest situation, the displacement on the surface of a spherical bubble may have the form of a single spherical harmonic or mode, such as was discussed in Chapter 1. Thus, if θ and ϕ are polar and aximuthal angles, respectively, the displacement may be of a form given by the product of cos $m\phi$ and $P_m^n(\mu)$, where $P_m^n(\mu)$ is a Legendre function and μ is cos θ. The mode for which the indices (m,n) are equal to (0,0), respectively, represents spherical symmetry while the (0,1) mode represents uniform translation along the axis of symmetry. The simplest mode representing perturbation of the surface from its spherical shape is the (0,2) mode; for this the displacement is given by

$$\xi_{02} = a_{02}(3 \cos^2\theta - 1) \cos \omega t, \tag{58}$$

where a_{02} is a constant. This mode is one in which the surface contracts at its equator ($\theta = \frac{1}{2}\pi$) while it expands along the polar directions ($\theta = \pm \pi$), and *vice versa*. The characteristic angular frequencies ω_n for modes (m=0) with symmetry about the axis have been given by Lamb[8] (Par. 275) for $n \geqslant 2$:

$$\omega_n^2 = (n - 1)(n + 1)(n + 2)\, \sigma/\rho_o R_o^3 \quad ; \tag{59}$$

where σ is the interfacial tension at the bubble surface, ρ is the liquid's density and R_o is the equilibrium bubble radius. For water at room temperature (σ = 73 dyn/cm) plots of ω_n against R_o are shown in Figure 6 for n = 2, 3 and 4, based on Eq. (59). For a bubble 3.3 µm in radius (approximately the size for volume resonance at 1 MHz) the frequencies $\omega_n/2\pi$ for modes 2, 3 and 4 are 0.78 MHz, 1.43 MHz and 2.15 MHz, respectively. For a bubble of radius 2 µm the corresponding frequencies are 1.67 MHz, 3.04 MHz, and 4.56 MHz.

That such waves should be set up is expected on the basis of scattering theory, such as was taken up in Chapter 1; this theory is based on solutions of a wave equation which is linear with constant coefficients. For a given "driving" (or "ambient" or "incident") field of pressure amplitude p_o and frequency ω, the characteristic amplitude a_{on} of a given mode is proportional to p_o and depends on both the driving frequency ω and the characteristic mode frequency ω_n. According to such theory the amplitude a_{on} tends to be relatively

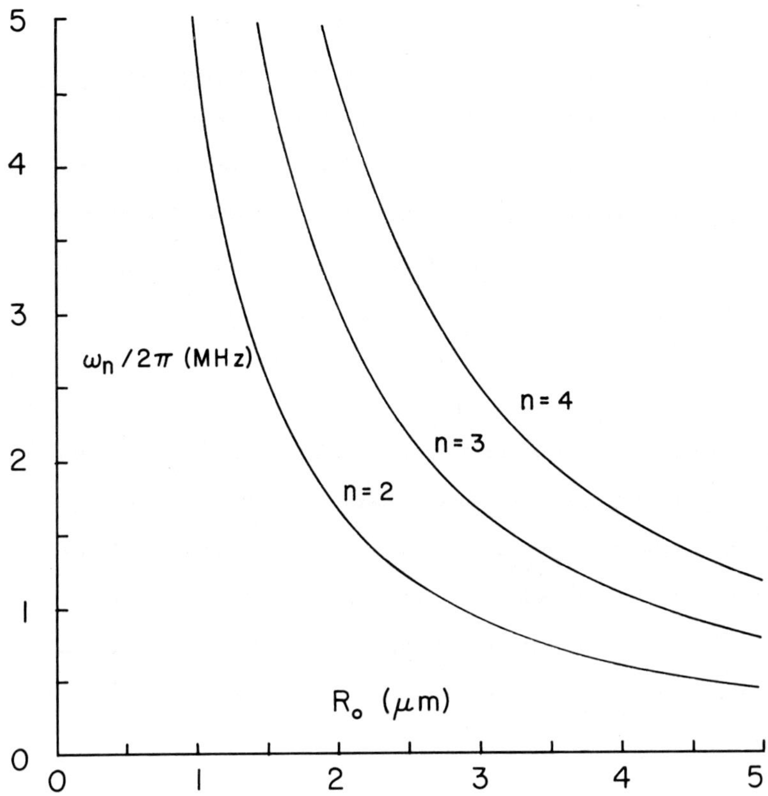

Fig. 6 Plot versus radius R_o of the characteristic frequency $\omega_n/2\pi$ for various surface modes n = 2, 3 and 4. From Eq. (59).

large (infinite, if damping is neglected) when ω is equal (or nearly equal) to ω_n. We shall, for convenience, refer to waves set up in this way as "linearly driven".

Alternatively, the waves may be set up by "parametric" excitation, a mechanism for which the theory was first given by Benjamin and Ursell.[28] For a plane liquid-gas interface vibrating with amplitude ξ along its normal with angular frequency ω they considered the growth/decay of a surface wave, assumed to have been initially set up in some arbitrary way. They found from hydrodynamic theory, that the amplitude of this wave satisfies

Mathieu's equation. Of particular interest is their prediction that the surface waves most likely to grow will be characterized by a frequency of $\omega/2$. Analogous theory for waves on a spherical surface has been discussed by Benjamin;[29] a surface mode is likely to be established when its characteristic frequency ω_n is equal to $\omega/2$, i.e., to half the driving frequency.

Unlike the situation with linearly driven surface waves the amplitude of a wave set up by parametric excitation does not vary linearly with the driving amplitude p_o. For a plane interface vibrating normal to itself with angular frequency ω and with amplitude ξ_o, Sorokin[30] and Eisenmenger[31] have shown that a sound wave is not set up parametrically until ξ_o exceeds a critical threshold value ξ_c given by $2b/k\omega$, where b is the damping coefficient and k the propagation constant for waves of frequency $\omega/2$. From Lamb[8] (Par. 349) b is given by $2\nu k^2$; also from Lamb[8] (Chap. 9) k^3 is given by $\rho_o \omega^2/4\sigma$ for waves of angular frequency $\omega/2$ omitting a gravitational term which is negligible at an air-water interface for frequencies above 1 kHz. Hence for the plane interface we obtain

$$\xi_c = \nu\ (16\ \rho_o/\omega\sigma)^{1/3}\ . \qquad (60)$$

It seems reasonable that this expression should hold also on a spherical surface, provided that the mode is of high order, such that n >> 1. With this in mind we compare the critical amplitude ξ_c with a representative value of the bubble radius, choosing for the latter purpose the resonant value, given approximately by R_o in Eq. (18). Designating the resonant radius by R_{res} and using constants for an air bubble in water we find that

$$\xi_c/R_{res} \sim 10^{-5}\ f^{2/3}\ . \qquad (61)$$

We expect this rough estimate to apply only to relatively large bubbles; such that the excited wave has a high mode number n and such that the resonant frequency is given by Eq. (18). We find from Eq. (61) that at 20 kHz the ratio ξ_c/R_{res} is 0.0074 and at 100 kHz the ratio is 0.022. At resonance, from Eqs. (13-15), the pressure ratio p/P_o is equal to $3\gamma\delta$ times the ratio ξ_{bo}/R_o. Equating the latter ratio to ξ_c/R_{res} and letting $\gamma = 1.4$ we find that p/P_o is equal to $4.2\delta\xi_c/R_{res}$. In water the damping coefficient δ at 20 kHz is 0.08 and at 100 kHz it is 0.11; hence, by this estimate, the driving amplitude required for parametric excitation of surface waves on resonant bubbles in water is only 0.0025 atm at 20 kHz and 0.010 atm at 100 kHz.

Experimental measurements have been made by Strasberg and Benjamin,[32] by Eller and Crum[33] and by Gould[16] on the threshold for wave excitation at the bubble surface, all for frequencies in the range 20-30 kHz. The driving pressure amplitude p_o required to produce the waves varies from 0.7 atm to less than 0.1 atm for bubbles ranging from 30 μm to 120 μm in diameter (all are small compared to resonant size). The expression in Eq. (61) for the displacement amplitude, together with the relation between displacement amplitude and p_o from Eqs. (13-15), provide a reasonable fit to data for the larger bubbles, but yields thresholds that are higher than experimental values for bubbles 50 μm or less in diameter. Better fit is obtained with a much more detailed theory developed by Eller and Crum.[33]

It was shown by Elder[23] that the microstreaming near a vibrating gas bubble is significantly altered when surface waves are set up. Gould[34] showed that the transport of heat from a small heat source can be speeded up as much as 10-fold by microstreaming from a nearby bubble, vibrating at 2 kHz on a fixed surface, and that this transport is sometimes affected by surface waves. Using freely suspended bubbles at 20 kHz Gould[16] has shown that the transfer of gas molecules between a vibrating bubble and the adjoining liquid is increased by as much as 20-fold in the presence of microstreaming generated by surface waves. At present there exists no theory for microstreaming which satisfies appropriate boundary conditions for surface waves on the gas bubble.

It was shown by Willard[35] that surface waves on a bubble surface at 18 kHz can give rise to profuse formation of clouds of much smaller bubbles or "microbubbles." Insight on the mechanics of microbubble generation by surface waves has come from studies by Nyborg and Rodgers[36] and Storm[37] at low frequencies. Photographs of microbubble generation at 20 kHz have been published by Neppiras and Fill,[38] together with a description of novel aspects of the process. The process of microbubble generation is important in that (i) this process sometimes leads to disintegration of the major bubble (Nyborg and Hughes)[39] and (ii) microbubbles act as sites for new cavitation activity (Neppiras and Fill).[38] Microbubble generation also gives rise to production of harmonics and noise (Neppiras and Fill)[38] a fact that must be kept in mind in interpreting spectra in a field of vibrating bubbles.

As noted above, parametrically excited surface waves tend to be characterized by subharmonic frequencies. A hydrophone in sonated liquid containing one or more bubbles typically reveals a complex pressure spectrum if the amplitude is sufficiently high to cause significant bubble action. It would not be surprising if subharmonic contributions to the spectra were to arise from surface waves. Strasberg[40] has pointed out that the pressure field radiated by a surface mode of high order drops off rapidly with distance from the bubble. However, appreciable contributions are to be expected in the vicinity of the surface. At audio-frequencies (750 Hz) Storm[41] has observed surface waves on resonant bubbles in water at driving pressure amplitudes p_o as low as 0.002 atm; with a small hydrophone near a bubble he has also observed subharmonic contributions to the pressure field at these low levels, and it is probable that the two phenomena are related. (See Hsieh).[42]

1.11 Superharmonics, Subharmonics and Noise

At low amplitudes the response of a gas bubble to a driving pressure $p_o \cos \omega t$ is governed by a linear differential equation with constant coefficients, Eq. (11). At higher amplitudes the applicable differential equation is complicated with nonlinear terms and time-dependent coefficients. As a consequence the pressure in the scattered field is no longer characterized by simple sinusoidal variation with time. Besides a fundamental component of angular frequency ω, the scattered pressure spectrum contains contributions of superharmonic frequency 2ω, 3ω, etc., as well as subharmonics of frequency $\frac{1}{2}\omega$, $\omega/3$, etc. In addition, other components appear with frequency equal to the sum or difference of the superharmonics and subharmonics. Theory shows that all of these contributions to the spectrum are to be expected even from a single bubble whose center remains fixed as it pulsates in a symmetrical volume mode. When bubbles are translated, or are set into surface modes of vibration, other mechanisms for sound production are present, giving rise to both discrete and continuous spectra.

The characteristics of sound fields generated by vibrating bubbles are important to bio-ultrasound, for acoustic spectra are frequently used as evidence for the presence or absence of cavitation. For example, Neppiras and Parrott[43] and Neppiras[44] have shown that the strength of the subharmonic ($\omega/2$) (as sensed by a hydrophone placed in the sonated liquid) can be a convenient and useful measure of cavitation activity. Investigations of bio-ultrasound are described (Section 2) in which the subharmonic and other acoustic indices are used as correlates of cavitation. While acoustic indices appear to be very useful, it should be realized that there is, as yet, no firm theoretical basis for interpreting acoustic spectra from cavitating fields. The subject is a complex one: an experimenter is faced with the task of specifying the parameters of a system containing many bubbles of varying size moving with varying speed in different directions; the theorist is faced with a problem in nonlinear mechanics for a many-body interacting system.

There is not space here to review the status of theory and experiment on sound spectra in cavitation fields. We shall only cite a few results of special interest. The theory is fairly straightforward for generation of the second harmonic component, i.e., the component of frequency 2ω. Consider a bubble of equilibrium radius R_o, subject to a driving pressure field $p_o \cos \omega t$, as in Section 1.3. Storm[45] has found that when nonlinearities are taken into account the displacement ξ has a component ξ_2 given by

$$\frac{\xi_2}{R_o} = \frac{(1 + 3\gamma + 5\Omega^2)(p_o/P_o)^2 \cos 2\omega t}{36\gamma^2 (1 - \Omega^2)^2 (1 - 4\Omega^2)} . \tag{62}$$

and that the pressure in the scattered field at the bubble surface has a component given by

$$p_2 = \frac{-\Omega^2[5 + 4(1 + 3\gamma)]p_o^{\,2} \cos 2\omega t}{12\gamma P_o \;(1 - \Omega^2)^2 \;(1 - 4\Omega^2)} \tag{63}$$

It is assumed that $\xi/R_o << 1$ and that changes in gas volume occur adiabatically. We see that the amplitude of either ξ_2 or p_2 is proportional to $p_o^{\,2}$, i.e., to the square of the driving pressure amplitude. To determine the pressure p_2 in the liquid at a distance r from the bubble center (for $r > R_o$) the expression on the right hand side of Eq. (63) should be multiplied by (R_o/r).

Available information from theory and experiment for subharmonic spectra in a cavitation field has been reviewed by Neppiras.[44] For a gas bubble in a field where the driving pressure is $p_o \cos \omega t$ contributions to the pressure of frequency $\omega/2$ arise partly from surface waves, and partly from volume oscillations. Surface waves are discussed in Section 1.10; it is seen there that the threshold for generation of an $\omega/2$ component by this means is sometimes very low, but the pressure amplitude of this component may also be low. For generation of subharmonic spectra by either surface or volume oscillations the governing differential equation is of the Mathieu type or Hill type. Neppiras[44] has shown that volume oscillations of frequency $\omega/2$ are especially likely to be excited in bubbles which are resonant to that frequency. For these relatively large bubbles the critical pressure amplitude p_{oc} for excitation of the $\omega/2$ volume mode is given by Eller and Flynn[46] as

$$p_{oc} = 6P_o\,\delta\ , \tag{64}$$

where δ is the damping coefficient for the bubble at the frequency $\omega/2$, to which it is resonant. According to Safar[47] and Storm[45] the expression for p_{oc} in Eq. (64) should be multiplied by γ. In a sound field of frequency 2 MHz, a bubble resonant to half this frequency, i,e., 1 MHz, is approximately of radius 3.3 μm and the corresponding damping constant in water is about 0.14; hence by Eq. (64) the pressure amplitude required for generation of a 1 MHz signal <u>via</u> the volume mode is 0.84 atm. If bubbles are not present of the appropriate size the threshold may be much higher. Much more extensive experimental tests of the theory are needed, especially in the megahertz frequency range. Important additions to the theory for subharmonic and other spectra has been published recently by Prosperetti.[47a,47b]

1.12 Radial Pulsation: Nonlinear Theory

In previous sections the theory we have presented has been based on approximations which are valid only when the radial displacement ξ_b of the bubble surface is small compared to the initial radius R_o. Such small-amplitude theory is very useful in that it leads to analytical expressions; from these we can see with relative ease the results of changing any parameters of the problem. However, the basic equations which govern bubble motion are nonlinear, and at high amplitudes the consequences of this nonlinearity cannot be avoided. Starting from theory originally given by Rayleigh,[48] Noltingk and Neppiras[49] arrived at a differential equation which has proved extremely useful. In a somewhat generalized form, essentially as given by Flynn,[4] the Noltingk-Neppiras equation is

$$2\rho R^2 \frac{d^2R}{dt^2} + 3\rho R\left(\frac{dR}{dt}\right)^2 + 4\sigma = 2R\left[p_o \sin \omega t - P_o + \left(P_o + \frac{2\sigma}{R_o}\right)(R_o/R)^b\right]. \quad (65)$$

Here R is the instantaneous radius, R_o the equilibrium radius, ρ the liquid density, σ the surface tension, p_o the driving pressure amplitude and P_o the pressure at infinity. The liquid is assumed incompressible, and the gas to be ideal; damping is ignored. Changes in state occur in such a way that pv^b is constant, where p is the pressure in the gas, v the bubble volume and b is a constant. For isothermal changes b = 1, and for adiabatic changes b is equal to γ, the ratio of specific heats for the gas; in general b has an intermediate value between 1 and γ.

Noltingk and Neppiras proceeded to obtain information from Eq. (65) by computational methods. They showed, as a consequence of this equation, that the pressure variation in the driving field during a sonic cycle may cause a bubble to expand somewhat, then contract to a small volume; during this contraction the temperature may reach very high values leading to "incandescence" of the gas. Flynn[4] proceeded further in the use of a computer to plot radius-vs.-time curves from equations such as Eq. (65). In his extensive review he presents many results obtained in this way. For example, he considers the response of a bubble with initial radius 26 μm, to a driving field of frequency 24.5 kHz at varying pressure amplitudes p_o. For p_o = 0.75 the bubble wall is found to reach a velocity of 0.02 c during collapse, where c is the velocity of sound in the liquid; at p_o = 0.80 atm the velocity reaches 0.045 c, at p_o = 0.90 atm it reaches 0.56 c and at p_o = 1.00 atm the bubble collapses supersonically, with velocity 1.6 c; under these latter conditions shock waves are expected in the surrounding liquid. Flynn shows that this dramatic change from subsonic to supersonic collapse results from increase of an "inertial function" representing the liquid motion. Under conditions of rapid collapse the radius-vs.-time curve is similar to one deduced by Rayleigh[48] for the shrinking of a spherical "void" created instantaneously in a liquid.

Flynn has introduced the useful terms stable and transient to distinguish two kinds of behavior of a bubble in response to a sound field. Transient cavitation occurs if the bubble collapse is very rapid and corresponds to the Rayleigh prediction; it is implicit that in this collapse the bubble disintegrates (perhaps by surface instability) and thus is "transient." Stable cavitation occurs if the bubble oscillates for a number of cycles without leaving the field. The theory presented in previous sections, applicable for relatively small amplitudes, is clearly for stable cavitation.

An experimental test of a modified Noltingk-Neppiras equation has been carried out by Lauterborn,[50] using high speed photography. Results are shown in Figure 7 where the curve is from theory, as determined by means of a computer. The points are experimental, for a bubble created by a pulsed ruby laser in a viscous silicone fluid. After the bubble is created it oscillates for a number of cycles with decreasing amplitude; its radius at rest is 0.8 mm.

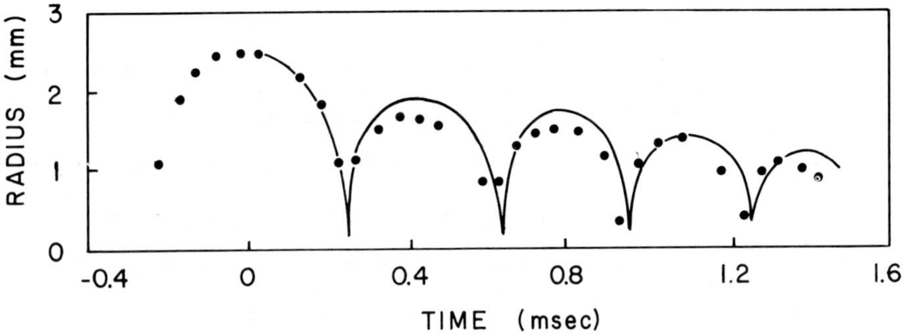

Fig. 7 Plot of radius versus time for a bubble created by a laser. Points are experimental; curve is theoretical, based on an equation similar to Eq. (65). From Lauterborn.[50]

Satisfactory agreement is demonstrated between theory and experiment. A characteristic of the radius-vs.-time curve is of much interest; the expansion phase of a cycle is broad, while its contraction phase is cusp-like. Considering the high temperatures which occur during contraction, we see that this bubble motion gives rise to a series of temperature pulses, the pulse-length being very short compared to the time between pulses.

1.13 Cavitation Complexes

Up to this point our discussion has been primarily directed to the dynamics of a single gas bubble. We now consider the much more complicated situation presented by a liquid, or other medium, in which cavitation is produced by irradiation in a typical manner with ultrasound. In such a situation there are usually hundreds or thousands of bubbles of varying size, moving rapidly and interacting with each other. While some aspects of complex cavitation fields can be understood in terms of single-bubble dynamics, novel features appear in these fields which are peculiar to multi-bubble systems. For these systems little theory has been developed; our treatment of this topic is mostly qualitative.

The simplest interaction is that between a pair of bubbles, say, A and B, which are near each other in a sound field where the ambient pressure is

p_o cos ωt. Each is set into vibration and generates a scattered wave. The total driving pressure seen by A, for example, is the sum of the ambient pressure p_o cos ωt and the field scattered by B. The situation has been analyzed by Nyborg and Storm;[22] one of their findings is that if both bubbles are smaller than the size for resonance (given in Eqs. (18-20)) they will each vibrate with greater amplitude than they would if the other were absent. This is reasonable in view of the fact, noted in Figure 4, that in the field near an isolated bubble of subresonant size the total pressure amplitude is greater than for the ambient field alone. Thus bubbles smaller than resonant size tend to reinforce each other. Different results are obtained near bubbles which are larger than resonant size, as may be expected from Figure 4; near a large bubble the total pressure amplitude tends to be smaller than the ambient value.

Because of these interactions it is reasonable that the threshold for onset of a phenomenon characteristic of cavitation (such as subharmonic production, sonoluminenscence or solid erosion) should be different in a multi-bubble field, than would be predicted on the basis of single-bubble dynamics. Evidence for such a difference is seen in the results of Howkins;[51] he found erosion of a solid surface to occur at a lower amplitude when a small pocket of gas was present. Surface waves occurred on this pocket, and a "ghost" was formed, evidently consisting of many small bubbles. The ghost was very effective in producing erosion, apparently because of bubble-bubble interactions.

Further evidence for interactions comes from the results of Saksena and Nyborg[52] in which sonoluminescence was observed from arrays of bubbles. Light was detected at rather low values of the ambient pressure amplitude from arrays containing as few as six bubbles, but not from smaller numbers. While the authors originally interpreted these results in terms of single-bubble dynamics it appears in retrospect that interactions may have been involved.

In other experiments Temple, et al.[53] studied sonoluminescence instigated by a single bubble suspended in a standing wave field. The arrangement was such that visual observations of the sonoluminescence could be correlated with visual observations of the bubble. It was found that sonoluminescence occurred at rather low driving pressure amplitudes, but not unless a multi-bubble complex, dubbed a "shuttlecock" because of its appearance, formed itself around the original bubble. It is believed that this shuttlecock was initiated by microbubbles shed from the original bubble through surface wave activity.

Howkins' "ghost" and Temple's "shuttlecock" may be related to the phenomena studied by Willard,[35] and mentioned previously. Willard prepared a motion picture showing the microbubble cloud generated near a vibrating bubble at 20 kHz. Under the conditions of these experiments the initiating bubble remained at its position in the field indefinitely without significant change in its size or activity, in spite of the continuous evolution of microbubbles. Evidently the microbubbles recombine with the initiating bubble, as a result of radiation forces such as were discussed in Section 1.8. A balance is achieved between loss of gas from the initiating bubble through microbubble generation, and gain through recombinations, leading to a kind of dynamic equilibrium.

In visual observations of a cavitating liquid it is common to see "streamers," i.e., narrow streaks formed of bubbles, which may be some millimeters in length. (See, e.g., D. E. Hughes and W. L. Nyborg).[54] Under some circumstances streamers may persist indefinitely in a given configuration, while in others they are unstable, intermittent and fluctuating. Usually a characteristic hissing noise accompanies a streamer, and the latter is the site of vigorous cavitation activity. Thus when the streamer is along a solid surface it often produces erosion of the solid. Also, in water, production of the free radicals H and OH may be produced along the streamers; this is made evident by localized concentration of color under conditions where free radicals produce colored reaction products. It was shown by Hughes and Nyborg[54] that a streamer consists of a large number of rapidly moving bubbles. The bubble motions are primarily along the direction of the streamer and appear to be governed by radiation forces. These forces govern the direction of the streamer; they also bring about interactions between bubbles within a streamer. Bubbles smaller than resonant size tend to combine with each other, thus producing larger bubbles. As a bubble grows and approaches resonant size it attracts smaller bubbles at an increasing rate. In a streamer bubble velocities are typically high, and may approach sonic speed. It was also shown by Nyborg and Hughes[39] that bubble annihilation occurs in streamers; presumably violent surface waves are set up, and these lead the bubble to fragment into microbubbles during the time of a few sonic cycles (at 20 kHz).

Another kind of cavitation field was first studied in detail by Willard[55] and has since been investigated further by Coakley and associates.[56,57,66] Here cavitation activity is produced in a body of liquid by means of a focused ultrasonic transducer. It appears that a single "cavitation center" is created in the liquid, proceeds outward from the transducer (as a result of acoustic streaming and radiation pressure), and grows as it approaches the focal region. At a critical point along its path an "explosion" occurs, resulting in the creation of an extensive and active cavitation complex. Acoustic streaming aids in gradual dissolution of the complex, after which another initiating "cavitation center" proceeds outward from the transducer, and grows until another "explosion" occurs, etc. These explosions have been referred to by Coakley as "Willard events." As is explained in later sections Coakley and co-workers have characterized these events by acoustic means, and have also investigated their effectiveness in bringing about chemical and biological effects.

2. EXPERIMENTAL STUDIES: LIQUIDS AND SUSPENSIONS

2.1 Introduction

In this section we review experimental findings from investigations of cavitation in water and other liquids, in chemical solutions, and in suspensions of biological cells and macromolecules. In these media most of the known effects of ultrasound appear to be brought about by cavitation. At frequencies around 20 kHz, where ultrasound is used routinely for the extraction of material from cells, there is interest in developing methods for assessing quantitatively the effectiveness of cavitation. Also, at these low ultrasonic frequencies there has been emphasis on seeking out the modes of behavior of the active bubbles, which are readily made visible in this frequency range. At megahertz frequencies resonant bubbles are small, of the order of microns in diameter, and are consequently relatively difficult to detect. At these frequencies, so useful in medical ultrasound, there is interest in learning which of the known effects in tissue, cell suspensions, and solutions of biopolymers are mediated by cavitation, and which may be attributed to some non-cavitational interaction between the specimen and the sound.

In many of the studies the need is apparent for better understanding of cavitation phenomena. This need is evident, for example, in the confusing terminology which has been introduced for classifying cavitation; it is also evident from the wide variation in criteria used for determining the absence/presence of cavitation. Flynn[4] has dealt thoroughly with the semantics of terms associated with cavitation. As stated in Section 1.1 "Acoustic Cavitation" is used as a very general term which encompasses the generation of bubbles by sound in liquids, the various kinds of motions of such bubbles, and the physical effects that they bring about. Flynn's use of the adjectives "stable" and "transient" is discussed briefly in Section 1.12. The phrases "cavitation event," "cavitation field" and "cavitation activity measure" were used by Flynn as very specific prescriptive terms limited to bubble fields in which transient cavities occur. The presence or absence of bubbles whose motions resemble those of transient cavities distinguishes two important kinds of bubble fields: one, a "stable bubble field," contains only bubbles resembling stable cavities, whereas the second, a "transient bubble field," contains bubbles resembling transient cavities. Flynn cautions that when a sound field generates bubbles in a liquid, a wide variety of motions is possible. He considered that the transient cavity and the stable cavity would be adequate models for certain limiting types of bubbles and that in an experiment there would be generated a range of bubbles between these limiting types.

Criteria for presence/absence of cavitation, and indices of cavitation activity, are based on various physical, chemical, and biological techniques. Acoustic indices have proven especially useful; these are based on the use of hydrophones in media of interest, to monitor spectra characteristics of cavitation. The physical basis for these spectra is understood, to some extent; theory for the generation of frequencies equal to (i) twice and (ii) half the driving frequency are discussed in Section 1.11. In general, however, the spectra are complex, including extended line structure as well as "white noise." Satisfactory theory for these complex spectra is not available. Correlation of acoustic indices with other indices of cavitation activity has been, so far, carried out primarily on an empirical basis.

2.2 Acoustical and Visual Manifestations of Cavitation

In this section we describe the findings of investigators on the acoustic emissions from cavitation fields, and also the visual appearance of these fields, under various circumstances. In all of these the fields can be characterized as "cavitation complexes," some aspects of which were discussed in Section 1.13.

Neppiras[58] has reviewed methods of detecting and measuring acoustic cavitation with particular reference to practical industrial processing systems. He concluded that for most purposes in that area interest centers on the effects which follow from the liquid shock pressure developed at the collapse of transient cavities and that a realistic assessment of cavitation activity can be obtained by basing a test on one of the following: (i) the white-noise portion of the acoustic output, (ii) erosion, (iii) dispersion. Neppiras also suggested that some of the small bubbles contained in the debris from collapsing transients may be in the size range for growth by rectified diffusion and that the growth and collapse cycle can repeat, perhaps many times. This is known as the gaseous cavitation cycle and is distinguishable as the visible streamer activity which accompanies the intense irradiation usual in practical processing systems. Therefore, a cavity which behaves as a transient cavity may trigger a series of events which lasts for many cycles of the sound field.

Blake[59] reported some early observations on gassy and degassed water exposed to a 60 kHz sound field. A hydrophone in the treated water monitored "noise" from the cavitating region. In gassy water bubbles appeared at the focus at low power inputs; as the power input was increased, lines and then sheets or streamers of very small bubbles of approximately radial resonant size appeared in the focal region and extended outward from it. The appearance of the streamers coincided with the detection of noise. Occasionally comet-like streamers with a bubble larger than resonant size at the head emerged and darted about the field. In "degassed" water Blake observed "ruptures" located at the focus when the sound pressure was near the threshold value. These ruptures were considered to be formed by single cavities moving rapidly in the water. They were accompanied by an audible snap and had a life time estimated at 1 to 10 msec.

Esche[60] described the appearance of, and examined the emissions from, cavitation bubbles in sound fields of fundamental frequency ranging from 10^2 to 3.3×10^6 Hz. A frequency band not containing any of the original frequencies, i.e., a white noise detector as described by Lange[61], was used for the sound fields of frequency 1.5 to 500 kHz. Spectra of subharmonic-emitting fields ranging from 3 to 3,300 kHz were also determined. Harmonics of the driving frequency were detected at the commencement of bubble formation. It did not prove possible to separate the onset of subharmonic emission from the onset of white noise when the cavitation "threshold" was reached. Overtones of the $f_o/2$ and $f_o/3$ subharmonics were also detected. Barger[62] also used white noise emission to determine cavitation thresholds and broadly confirmed Esche's findings. In gassy water at an ambient pressure of 1 atm the threshold below 200 kHz was for the formation of white noise emitting streamers of "gassy" or "stable" bubbles while at higher frequencies the threshold was for randomly occurring transients of less than 100 msec duration, which emitted a stronger white noise signal. Willard[55] has photographed cavitation activity, lasting for milliseconds, arising from a single nucleus at the focus of a 2.5 MHz field. Flynn[4] considers that Willard's events are different from Blake's "comments" but the postulation of a different explanation for the two observations may be unnecessary. By means of acoustic techniques Coakley and his colleagues[56,57,66] have detected the focal events described by Willard and have also identified non-focal regions in the sound field, where the pressure is lower than the focal level as sites of bubble activity. Two examples of Willard type events are shown in Fig. 8.

Subharmonic and other low frequency emissions from bubbles in sound irradiated liquids have been examined in detail by Neppiras[44] in experi-

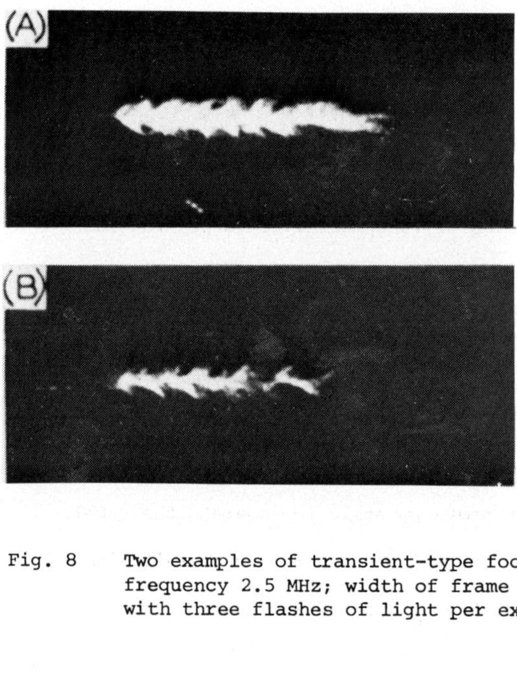

Fig. 8 Two examples of transient-type focal events in water from Willard;[55] frequency 2.5 MHz; width of frame 17.5 mm; 22 msec exposure; illuminated with three flashes of light per exposure.

ments in the frequency range 18-46 kHz. In agreement with the theoretical treatment of Eller and Flynn[46] (see Section 1.11) Neppiras observed that if a liquid was seeded with bubbles of a size twice the resonant radius, then a first subharmonic signal was detected well below the sound pressure necessary for white noise emission. In unprepared liquids a very low signal at the first subharmonic was sometimes detected at even lower driving pressures. The level of first subharmonic increased rapidly with the onset of white noise. The $f_o/3$ subharmonic was also detected. Emissions corresponding to frequencies of $2^{-1/3} f_o, 3^{-1/3} f_o$, etc. were explained as arising from shock excitation of bubbles of sizes resulting from agglomeration of bubbles of resonant size. As in Esche's case, harmonic emissions preceded subharmonic radiation. In a more recent theoretical study Eller has shown that the first subharmonic can be excited over a broad frequency range, and not just for a driving frequency of twice the bubble resonance frequency.[63] Eller and Flynn's theoretical predictions of the threshold sound pressures for subharmonic emission, and the effects of dissolved gases on the threshold[46] have been broadly confirmed quantitatively by recent studies of Ceschia and Iernetti, working at 700 kHz.[64]

Some recurring features in the description of optical observations of bubble activity have been (i) The growth of bubbles by rectified diffusion (see Section 1.7) and agglomeration via radiation forces (Section 1.8),[59,62,44] (ii) Occasional changes in the appearance of the bubble surface both at radial and non-radial resonant frequencies to a form which was "frosted,"[62] "milky and irregular,"[59] "crispated;"[65] (iii) The shedding of microbubble clouds by bubbles.[55,44,39,36] The behaviour reported by the above authors in (ii) and (iii) has been ascribed by some of them to surface wave activity; see also Section 1.10. Meyer[65] forced bubbles to vibrate in a 14.6 kHz field and found that the active bubbles preferably showed diameters for which one of the many modes of surface oscillation had a relationship to the radial natural frequency. Willard[35] reported microbubble production from bubbles driven at 18 kHz in the presence of surface wave activity and suggested that the same phenomenon might explain his previous observations of focal transient cavitation at 2.5 MHz. Strasberg[40] concluded that surface activity by itself would radiate little energy compared to a volume oscillation and indeed Nepirras[44] reported that he was unable to detect any subharmonic which could clearly be assigned to surface activity, even though shape oscillations have an angular frequency of $j\omega/2$ where $j = 1,2,3,$--- and ω is the driving angular frequency.[33] Hsieh has examined the association between surface waves and subharmonic emission and decided that unless the transducer is placed near the bubble the subharmonic generated by the surface waves may not be detectable[42] in agreement with the observations of Storm.[41] Neppiras and Coakley[66] filmed cavitation in a focused field of 1 MHz and found clear evidence for microbubble production by surface wave activity on bubbles larger than radial resonant diameter. These large bubbles were involved in non-focal cavitation activity and the authors concluded that bubbles larger than resonant size also provided the initiating bubble for the transient-type focal events they observed.

To summarize the information available from experiments on emissions in water: harmonics of the fundamental may be detected first,[60] while subharmonics will be observed if bubbles of a size whose resonant radial frequency is a subharmonic of the driving frequency are present or grow;[44] sometimes harmonic, subharmonic and white noise emission become detectable at around the same sound field pressure.[66] A low level of white noise can accompany periods of the life time of stable bubbles.[38,62] White noise and microbubble clouds are considered to be associated with the collapse phase of a transient cavity while microbubble clouds may also accompany surface wave activity.[38] Events described as transients may have a life time from one to thousands of cycles of the sound wave but rarely longer than 10 msec. Emissions arising either from transient cavity collapse or from surface-wave activity may truly be said to require a "threshold pressure" while in theory all finite excursions of a bubble surface are non-linear and emit harmonics of the driving frequency.

There is a need for greater understanding of the development phases a bubble needs to pass through before resulting in a particular form of emission in water. This is particularly true if interferences are to be drawn about the occurrence of cavitation in tissue. There would be particular significance in knowing if the absence of a particular emission can be taken as evidence for the absence of non-linear bubble motion in the tissue.

2.3 Thresholds for Cavitation

Published values of "cavitation thresholds" display a large degree of variation. It is generally agreed that the threshold is frequency dependent. However, even at the fixed frequency of 1 MHz, Iernetti has pointed out that the reported threshold values range from 1.75 to 240 atmospheres[67] while a value as low as 0.7 atm has been published.[68] Several factors will affect the threshold in water. The gas content and degree of preparative cleaning of the water are important. With gassy water the length of time the sample has been standing will determine the cavitation nuclear size distribution in the liquid. The presence of solid-liquid interfaces in parts of the field where the sound pressure is high may provide a rough surface where bubble nucleation and growth may be enhanced. The presence of standing waves in a sound field will have two effects on the bubble population. In the first case they are likely to produce regions in the field where the sound pressure is higher than would exist in a free field. Secondly, bubbles smaller than resonant size are attracted to standing wave pressure antinodes and, being held there, have an opportunity to grow to resonant size in a manner which might not have been possible in a streaming free field. Iernetti ascribes the lower experimental thresholds for cavitation in large volumes to the higher probability that a nucleus capable of developing into a cavitation event will be present in the large volume.[67] Since the treatment of larger volumes generally involves plane wave irradiations the greater likelihood of a high standing wave ratio may also influence the above result. The influence of liquid properties such as surface tension, viscosity and thermal conductivity, on cavitation behaviour is discussed by Flynn.[4] A sharp fall in threshold occurs in the temperature range 5.0 to 0.5°C.[69]

When stating cavitation threshold pressure values, every effort should be made to determine the site in the field at which cavitation is occurring. Some of the active sites in focused fields are not located at the focus, therefore a statement of the focal intensity would lead to too high a threshold value.[57] In order to protect a measuring hydrophone the sound pressure at the cavitation site is best measured when the acoustic pressure is below the cavitation level and the pressure at which cavitation sets in should be inferred from that measurement. Unfortunately some threshold pressure values have been determined by the inappropriate procedure of converting the average intensity at the face of the transducer to sound pressure. Further information on the type or types of cavitation present may be gleaned if observations on the bubble behaviour of the type described above are reported. It is also sometimes not sufficiently clear if reported threshold values are real thresholds for the bubble activity and not the lower limit of resolution of the detecting system.

2.4 Sonochemistry and Sonoluminescence

2.4.1 Association with Cavitation

A sound field may accelerate chemical reactions which would take place in the absence of sound. Chemical reactions which require the presence of cavitation may be initiated in a sound field. These latter reactions are referred to as sonochemical. Richards and Loomis[70] first reported the initiation of chemical reactions by ultrasound in 1927. Frenkel and Schultes observed a very faint light, visible to the dark-adapted eye, from a cavitating liquid in 1934.[71] They called this phenomenon "sonoluminescence." Flynn[4] concluded that there are enough coincidences between the yields of chemical reactions and the intensity of luminescence in cavitation fields to justify the assumption that what causes one probably causes the other. In reviewing the information available up to 1964 he concluded that the "hot-spot" theory of Noltingk and Neppiras[49,72] or a variant thereof, involving high temperatures in contracting bubbles offered the most likely explanation for the effects observed. Prudhomme[73] interpreting in a different manner some of the experimental evidence quoted by Flynn in support of a thermal mechanism, concluded that an electric discharge in the active bubble is a more likely cause of the phenomena of sonochemistry and sonoluminescence. A new theory to explain the initiation of sonochemical effects and sonoluminescence by a microscopic discharge in the bubble has been recently proposed[74] but critical experimental evidence to support this hypothesis against the thermal one is still lacking.

Water may be decomposed by cavitation and give rise to free radicals primarily by the following reaction

$$H_2O \rightarrow \dot{H} + \dot{O}H$$

Del Duca, Yeager, Davies and Hovorka,[75] in an important paper, have listed ten possible secondary reactions which would involve the production of H_2, H_2O_2, O_2 and HO_2. Solvated electrons have also been observed as products of ultrasonic cavitation.[76] The extent to which the radical recombinations and secondary processes occur in the liquid or in the gas phase is still a matter of uncertainty.[77] Interactions between active species and biopolymers must of course occur in the liquid phase. In common with other authors Margulis[76] points out that free radical formation from water lysis by cavitation gives rise to the same products as ionizing radiation lysis of water. The wealth of information available from radiochemical studies may, therefore, serve as a guide in sonochemical work. An essential difference between sonochemical and radiochemical reactions are that solutes in the latter case may be affected by direct interaction with the radiation (a direct effect as well as by the products of the interaction of the radiation with water (indirect effect) while the sonochemical reactions depend entirely on indirect effects. Also gases influence that happens in a cavitation field in a more dramatic way with ultrasound than with ionizing radiation.

The density of ionization associated with the path of an ionizing particle is characterized by the linear energy transfer coefficient (LET). Particles with a high LET value in water have relatively higher densities of active species along their paths, and interactions between the species, to form molecular products like H_2, H_2O, H_2O_2, are more likely. In aqueous solutions exposed to ionizing radiation the oxidation of ferrous ion depends largely on the free radicals produced while the reduction of ceric ion mainly reflects the presence of molecular products. The ratio of ferrous oxidation to ceric reduction decreases to a limiting value of 1.3 as LET increases.[78] Weissler[79] calculated that this ratio was 1.6 in a cavitation field. From a more comprehensive study Todd[78] established a value of 1.29 $\pm$ 0.04. Both the above values suggest that sonochemically active cavitation events have a high density of active products associated with them.

Sonochemical dosimeters have proved useful as indicators of bubble activity. Weissler[79] compared the sensitivities of several different systems and concluded that the production of molecular iodine from a potassium iodide, carbon tetrachloride solution was the most sensitive indicator. Clarke and Hill[80] found that this test was twenty times more sensitive at 1.0 MHz than the production of iodine from Kl in the absence of carbon tetrachloride.

Much less effort has been applied to establishing the forms of bubble behaviour associated with sonochemical and sonoluminescence activity than has been the case with emission of sounds or noise. Since the review by Flynn[4] which associated sonochemical and sonoluminescent phenomena with high temperatures reached during bubble collapse, Vaughan[81] found at 27 kHz that the first peak in subharmonic output was coincident with visual signs of cavitation and preceded slightly the liberation of iodine from Kl. Saksena and Nyborg[52] measured sonoluminescence from an array of bubbles at 30 kHz. They concluded that light generation was possible even from stable cavitating bubbles having periodic volume oscillations. Evidence was presented for a thermochemical mechanism. Later work (see page 120) suggests that bubble-bubble interactions may have played a role in the phenomena observed here. Clarke and Hill found that on rotating their sample holder in a 1.0 MHz sound field sonochemical activity and biological cell death had essentially the same threshold[80] and that the subharmonic signal was detectable at slightly lower intensities. Sonoluminescence could not be detected using either the dark-adapted eye or a sensitive photomultiplier system which had been successfully used for the measurement of luminescence produced by ionizing radiation. This observation led the authors to suggest that the sonochemical activity arose from "stable" rather than transient cavitation. The finding is an exception to Prudhomme's generalization that sonoluminescence and sonochemical reactions are concurrent.[73] Iernetti and his colleagues have studied sonoluminescence and sound emissions in a variety of cavitating fields at 0.7 MHz.[69,82-84] In gas saturated water the threshold for subharmonic emission was much lower than the threshold for sonoluminescence. The onset of sonoluminescence was considered to occur at the threshold pressure for imploding bubbles. The trend in the dependence of the measured sonoluminescence threshold on pulse regimes and bubble size was shown to be similar to the trend in the theoretical threshold for bubble implosion.[84]

Coakley and Sanders[57] found that a sonochemical yield was detectable in a focused field at 1.0 MHz from both Willard cavitation events and from stabilized non-focal regions of cavitation activity. The non-focal activity was present at low white noise levels before the rapid growth of the white noise amplitude occurred.[57,66]

It would appear then that there are exceptions to the generalization that sonochemical and sonoluminescent activity are concurrent.[80] Both are most noticeable when strong white noise emission occurs although sonochemical reactions can take place at megahertz frequencies when the white noise signal is low,[57,41] and sonoluminescence has been observed without bubble collapse.[52] It may be then that the two types of white-noise emitting cavities described by Barger[62] as well as imploding bubbles are capable of sonochemical and sonoluminescent activity. It may also be said that most authors favour the thermal mechanism for both phenomena.

When attempting to discover if cavitation could have occurred in an experimental arrangement in which a biological effect has been observed some authors have replaced their biological samples by a chemical solution and have examined whether a change has occurred in the system after exposure to sound. This procedure is of doubtful validity since the chemical solution usually had a different history, and therefore a different cavitation nucleus distribution to the biological sample. The cavitation "threshold" may also be altered by the presence of a volatile component like carbon tetrachloride in the chemical solution. Little information is available on the relative sensitivities of chemical tests and biological changes in a cavitating sound field. Furthermore a resonant bubble may not be oscillating with sufficient amplitude to give rise to chemical products and yet may have sufficient microstreaming associated with its motion to bring about changes in cell behaviour.

A chemical test requires that the products accumulate to a detectable level and may well require the integrated products of a number of cavitation events before becoming detectable. Taking the lowest resolvable optical density change in a KI, CCl_4, starch solution to be 0.01 Coakley and Sanders[57] showed that this change could be brought about in a 3 ml volume at the focus of a 1 MHz field by about 20 Willard events when the focal intensity was 515 W/cm^2. The average number of molecules of iodine produced by a Willard event per cycle of the sound field (1 μsec) was estimated to be 3×10^{10}, while from a non-focal cavitation region the yield was estimated to be about 10^9 molecules/μsec when the focal intensity was 140 W/cm^2. Table 2 shows the yield of chemical products per cycle

Table 2

Comparison of Sonochemical Yields per Cycle in 1.0 MHz Cavitating Fields

Wave	Sample Volume (ml)	Intensity (W/cm^2)	Test Solution	Molecules of Product/μ/sec.	Ref.
Plane	3	7.2	$KI+CCl_4$	4×10^7	(80)
Plane	200	600 watts, power	$KI+CCl_4$	3×10^{11}	(85)
Plane	4.0		$Fe^{++}SO_4$	3×10^9	(78)
Focused	3.3	515	$KI+CCl_4$	3×10^{10}	(57)
Non-focal	385	140	$KI+CCl_4$	10^9	(57)

of the sound field from different authors. The order of magnitude of the yields is comparable with the exception of the work of Clarke and Hill[80] which may be taken as support for their conclusion that the effects which they observed arose from the volume oscillations of bubbles driven at twice their resonant radius.

2.4.2 Influence of Gases

The nature and concentration of free and dissolved gases in a cavitating liquid significantly influence the mechanical and chemical effects of the cavitation. Cavitation threshold sound pressures increase as the gas content of the water decreases.[60] The presence of CO_2 in a treated biological cell suspension reduced the mechanical forces associated with the cavitation.[86] Sonochemical and sonoluminescence activity is greatest when the dissolved gas has a low thermal conductivity. Presumably gases with lower conductivities produce higher temperatures during the adiabatic compression of the bubbles. Likewise, solutes with a low value for γ, the ratio of specific heats inhibit sonochemical reactions.[4] Hydrogen has a high thermal conductivity and may be used as the saturating gas when sonochemical effects are undesirable. Helium also functions well in this respect. Argon may be used to increase sonochemical yields. Margulis[76] notes that the experimental results obtained by different workers on the kinetics of sonochemical reactions are conflicting in many cases. He considers that this difficulty arises because of the failure of many workers to control or report the gas concentration in the treated samples.

2.4.3 Sonochemical Effects on Cell Suspensions

Cells in cavitating fields are exposed to the shock waves and shear stresses associated with the bubble motion and may also be influenced by sonochemically induced changes in the environment. Hughes[86] when exposing yeast cells to cavitation in a 20 kHz field found that disintegration of the cells and sonochemical activity varied independently. He concluded that the chemical products of cavitation made little if any contribution to cell breakage. This result was confirmed by Bronskaya, et al.[87] who found that increasing the ambient pressure on a sonicated suspension of *Chlorella pyrenoidosa* increased the cell breakage, yet the oxidation of solutions of α-keto-glutaric acid decreased under increased ambient pressure. Effects of pH changes on bone marrow cells in suspension, arising from the production of nitrous acid from nitrogen in the air of cavitating bubbles, were examined by Hrazdira.[88] He concluded that loss of viability due to sonochemical effects was only 2-3% of the loss due to mechanical effects. In their study, Clarke and Hill[80] considered the possibility that short lived radicals might have influenced their mouse lymphoma cells when exposed to cavitation in a 1 MHz field. They concluded that while prolonged sonication of the medium can eventually reduce its ability to support cells, such indirect effect was insignificant in the relatively short irradiation time required to kill 99% of the cells. The common experience of the above authors, therefore, was that in cavitating conditions where the bubble activity was intense enough to result in sonochemical activity the associated mechanical forces were such as to disintegrate the cells without leading to significant damage by chemical agents. This was especially true where the medium acted as a free-radical scavenger.[80]

2.4.4 Sonochemical Modification of Biological Molecules

The chemical modification of many molecules of biological interest has been reported. These reports have been comprehensively reviewed by Él'piner.[89] The treatment of very alkaline carbohydrate solutions and aqueous solutions of polyhydric alcohols results in new products which absorb in the ultraviolet. Reducing substances, probably aldehydes, are also formed. Treatment of alkaline aqueous solutions of some aliphatic acids also results in the formation of ultraviolet absorbing material. On the other hand, aromatic and heterocyclic compounds, including benzene, treated in aqueous solution undergo changes which result in the disappearance of the characteristic band in the ultraviolet. Él'piner notes that the same decomposition products of sonicated aqueous solutions of benzene have been found as a result of the action of x-rays on aromatic compounds. The presence of a side chain in an aromatic compound protects the benzene ring from destruction in an ultrasonic field. In such cases only decarboxylation and deamination of the side chain with the formation of the corresponding aldehyde are observed. When solutions of the nucleic acid nitrogenous bases adenine, uracil and guanine were exposed to cavitation their characteristic ultraviolet absorption pattern was altered. The position of the absorption minimum moved to higher wavelengths and the amplitude of the absorption maximum decreased. The reduction in the absorption maximum probably resulted from the breakage of the pyrimidine or purine ring in the bases. Chromatograms of the irradiated bases contained a series of spots which fluoresced under ultraviolet light, showing that new products were formed.

A recent study by Staas and Spurlock[90] confirmed earlier findings of degradation of amino acids exposed to ultrasound. Ammonia, carbon monoxide, and formaldehyde were produced in all treated samples and some new products were observed. These authors suggest a scheme for general pathways of deamination, decarbonylation, and production of formaldehyde.

Since the above basic constituents of biopolymers are degraded in cavitating fields it is not surprising that the biopolymers themselves are affected.

Enzymes irradiated in the presence of oxygen are usually inactivated, i.e., biocatalytic activity is lost. Oxidases were found to be more sensitive to ultrasound than reductases and amylases, catalase was not inactivated. In particular, enzyme activity was lost or reduced in treated samples of trypsin, pepsin, DNA-ase, diastase and hyaluronidase.[89] The finding that catalase was not inactivated has recently been confirmed[91] and it was suggested that its survival may be due to the fact that its substrate is hydrogen peroxide, one of the products of cavitation. MacLeod and Dunn[92] confirmed the inactivation of trypsin by ultrasound and examined the inactivation of lactate dehydrogenase in some detail. They found that the enzyme inactivation rate was constant and that after 15 min. the sedimentation coefficient had decreased from 5.78s for untreated material to 3.94s for the treated material, which retained 37% of its original activity. The reduction in s value indicated either fragmentation of the protein or an unfolding that increases the frictional coefficient of the molecule. The limited results for α-chymotrypsin inactivation reported by MacLeod and Dunn[92] suggested that the inactivation of this enzyme was not a first order reaction. In a different study of the effects of 20 kHz cavitation the inactivation of lysozyme, and alcohol dehydrogenase has been shown to be exponential.[91] The protective effect of enzyme concentration or presence of other protein has been shown to be due to competition for inactivating species, and the observed inactivation rate was similar to that expected in an ionizing radiation field producing similar amounts of free radicals. MacLeod and Dunn[92] did not detect amino acids or small peptides in treated solutions while Coakley, et al., using gel electrophoresis, found no evidence for protein fragmentation.[91] These findings appear to conflict with Él'piner's results[89] but it is worth noting that, in general, Él'piner's treatments of biological solutions were of a number of hours duration and effects which might not be detected in the shorter treatments by other workers were more likely to be noticeable.

Alcohol dehydrogenase has been inactivated in the presence and absence of its cofactor NAD^+.[93] It was found that the cofactor protected against ultrasonic inactivation while it served as a sensitizer to γ-ray and heat inactivation. Studies of the inactivation of lysozyme in dilute aqueous solution by γ-rays in the presence of nitrous oxide (to convert solvated electrons to hydroxyl radicals) and t-butanol (to scavenge OH radicals) have clarified the mechanism of the inactivation of this enzyme.[94] There have been few similar critical examinations of ultrasonic inactivation of enzymes resulting in identification of the inactivating species, possibly because of difficulty in assigning only one function to any additive in a sonic field. Changing a gas or adding a solute which might vaporize could effect the cavitation bubble chemical activity. The main practical interest in sonochemical enzyme inactivation rates lies in the degree to which enzymes may be inactivated during extraction from cells by conventional ultrasonic cavitation cell disintegration. The above discussion dealt only with soluble enzymes, membrane-bound systems will be discussed later (see page 149). Él'piner[89] considers that, in general, mechanical forces associated with cavitation will not degrade a globular protein of molecular weight less than 20,000 - 25,000. However, he has observed that individual amino acids or low molecular weight polypeptides, comprising a small number of peptides, may be broken off protein molecules by chemical attack. The location of the break in the polypeptide chain and the nature of the chemical transformations of the end amino-acid residues depend on the gas present. In general,

El'piner found that prolonged treatment of proteins in the presence of oxygen led to reduction in the size of the polymer while treatment in the presence of hydrogen led to increased molecular weight. The molecular weight increase was not necessarily accompanied by a loss of biocatalytic activity of treated enzymes.

Inactivation of enzymes may arise if a conformational change occurs in an enzyme or if an amino acid residue in the active site interacts with a radical,[94] thereby interfering with the active-centre-substrate complex formation. Determinant sites, specific sequences of amino acids, determine the antigenic properties of a protein and enable it to bind to its antibody by forces similar to those involved in enzyme-substrate binding.[95] Edwards and his colleagues[96] have found that cavitation reduced the ability of RNA-ase to combine with its antibody. The rate of loss of antigenicity was one-seventh of the rate at which enzymic activity was lost. Fragments antigenic both to light and heavy chain antibody have been found in solutions of rabbit γ-globulin exposed to cavitation.[97]

Fibrous proteins, e.g., collagen[89] and keratin[98,99] have been treated with cavitation and the molecular weights of the products measured in an attempt to isolate fundamental subunits of the proteins. Changes in the biological activity of the hormone insulin have been observed after treatment.[89] Ultrasound is used in pharmacy for the extraction of alkaloids and glycosides. Skauen[100] has reviewed the occasional reports of damage to the extracted material and suggested that further work is needed to help avoid such degradation. The four bases of DNA, i.e., quanidine, cystosine, adenine and thymine, have been broken down chemically by cavitation. Samples of the pyrimidine bases were not only broken down but some of them underwent chemical transformation and produced new fluorescent compounds.[89] Since DNA bases have been affected by cavitation, then when DNA itself is exposed for a long time, effects like production of ammonia would be expected. These effects have been observed.[89] On assaying the biological integrity of DNA, Combes found that in vitro exposure to cavitation in a 20 kHz field was unable to produce mutagenic lesions that could be detected by transformation. Transforming activity could be reduced without a corresponding decline in linkage between two contiguous markers. This suggests that no significant mutagenic modification of transforming DNA had occurred. The decrease in transforming ability was ascribed to a reduction in DNA molecular weight arising from the mechanical effect of cavitation. When nucleoproteins *in vitro* were treated by Hrazdira[102] he was able to show that the release of polydeoxyribonucleotides arose from the mechanical effects of cavitation and that a second effect, the sensitization of the nucleoprotein to the action of heparin was a sonochemical effect. On comparatively short exposure to cavitation the main effect on DNA is reduction of molecular weight by mechanical cleavage.

2.5 Mechanical Effects of Cavitation on Biopolymers

Degradation of biopolymers in solution is just a particular example of the degradation of macromolecules in ultrasonic fields. It is usually observed that the molecular weight falls rapidly to a limiting value. This observation has led many authors to conclude that the degradation is mechanical in nature[103] although Alexander and Fox[104] have stated that 30% of the breakage of DNA in aqueous solution in the presence of cavitation could be due to chemical effects. As previously pointed out (see page 139) Él'piner and his colleagues[89,97] have observed small polypeptides in solution after long term exposure of proteins to cavitation. Kinetic studies of the approach of molecular weight to a limiting value have been reviewed by Peacocke and Pritchard.[105] It is likely in the early stages of macromolecular degradation by cavitation that mechanical forces produce the rapid decrease in molecular weight which is commonly observed while on longer irradiation modification of the polymers by chemical attack may become apparent.

On ultrasonic irradiation of synthetic polymers at intensities from 5 to 700 W/cm^2 over the frequency range 210 kHz to 2 MHz several investigators have detected polymer degradation which was not observed when steps were taken to suppress cavitation.[106-110] A similar result was obtained by Hill et al. on exposing DNA to ultrasound.[111] Some investigators[112,113] have reported observing continued degradation of synthetic polymers, though at a reduced rate, with suppression of cavitation by irradiating samples under reduced ambient pressure. This method of cavitation suppression has not been widely used and no quantitative measurements are available on the efficiency of this technique.

Electron micrographs of DNA fragments produced by cavitation at 9 kHz have shown that the molecule breaks by double back-bone scission, i.e., the ends of the fragments were double rather than single stranded.[114,115] An investigation of the viscosity of sonic fragments from a cavitating field at 800 kHz and 15 W/cm^2 supports the view that double back-bone scission, rather than an accumulation of single strand breaks, is the cause of DNA degradation.[116] The chemical work of Richards and Boyer[117] suggests that the main bond involved in back-bone scission is the C-O bond (90%) with 10% P-O rupture and no appreciable C-C damage. The broken molecules are macro-radicals[89] and these have been successfully labelled with radioactive iodine.[118] Bubbles trapped in holes in a solid surface and driven at 20 kHz have been used to degrade DNA under conditions of stable cavitation.[119] Attempts were made to relate the kinetics to the degradation of DNA in non-sonic laminar shear devices but it has been pointed out[120] that the stable cavitation condition present during the above experiments probably included surface wave activity and that available microstreaming theory might not be adequate. Transient Willard type cavitation events have been shown to be responsible for DNA breakage in small containers placed at the focus of a 1 MHz bowl, when the sound intensity was of the order of 500 W/cm^2. The volume of degraded DNA, i.e., the sphere of influence of the cavitation event, had dimensions of the order of 1 mm.[121] At lower intensities the breakage did not depend on the number of Willard events detected and it was suggested that a non-cavitation interaction between the molecule and the sound may have stressed the molecules. Subsequent experiments with sonochemical reagents in the same system showed that non-transient stabilized cavitation centres could exist in the treatment vessels and these may have been responsible for the degradation of the DNA at the lower intensities.[57]

The most noteworthy attempt to detect a mechanism for DNA breakdown which does not involve the effects of cavitation was that of Hawley, MacLeod and Dunn.[122] Several methods were employed simultaneously to detect the presence of cavitation when degassed DNA was degraded at 25-31 W/cm^2 and 981 kHz. These methods were (i) a thermoelectric probe mounted behind the sample to monitor transmitted intensity, (ii) ultraviolet light was passed through the sample holder and monitored to detect changes in the light transmitting properties of the medium, (iii) when cavitation occurred the acoustical impedance of the liquid changed and caused a movement of a meter needle in a voltage feed-back circuit. While all the above tests proved negative the absolute sensitivity of the detecting methods is unknown and the possibility that a small number of stable resonant bubbles were present cannot be ruled out.

Because of the great biological significance of DNA some authors have made the fact that the molecule may be degraded *in vitro* by ultrasound the basis for suggestions that ultrasound in medicine may be hazardous, without taking note of whether or not cavitation was occurring in the treated solutions or noting the difference between the conformation of DNA in solution compared with DNA in the cell nucleus.

2.6 Mechanical Effects of Cavitation on Cells

It has previously been pointed out (see page 136) that cells in suspension exposed to ultrasonic cavitation are broken, mainly by the mechanical forces associated with cavitation. Él'piner[89] refers to almost 200 reports of studies of the effects of ultrasound on micro-organisms and viruses. Much of this work was concerned with the release of intracellular components from micro-organisms treated with cavitation in sound fields of frequency around 20 kHz. The reports include disintegration of Gram-positive, Gram-negative, aerobic and anaerobic, pathogenic and non-pathogenic micro-organisms. Eucaryotes as well as procaryotes have been disrupted.

The rate of cell disintegration has usually been found to be constant and the surviving cell population has decreased exponentially. Thacker has examined cell survival in haploid and diploid yeast cells *Saccharomyces cerevisiae* and shown that the larger, i.e., dividing, cells are most likely to break.[123] He suggests that since non-synchronous cell population, therefore, possess cells of differing susceptibility to cavitation that the survival curves may be bi- or multi-phasic. His experimental results, when working near the cavitation threshold, show a departure from the usual exponential cell survival curves. However, for most users of ultrasound at levels well above cavitation threshold, the breakage rate will be constant, at least until the surviving population falls below 10% of the initial cell population.

The structure of the outer envelope of the cell has a great influence on its susceptibility to sound. The breakage rate constant for a polysaccharide walled budding yeast or a bacterium like *Escherichia coli* is about one-hundredth of the constant for the disruption of a membraneous outer envelope cell like a red blood cell. Erythrocytes have been disrupted by microstreaming about a single bubble vibrating in a viscous medium to minimize the possibility that bubble surface activity could arise,[124] and also by arrays of bubbles where surface activity may have been present.[119] Hughes and Nyborg[54] found that protein release, which in most cases is a reliable index of cell breakage, was detected when suspensions of the bacterium *E. coli* were exposed to an array of stable bubbles vibrating at 20 kHz. The protein release occurred below the amplitude threshold for free iodine production. This led the authors to suggest that transient, i.e., collapse-type cavitation may not be necessary for cell breakage but that its presence during cell disintegration with 20 kHz sonication coincides with the increased activity of the stable bubbles which disrupt the cells. At megahertz frequencies it has been suggested that *in vitro* mammalian cell breakage has resulted from the activity of stable bubbles resonant at half the driving frequency.[80] A search for emissions in addition to the first subharmonic would provide more information on the exact type of types of bubble activity which affected the breakage. One Willard type cavitation event is capable of breaking all the ameobae *Hartmanella castellanii* in a volume with typical dimensions of 1 mm.[125] The degree of erythrocyte breakage found at 20 kHz in a commercial disintegrator and at 1 MHz in the presence of stabilized cavitation centres were comparable for similar power inputs.[57] Cavitation activity, however, is more dependent on gas content at the higher frequency.

Cell suspensions examined after exposure to cavitation display a range of degrees of damage from complete disruption to apparently undamaged cells. An examination of treated yeast under the electron microscope showed that in the absence of envelope rupture the cells appeared to be undamaged intracellularly.[123] When suspensions of *E. coli* were exposed to "soft" cavitation by treating the cells in a polypropylene tube in a 20 kHz sound field the rate of cell disintegration was considerably reduced compared with cells exposed directly to the disintegrator horn.[126] A range of degrees of cell damage was observed on electron microscope examination. The objective of the work was to produce a bacterium where the protein synthesizing systems remained functional and yet its envelope would be permeable to macromolecules. The effect of molecules like actinomycin D, adenosine triphosphate, trypsin, ribonuclease and polyuridylic acid on viability and protein synthesis of the permeable, slightly ruptured cells was examined.

When cavitation occurred in cell suspensions which were being treated with a view to determining if any non-cavitational interaction, damaging to the cell, had occurred between the cell and the sound field, a number of workers have examined the surviving cell fraction for any changes.

No mutagenic effects on Bacillus subtilis have been detected on examination after exposure of the cells to cavitation at 20 kHz.[101] Thacker tested four different genetic systems for mutations in treated yeast cells. Only the system assaying for genetic loss of mitochondrial DNA function showed an increase of mutant frequency in samples exposed to cavitation. The mutation increase was present under experimental conditions which led to a temperature increase or a build up of sonochemical products.[127] No chromosome aberrations were observed in lymphocyte cultures after exposure to cavitation at megahertz frequencies.[128] Schnitzler[129] and his colleagues, Kim, et al.[130] have, however, reported a high incidence of single chromatid breaks when stable cavitation occurred in an in vitro suspension of proliferating ascites carcinoma cells at 810 kHz. These chromatid breaks indicated either that the mitotic apparatus had been damaged during anaphase or that the centromeres had been broken. It had been suggested also that resonant microbubbles may have caused chromosome aberrations in suspensions of human lymphocytes treated at 2.25 MHz[131] but this suggestion has now been withdrawn since the effect was not reproduced.[132] The difficulties associated with deciding, by subsequent growth studies, if cell death in a culture exposed to cavitation has resulted from cavitation, or some other interaction, are discussed by Coakley, Hampton and Dunn.[125] Clarke and Hill concluded that there was no effect on the progress of mouse lymphoma cells through the cell cycle after exposure to cavitation.[133] Dvorak and Hrazdira describe in detail the changes in the ultrastructure of rat bone marrow cells after exposure of the cells in vitro to what was apparently a cavitating field. Membrane lesions, mitochondrial swelling, and nuclear changes were observed.[134]

Erythrocyte lysis,[135] and electrophoretic mobility changes in a mouse ascites tumour cell exposed to ultrasound,[136] have been claimed as effects on cells which are not explainable in terms of cavitation. The electrophoretic motility change in the mouse ascites cells was observed in the absence of free iodine liberation and without detectable subharmonic emission. An attempt by other workers to detect an electrophoretic motility change in Chinese hamster cells was unsuccessful unless cavitation was present.[137]

When a biological effect is observed in aqueous cell suspensions after exposure to ultrasound even at megahertz frequencies, the presence of cavitation must always be suspected. The low cavitation threshold of around 200 mW/cm^2 reported by some workers for cavitation and erythrocyte lysis at 1 MHz highlights the above point.[68] It is not always practical to treat cells in degassed suspension since the cells may be affected by the absence of oxygen. Visible bubble production is a useful guide to the presence of cavitation although on occasions the rise in temperature on placing a cold sample saturated with gas in a warm treatment tank can lead to bubble formation in the absence of sound. The absence of visible bubbles in a treated sample is no guarantee that cavitation has not taken place.[57,138] There are several reports of biological damage to cells in a bubble field achieved before sonochemical activity was observed so the absence of a chemical yield cannot guarantee the absence of cavitation.[54,139] Failure to detect a significant first order subharmonic emission signal has been recorded under conditions where bubble activity could reasonably be suspected of being responsible for effects observed.[57,121] In the above cases no increase in detected subharmonic above the level when the sample container was not in the field could reliably be detected. Problems arising from background emissions coming from even carefully "degassed" water may be overcome by coupling samples to the transducer by liquids with a higher cavitation threshold than water.[140] The published work on cavitation bubble emission suggests that the emissions most likely to be detected as the first sign of non-linear bubble activity are harmonics of the driving frequency.[44] Increasing ambient pressure on a sample will suppress some aspects of bubble activity but cannot guarantee the absence of all forms of non-linear bubble motion. Increasing the ambient pressure on DNA solutions treated at megahertz frequencies has been shown to reduce and sometimes suppress cavitation activity as measured by DNA breakage, sonochemical activity, and subharmonic emission.[140] Increased ambient pressures have, in other cases, enhanced cavitation activity [141,142] possibly because transient collapse bubbles rather than stable bubbles were active in these fields. The suppression of cavitation between a sample and the transducer in a focused field, under increased pressure will lead to higher intensities reaching a sample.[143] The effects of increased ambient pressure also depend on the dissolved gas concentration and on whether the pressure increase is achieved by mechanical compression of the load without a gas-liquid interface, or by increasing the gas pressure over the liquid surface.[109]

2.7 Optimization of the Release of Undegraded Material from Cells by Cavitation

The disruption of cells for the release of intracellular materials by methods such as mechanical stress, chemolysis, freeze thawing, and enzyme extraction has long been a standard procedure in biology.[144] Ultrasound has played an increasing role in cell disruption since the late 1950's because of the availability of commercial equipment and because of the relative simplicity of the method. One of the earliest critical examinations of the method was that of Hughes and Neppiras[145] working with a 60 watt disintegrator of a potential sonic efficiency of 71%. They found, among other things, that the cells were being broken at an exponential rate by a mechanical stress associated with cavitation, and also that the distance between the end of the transducer and the bottom of the treatment vessel should not be a quarter wavelength. Best reproducibility was achieved by having the end of the transducer a few millimeters from the base of the sample holder. Other experiments with a higher powered generator showed that the product of the release rate of protein (k_1) from brewers yeast, and the volume of the treatment vessel (V) was constant over a volume range from 70 to 300 ml and over a cell concentration range from 1 to 60%.[146] Since the product k_1 V was constant for a given cell at a given power level, then this product may be a useful means of characterizing relative biological cell susceptibilities to cavitation.

The rate of erosion of transducer tips by cavitation becomes marked at intensities of the order of 150 W/cm^2. In an unreported brief study one of us (W.T.C.) found that cavitation unloading, i.e., the growth of large air bubbles on the surface of the transducer, which prevents the transmission of sound and causes splashing of the cell suspension, occurs when the acoustic power density in the liquid is around 3 Watts/ml.

Flow systems are now frequently employed with ultrasonic cell disintegrators. The steady-state fraction x_{eq} of the cell population disrupted when the flow rate is Y ml/min is given by

$$x_{eq} = (1 + Y/k_1V)^{-1} \tag{66}$$

when the input power is constant.[147]

When cells are disrupted by any method to release mitochondria, nuclei, lysosomes, intracellular endoplasmic reticulum membrane fragments, soluble or membrane bound enzymes, or enzyme complexes, ribosomes, DNA or RNA, it is often desirable to avoid degradation of the material of interest by the method used to break the cell. Cell organelles and high molecular weight ($M > 10^6$) polymers like DNA if broken by ultrasound on extraction may be considered to be broken by the mechanical effects of cavitation. Biological activity of smaller polymers like globular proteins, e.g., enzymes, will in general be most susceptible to sonochemical attack. Mechanical breakdown of released product, and its rate constant k_2, is largely independent of the concentration of material released, whereas the rate of chemical attack on soluble enzymes will be concentration dependent.[91]

For systems where material such as organelles are to be released at a rate k_1 and subsequently damaged by mechanical forces at a rate k_2 the resultant fractional yield D is given[147] by

$$D = [\exp(-Bk_1t) - \exp(-k_1t)]/(1-B) \quad (67)$$

where $B = k_2/k_1$. Eq. (67) is shown in Fig. 9. Computer solutions for the case where k_2 is dependent on the concentration of released material have been obtained and the general trends in the solutions are similar to Fig. 9.[147] The fractional yield of active material from a flow system has also been determined. It has been shown that under optimum conditions the fraction of active material releasable in a batch system is higher than the case for flow systems. An interesting observation is that in a flow system the active fraction increases to a maximum value after flow begins and then decreases to its equilibrium value. The equilibrium amount of undegraded product obtainable from a cell suspension by sonication expressed as a fraction D_{eq} of the total pool initially in the cells is shown in Fig. 10 as a function of a flow rate parameter when k_2 is constant. If the ratio k_2/k_1 is large, Figs. 9 and 10 show that ultrasound would not be indicated for extraction in these circumstances.

Soluble enzymes may be protected against sonochemical attack by treatment of high cell concentrations or even by extraction in a growth medium which is rich in free-radical scavengers. A more complicated situation than those discussed above may arise where a membrane bound enzyme chain is being extracted. The mechanical forces of the cavitation will disrupt the intracellular membranes, possibly breaking the chain. Vesicles may also be formed which may present a permeability barrier to the substrate, thus reducing the enzyme rate. This effect has been demonstrated as follows. The action of the antibiotic D-cycloserine on intact *Pseudomonas aeruginosa* inhibits transport of the substrate of D-alanine oxidase across the cell membrane. When an isolated membrane preparation was obtained by extraction with the Hughes Press D-alanine oxidase was free to oxidise its substrate. Sonication of the membrane preparation formed closed vesicles which re-established the enzyme inhibition by

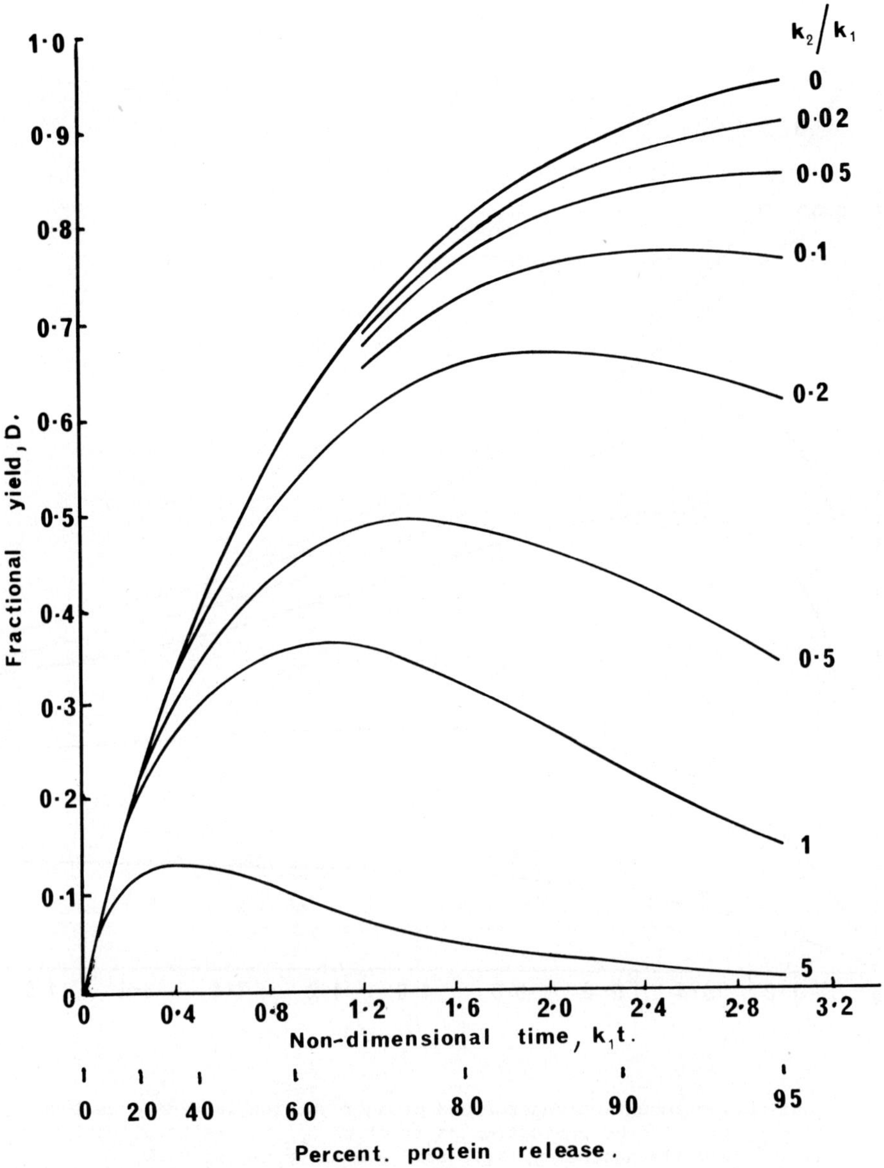

Fig. 9 D for batch sonication against k_1t for several values of k_2/k_1. The lower numbers on the abscissa scale refer to the percentage of protein release or cell breakage associated with the corresponding value of k_1t.

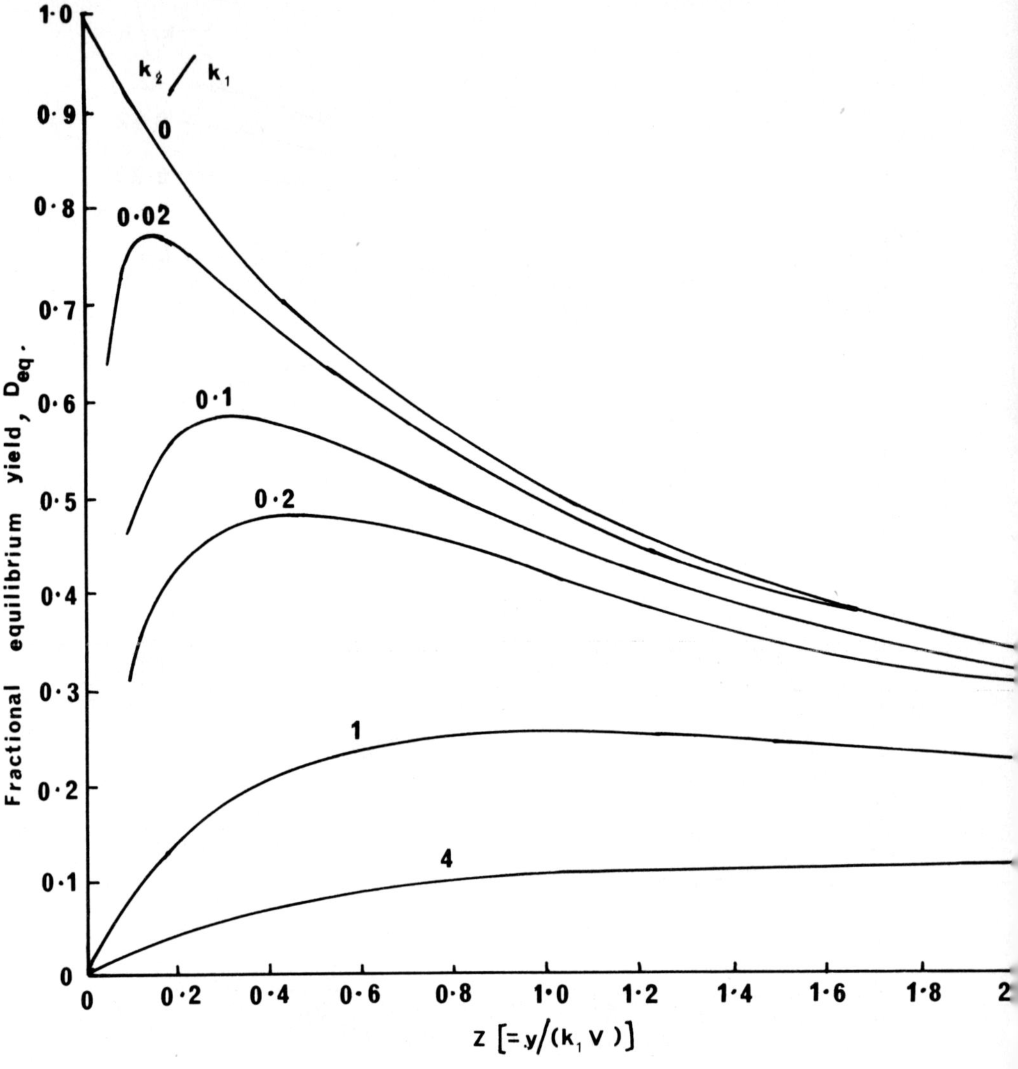

Fig. 10 Equilibrium amount of undergraded product obtainable from a cell suspension by flow system sonication as fraction D_{eq} of the total pool initially in the cells against $y/(k_1V)$ for several values of k_2/k_1.

D-cycloserine. Cavitation had, therefore, produced closed vesicles with functional transport and oxidation systems. (R. C. Brown, personal communication). Sub-mitochondrial particles of diameter ranging from 200 to 5000Å are produced when intact mitochondria are exposed to cavitation.[148] The inside-outside order of the membrane is reversed when vesicles are formed from the mitochondrial inner membrane.[149] When mixtures of bacterial membranes labelled with a density-labelled fatty acid and unlabelled membranes are sonicated in a cavitating field of 20 kHz a hybrid membrane vesicle of intermediate density results.[150] Apparently the sonication gave rise to fragments smaller than those finally observed. These small fragments aggregated to form vesicles which exhibited normal structure by electron microscopy. Chance, *et al*. found that the oxidation of cytochrome *c* was reduced in submitochondrial particles compared with unsonicated intact mitochondria and they concluded that sonication had produced a permeability barrier to the oxidants.[151] They also suggested that submitochondrial particles produced in intense cavitation fields are reassembled from smaller fragments which resulted from the disruptive action of the cavitation. Sonicated aqueous dispersions of phospholipids have gained importance because of their use in model membrane, lipoprotein and enzyme studies. The spherical vesicles resulting from continued sonication of egg - lecithin have diameters of the order of 250Å . On studying the chemical degradation of the membranes Hauser found that sonication under a nitrogen atmosphere eliminated oxidation of the unsaturated hydrocarbon chain but lysolecithin and fatty acids were among degradation products resulting from long sonication even under a nitrogen atmosphere.[152]

3. ACKNOWLEDGEMENTS

We are grateful to a number of individuals for helpful comments and suggestions on this manuscript, especially Drs. A. I. Eller, E. A. Neppiras, D. L. Storm, M. Strasberg and D. L. Miller.

REFERENCES

1. P. S. Epstein and M. S. Plesset, *J. Chem. Phys.*, 18 (1950) p. 1505.
2. E. N. Harvey, in *Decompression Sickness*, Ed., J. F. Fulton, W.B. Saunders, Philadelphia, 1951, Chap. IV.

3. F. E. Fox and K. F. Herzfeld, J. Acoust. Soc. Amer., 26 (1954) p. 984.

4. H. G. Flynn, in Physical Acoustics, Vol. 1B, ed., W. P. Mason, Academic Press, New York, 1964, pp. 57-172.

5. J. F. Fulton (Ed.), Decompression Sickness, W. B. Saunders Co., Philadelphia, 1951.

6. R. Gramiak and P. M. Shah, Radiology, 100 (1971) p. 415.

7. C. Devin, Jr., J. Acoust. Soc. Amer., 31 (1959) p. 1654.

7a. D. L. Miller, J. Acoust. Soc. Amer., publication in 1977.

8. H. Lamb, Hydrodynamics, Dover Publications, Inc., New York, 1945.

8a. A. I. Eller, J. Acoust. Soc. Amer., 47 (1970) p. 1469.

9. E. N. Harvey, D. K. Barnes, W. D. McElroy, A. H. Whitely, D. C. Pease and K. W. Cooper, J. Cellular Comp. Physiol., 24 (1944) p. 1.

10. D-Y Hsieh and M. S. Plesset, J. Acoust. Soc. Amer., 33 (1961) p. 206.

10a. M. H. Safar, J. Acoust. Soc. Amer., 43 (1968) p. 1188.

11. M. Strasberg, J. Acoust. Soc. Amer., 33 (1961) p. 359L.

12. A. I. Eller and H. G. Glynn, J. Acoust. Soc. Amer., 37 (1965) p. 493.

13. A. I. Eller, J. Acoust. Soc. Amer., 52 (1972) p. 1447.

14. M. Strasberg, J. Acoust. Soc. Amer., 31 (1959) p. 163.

15. A. I. Eller, J. Acoust. Soc. Amer., 46 (1969) p. 1246.

16. R. K. Gould, J. Acoust. Soc. Amer., 56 (1974) p. 1740.

17. D. E. Goldman and W. W. Lepeschkin, J. Cell. Comp. Physiol., 40 (1952) p.255.

18. F. G. Blake, Jr., J. Acoust. Soc. Amer., 21 (1949) p. 551L.

19. A. Eller, J. Acoust. Soc. Amer., 43 (1968) p. 170L.

20. R. K. Gould, J. Acoust. Soc. Amer., 43 (1968) p. 1185L.

21. L. A. Crum and A. I. Eller, J. Acoust. Soc. Amer., 48 (1970) p. 181.

21a. P. J. Westervelt, J. Acoust. Soc. Amer., 29 (1957) p. 26.

22. W. L. Nyborg and D. L. Storm, Eighth International Congress on Acoustics, London, 1964, Goldcrest Press, Trowbridge, Wilts, England, p. 470.

23. S. A. Elder, J. Acoust. Soc. Amer., 31 (1959) p. 54.

24. W. L. Nyborg, J. Acoust. Soc. Amer., 30 (1958) p. 329.

25. J. A. Rooney, Science, 169 (1970) p. 869.

26. B. J. Davidson and N. Riley, J. Sound Vib., 15 (1971) p. 217.

27. W. L. Nyborg, D. L. Miller and A. Gershoy, Seventh Rochester International Conference on Environmental Toxicity, June 1974, Plenum Press.

28. T. B. Benjamin and F. Ursell, Proc. Roy. Soc. (London), A225 (1954) p. 505.

29. T. B. Benjamin, in Cavitation in Real Liquids, Ed., R. Davies, Elsevier Publishing Co., Amsterdam, 1964, pp. 164-180.

30. V. I. Sorokin, Soviet Phys.-Acoust., 3 (1957) p. 281.

31. W. Eisenmenger, Acustica, 9 (1959) p. 327.

32. M. Strasberg and T. B. Benjamin, J. Acoust. Soc. Amer., 30 (1958) p. 697A.

33. A. I. Eller and L. A. Crum, J. Acoust. Soc. Amer., 47 (1970) p. 762.

34. R. K. Gould, J. Acoust. Soc. Amer., 40 (1966) p. 219.

35. G. W. Willard, J. Acoust. Soc. Amer., 26 (1954) p. 933A.

36. W. L. Nyborg and A. Rodgers, Biotechnology and Bioengineering, 9 (1967) p. 235.

37. D. L. Storm, in Finite-Amplitude Wave Effects in Fluids, Ed., L. Bjørnø, IPC Science and Technology Press, Ltd., Guildford, Surrey, England, 1974, pp. 234-239.

38. E. Neppiras and E. Fill, J. Acoust. Soc. Amer., 46 (1969) p. 1264.

39. W. L. Nyborg and D. E. Hughes, J. Acoust. Soc. Amer., 42 (1967) p. 891.

40. M. Strasberg, J. Acoust. Soc. Amer., 28 (1956) p. 20.

41. D. L. Storm, J. Acoust. Soc. Amer., 52 (1972) p. 152A.

42. D-Y Hsieh, J. Acoust. Soc. Amer., 56 (1974) p. 392.

43. E. A. Neppiras and J. Parrott, Fifth International Congress on Acoustics, Liege, 1965, Paper D51.

44. E. A. Neppiras, J. Acoust. Soc. Amer., 46 (1969) p. 587.

45. D. L. Storm, Ph.D. Thesis, University of Vermont, 1975.

46. A. Eller and H. G. Flynn, J. Acoust. Soc. Amer., 46 (1969) p. 722.

47. M. H. Safar, J. Phys. D.: Appl. Phys., 3 (1970) p. L48.

47a. A. Prosperetti, J. Acoust. Soc. Amer., 56 (1974) p. 878.

47b. A Prosperetti, J. Acoust. Soc. Amer., 57 (1975) p. 810.

48. Lord Rayleigh, Phil. Mag., 34 (1917) p. 94.

49. B. E. Noltingk and E. A. Neppiras, Proc. Phys. Soc. (London), B63 (1950) p. 674.

50. W. Lauterborn, in Finite-Amplitude Wave Effects in Fluids, Ed., L. Bjørnø, IPC Science and Technology Press, Ltd., Guildford, Surrey, England, 1974, pp. 195-202.

51. S. D. Howkins, J. Acoust. Soc. Amer., 39 (1966) p. 55.

52. T. K. Saksena and W. L. Nyborg, J. Chem. Phys., 53 (1970) p. 1722.

53. P. Temple, R. W. Detenbeck and W. L. Nyborg, J. Acoust. Soc. Amer., 50 (1971) p. 112A.

54. D. E. Hughes and W. L. Nyborg, Science, 138 (1962) p. 108.

55. G. W. Willard, J. Acoust. Soc. Amer., 25 (1953) p. 669.

56. W. T. Coakley, J. Acoust. Soc. Amer., 49 (1971) p. 792.

57. W. T. Coakley and M. F. Sanders, J. Sound Vib., 28 (1973) p. 73.

58. E. A. Neppiras, IEEE Trans., SU-15 (1968) p. 81.

59. F. G. Blake, Jr., The Onset of Cavitation in Liquids. 1, Tech. Mem. No. 12, Acoustics Research Lab., Harvard University, Cambridge, Mass., 1949.

60. R. Esche, Acustica, 2 AB (1952) p. 208.

61. T. Lange, Acustica, 2 AB (1952) p. 75.

62. J. E. Barger, Thresholds of Acoustic Cavitation, Tech. Mem. No. 57, Acoustics Research Lab., Harvard University, Cambridge, Mass., 1964.

63. A. I. Eller, J. Acoust. Soc. Amer., 55 (1974) p. 871.

64. M. Ceschia and G. Iernetti, J. Acoust. Soc. Amer., 56 (1974) p. 369.

65. E. Meyer, J. Acoust. Soc. Amer., 29 (1957) p. 4.

66. E. A. Neppiras and W. T. Coakley, J. Sound Vib., 45 (1976) p. 341.

67. G. Iernetti, Acustica, 25 (1971) p. 191.

68. D. Doynon and Y. Simont, Comp. Rend., 232 (1951) pp. 2011 and 2411.

69. G. Iernetti, Acustica 26 (1972) p. 112.

70. W. T. Richards and A. L. Loomis, J. Amer. Chem. Soc., 49 (1927) p. 3086.

71. H. Frenzel and H. Schultes, Z. Physik. Chem., 27 (1934) p. 421.

72. E. A. Neppiras and B. E. Noltingk, Proc. Phys. Soc. (London) B34 (1951) p. 1032.

73. R. D. Prudhomme, Vijnana Parishad Anusandhan Patrika 15 (1972) p.3.

74. M. DeGrois and P. Baldo, Ultrasonics 12 (1974) p. 25.

75. M. Del Duca, E. Yeager, M. O. Davies and F. H. Hovorka, J. Acoust. Soc. Amer., 30 (1958) p. 301.

76. M. A. Margulis, Sov. Phys. Acoust., 16 (1971) p. 361.

77. J. H. Todd, Ph.D. Thesis, London University, 1970, p. 4.

78. J. H. Todd, Ultrasonics, 8 (1970) p. 234.

79. A. Weissler, J. Acoust. Soc. Amer., 32 (1959) p. 283.

80. P. R. Clarke and C. R. Hill, J. Acoust. Soc. Amer., 47 (1970) p. 649.

81. P. W. Vaughan, J. Sound Vib., 7 (1968) p. 236.

82. I. Gabrielli, G. Iernetti and A. Lavenia, Acustica 18 (1967) p. 173.

83. G. Iernetti, Acustica 23 (1970) p. 189.

84. M. Ceschia and G. Iernetti, Acustica 29 (1973) p. 127.

85. A. Weissler, H. W. Cooper and S. Snyder, J. Amer. Chem. Soc., 72 (1950) p. 1769.

86. D. E. Hughes, J. Biochem. Microbiol. Technol. Eng., 3 (1961) p. 405.

87. L. M. Bronskaya, V. S. Vigderman, A. V. Sokol'skaya and I. E. Él'piner, Sov. Phys. Acoust., 13 (1968) p. 374.

88. I. Hrazdira, Folia Biol. (Prague), 11 (1965) pp. 330-333.

89. I. E.Él'piner, Ultrasound: Physical, Chemical and Biological Effects, Consultants Bureau, New York, 1964.

90. W. H. Staas and L. A. Spurlock, J. Chem. Soc., (Perkins Trans. I) Part II, (1975) p. 1675.

91. W. T. Coakley, R. C. Brown, C. J. James and R. K. Gould, Arch. Biochem. Biophys., 159 (1973) p. 722.

92. R. M. MacLeod and F. Dunn, J. Acoust. Soc. Amer., 42 (1967) p. 527.

93. A. Wiseman, P. H. Leslie and G. C. Granville, Biochem. Soc. Trans., 1 (1973) p. 606.

94. G. E. Adams, R. L. Willson, J. E. Aldrich and R. G. Cundall, Int. J. Rad. Biol., 16 (1969) p. 333.

95. I. M. Roitt, Essential Immunology, Blackwell Scientific Publications, Oxford (1971) p. 8.

96. J. Edwards, C. J. James, W. T. Coakley and R. C. Brown, J. Acoust. Soc. Amer., 59 (1976) p. 1513.

97. O. M. Zorina, R. S. Nezlin, T. I. Vengerova and I. E. Él'piner, Biofizika, 13 (1968) p. 595.

98. J. Catherall, Ph.D. Thesis, University of Leeds, 1972.

99. J. Catherall and P. T. Speakman, Nature, 219 (1968) p. 610.

100. D. M. Skauen, Chem. Eng. Prog. Sym. Series, 67 (1972) p. 35.

101. R. D. Combes, Ph.D. Thesis, University of London, 1972.

102. I. Hrazdira, Scripta Medica, 42 (1969) p. 5.

103. G. Gooberman, J. Polymer Sci., 42 (1960) p. 25.

104. P. Alexander and M. Fox, J. Polymer Sci., 12 (1954) p. 533.

105. A. R. Peacocke and N. J. Pritchard, Prog. Biophys. and Mol. Biol., 18 (1968) p. 185.

106. W. Roberts, E. Yeager and F. Hovorka, Tech. Report No. 18, Western Reserve University (1957).

107. G. Schmid, G. Paret and H. Freifelder, Colloid Zeit., 124 (1951) p. 150.

108. P. Grabar and R. O. Prudhomme, J. Chem. Phys., 46 (1949) p. 667.

109. H. W. Brett and H. H. G. Jellinek, J. Polymer Sci., 21 (1956) p. 535.

110. A. Weissler, J. Appl. Phys., 21 (1950) p. 171.

111. C. R. Hill, P. R. Clarke, M. R. Crowe and J. W. Hammick, Ultrasonics for Industry Papers, Iliffe Ind. Pub., London, 1969, p. 26.

112. H. W. Melville and A. J. R. Murray, Trans. Faraday Soc., 46 (1950) p. 996.

113. M. A. K. Mostafa, J. Polymer Sci., 33 (1958) p. 311.

114. C. E. Hall and P. Doty, J. Amer. Chem. Soc., 80 (1958) p. 1269.

115. C. E. Hall and M. Litt, J. Biophys. Biochem. Cytol., 4 (1958) p. 1.

116. N. I. Ryabchenko, F. I. Braginskaya and I. E. Él'piner, Biofizika, 9 (1964) p. 31.

117. O. C. Richards and P. D. Boyer, J. Mol. Biol., 11 (1965) p. 327.

118. A. R. Williams, Biopolymers, 8 (1969) p. 555.

119. N. J. Pritchard, D. E. Hughes and A. R. Peacocke, Biopolymers, 4 (1966) p. 259.

120. W. L. Nyborg, Proc. 1973 Symposium on Finite Amplitude Waves in Fluids, Ed., L. Bjørnø, IPC Science and Technology Press, Ltd., Guildford, Surrey, England, 1974, p. 245.

121. W. T. Coakley and F. Dunn, J. Acoust. Soc. Amer., 50 (1971) p. 1539.

122. S. A. Hawley, R. M. MacLeod and F. Dunn, J. Acoust. Soc. Amer., 35 (1963) p. 1285.

123. J. Thacker, Biochem. Biophys. Acta, 304 (1973) p. 240.

124. J. A. Rooney, J. Acoust. Soc. Amer., 52 (1972) p. 1718.

125. W. T. Coakley, D. Hampton and F. Dunn, J. Acoust. Soc. Amer., 50 (1971) p. 1546.

126. H. Amos, A. Kuhn and J. Andre-Schwartz, J. Bacteriol., 94 (1967) p. 232.

127. J. Thacker, Brit. J. Radiol., 47 (1974) p. 130.

128. J. Braeman, W. T. Coakley and R.K. Gould, Brit. J. Radiol., 47 (1974) p.158.

129. R. M. Schnitzler, Interactions of Ultrasound and Biological Tissues, DHEW Publication (FDA) 73-8008, BRH/DBE 73-1 (1973) p. 69.

130. A. Kim, S. Pavlovic, R. M. Schnitzler and K. H. Woeber, (A) First Nat. Rumanian Conf. on Biophysics, Bucharest, 1971, p. 67.

131. I. J. C. Macintosh and D. A. Davey, Brit. J. Radiol., 48 (1975) p. 230.

132. I. J. C. Macintosh, R. C. Brown and W. T. Coakley, Brit. J. Radiol., 48 (1975) p. 230.

133. P. R. Clarke and C. R. Hill, Exptl. Cell. Res., 58 (1969) p. 443.

134. M. Dvorak and I. Hrazdira, Zeitschrf. Mikroanat. Forschung., 75 (1967) p.451.

135. C. C. Connolly, Ph.D. Thesis, University of London, 1969.

136. K. J. W. Taylor and D. L. Newman, Phys. Med. Biol., 17 (1972) p. 270.

137. G. P. Joshi, C. R. Hill and J. A. Forrester, Ultrasound in Med. and Biol., 1 (1973) p. 45.

138. W. T. Coakley, Ph.D. Thesis, University of Wales, 1971.

139. J. A. Rooney, Interactions of Ultrasound and Biological Tissues, DHEW Publication (FDA) 73-8008, BRH/DBE 73-1 (1973) p. 93.

140. C. R. Hill, J. Acoust. Soc. Amer., 52 (1972) p. 667.

141. M. G. Sirotyuk, Sov. Phys. Acoust., 12 (1966) p. 199.

142. F. A. Angona, J. Acoust. Soc. Amer., 50 (1971) p. 277.

143. B. A. Agranat and V. I. Bashkivrov, Sov. Phys. Acoust. 15 (1970) p. 520.

144. D. E. Hughes, J. W. T. Wimpenny and D. Lloyd, in Methods in Microbiology, Vol. 5B, Ed., J. R. Norris and D. W. Ribbons, Academic Press, London (1971) p. 1.

145. E. A. Neppiras and D. E. Hughes, Biotechnol. Bioeng., 6 (1964) p. 247.

146. C. J. James, W. T. Coakley and D. E. Hughes, Biotechnol. Bioeng., 14 (1972) p. 33.

147. W. T. Coakley, R. C. Brown, C. J. James and R. K. Gould, Biotechnol. Bioeng., 16 (1974) p. 659.

148. G. D. Greville, E. A. Munn and D. S. Smith, Proc. Roy. Soc., B, 161 (1965) p. 403.

149. J. M. Fessenden and E. Racker, J. Biol. Chem., 241 (1966) p. 2483.

150. N. Tsukagoshi and C. F. Fox, Biochemistry, 10 (1971) p. 3309.

151. B. Chance, M. Erecinska and C. P. Lee, Proc. Natl. Acad. Sci., 66 (1970) p. 928.

152. H. D. Hauser, Biochem. Biophys. Res. Comm., 45 (1971) p. 1049.

CHAPTER III

DOPPLER PRINCIPLES AND TECHNIQUES

Donald W. Baker, Fred K. Forster and Ronald E. Daigle

1. INTRODUCTION

The detection and measurement of blood flow is a goal which researchers have been pursuing for many years. This topic of investigation, while appearing deceivingly simple, is quite complex and touches on many fields of endeavor, ranging from clinical medicine to basic engineering sciences. Interestingly, there may be as many possible solutions to the basic problems as there are clinical applications or needs. A challenging engineering and clinical problem is to select from all the possible combinations and permutations the best approach to the eventual transcutaneous, quantitative measurement of blood velocity and volume flow rate. The "grand design" for some ultimate, practical, clinically useful instrument is just beginning to emerge. Research under way at the University of Washington, in Seattle, would suggest that such an ultimate system would consist of an integrated concept comprised of real time imaging, pulsed Doppler flow sensing, a transducer locating system for determining the orientation of the sound beam, and appropriate computer processors and displays to reduce the primary measurements into flow velocity or volume rate information. The development of highly dedicated systems for specific measurement applications does not seem to be a practical development philosophy at the present time, since no large volume of procedures requiring specific protocols have been identified to date. As a consequence, blood flow quantification will probably emerge slowly on a broad front as the comprehensive understanding of the instrumentation concepts become a central part of this ultimate measurement scheme.

This chapter deals with what we consider to be the more significant aspects of this increasingly important modality. Following a brief history of some important milestones achieved, the material goes on to deal with fundamental physical principles of the Doppler effect and the interaction of ultrasound with blood and tissue. This is followed by a discussion of the important instrument concepts involved in the detection of the Doppler signal. Various signal analysis methods are dealt with and then there is a fairly detailed discussion of the spatial and velocity resolution characteristics. Several approaches to quantitation are covered in a later section and the chapter closes with selected examples of diagnostic clinical applications using the Doppler effect instrumentation.

1.1 Chronology of Important Developments in Blood Flow Detection

A number of significant milestones have been reached over the last twenty years toward the ultimate development of Doppler techniques. The first demonstrated use of continuous wave Doppler ultrasound for the transcutaneous detection of blood flow was by Satomura.[1] His early vacuum tube devices appeared to have sufficient sensitivity to detect signals from superficial vessels. Whether these signals were vessel wall motion or actual blood flow is not clearly known. Satomura and Keneko[2] felt that the signals that they were detecting were due to flow disturbances within the vessels themselves and therefore apparently did not fully appreciate the concept of scattering of the sound by the red blood cells themselves. These early devices recorded the intensity of the Doppler signal rather than a voltage proportional to frequency as an indication of velocity.[3,4]

Franklin, et al.[5] independently developed a continuous wave Doppler device in which the transducer was implanted within an animal using a cuff around the vessel. These transducers used the then newly available barium titanate piezoelectric elements which were more sensitive than quartz. This new sensitivity permitted a significant step in the detection of the Doppler shift signal from moving blood. This early system, which operated at a frequency of 5 MHz, detected the Doppler shift frequency using a homodyne principle in which the transmitted and received signals are mixed in the receiving crystal and receiving amplifier. The early devices were non-direction sensing and used two types of signal processors. The first method used a simple zero crossing type frequency meter which was unable to detect the direction of flow velocity. The second method developed involved taking the derivative of the Doppler signal using a simple analog amplifier. The amplitude of the output of this device is proportional to the Doppler shift frequency. This approach was simple, but had the disadvantage that the output analog waveform was also dependent on the amplitude of the backscattered signal. As a result, it was difficult to separate the Doppler signal amplitude from Doppler shift frequency.

McLeod[6] started development of a flow direction sensing readout which led to the quadrature phase detection scheme based on single sideband detection method used in amateur radio communications.[7] This detection process determined the Doppler shift frequency by using a pair of phase detectors. The shift was determined by comparing the back-scattered signal frequency with a reference frequency supplied in quadrature format so that the resultant Doppler shift signal also had quadrature components. He designed a method for comparing the relative phase shift between the two signals using a zero crossing concept. This idea was based on the fact that flow towards the transducer caused one quadrature channel to lead the other by a 90 degree phase shift. When flow occurred in the opposite direction, these phase relationships are reversed. He accomplished this detection using a pair of cross-coupled, one-shot multivibrators. These formed a zero crossing type frequency meter which determined the direction of flow, based on the relative phase shift between the adjacent quadrature channels. His device was initially used with implanted flowmeter cuffs in experiments involving sheep.

Baker[8] demonstrated the first miniaturized, solid state, continuous wave Doppler for transcutaneous use on man. This device became practical through the development of the silicon planar, low noise transistor. Previous Doppler devices had depended on the use of low noise vacuum tubes. All of these active elements required high plate voltage and filament currents to operate them. Since they were not easily decoupled from the 60 Hz electrical power supply source, these electrical leakage currents often produced audible outputs larger than the desired Doppler signals. The development of low noise, silicon planar transistors permitted the design of electrically isolated, battery operated Doppler devices which solved this problem.

Franklin, et al[9] completed development of a miniaturized modular continuous wave Doppler flowmeter and telemetry system under a NASA contract, which formed the basis for several physiologically oriented continuous wave flowmeter designs. The first commercial Doppler devices for clinical use were derived from the University of Washington development and were marketed as the SKI "Doptone." Parks*, in 1964, introduced a 10 MHz continuous wave Doppler for use with peripheral vascular applications based on the original Franklin design. All of these early devices were intended either for use with a cuff placed directly around the vessel or for transcutaneous detection of flow velocity in superficial vessels. One exception was the use of the "Doptone" for detection of fetal life and placental blood flow in pregnant mothers. These devices used 2.25 MHz transducers instead of 5 or 10 MHz and were able to detect signals from depths as great as 15 cm, though often of relatively poor signal-to-noise ratio.

There was early interest in using Doppler techniques for detecting flow signals from the heart and great vessels. This was a difficult objective, since the continuous wave devices were sensitive to any and all motion in the path of the sound beam. This characteristic resulted in Doppler signals from heart valves, myocardial wall motion, as well as from blood flow, all of which were difficult to separate audibly and even more difficult to separate electronically. The first attempts to overcome this problem for the detection of aortic flow came in 1965-66 by a number of groups functioning independently. These included D. Baker, W. Martin and E. McCutcheon at the University of Washington and F. Duck at the University of Newfoundland.** Both of these approaches involved placing piezoelectric transducers on the tip of a cannula,

---

* Parks Electronics, Beaverton, Oregon.

**To the best of our knowledge, none of these early efforts was ever published.

which was introduced via the esophagus in animals and in man, so as to position the transducer near the descending aorta. While it was possible to detect flow signals in these locations, there were numerous problems. Motion of the heart wall still produced large amplitude signals which were removed by a high pass filter amplifier, designed to reject these low frequency components of the Doppler signal. Unfortunately, this filtering also removed flow signals as well. Another difficulty was that the transducer tended to move about with respect to the vessel due to peristaltic motion of the esophagus. And finally, it was still not possible to quantify the flow velocity or volume flow rate since the angle between the sound beam and the flow axis are unknown and the dimensions of the aorta were unknown.

The next steps taken toward determining great vessel blood flow were via the development of catheter tip continuous wave Doppler flow detectors. These early devices developed by Stone, et al[10] in Texas and Reid, et al.[11] in Seattle involved placing a ceramic transducer at the top of a small catheter, which was introduced directly into the blood stream. Stone constructed the catheter by placing a single transducer disc at the tip of the catheter, which was then split to form separate D-shaped transmitting and receiving elements. In this way, he was able to use the pre-existing electronic circuitry designed for continuous wave Doppler devices. Reid's design of the transducer placed a single element at the tip of the catheter and then used a specially balanced decoupling transformer, or balun, which permitted transmitting and receiving the continuous wave signal on the same transducer. This decoupler used a bridge-type circuit which permitted tuning out the continuous wave excitation from the receiver input. Both methods were able to detect flow velocity signals. The major difficulty with these approaches was that it was difficult to know the location of the transducer within the flow lumen. Usually, a fluoroscope table was used to help localize the catheter. Even so, the output was still only an indication of velocity directly in front of the transducer and it was not feasible to accomplish a calibration.

Pulsed Doppler developments designed to overcome limitations in the continuous wave approach began independently in two laboratories on opposite sides of the world.[12,13] These systems were similar in that they were both phase coherent pulsed Doppler devices operating in the range between 4 and 8 MHz; Peronneau using 4 and 8 MHz and Baker using 5 MHz. The Doppler device by Peronneau, et al.[14] used receiver amplifier gating preceding the quadrature phase detectors for detection of the Doppler signal. The Baker[15] system used quadrature phase detectors followed by sample and hold amplifiers for detection. The principal difference between the two approaches is that the latter method was directed toward the transcutaneous measurement of blood flow requiring much higher signal-to-noise ratios. The Peronneau system was primarily used for physiologic blood flow measurements on experimental animals. Both these devices were developed to overcome the range limitation of the continuous wave Dopplers. In principle, the pulsed Dopplers could detect flow signals within the heart or great vessels and exclude the masking effects of other nearby vessels or moving structures. This capability has generally been demonstrated by both systems either in peripheral vascular or cardiac blood flow applications. In terms of an ultimate flow quantitation objective, these devices still only contributed velocity information. Still unknown was the flow angle and the dimensions of the vessel.

Hokanson[16] demonstrated the use of pulsed Doppler techniques for making flow images of the vessel lumen. This approach, using the previously developed pulsed Doppler systems, mapped onto a cathode-ray tube storage oscilloscope an image of the vessel where flow exceeded a certain threshold velocity. The major contribution of this work was the possibility of using the display to determine the angle between the sound beam and the flow vector. The major limitation of this approach was that it required many cardiac cycles to produce the flow lumen image. Also, the angle measurement assumed that the flow vector was parallel to the vessel wall.

Barber,[17] at the University of Washington, demonstrated the first "Duplex" scanner. This device combined real time, two-dimensional imaging with simultaneous pulsed Doppler flow detection. This step brought the solution of the problem of orientation between the Doppler beam and the flow axis one step closer. In this device, the pulsed Doppler sound beam was constrained to scan in the same plane as the real time image. This relationship between the two sound beams has been effective for most peripheral vessel applications, but may not be effective for cardiac flow measurements. Basic to all combined Echo-Doppler devices is that imaging and flow detection are not optimal using the same beam orientations. This is a consequence of the fact that echo imaging is basically a specular process requiring nearly a 90 degree angle between the interface and the sound beam axis, whereas, flow detection prefers the sound beam to be on the flow axis. This situation suggests that the Doppler and Echo systems should always be functioning orthogonally to each other in space. Therefore, for example, cardiac blood flow detection may require that the echo scanner and the Doppler system be detached and used as orthogonally as possible. This procedure will require knowledge of the location of the scan image with respect to the detached Doppler transducer beam.

Moritz[18] developed a sonic locator system which permits unconstrained movement of a transducer while periodically computing its location and orientation with respect to an arbitrary coordinate system. This locating system is the key link required to bring together the coordinates of the real time image with the Doppler beam for eventual calculation of the orientation angle between the flow vector and the sound beam axis.

It would appear that the achievement of the long range goal of flow quantitation will necessitate the continuation of this logical sequence to integrate these systems into an effective unit. Further development of the basic understanding of each of these modalities must continue in parallel so that a quantitative result can ultimately be achieved. This chapter attempts to bring into focus some of the essential considerations for the detection and quantification of flow following a line of logical development.

1.2 Basic Physical Principles

Doppler flow detectors depend on the interaction of impinging sonic energy with moving blood. The early work of Reid, et al[19] shows that the primary source of ultrasonic scattering is from the red blood cell. Measurements of scattering from platelets indicated that the levels were considerably lower than that from red blood cells. This is explained by the fact that the scattering cross section for platelets is a factor of 1×10^{-3} lower than for red cells, except in cases of concentrated solutions. In normal blood, platelet scattering would be considered to be negligible.

At low concentrations of red blood cells, less than 10%, the scattering is a linear function of hematocrit. At higher concentrations, Shung, et al.[20] described a more complex relationship using a frequency of 5 MHz for his studies. For hematocrits below 8%, the red cells behave as incoherent randomly-distributed scattering centers. About these levels the blood appears to gradually take on a crystaline characteristic reaching a nearly perfect crystaline character with hematocrits above 58%. In his experiments, the maximum scattering coefficient occurred for a hematocrit of 26%. Figure 1 summarizes Shung's work. The strong evidence for particle interaction in blood for hematocrits above 8% may require reappraisal of previous concepts on the subject and will call for new work to evolve an adequate analytical model.

The work of Roevros[21] proposed that the total backscattered power can be determined by simply summing the contribution of individual scatterers.

He showed that the total power of the received signal from the moving blood, P, is defined by the following equation.

$$P = pCL\pi r^2 \quad (1)$$

where C = particle concentration

p = sound power scattered per red cell

L = length or sound beam volume intercepting blood vessel

r = radius of sound beam

This expression shows that the total power of the synchronously detected signal is only dependent on the magnitude of the measuring volume, particle

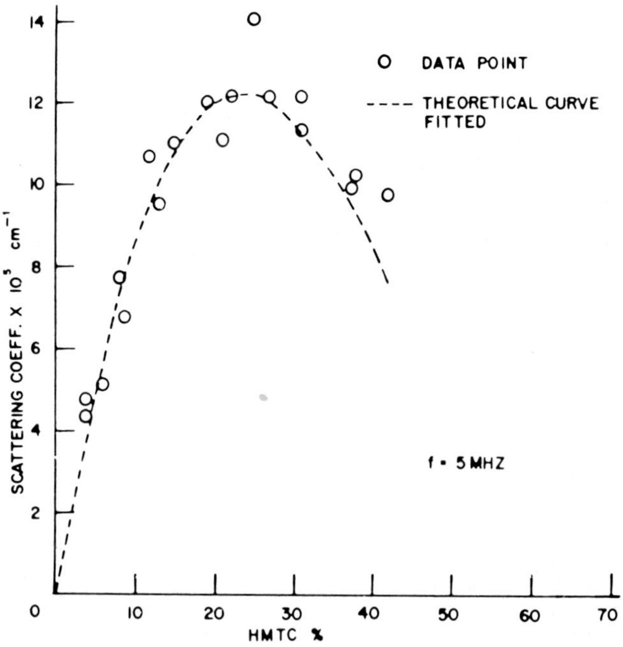

Fig. 1 The results of recent experimental work show a nonlinear relationship between backscattered intensity and hematocrit. This data, from experiments using 5 MHz ultrasound, is based on the work of Shung, et al.[20]

concentration, and the power scattered per red cell. From Shung's work, this concept may only be useful for low hematocrits, i.e., less than 8 - 10%.

Measurements at different ultrasound frequencies show that the higher the frequency, the stronger the scattering, Shung, et al.,[20] Figure 2. The graph shows the theoretical fourth power dependence upon frequency which would be expected from scatterers which are small with respect to wavelength. Tests made with red cell concentrations of 30% or less by Shung, et al[20] showed some angle dependence in the relative scattered intensity. His data shown in Figure 3 shows a variation of $\pm$ 2 to 3 db for angles of 50° to 170° between the transmitting and receiving transducer beams. When all these physical effects are taken into consideration, the interaction, for hematocrits less than 10%, can best be described in terms of Rayleigh scattering theory. Each cell acts as

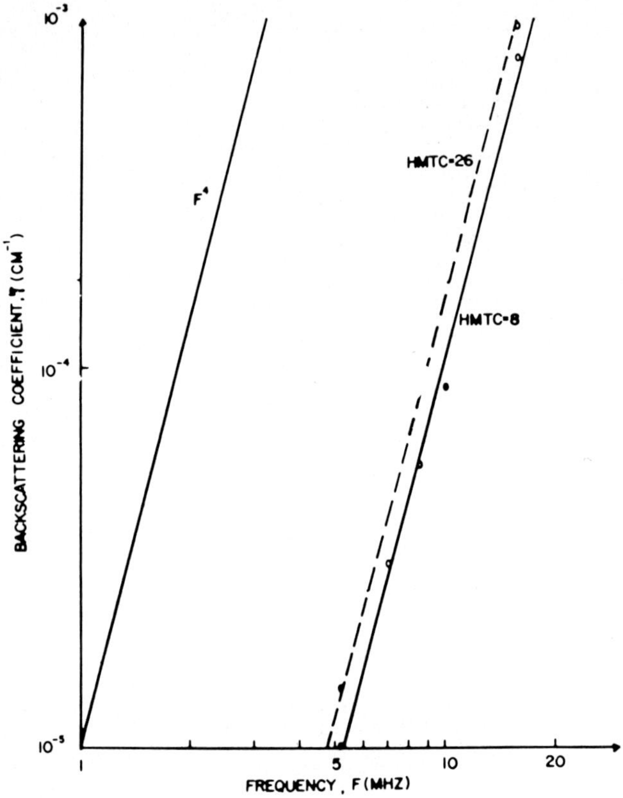

Fig. 2 Experiments with hematocrits of various concentrations show that the backscattered signal intensity closely follows the fourth power of the frequency.

an isotropic scatterer, scattering a small fraction of the incident power equally in all directions. Assuming single scattering, the total scattered power from a group of cells is thus simply the sum of power scattered by each cell. New analytical models may be required for hematocrits above 10%.

1.3 Doppler Effect

From knowledge about the scattering of ultrasound, we can build up a fairly complete picture of the Doppler effect produced by a single moving scatterer. The Doppler frequency shift associated with scattering from an interface, or a single moving cell, can be determined under certain conditions.

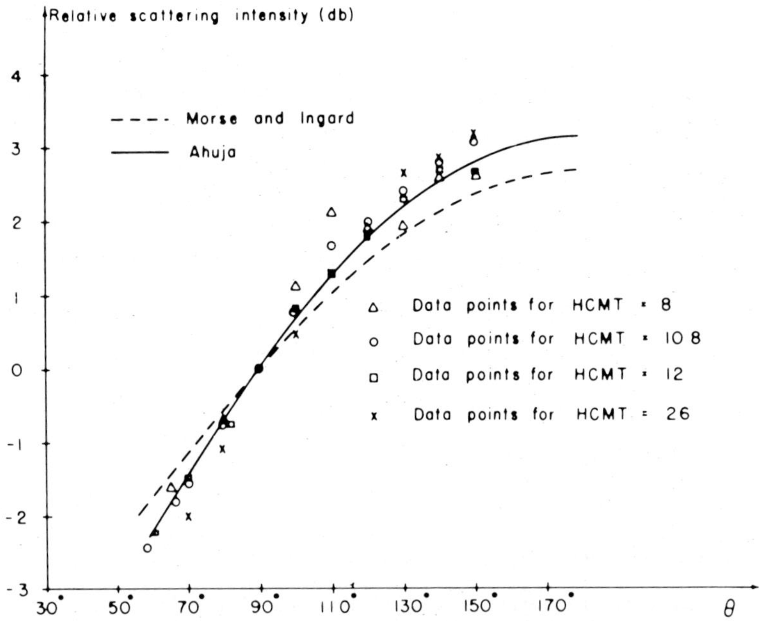

Fig. 3 Comparison of measured values of scattering intensity with analytical predictions of Morse and Ingard (1968) and Ahuja (1972) as the angle between the transmitting and receiving transducer is varied. Taken from Shung.[20]

For example, consider a single scatterer within a blood vessel moving at a velocity, v, through an infinite sound field. The sound field is generated by an ultrasonic transducer which contains elements for receiving, as well as transmitting. Assume that both elements intercept the vessel with the same angle. The total transmission path, x, from transmitter to scatterer (a red cell) and back to the receiver will be changing as a linear function of time. This produces a phase shift between the excitation and received signal which increases linearly. Since frequency is defined as the rate of change of phase, the frequency of the backscattered energy is changed by a constant proportional to the velocity of the scatterer. This frequency change, the difference between the transmitted and the received frequencies, is termed the Doppler shift. The magnitude of the Doppler shift can be easily calculated in this situation. If the transmitted wave is $\cos 2\pi ft$, then the time dependence of the received wave is

$$\cos 2\pi f_o (t + \frac{2x}{c}) \tag{2}$$

where c is the velocity of propagation. If x increases at a rate corresponding to the velocity, v, Equation 2 becomes

$$\cos 2\pi f_o t(1 \pm \frac{2v}{c}) \tag{3}$$

The Doppler shift frequency is given by the second term within the brackets in Equation 3, $\Delta f = 2f_o v/c$. When the transducer beam intercepts a vessel at an angle, the Doppler shift detected corresponds to the component of velocity acting along the sound beam axis, Figure 4a. The Doppler shift can then be defined in the following form,

$$\Delta f = \frac{2vf_o \cos\theta}{c} . \tag{4}$$

Unfortunately, this simple description only describes the basic concept. In a real flow sensing situation, there may be many velocities within the vessel or tube according to the velocity profile. Also, the actual ultrasound transducer has a finite beam width which intercepts the vessel at varying angles. The example shown in Figure 4b demonstrates the orientation of a typical transducer. The final distribution of frequencies in the Doppler spectrum shown in Figure 4c is due to an effective change in the Doppler angle as a particle moves through the sound field. In addition, the spectrum is affected by the total time the particle spends in the sound field (transit time broadening). This will be discussed in more detail later. The area under the frequency spectrum is proportional to the power of the received signal.

1.4 Phaser Description of Doppler Signal

The easiest way of characterizing the backscattered Doppler signal is via a vector or phaser type of description. This electrical phaser approach will be helpful in understanding the relationships between the various components of the returned signal and its amplitude and modulation format. It will also be helpful in understanding the types of signal detection which can be used to derive the Doppler information.

The continuous wave Doppler device is a simple case which is easily

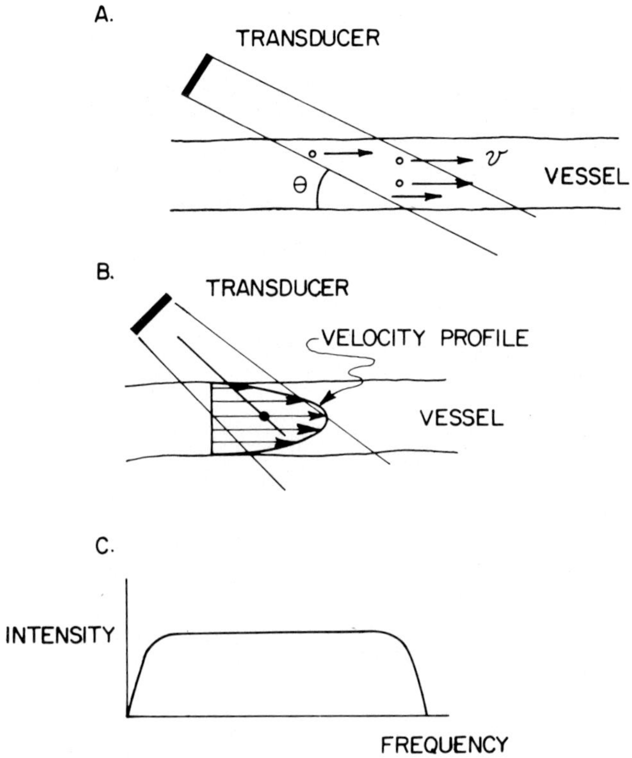

Fig. 4 The detailed character of the Doppler shift spectrum shown in (C) depends on the velocity of the scatterers in (A), the angle between the sound field axis and the flow vector, and the range of velocities present within the vessel, (B). The resultant spectrum will consist of the summation of all of these effects.

analyzed in phaser notation. This simple device, comprising separate transmitting and receiving transducers, projects ultrasound into the direction of the vessel from the transmitting element. There are two primary components of this signal which are detected by the receiving element: the leakage phase, ϕ_L, and the Doppler shift component, ϕ_D, backscattered from the moving red blood cells as shown in Figure 5. Both of these signals are picked up by the receiving transducer and summed together to form the receiver input, $\phi_L + \phi_D$, Figure 6. When these two components combine at the receiving transducer, amplitude modulation and phase modulation occur. The rate of change of the

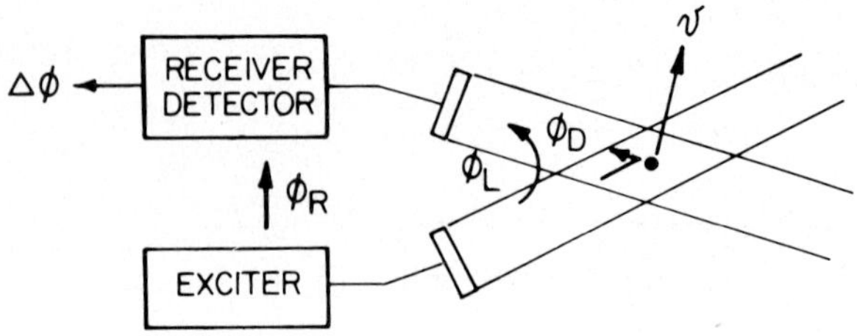

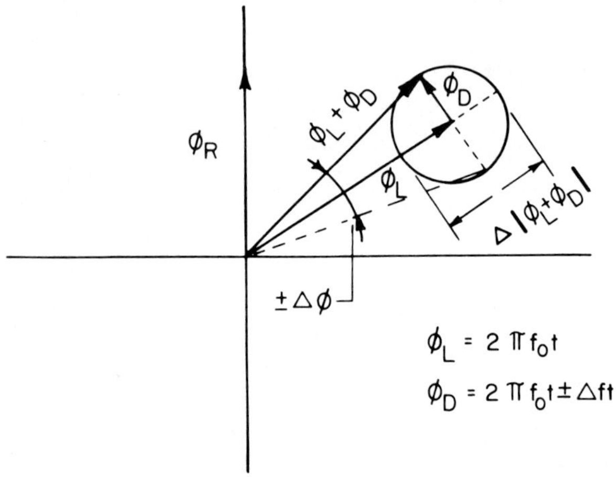

Fig. 5 The signal picked up by the receiver detector amplifier in a continuous wave Doppler consists of a Doppler shifted component, ϕ_D, and a leakage component, ϕ_L. The Doppler shift is derived by comparing the sum of $\phi_D + \phi_L$ with ϕ_R, the reference signal.

Fig. 6 The phase relationships between the Doppler component, ϕ_D, the leakage component, ϕ_L, and the reference signal, ϕ_R. The sum of $\phi_D + \phi_L$ produces both phase and amplitude modulation.

phase of the sum, with respect to a reference phase, ϕ_R, corresponds to the Doppler shift rate. The magnitude of the sum also varies at the Doppler shift frequency, giving a form of amplitude modulation. The ratio of the two signals, ϕ_D/ϕ_L, determines the overall sensitivity of the Doppler device. The larger the backscattered component, ϕ_D, is with respect to the leakage component, ϕ_L, the more sensitive the device will be, since a larger change in phase, $\pm\Delta\phi$, will occur. The ideal device would minimize the leakage phase, ϕ_L, to zero. In a typical Doppler device, the signal ϕ_D will be some 40 to 50 db below ϕ_L, giving an amplitude modulation on the order of 1%. The corresponding magnitude of the phase modulation for this signal relationship gives a $\pm\Delta\phi = \pm 0.57^{o}$ at the carrier frequency, f_o.

This same type of phaser notation can also be used to describe the return signal in a pulsed Doppler device. In contrast to the continuous wave Doppler, a pulsed Doppler has a region of flow sensitivity which is determined by the duration and delay, from the transmitted burst, of a receiver range gate. The continuous wave Doppler, having no range gate, is sensitive in the region of beam intersection, Figure 5. In contrast to this, the pulsed Doppler is sensitive only to velocities occurring within the sample volume. In the example in Figure 7a, the signal consists of two components: the first, ϕ_D, the Doppler signal, returning from the common volume between the sample volume and the vessel, and secondly, a larger component, ϕ_L, is backscattered from the region within the sample volume but outside the vessel lumen. In this case, the ratio ϕ_D/ϕ_L will be small. In Figure 7b, a much larger vessel is being scanned; the sample volume is small with respect to the vessel lumen, and the ratio ϕ_D/ϕ_L will be much larger. These two cases are summarized in Figure 7c. In case B, a large vessel with small sample volume, the ratio is low when the sample volume is outside the vessel and then rises to a fixed value. The output stays at that level while the sample volume is entirely within the vessel and then falls on the far side. In case A, with the small vessel and large sample volume, the ratio is low, reaches a lower value and returns to a low level beyond the vessel. The ratio will always be smaller in the case where the sample volume is equal to, or smaller than, the vessel diameter. The conclusion is consistent with Roevros's[21] expression that the total backscattered power is only dependent on the magnitude of the measuring volume; the particle concentration assuming single scattering and the power scattered per particle. The effects described here are primarily dependent on the fact that the measuring volume is changing as the sample volume moves through the vessel, except in the case where the sample volume is smaller than the vessel, in which case the relative ratio between ϕ_D/ϕ_L becomes con-

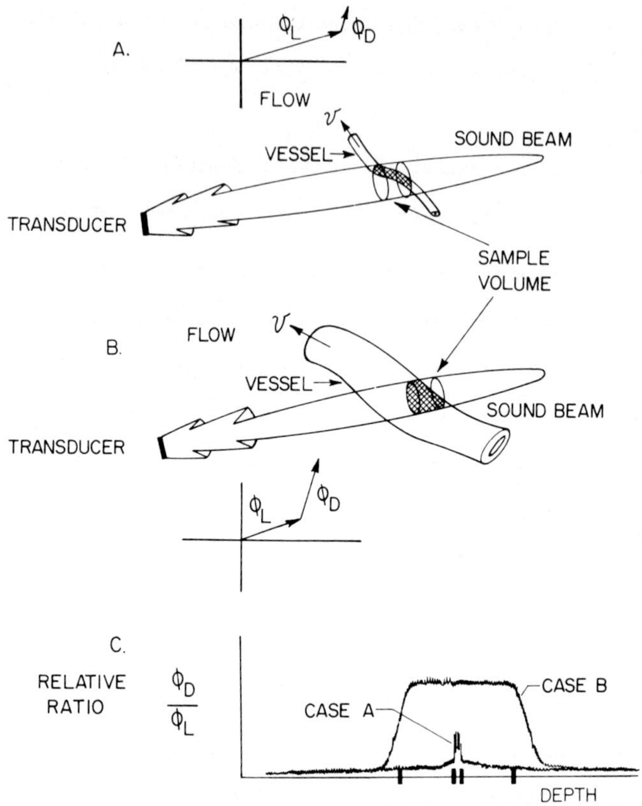

Fig. 7 In a pulsed Doppler device, the region of sensitivity can be represented by a small cylinder along the axis of the sound beam, (A). The signal detected from this region will consist of two components, the leakage phase, ϕ_L, and the Doppler shift phase, ϕ_D. The magnitude of the Doppler component, ϕ_D, will depend on the dimensions of the common volume between the vessel and the sample volume as well as the scattering characteristics of blood. This ratio will change significantly when the sample volume is small with respect to the vessel dimensions, (B). The maximum sensitivity is achieved when the sample volume is entirely within the lumen of the vessel. The relative ratio of the Doppler shift component to the leakage component determines the overall sensitivity of the pulsed Doppler device, (C). In case (B), the amplitude of the backscattered signal will reach a constant level if the sample volume is smaller than the vessel lumen. In case (A), the intensity of the backscattered signal will be smaller and never reach the same magnitude when the vessel is smaller than the sample volume.

stant for as long as the sample volume is within the vessel. Both of these examples assume that there is no filtering of the detected Doppler signal. Had the velocities varied over a wide range and had there been low frequency filtering causing discrimination against the low velocities of the detected Doppler signal, then the amplitude may have varied with velocity as the sample volume moves across the vessel.

The sensitivity of a Doppler device will depend on the sound beam intensity and the dimensions of the common volume between the sound beam and the vessel. The Doppler sensitivity in the CW case, as demonstrated by the ratio ϕ_D/ϕ_L, will depend primarily on the dimensions of this common volume. This is not the situation in the case of the pulsed Doppler example shown in Figure 8. In this example, the leakage phase, ϕ_L, will vary in amplitude following the echo amplitude along the beam as received from the walls of the vessel. The

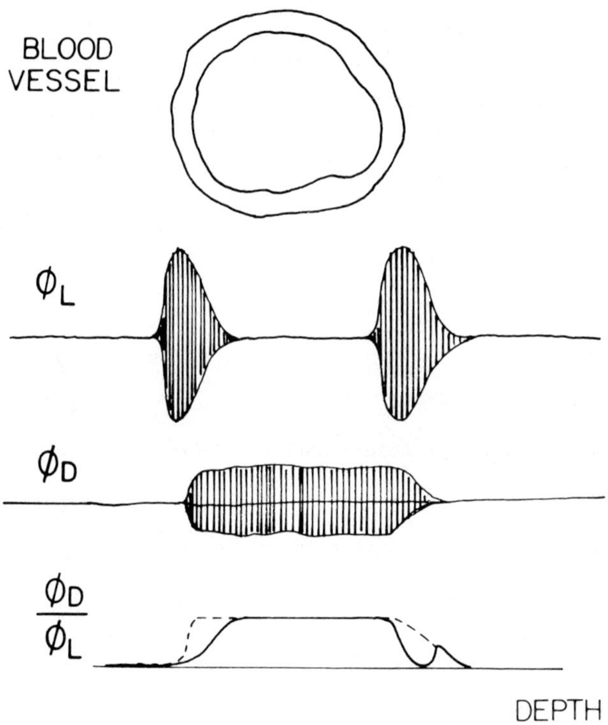

Fig. 8 In a pulsed system, the magnitude of the leakage phase, ϕ_L, will depend on the magnitude of the reflected signals from major interfaces. The magnitude of the Doppler shift component, ϕ_D, will depend on the scattering cross section and the sound field intensity. The ratio of the two determines the overall sensitivity as a function of depth. Wall echoes have the effect masking the Doppler signals near the vessel wall.

backscattered Doppler component, ϕ_D, only exists in the region along the sound beam which is bounded by the lumen of the vessel with some extension beyond the lumen due to the finite size of the sample volume. The sensitivity of the Doppler device, in this case, depends once again on the ratio ϕ_D/ϕ_L. In the example, it is easy to see that this ratio will vary as a function of the distance along the beam across the axis of the vessel. The dotted line shows the pulsed Doppler sensitivity based upon ϕ_D alone. The actual sensitivity is skewed due to near wall echoes, is constant as the measurement proceeds into the vessel, and then decreases at the far wall due to the large wall echoes.

1.5 Signal-to-Noise Optimization

Ultrasound traveling through the intervening tissue to the region of the blood vessel is absorbed at a rate proportional to the frequency[22] while the backscattered signal intensity varies with the fourth power of the frequency. Based on this, it would appear that there is some optimal relationship between the ultrasound frequency and the depth of penetration for an optimal signal-to-noise ratio for a Doppler device. McLeod[23] derived a set of general equations which show the factors and principal relationships influencing the signal-to-noise ratio. While actual values calculated will depend on the particular situation, and on the constants involved, the equations serve to provide some insight into the ultrasound interaction process. In the following equation, based on McLeod's formulation, are shown the relationships between scattering cross section, depth of vessel, absorption coefficient, emission power, sound beam size, and system bandwidth, as well as system noise characteristics,

$$\frac{S}{N} \propto \frac{V_v \sigma \varepsilon^{-2\alpha R}}{V_B R^2 \eta BW} P_a \tag{5}$$

where σ = scattering cross section of blood

R = range depth

α = tissue absorption coefficient

P_a = average emission power

η = electronic noise component

V_v = volume of vessel lumen falling within sound beam

V_B = volume of sound beam within boundary set by system noise threshold

BW = system bandwidth.

The ratio $\frac{V_v}{V_B}$ has been added to cover the case of a blood vessel with dimensions smaller than the sound beam sample size. This equation is shown to be equally effective whether it is applied to a continuous wave Doppler or a pulsed Doppler, since it is independent of the duration of the emission, the emission repetition rate, and the receiver sampling time. It states that for a given average emission power, the usable range of penetration for the pulsed Doppler and continuous wave instruments are optimally identical.

Seeking to maximize the signal-to-noise ratio by proper selection of the center frequency calls for a compromise between maximizing the scattering cross section, minimizing attenuation losses, and adjusting system bandwidth.

Recognizing that the wavelength of ultrasound is orders of magnitude greater than the dimensions of the scattering blood cells, the power level of the scattered sound follows the Rayleigh fourth power law assuming single scattering. In biological soft tissue the attenuation coefficient varies linearly with frequency. The rate of attenuation depends on the tissue type and ranges from 0.2 db/MHz/cm to more than 2 db/MHz/cm. The system bandwidth appearing in the noise term is dictated by the maximum expected Doppler shift for a given set of velocities and is proportional to the center frequency, f_0. The signal-to-noise ratio expression, Equation 5, can be rewritten in the following form if only the frequency dependent terms are included.

$$\frac{S}{N} \propto \frac{f_0^4}{f_0 R^2} e^{-2\alpha f_0 R} \qquad (6)$$

where α is the soft tissue attenuation coefficient, R is the depth, and f_0 is the center frequency. Variations in the sound field directivity pattern may require different terms and signal-to-noise effects in the near field of the transducer might also alter this formulation. When this equation is differentiated with respect to frequency and the result set to zero, the optimal frequency, f_0 equals $\frac{3}{2\alpha R}$. If the attenuation is γdb/MHz/cm, it follows that

$$f_0 = \frac{15 \log e}{\gamma R} \text{ MHz.} \tag{7}$$

For $\gamma = 0.5$, f_0 is approximately $\frac{13}{R}$ MHz. Figure 9 shows the range of optimal frequencies concluded by McLeod,[23] Reid and Baker,[24] and Gill.[25] While very little experimental work has been done to confirm these conclusions, impressions, based on pulsed Doppler systems developed at the University of Washington, would indicate that the optimal signal-to-noise ratio constant may actually be as high as $\frac{30 \text{ MHz}}{R}$. Stated in another way, this constant could be regarded as indicating the highest frequency, f_0, that can be utilized for a given depth in centimeters which will provide an optimal signal-to-noise ratio. Further work based on data by Shung, et al.[20] shows that the back-

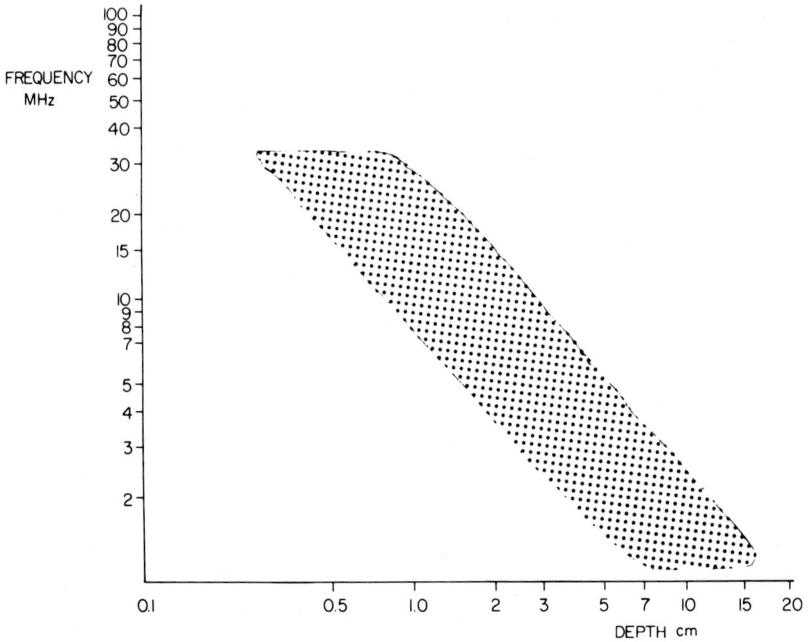

Fig. 9 Calculations from several laboratories to determine the optimal frequency for maximum signal-to-noise ratio versus depth all fall within the dotted area. The precise value is influenced by a number of complex trade-offs concerning the design of specific instruments. New instrument developments will probably lead to improved signal-to-noise ratios at greater depths.

scattered signal intensity may not strictly follow the fourth power of frequency. The exact value of these constants could be influenced by the noise power constant, η, in Equation 5, the emission power, P_a, incomplete or poor information on scattering cross section, σ, and the attenuation characteristics, α. Since many of these factors are quite variable, specific calculations are difficult. As advanced systems are designed to provide more optimal performance, it will, no doubt, be necessary to investigate these fundamental interactions much more completely.

1.6 Emission Power Levels

It has been demonstrated that the performance of either a pulsed or continuous wave Doppler system is proportional to the average power level transmitted into the vessel of interest. Continuous wave Dopplers utilize power levels ranging from 10 mw/cm^2 at 2.25 MHz for obstetrical applications, to as high as 600 mw/cm^2 at 8-10 MHz for shallow penetration peripheral vascular Dopplers. These latter devices have extremely good signal-to-noise ratios in the range of 20 to 40 db. Obstetric devices, which are purposely operated at low power levels for safety reasons, often have marginal signal-to-noise ratios. The audible signal output from these devices is usually interpretable only because of human aural analytical abilities. These low power level devices generally have difficulty producing adequate signals for electronic determination of velocity.

The pulsed Doppler devices that have been built to date have average power outputs in the range of 20 to 300 mw/cm^2. Since pulsed Doppler devices have a requirement for a variable pulsed repetition rate to avoid range velocity ambiguities, their average power levels will vary with the pulsed repetition rate. Most of the pulsed devices described to date utilized fixed duration, fixed excitation bursts for the transmitted signal. If one accepts 50 mw/cm^2 as a safe spatial average power level, then in a pulsed Doppler device for a fixed frequency, it is conceivable that the peak power should vary with the depth of penetration desired for the measurement. Based on this objective, a further modification of Equation 5 for maintenance of high signal-to-noise ratios in a pulsed Doppler system would be in order. The average power in a pulsed system is equal to the following equation,

$$P_a = P_p \frac{\tau_e}{\tau_p} \quad . \tag{8}$$

Substituting $\tau_p \geq 13R$ (see Equation 11 below), then

$$P_a \leq P_p \frac{\tau_e}{13R} \tag{9}$$

where P_p equals the peak power, τ_e equals the emission duration, τ_p equals the period between emission pulses, R equals the range, and 13 is the range constant of 13 μsec/cm depth. The signal-to-noise ratio defined by Equation 5 is proportional to the average emission power. If the formula for average power in a pulsed system is inserted into Equation 5, the signal-to-noise ratio is now proportional to the following formulation,

$$\frac{S}{N} \propto \frac{V_v \sigma \varepsilon^{-2\alpha R}}{V_{sv} R^3 \eta BW} P_p \tau_e \tag{10}$$

In this new form, the signal-to-noise ratio is proportional to the peak power but is further reduced by the depth, so that the formulation now includes an additional R term in the denominator.

Measurements on pulsed Doppler systems developed at the University of Washington indicate a transducer efficiency on the order of 10 - 15%. These elements are excited with peak powers in the range of 30 w, which provides a peak acoustical output of 3.0 w, which at a pulse repetition rate of 15 kHz and a burst duration of 1.0 μsec, produces an average power level of 50 mw/cm^2. As these devices are currently designed, the average power level will fall as the pulse repetition rate is reduced depending on the depth of the vessel of interest. If 50 mw/cm^2 is acceptable as a safe operating level, then it may be desirable in an instrument of this type to increase the peak power in direct proportion to the desired depth of penetration in order to maintain the signal-to-noise ratio constant over wide ranges.

1.7 Characteristics of Doppler Devices

The simplest form of an ultrasonic Doppler device is shown in the block diagram of Figure 10. This simple device consists of a transducer, an exciter, a receiver, a simple detector, and audio amplifier for auscultatory analysis of the Doppler signal. Various signal processing readout methods can be added to convert the raw Doppler signal into an analog signal for flow velocity analysis. These devices, functioning in a continuous wave (CW) mode, can only be used for velocity detection since they have no ranging capability and the only spatial resolution available is determined by the lateral beam width.

The transducers used with CW Doppler devices utilize the same barium titanate or lead zirconate titanate ceramic elements as used in conventional echo instruments. These transducers have been built to function in the range from 1 - 10 MHz using high Q, (A = 10 - 15), airbacked transducer construction. The transducer Q is defined by the logarithmic decrement measured from the rate

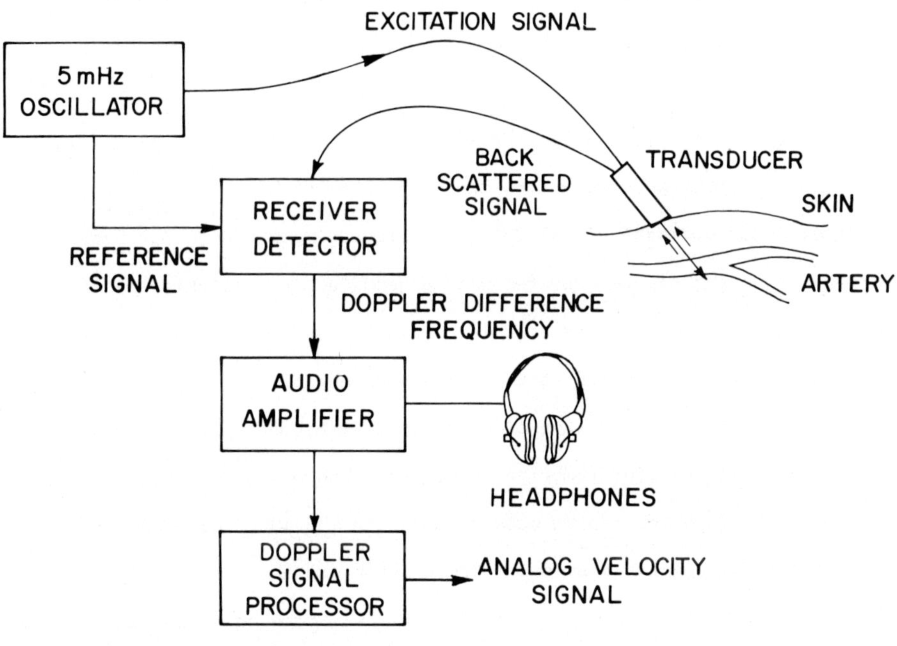

Fig. 10 A simple continuous wave ultrasonic flow detector usually has only an audible output and is manufactured in a number of hand-held versions. The output can be analyzed for recording as an analog waveform or for frequency spectral plots.

of decay of the damped natural oscillation of the crystal element. Relatively high Q's are utilized to improve the sensitivity of the instrument since very modest bandwidths, 10 to 20 kHz, are required. These transducers are highly capacitive reactive and have relatively low electrical impedance, in the range of 30 - 100 ohms, depending upon the dimensions and frequency of the element.

Simple transducer excitation circuitry is required for these instruments. The primary requirement is to have a very low output impedance, 10 ohms or less, in order to drive the highly capacitive load. The voltage level that is used to drive these elements is in the range of 3 - 10 volts, peak to peak, which will produce average power levels in the range of 10 - 600 mw/cm^2, depending on element size, shape and frequency.

The receiving amplifier used to detect the backscattered signal from blood requires a bandwidth on the order of 20 - 50 kHz maximum with voltage gains in the range of 50 db. The signal-to-noise ratio for this amplifier is very critical since the Doppler shifted component of the returned signal is some 40 - 60 db below the leakage component, ϕ_L. In a typical device, the amplitude of the Doppler signal sidebands are in the range of 1 - 10 μvolts at the input to the receiving amplifier. Because of these low signal levels, proper impedance matching between the transducer element and the input transistor is required for optional signal-to-noise ratios. From 0.2 to 0.4 μvolts equivalent noise input can be expected with a good design. If this specification is met, the final detected signal-to-noise ratio will be in the range of 20 - 30 db for vessels in the range of 1 - 5 cm depth.

Continuous wave Dopplers have been designed with two types of signal detection. Square law transfer function detectors using either collector detection or diode rectification can be used to separate the amplitude modulation component of the backscattered signal. These detectors are very simple, inexpensive, and for most current clinical devices, are more than adequate. Phase detection of the backscattered Doppler signal can also be accomplished in a CW device and is sometimes used when it is desirable to have a fixed off-set frequency corresponding to zero velocity. When there is flow or motion towards the transducer, the Doppler signal frequency is plotted above the off-set baseline level and for flow away it is indicated below the baseline level.

2. INSTRUMENT CHARACTERISTICS UNIQUE TO COHERENT PULSED DOPPLERS

By pulsing the acoustic emission and detecting the Doppler shift of the echo at selected depths, it is possible to determine both the velocity and the position of moving interfaces. Within certain limitations, this approach overcomes the lack of range resolution in the continuous wave Doppler devices. Such instruments may permit quantification of flow by being able to measure the average velocity from the velocity profile, determine the vessel diameter, and to also have the possibility of triangulating the orientation of the flow vector with respect to the sound beam axis.

The design of the pulsed form of a Doppler device must meet certain requirements to achieve this goal. A pulsed Doppler is basically a sampling method for detection of Doppler shift frequencies, which must follow certain rules to unambiguously detect both velocity and position.

The maximum depth for which a pulsed device can measure the position of an interface is determined by the pulse repetition rate. The following expression shows the relationship between the pulse repetition rate, PRF, and depth.

$$PRF_{max} = \frac{c}{2R_{max}} \tag{11}$$

This equation simply states that all the echoes from the range of interest (R_{max}) must return before the next transmission.

The sampling theorem says that to adequately describe a signal it must be sampled at a frequency equal to, or higher than, twice the highest frequency present. This requirement places a further restriction on a pulsed Doppler by limiting the maximum detectable Doppler shift. The following expression shows the relationship between the maximum detectable Doppler shift and the range or depth.

$$\Delta f_{max} = \frac{c}{4R_{max}} \tag{12}$$

When this restriction is incorporated into the basic Doppler shift equation, we can determine the maximum velocity which can be detected for a given depth in terms of the basic defining parameters. If we substitute Δf_{max} into Equation 4, we have the following expression.

$$V_{max} = \frac{c^2}{8f_0 R_{max} \cos\theta} \qquad (13)$$

The graph shown in Figure 11 plots out the V_{max} which can be detected using a 3 MHz pulsed Doppler and a range of PRF frequencies over a range of depth from 0 to 20 cm. The example curves apply for the case where cosθ is equal to 0°, 45°, and 60°.

In the totally phase coherent, pulsed Doppler, the pulse repetition frequency is derived from the master oscillator frequency using a digital frequency divider network to produce an integer ratio. The transmitted signal is formed by gating the master oscillator at the PRF so that the output burst is always in phase with the original master oscillator frequency. Later, when the backscattered signal is compared with the original emission frequency in the phase detector, the associated PRF signal component is locked into a null, zero beat, situation such that no mixing occurs with Doppler signal frequencies.

A block diagram for this type of pulsed Doppler is shown in Figure 12. This device consists of a master oscillator at the emission frequency f_0, a PRF divider which counts down the master oscillator frequency to the desired pulse repetition rate, a linear transmission gate located between the master oscillator and the transducer exciter, a range gate and associated delay circuitry, a receiver and quadrature phase detector, a sample gate, or sample and hold circuit, followed by appropriate bandpass amplifiers, and finally, display circuitry. Each of these major subsections will be discussed individually.

2.1 *Master Oscillator*. The acoustic emission frequency is set by this circuit and can be developed using either a crystal controlled oscillator or an LC type circuit. The main requirement of this oscillator is that it have very good phase stability and low center frequency drift. Precise control of

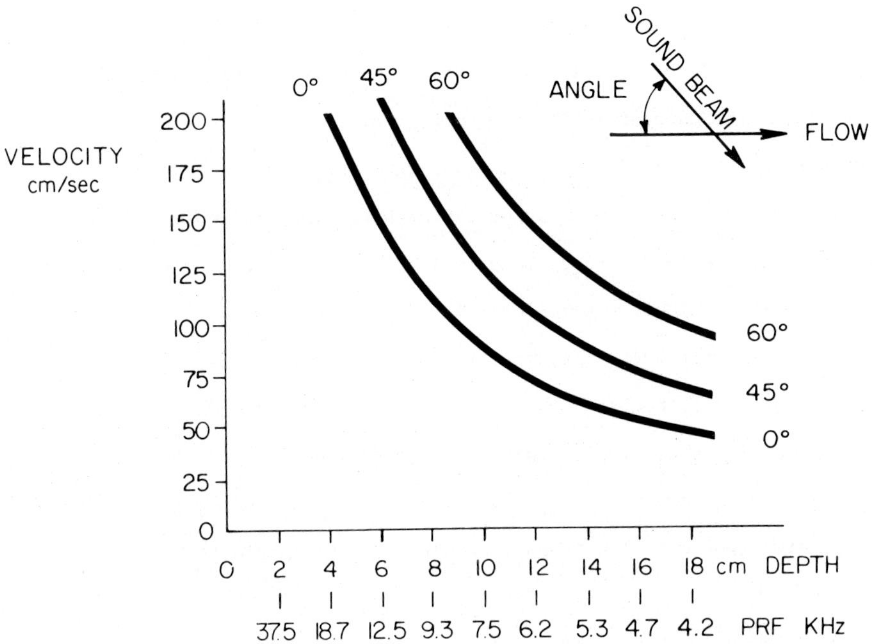

Fig. 11 The maximum velocity that can be detected with a pulsed Doppler depends on the distance between the transducer and the vessel or location of the sample volume. The greater the depth, the lower the pulse repetition rate and the lower the Doppler shift that can be detected. This limitation is maximum when the sound beam and the flow vector are parallel. As the angle increases from 45 to 60 degrees, it is possible to detect greater Doppler shifts at the same depth.

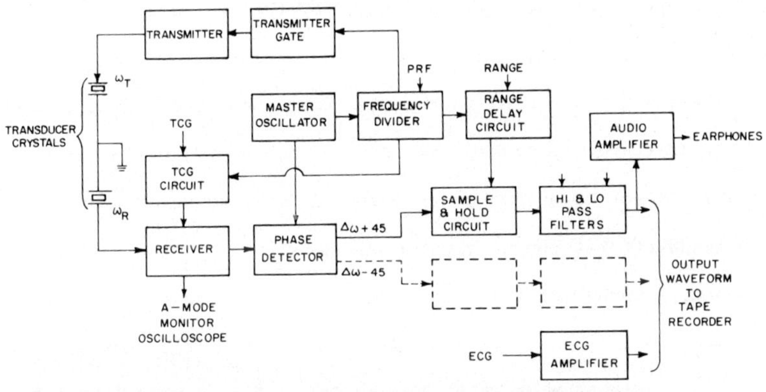

Fig. 12 The block diagram for a pulsed Doppler system shows the major functional elements of this device.

the frequency is not as important as phase stability. Drifts on the order of a few hundred Hz are insignificant since this change is translated on a percentage basis to the Doppler shift frequency.

2.2 *PRF Divider*. Standard TTL digital dividers are adequate for this application. A divide by N counter which can be programmed by logic levels permits the pulse repetition rate to be easily adjusted according to the depth over which the flow velocity measurement is to be made.

2.3 *Transmitter Exciter and Transmitter Gate*. A sample of the master oscillator frequency is derived by a linear transmission gate which in turn drives the transmitter exciter circuit. The transducer exciter amplifies the signal into the range of 40 - 80 volts, peak to peak, and it is coupled through a transformer with low output impedance to the transmitting transducer.

2.4 *Range Gate, Range Gate Delay*. The performance of the range gate, range gate delay is critical in terms of the overall signal-to-noise ratio in pulsed Doppler systems. The delay required will vary from approximately 10 μsec to 150 μsec based on a delay of 13 μsec/cm depth. This delay has to be accomplished with an overall jitter of less than 10 nsec with respect to the transmitted burst time.

2.5 *Transducer*. As shown previously, the resolution-sensitivity characteristics of a pulsed Doppler are largely determined by the characteristics of the transducer itself. Typically, these transducers will have Q's in the range of 4 to 14. The low Q's are used in the high frequency, short range, high resolution pulsed Dopplers and the high Q transducers are used with the long range, deep penetration lower frequency units designed for cardiac or abdominal applications. The trade-off between resolution and sensitivity is an intrinsic one which still requires further optimization. The Q of these transducers is controlled by the type and characteristics of the backing material. Low Q transducers utilize a backing of epoxy and aluminum or lead powder usually not centrifuged. The high Q transducers are fabricated with an air backing, very similar to the type used with the continuous wave Doppler. Most of these transducers use concave lenses molded to the crystal face to produce some focusing effect in the sound field. The position of this focal zone may vary from 2 cm, for peripheral vascular applications, to 5 - 8 cm for cardiac applications. The diameter of the transducer element may vary from 2 to 15 mm.

2.6 *Transmit-Receive Switch and Receiver.* The pulsed Doppler devices usually employ a common transmit-receive transducer element which necessitates a decoupler. Overload of the receiver usually is prevented by diode decoupling circuitry at the input. This decoupler presents a high impedance path to the receiver in transmit mode and a low impedance path to the receiver in receive mode.

The receivers used with the pulsed devices have a wide range of gains and bandwidths depending upon the particular application. High frequency, short range, high resolution pulsed Dopplers use receivers having gains up to 40 db with Q's in the range of 2 to 4. Long range, deep penetration cardiac or abdominal pulsed Dopplers will use receivers having gains as high as 90 db with Q's in the range of 5 to 8. All receivers need to be designed for an optimal noise figure commensurate with the bandwidth of the system. These receivers also require time compensated gain controls to correct for tissue attenuation effects. This is important since the outputs are often used for displaying the echoes produced by interfaces along the path of the beam to help in determining the location of the Doppler sample point. The compensation of the receiver gain with depth also allows adjustment to prevent overload of the receiver due to large amplitude stationary interfaces, such as vessel walls. The time compensated gain function usually permits a dynamic adjustment of the gain over a range of 40 to 60 db within the interval of the PRF period.

2.7 *Signal Detection in Pulsed Dopplers.* Phase detection is the most frequently used detection scheme in pulsed devices. The transfer function for a typical phase detector is shown in Figure 13a. Using the phaser notation shown in 13b, the sum of the leakage phase, ϕ_L, and the Doppler shift phase, ϕ_R, varies according to the Doppler shift frequency. In a typical phase detector, the output voltage will be zero when the phase relationship between

A.

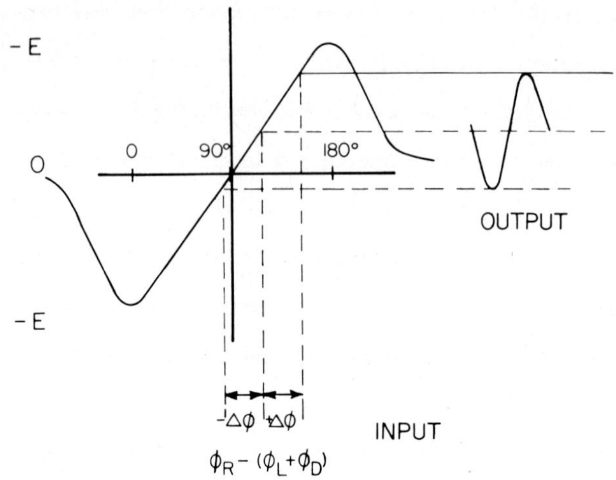

B.

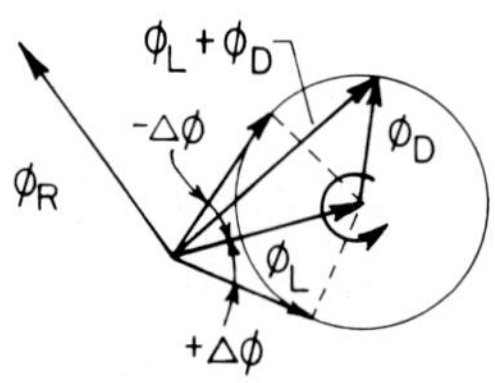

Fig. 13 The typical transfer characteristics for a phase detector are shown in A. The output voltage will be zero when the phase relationship between the reference and the unknown is 90 degrees. Variations about an operating point $\pm\Delta\phi$ produces an output signal. The differences in phase, $\pm\Delta\phi$, are equal to the difference in phase between the reference signal, ϕ_R, and the sum of the leakage and Doppler shift phase.

the reference and the unknown is 90 degrees. In most pulsed Doppler devices, the phase relationship between the reference vector and the resultant of the Doppler shift and leakage vectors cannot be readily controlled, therefore, the operating point on the phase detector transfer curve is randomly determined as shown in Figure 13a. Phase shifts about this random operating point produce a corresponding output voltage. The phase excursions $\pm\Delta\phi$ typically will fall in the range of a few degrees or less with a center frequency of 5 MHz. If the phase detector transfer function is reasonably linear over a range of 140 to 150 degrees, and the operating point happens to fall in this range, then the gain of the phase detector will be fairly uniform. However, if the vessel moves or a nearby interface moves displacing the vessel such that the range phase or number of cycles between the transducer and the leakage interface shift, then the operating point will shift along the transfer function curve producing intervals of high sensitivity and low sensitivity, Figure 14. The regions of low sensitivity occur when the operating point is at the peak or valley of the transfer curve, point b on lines A and B of the figure. In some situations, the phase detector may even become a displacement, instead of frequency detector. This occurs when the Doppler component is very small with respect to the leakage component and when there are displacements of interfaces producing the leakage component. The output of the phase detector will then be a displacement curve which is linear, provided that the displace-

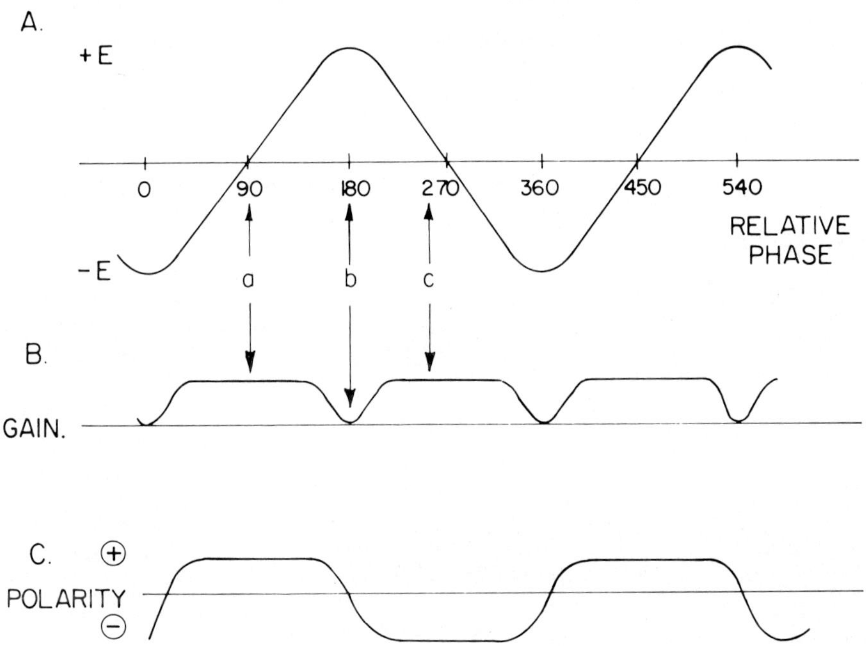

Fig. 14 The transfer characteristics for a phase detector operated over a range greater than 180 degrees are shown in (A). The gain of the transfer function will minima at the points of maxima and minima of the transfer function, (B). The polarity of the output switch as the operating point moves from a region of positive slope to a region of negative slope on the transfer function, (C).

ment falls within the range of 100 degrees phase shift or less. If this displacement produces phase shifts larger than this, then the output of the phase detector will be a periodic signal of little meaning unless the number of times the phase excursion exceeds 180° is recorded, Figure 15.

2.8 *Range Gating*. In a phase coherent pulsed Doppler, the output of the phase detector in any one PRF period will be a function of the instantaneous phase difference between the backscattered return and the reference or master oscillator frequency. This function might be called the range phase. To detect the Doppler shift requires that this function be sampled at a particular delay time repeated regularly over many PRF cycles. There are two slightly different methods currently in use for sampling the Doppler shift frequency. The most commonly used method utilizes a sample-and-hold (S-H) approach. The second involves a sample-and-average (S-A). Each has distinct advantages and limitations, and both require some type of bandpass filtering following the S-H or S-A step. The conceptual process for the sample-hold method can be easily followed by referring to Figure 16.

The output from the phase detector is shown in Figure 16a along with the output of the sample and hold circuit which immediately follows it. In the

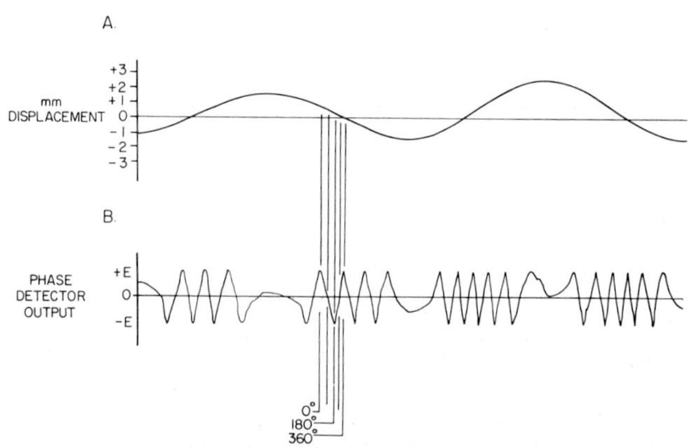

Fig. 15 The phase detector used in Doppler devices can become a displacement detector if the phase shifts exceed 360 degrees. When this occurs, the output of the phase detector will be a periodic signal resembling the transfer function characteristic, (B).

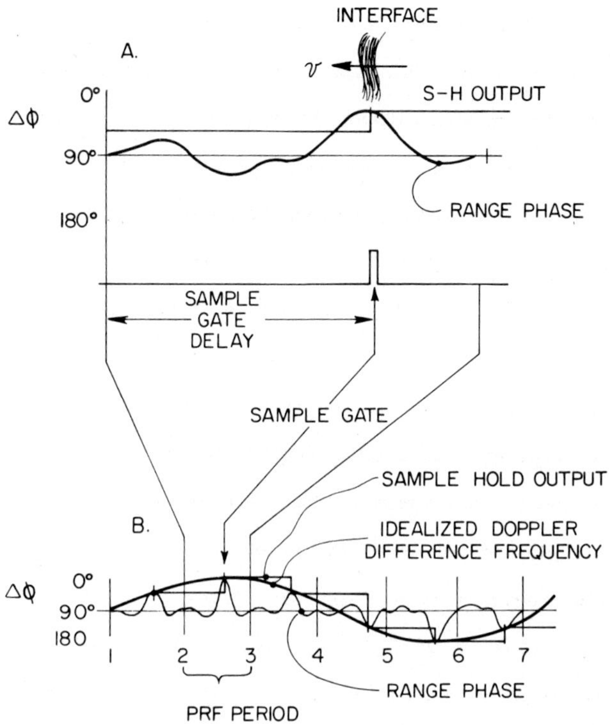

Fig. 16 In a pulsed Doppler device using the sample-and-hold type detector, the range phase, $\Delta\phi$, is sampled by the sampel gate. The value sensed is held in a sample-hold circuit, (A). If the interface is moving towards the transducer, the relative phase between the reference and backscattered signal will shift. The idealized Doppler shift difference frequency is shown in (B) with the successive sample-and-hold outputs derived from the shifting range phase signal.

example an interface is presumed to be moving toward the transducer at a constant velocity and Doppler shift. As the interface moves the relative phase between transmitted and received frequency will vary in a periodic fashion which acts to modulate the range phase function at the point corresponding to the interface depth. The amplitude of the range phase modulation corresponds to the amplitude of the Doppler shift frequency, Figure 16b. The frequency of this modulation over many PRF intervals will be the Doppler shift frequency. If the modulated range phase signal is sampled and held during each PRF cycle, then the resulting envelope of this new function will be the

Doppler shift signal. The principal disadvantage of the sample-hold approach is its relatively poor frequency response. This sampling process is a type of low pass filter with a sin x/x response which means the output will fall gradually as the input frequency approaches one-half the sampling frequency. The sample-and-hold output will also contain varying amounts of phase shift between it and the idealized or original Doppler signal as seen in Figure 16b. This shift will depend on the relationship between the sample frequency and the Doppler frequency.

Further filtering follows the sample-hold circuits to remove the sample (PRF) transients and to remove large amplitude low frequency Doppler signals associated with blood vessel or heart wall motion. This bandpass filter may have an overall gain of 20 to 40 db and will have adjustable high and low frequency cut-off points. The low frequency cut-off usually ranges in steps, i.e., 50, 100, 200, 400, 800 Hz. This is accomplished using 3 to 6 pole high pass active filters. The high frequency points are usually set just below the half PRF frequency using 4 to 8 pole low pass filters. Common cut-offs used in a 3 MHz deep penetration cardiac pulsed Doppler are, for example, 1.2, 2.5, and 5 kHz.

The principal advantage of the sample-and-hold method is its relatively high sensitivity for detecting the Doppler modulation on the range phase signal compared to the sample-and-average approach. Also, the sensitivity is constant independent of the sample gate duration and only depends on the actual closing or turn-off rate at the gate.

The sample-and-average method has two principal advantages over the sample-and-hold approach. The primary of these is the slightly increased frequency response out to the half PRF frequency limit. A second advantage is that the sample gate duration can be made variable to adjust the effective size of the pulsed Doppler sample volume. Very little analysis or application of this approach has been carried out to date. The principal disadvantage of this method is that the sensitivity is less because the output of the sample gate is averaged over the PRF interval as is shown in Figure 17 a and b.

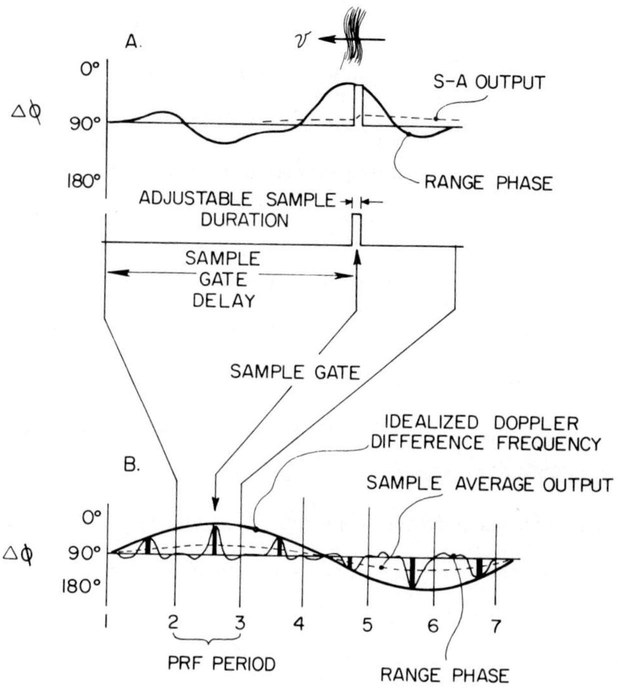

Fig. 17 A pulsed Doppler with a sample-average detector derives the Doppler shift frequency by sampling the range phase at the repetition rate (PRF) interval. These periodic samples are averaged over the PRF period to produce the Doppler difference frequency signal.

Essentially, the same type of bandpass filtering is used following the sample-average method as is used with the sample-hold approach. The low pass filter used to remove the sample PRF harmonics is generally of a higher order with this method. The output from the bandpass filter amplifiers will be the Dopper velocity signal now ready for further analysis.

3. RANGE AND VELOCITY RESOLUTION CHARACTERISTICS

3.1 General Characteristics of Doppler Systems

In virtually all applications of ultrasonic Doppler techniques to cardiovascular measurements, one is interested not only in blood velocity (or volume flow rate), but also the dimensions of the region over which the velocity estimate is made. In this sense, the overall performance of an ultrasonic Doppler velocimeter is strongly related to its simultaneous velocity and spatial resolution. There are numerous system characteristics that influence the resolution of a given Doppler instrument. However, the fundamental parameters that determine the space/velocity resolution, common to all types of Doppler instruments, are the transducer sound beam pattern and the waveform of the transmitted signal.

3.1.1 *ACOUSTIC BEAM EFFECTS.* To investigate the effect of the acoustic beam pattern on space/velocity resolution, consider a CW system transmitting a sinusoidal waveform. Also consider an acoustic beam pattern that is of uniform intensity and of infinite extent. A particle moving with a velocity, v, through such a sound field is shown in Figure 18a. The Doppler signal is obtained by mixing (multiplying) the transmitted waveform with the backscattered Doppler shifted signal. As both of these signals are continuous sinusoidal

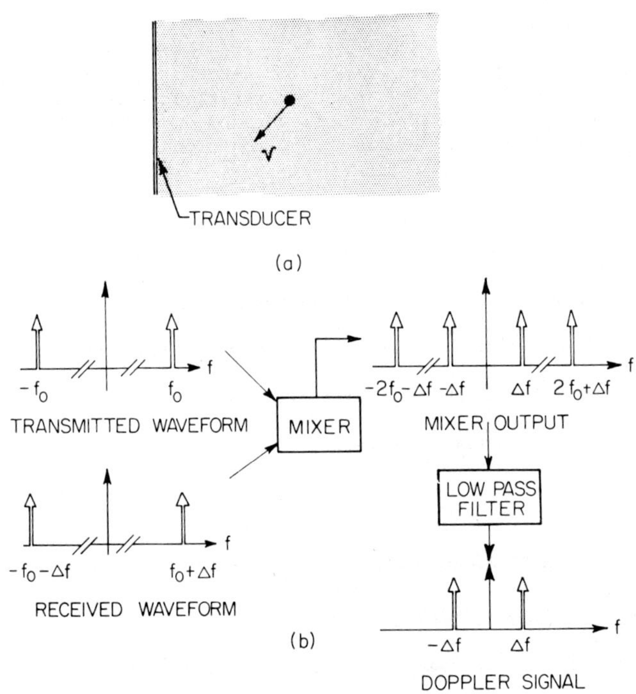

Fig. 18 CW Doppler system with an acoustic sound field of <u>infinite</u> extent showing the acoustic sound field (a) and the frequency domain representation of the signal processing, (b).

waves, they are represented by Fourier transforms on the frequency domain as impulses. Multiplication in the time domain is equivalent to convolution in the frequency domain. The result of this process is shown in Figure 18b. Note that the frequency spectrum of the doppler signal, after filtering off the higher frequency component, consists of a single impulse at Δf, a frequency directly related to the particle velocity, v, by the equation

$$\Delta f = \frac{2f_o}{c} \left| v \cos\theta \right| \qquad (14)$$

where f_o is the transmitted frequency, c is the speed of sound, and θ is the angle between the velocity vector and wavefront direction. Since the Doppler spectrum is an impulse in this case, the measurement of $v \cos\theta$ is unambiguous. This case represents perfect velocity resolution, and is related to the fact that the particles backscatter sound energy for an infinite length of time.

Now we consider a more realistic case in which the particle moves through a finite width acoustic beam as shown in Figure 19a. The particle backscatters acoustic energy only for the length of time it is in the sound beam. Thus, the returning signal is a finite sinusoidal waveform. On the frequency domain such a signal is ideally represented as a $(\sin\chi/\chi)^2$ function. The Doppler signal resulting from the mixing and filtering process described above is shown in Figure 19b. The resulting Doppler spectrum has a frequency spread which is inversely proportional to the time the particle spends in the sound beam. The spread can be considered an ambiguity in the velocity measurement since the ability to resolve two different velocities degrades with the width of the spectrum. These same results have been obtained by Brody[26] for the case of a large number of randomly distributed scatters in the sound beam.

From the above examples we have seen that velocity resolution is related to the time available for the measurement. We thus see the trade-off between spatial and velocity resolution. Decreasing the beam width improves the lateral resolution of the system at the expense of velocity resolution. This trade-off is evident in numerous aspects of Doppler instrumentation. Although not a direct result of the uncertainty relation of quantum mechanics,

$$\Delta f \, \Delta t \simeq 1, \qquad (15)$$

the trade-off is related to the fact that to obtain good spatial resolution the signal must be concentrated in the time domain, while for good velocity resolution the signal must be concentrated on the frequency domain. Fourier transform theory, however, tells us that these conditions cannot be met simultaneously.[27]

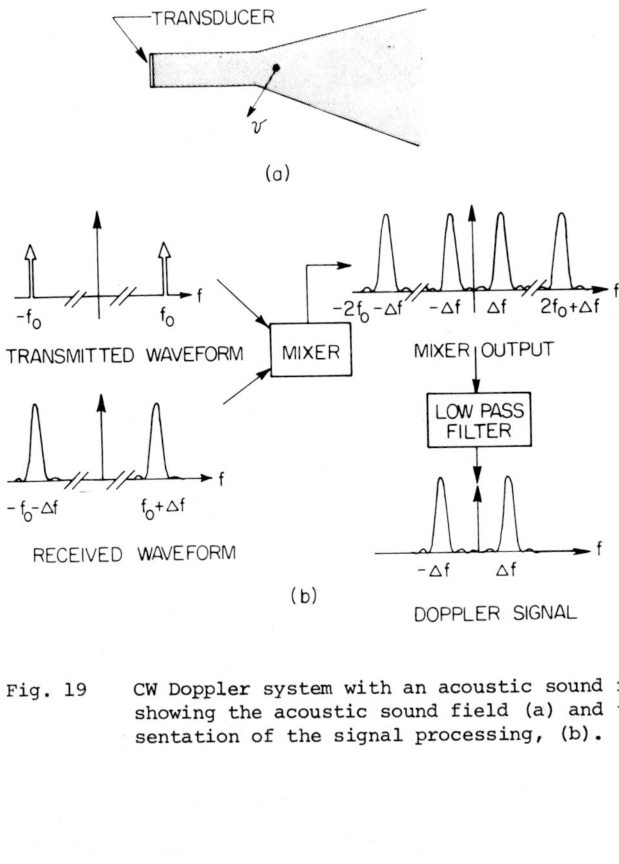

Fig. 19 CW Doppler system with an acoustic sound field of finite extent showing the acoustic sound field (a) and the frequency domain representation of the signal processing, (b).

The discussion above for CW Doppler velocimeters can be extended to realistic acoustic beam geometries where separate transmitting and receiving crystals are used. Figure 20 shows a number of such configurations. Note that the configuration shown in Figure 20c is capable of high spatial resolution.

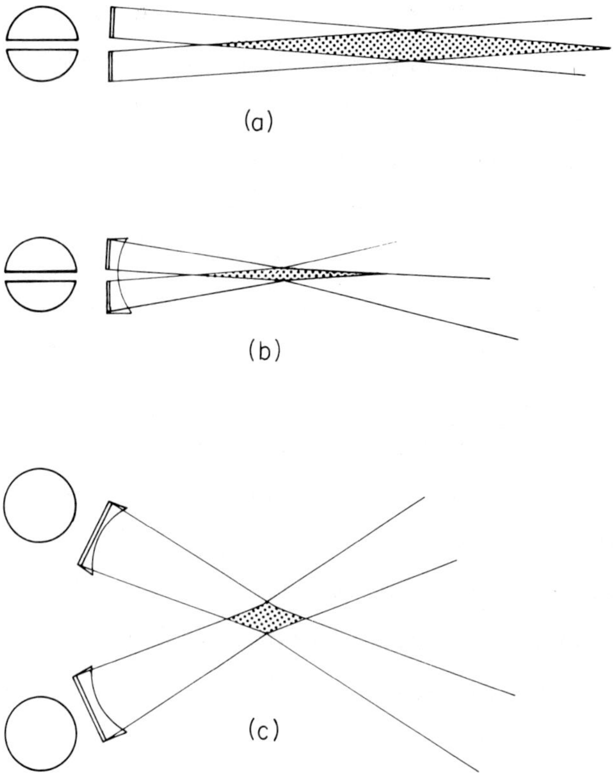

Fig. 20 Possible configurations of the receive and transmit transducers. (a) Unfocused split "D"; (b) Focused split "D"; (c) Focused crossed beams.

3.1.2 *TRANSMITTED SIGNAL EFFECTS.* Previously, we considered continuous sinusoidal transmitted signals, i.e., signals of zero bandwidth. In those cases the spatial resolution was completely determined by the sound field. However, as is well known from radar theory, range resolution can also be achieved by transmitting a waveform of nonzero bandwidth. Since we have shown above that velocity resolution was directly related to temporal length of the returning signal, one might conclude that for simultaneous range and velocity resolution the time/bandwidth (TB) product should be maximized. However, numerous other factors also influence resolution. The problem is too complex to be completely described by the TB product. In fact, Rihaczek[28] (Chapter 5) has shown that the upper bound for a measure of error in simultaneous range and velocity estimations is directly proportional to the TB product. Thus, large TB products do not guarantee accuracy.

Due to the importance of the transmitted waveform with respect to simultaneous range/velocity resolution, a large number of possibilities have been investigated in radar applications, and some of these have also been used for ultrasonic applications in medicine. It is important to note, however, that there is no "perfect" waveform. The optimum transmitted waveform depends on the characteristics of the target and the environment in which the target is located. In this section we will briefly discuss the general characteristics of a number of different transmitted waveforms that have been implemented or proposed for medical ultrasonic Doppler systems.

a. *Pulsed Phase Coherent Signal.* In the phase coherent pulsed system, a constant frequency carrier is amplitude modulated to obtain short harmonic bursts which are then phase coherent from pulse to pulse.[15] The backscattered signal is demodulated and sampled at time, τ, from the transmission of the last burst. The sampled signal, then, corresponds to backscattered energy originating at ranges

$$\frac{C}{2}\left(\tau + \sum_{n=1}^{\infty} \frac{n}{f_r}\right) \tag{16}$$

where f_r is the pulse repetition rate. Normally, one relies on the fact that no targets are located at the ambiguous ranges represented by the summation term in the above equation, or that attenuation of the transmitted signal is sufficient to rule out the possibility of detectable signals coming from the ambiguous region. There are similar discrete velocity ambiguities associated with the fact that the pulse repetition interval is regular. However, one normally picks the pulse repetition frequency at least twice the highest expected Doppler frequency to satisfy the Nyquist sampling theorem. By so doing, the complete spectrum of the Doppler signal will be contained in the frequency range,

$$o \leq f \leq f_{r/2}. \tag{17}$$

The lower bound for the pulse repetition frequency given above unfortunately leads to the major disadvantage of the pulsed coherent system. For the signal to return from the desired region before the next burst is transmitted, the maximum range is limited to

$$R_{max} = \frac{c}{2}\,\frac{1}{f_r}\ . \tag{18}$$

Otherwise, at the time signals returned from the desired range, signals from ranges nearer the transducer could also arrive. Under normal conditions, these signals would be stronger than the desired signal due to less tissue attenuation, for example.

b. *Pulsed Noise Signal*. In the Doppler system proposed by Bendick and Newhouse,[29] the harmonic signal of the pulsed coherent system is replaced with bursts of random noise. A block diagram of this system is shown in Figure 21. As previously discussed, the pulse repetition frequency still must be chosen to satisfy the sampling theorem. However, with random noise there are not the discrete range ambiguities at the intervals

$$\frac{c}{2}\,\frac{n}{f_r} \quad n = 1, 2, \ldots. \tag{19}$$

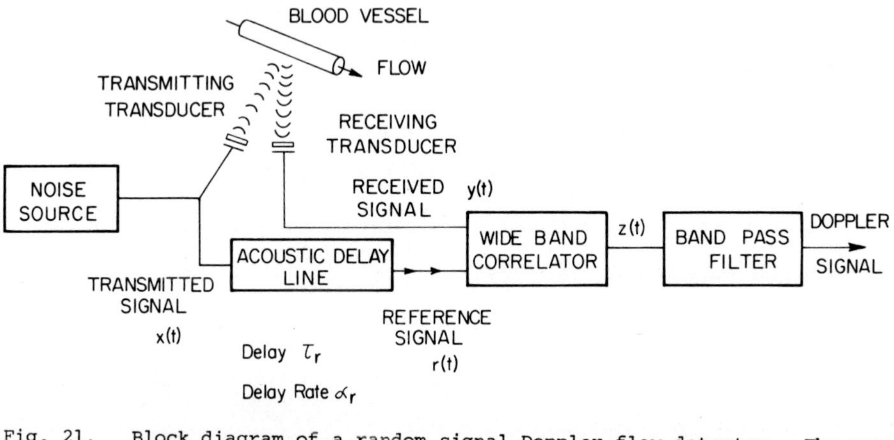

Fig. 21. Block diagram of a random signal Doppler flow detector. The received signal is correlated with a delayed sample of the original transmitted signal. The bandpass filter and output Doppler signal is essentially identical to the coherent type pulsed Doppler.

as shown in Equation 16 for the pulsed coherent system. This is due to the fact that each burst is different from all other bursts so that the echo from any one burst will not correlate with any other burst. Thus, there is not a range limitation for a given pulse repetition frequency as shown in Equation 18.

Another limitation of the pulsed coherent Doppler systems is caused by the fact that the duration of the transmitted burst must be kept short to maintain fine range resolution. This requirement imposes very high ratios of peak-to-average transmitted power. Since the echo from a target at range R gives a range delay or time of flight,

$$\tau = 2R/c, \tag{20}$$

it can be seen by differentiation that the range resolution ΔR for a transmitted burst of duration τ_e is at best

$$\Delta R = c\tau_e/2 \ . \tag{21}$$

The ratio of peak-to-average transmitted power can be written from Equation 8 as

$$P_p/P_a = \tau_p/\tau_e \tag{22}$$

where τ_p is the period of the transmitted signal, equal to the inverse of the pulse repetition rate. Using Equations 18 and 21, this gives

$$P_p/P_a = R_{max}/\Delta R \tag{23}$$

which is equal to at least 10^2 in any high resolution coherent system such as those used to observe deep-lying structures in the body.

It can be shown that in a random signal system, range resolution is not determined by the duration of the transmitted burst but by the bandwidth, B, such that $\Delta R = c/2B$. Thus, the transmitted power can be spread over a longer time period, reducing the ratio of peak-to-average transmitted power required for a given sensitivity.

From the foregoing discussion, one may conclude a number of possible advantages of the random signal system described.

1. Random signal processing can potentially extend limitations in velocity measurements by overcoming the discrete range ambiguities associated with coherent systems.
2. The same average power can be maintained with improved signal detectability due to the large duty factor possible with random signal ultrasonic pulses without sacrificing the range resolution. However, signal detectability of the random signal and conventional coherent pulsed Doppler systems using the same average power and spatial resolution are theoretically equivalent. In such cases, the random signal technique may have the advantage of simpler implementation and freedom from ambiguous responses.
3. Separate instruments using independent noise signals could be used to monitor different flow sites on the same subject at the same time without mutual interference. This scheme may not be possible with a conventional coherent pulsed Doppler system or with a system using a pseudo-random binary sequence.

A trade-off is made, however, to achieve these advantages. Uncorrelated clutter (received power from ranges other than the range of interest) results in an additional band limited white noise component in the Doppler signal.[30] The effect can be noted in the results of Bendick and Newhouse.[29] This can be interpreted as a degradation of velocity resolution as the additional frequency content masks the true Doppler signal. Again, a trade-off between range and velocity resolution is the result.

c. *Random Pulsed Noise Signal*. A variation of the noise system described above has been proposed by Jethwa.[31] In this system the transmitted pulses are randomly staggered. This essentially relaxes the requirement that the pulse repetition rate must be greater than twice the maximum expected Doppler frequency. The author demonstrates that the peaks in the ambiguity function,

$|A(\tau,\nu)|$* , which occur at intervals of f_r for periodic pulses will occur at Lf_r, where L is the total possible number of different intervals which are possible in the pulse repetition code (i.e., the probability of any one position is 1/L). It must be remembered that decreasing the velocity ambiguities must result in a resolution trade-off. The total volume under $|A(\tau,\nu)|$ is constant. Thus, diminishing the number of peaks associated with the Doppler ambiguity will result in increasing the function elsewhere, such as by broadening of the widths of the remaining peaks (close target separability) or increasing the height of the pedestal (wide target interaction or self-clutter).

d. *Continuous Wave Phase Modulated Signal.* Two more variations deserve mention. These methods are different from the above methods in that the transmitted signal in continuous. However, over constant intervals the phase or the frequency is changed. Waag, et al[32] proposed a phase modulation with a pseudo random binary sequence. For efficient phase modulation codes, ambiguities in range can be minimized to 1/N of the main response, where N is the total number of elements in the code. However, Rihaczek[28] demonstrated that phase reversal codes, in general, can result in rather large Doppler ambiguities. Although the variance in Doppler side lobes can be made small for large N, a low variance does not guarantee that some side lobes would not be large. The particular choice of code, therefore, is extremely important in determining system performance.

Lastly, McCarty and Woodstock[33] have reported the implementation of a continuous wave Doppler flowmeter in which the transmitted frequency is varied linearly, in a saw tooth fashion. Linear FM has a peculiarity in that there is strong coupling between range and Doppler shift. This effect is seen in Figure

---

*$A(\tau,\nu)$ is defined as the complex envelope of the output of a matched-filter Doppler system as a function of mismatch in delay and Doppler shift. It is defined for single targets moving at constant velocity and is obtained by correlating a returning signal with a Doppler-shifted time-delayed version of the transmitted signal. The output is maximum when the Doppler shift and delay match that of the returning signal.

22. Even though the received signal is due to a particular Doppler shift, ν_o, and range delay, τ_o, an infinite number of other combinations could also give the same received signal. This can create problems in determining simultaneous range and velocity measurement. The problem is compounded when the target consists of many particles, as in the case of blood. In that case, an FM system consisting of both up-sweep and down-sweep segments would be advantageous.[28] At this time no one has yet demonstrated the ability to measure simultaneously velocity and range of blood, i.e., determine velocity profiles with such a system.

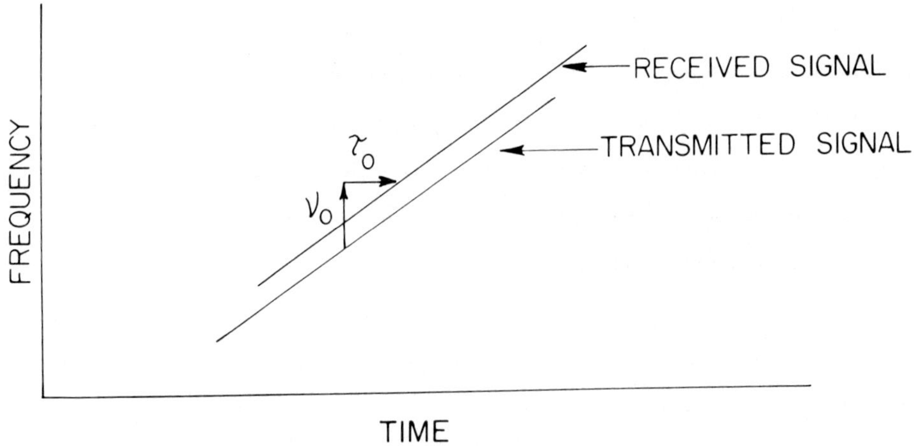

Fig. 22 The relation between frequency and time for an FM Doppler system.

The fact that there are numerous ultrasonic Doppler systems capable of simultaneous range and velocity resolution demonstrates that there is not complete agreement as to which is the best approach. Aside from the resolution characteristics described above, many other relative advantages and disadvantages exist between the various methods. These are characteristics such as acoustic intensity levels necessary, transducer technology required, signal processing complexity, and cost. It must also be remembered that there is a broad spectrum of Doppler applications from implantable transducers to transcutaneous measurements of flow in deep vessels. There are also vessels of interest which are in close proximity to other vessels and vessels which exhibit relative motion. Target environments such as these may optimally require different systems.

3.2 Specific Characteristics of Doppler Systems

In the previous section the fundamental effects of beam geometry and transmitted waveform on range and velocity resolution were described. In this section the specific system characteristics which influence performance will be discussed. Although much of the text below applies to a number of different types of Doppler velocimeters, the pulsed coherent system is taken as the specific example.

3.2.1 *SPATIAL RESOLUTION.* In a pulsed coherent system, short bursts of acoustic energy are transmitted by amplitude modulation of a carrier frequency. The returning signal is sampled at a specific delay time from transmission of each pulse. As a consequence, the sampled portions of the returning signal correspond to a backscattered acoustic energy originating from a specific region of space called the sample volume, whose range from the transducer is determined by the delay time, Figure 23. Note, we are neglecting any effects of previous transmitted bursts as discussed above.

The spatial resolution of a Doppler system is determined by the size of the sample volume. Typical dimensions of the sample volume are 1 mm diameter by 2 mm long in a 5 MHz system with 1 cm diameter transducer and a 3 cm focal length. This corresponds to a volume containing on the order of 10^7 red blood cells, each contributing to the backscattered signal. The amplitude and phase of the returning signal from each particle is determined by the effective intensity distribution throughout the sample volume and the location of each particle within it. In blood flow it is normally assumed that the blood cells represent a random distribution of scatterers.

The effective intensity distribution within the sample volume, and hence its spatial extent, can be determined from the form of the transmitted signal envelope and other system parameters. Figure 24a shows the excitation signal

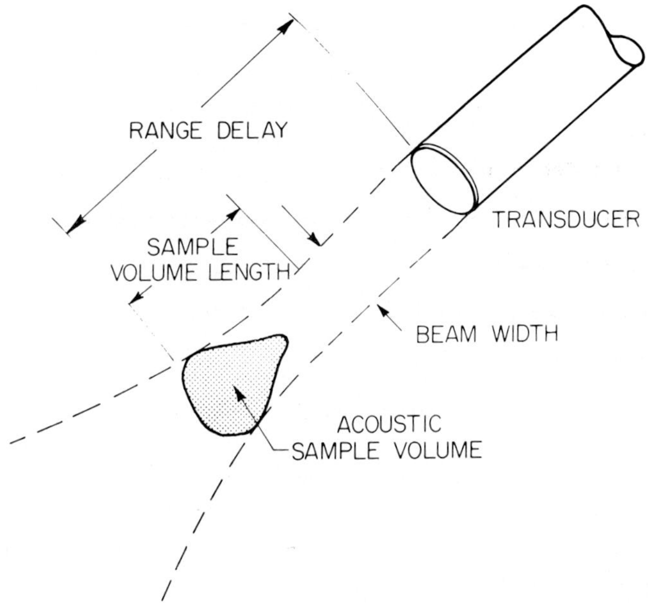

Fig. 23 The acoustic sample volume for a pulsed Doppler system.

applied to the transducer and the signal received from a stationary reflector. The envelope of the signal is determined by the exponential rise as the transducer is excited and the exponential decay in amplitude as the crystal oscillates in free response. Normally the transmitter and receiver have bandwidths equal to, or greater than, that of the transducer, so that Figure 24b reflects the actual transducer response. Although the exponential envelope reflects the effective axial intensity distribution of the sample volume, the length of the sample volume corresponds to the distance sound travels in one-half the transmitted burst duration, Δt_b,[34] i.e.,

$$\text{Sample volume length} = \frac{c}{2} \Delta t_b \ . \tag{24}$$

The length of the sample volume is also related to the method of sampling the returning signal. There is an additional contribution to the length equal to

$$\frac{c}{2} \Delta t_g \tag{25}$$

where Δt_g is the effective observation gate time.[34] For ideal sampling, $\Delta t = 0$. For sample-and-hold circuits, which are commonly used, Δt is approximately equal to zero.

The acoustic beam pattern determines the intensity distribution in the lateral direction. The changes in field strength in the range direction are normally small over the sample volume length, as determined above. However, for highly focused transducers the field pattern could also influence the intensity distribution in the range direction.

If the output of a frequency meter is to be quantitatively related to

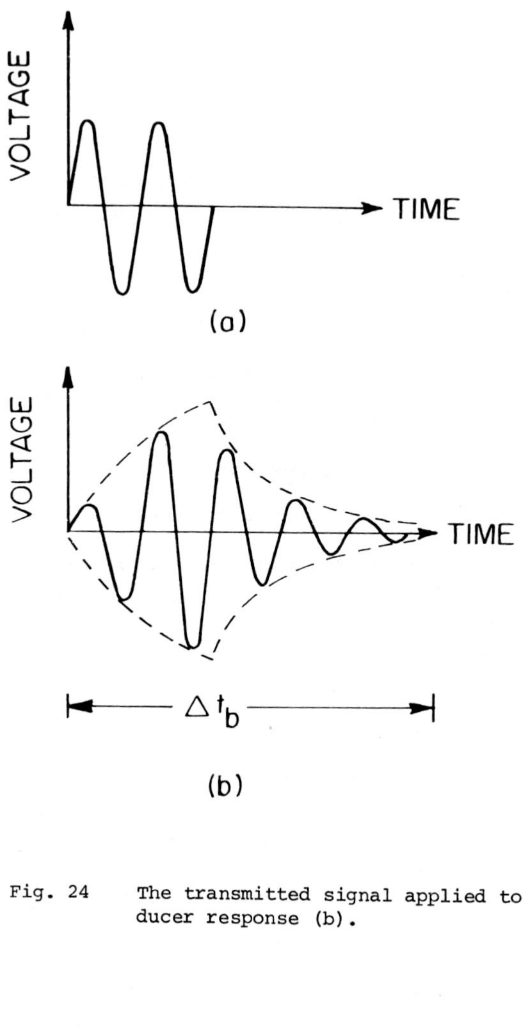

Fig. 24 The transmitted signal applied to the transducer (a) and the transducer response (b).

the velocity field, additional instrument parameters must also be known. The trigger level of zero-crossing detectors effectively causes truncation of the sample volume at intensities below the level set. In addition, the nonlinear behavior of such frequency meters with signal-to-trigger level, as shown in Figure 25, must be included.[35]

Knowledge of the system parameters discussed above make possible accurate measurements of velocity fields, such as is necessary for determining flow rates from velocity profiles in small vessels or constructing jet images to determine the degree of valvular stenoses. Normally, the velocity field is distorted by sample volume size effects, but corrections can be made using deconvolution techniques.[36] This will be discussed in detail in a later section.

3.2.2 *VELOCITY RESOLUTION*. The concept of velocity resolution is closely related to the width of the Doppler spectrum. The broader the spectrum, the more difficult it would be to measure a small shift of the spectrum (change in

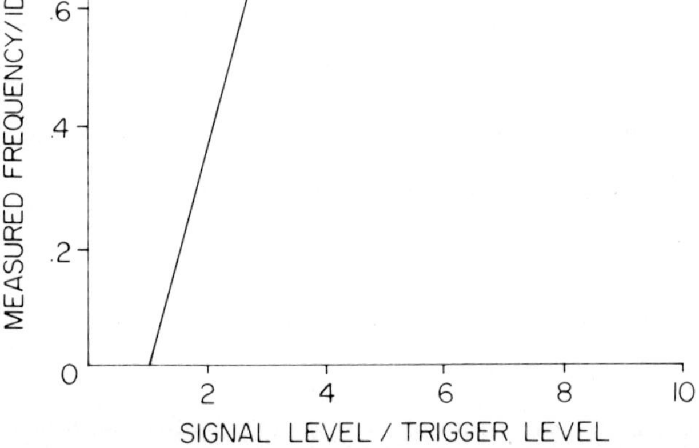

Fig. 25 The effect of the trigger level and signal amplitude on the output of a zero crossing detector.

velocity). Factors influencing Doppler spectral broadening can be classified into instrument effects and flow-related effects. Even if all scatterers move through the sample volume at the same constant velocity, the spectrum is composed of a range of frequencies. This effect of the instrument itself is demonstrated in Figure 26, which shows a time history and power spectral estimate for the Doppler signal taken in laminar, steady plug flow measured at the centerline of a circular pipe. In addition to these instrument effects, spatial and temporal velocity fluctuations occurring within the sample volume also contribute to spectral broadening. Each of these effects will be discussed below.

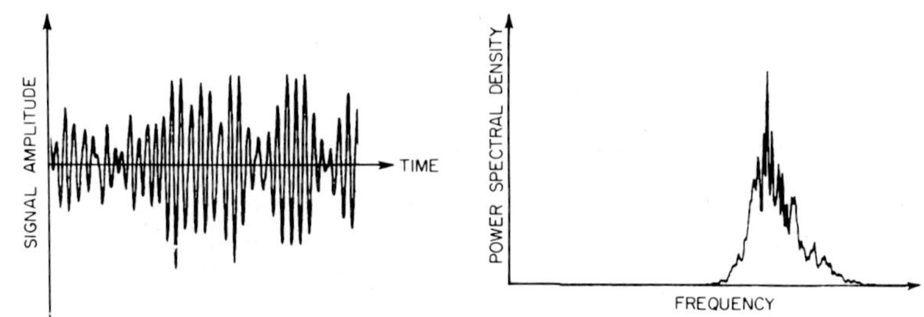

Fig. 26 A typical Doppler signal in both time and frequency domains for a uniform steady velocity field.

The purpose of this section is not to investigate improvements in velocity resolution without limit, because any improvement in velocity resolution (narrower Doppler spectrum) would lead to a degradation in spatial resolution. Rather, we are interested in understanding the causes of spectral broadening. In fact, it will be shown that a measure of spectral broadening can be used as an important parameter characterizing blood flow.

a. *Spectral Broadening - Instrument Effects*. Broadening of the Doppler spectrum can be attributed to instrument effects and flow effects. The instrument effects, shown in Figure 26, can be described by considering the movement of a single particle through the acoustic field at constant velocity. This situation yields a qualitative description of the actual situation which has been modeled for CW systems as the movement of random distribution of particles through the sound field.[26] It will be seen that the form of the Doppler spectrum is dependent on the sample volume dimensions (beam width and burst duration) and the method of signal sampling used.

For simplicity, we consider a rectangular sample volume and a single particle whose path of motion intersects the sample volume center, as shown in Figure 27. The Fourier transform of the received signal, $U_r(f)$, from a stationary particle is shown in Figure 28. The returning Doppler shifted signal, $U_s(f)$, from a moving particle is given by

$$U_s(f) = \frac{1}{(1 + \alpha)} \; U_f \left(\frac{f}{1 + \alpha}\right) \tag{26}$$

where $\alpha = 2v\cos\theta/c$. In addition, the returning signal exists only for the time the particle traverses the sound field given by the time,

$$\frac{1}{f_t} = \frac{w}{v\,\sin\theta} \tag{27}$$

This amplitude modulation is taken into account by multiplying $U_s(t)$ with the

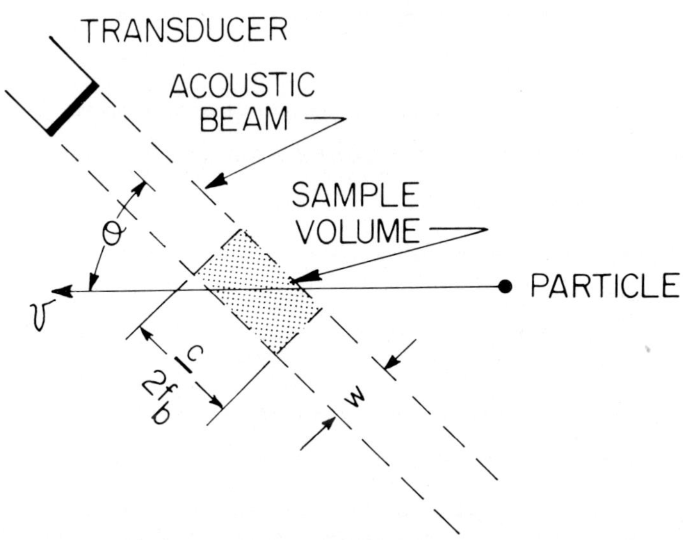

Fig. 27 Idealized configuration of the acoustic sample volume.

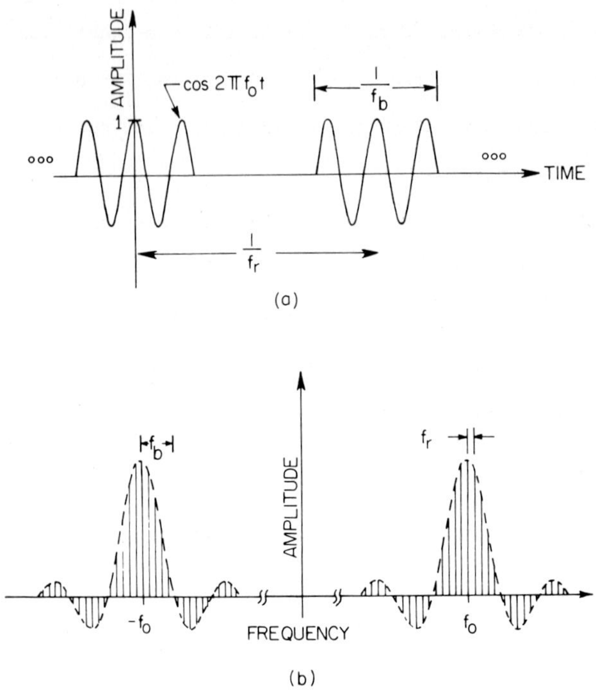

Fig. 28 The transmitted signal for an idealized Doppler system in the time domain (a) and the frequency domain (b).

rectangular function shown in Figure 29a, or equivalently, by convolving $U_s(f)$ with the transform shown in Figure 29b with f_o replaced by $f_o(1+\alpha)$, f_r replaced by $f_r(1+\alpha)$ and each impulse replaced by the sinc function shown in Figure 29b (Note that $f_t << f_r$). We call the resulting function $U_{st}(f)$.

The Doppler signal is obtained by mixing $U_{st}(f)$ with $\cos 2\pi f_o t$ and then sampling. The output of the mixer is given by

$$U_m(f) = \frac{1}{2}U_{st}(f-f_0) + \frac{1}{2}U_{st}(f+f_o) \tag{28}$$

This is equivalent to translating $\frac{1}{2}U_{st}(f)$ to the right by f_o and adding that function to $\frac{1}{2}U_{st}(f)$ shifted by f_o to the left. The contributions centered at $\pm 2f_0(1+\alpha)$ are then filtered off. Ideal sampling is equivalent to convolving $U_m(f)$ with the sampling function $S(f)$, i.e.,

$$U_d(f) = U_m(f) * S(f) \tag{29}$$

where

$$S(f) = \sum_{m=\infty}^{\infty} \delta\ (f - mf_r) \tag{30}$$

The final result is shown in Figure 30 in terms of the frequency spectrum $|U_d(f)|^2$ where $U_d(f)$ on the interval $0 \leq f \leq f_r/2$ is given by

$$U_d(f) = \text{const.} \sum_{n=\infty}^{\infty} \frac{\sin\left(\frac{\pi}{f_b}\ f_r n\right)}{\pi n}\ \frac{\sin \frac{\pi}{f_t}\ (f - f_o\alpha + f_r\alpha n)}{\pi(f - f_o\alpha + f_r\alpha n)} \tag{31}$$

if contributions from outside the interval $0 \leq f \leq f_r/2$ are neglected. Note, $U_d(f)$ is repeated every f_r and is mirrored around $f_r/2$, which represents the Nyquist frequency. Normally, the audio Doppler signal refers to this signal filtered at $f_r/2$. The effect of the burst duration (source broadening)

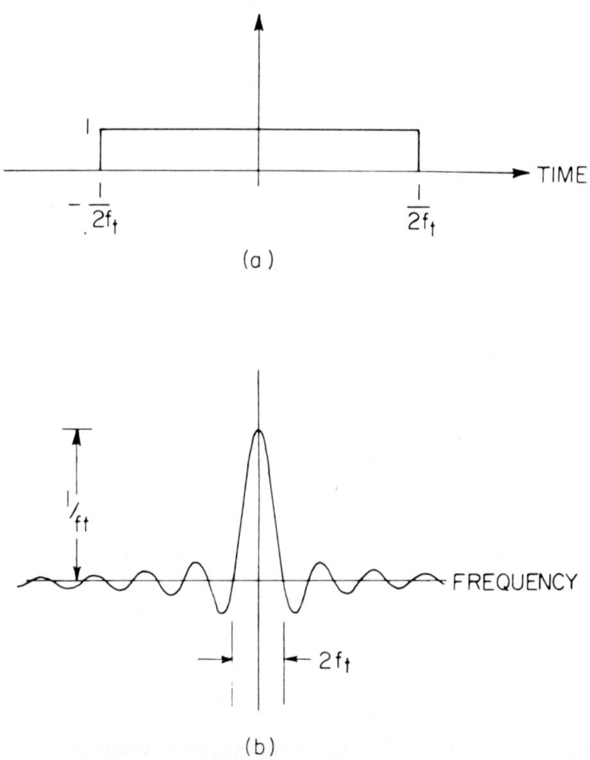

Fig. 29 The transit-time window in the time domain (a) and frequency domain (b).

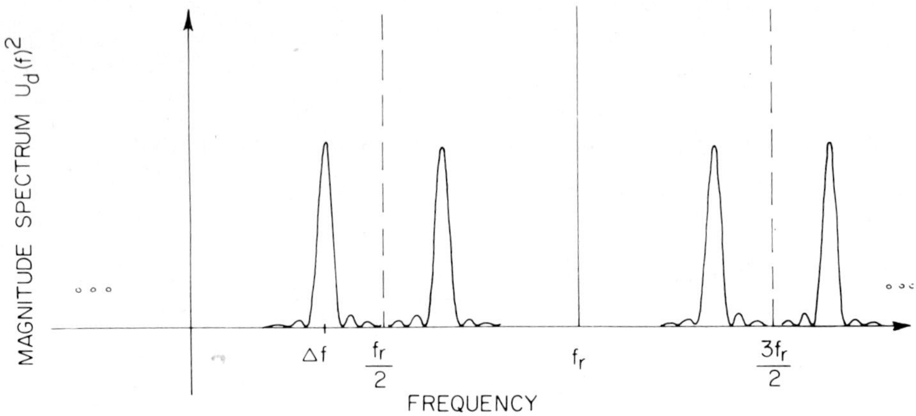

Fig. 30 The Doppler spectrum for an idealized system.

occurs through the term containing f_b. The effect of the acoustic field (transit time broadening) occurs through the term containing f_t. The overall spectrum is referred to as the baseline ambiguity. It should also be noted that the width of the baseline ambiguity is directly proportional to the velocity of the particle traversing the sample volume. Thus, the fractional broadening, defined as a measure of spectral width divided by f_d, is independent of velocity.

For the ideal sampling function used above, the mean frequency of the Doppler spectrum on the interval $0 \leq f \leq f_r/2$ is

$$\Delta f = \frac{2f_o}{c} \, |v \cos\theta| \, . \tag{32}$$

Thus, a frequency meter that measures the mean frequency of the spectrum yields an output directly related to the particle velocity. However, an idealized sampling function is not realizable. To obtain an instantaneous sample, the amplitude of the sampling impulse would have to be infinitely large. Instead, sample-and-hold circuits are employed. Such circuits are low-pass filters. The Doppler spectrum is therefore distorted, and the mean value (first moment) is lowered. The distortion can be large as the equivalent filter function is[23]

$$\left[\frac{\sin\left(\frac{\pi f}{f_r}\right)}{\frac{\pi f}{f_r}}\right]^2 \qquad (33)$$

which corresponds to an attenuation factor of $4/\pi^2$ on the Doppler spectrum at $f_r/2$. Thus, substantial distortion will result as the Nyquist sampling frequency is approached.

To summarize, this section has dealt with how the instrument itself contributes to the resolution in velocity of a moving particle as measured by the width of the Doppler spectrum. The spectral broadening, in this case, is due primarily to the burst duration (bandwidth) of the transmitted signal and the transit time of the target through the acoustic field. In addition, the resulting Doppler spectrum, called the baseline ambiguity, is affected by the method used to sample the returning signal.

b. *Spectral Broadening - Flow Effects.* In the previous section it was shown that if all the scatterers move through the sample volume with the same constant velocity, there is broadening of the Doppler spectrum, referred to as the baseline ambiguity. These effects are due to characteristics of the instrument. In addition, various characteristics of the flow lead to additional spectral broadening. These can be referred to as spatial effects and temporal effects.

Spatial effects are simply due to the fact that if velocity gradients exist in the flow field, a range of velocities of the particles moving through the sample volume will exist. This range of velocities, Δv, will result in additional frequency components in the Doppler spectrum. It is important to note that Δv is not a function of the velocity gradient alone, but is also a function of the sample volume size. Thus, gradient broadening can be reduced by reducing the sample volume size.

Temporal effects refer to the fact that over a given period of time the velocities of the particles moving through the sample volume may change. If the Doppler signal were viewed on a spectrum analyzer whose integration time was so long as to exceed the period of the heart cycle, the Doppler spectrum would contain the entire range of Doppler shifts present throughout the flow cycle. The spectrum would also be constant in time (stationary). However, the frequency components in the flow pulse are much lower than the Doppler shift frequencies. As a result, one can observe quasi-steady spectra at various stages of the pulsatile flow cycle. In other words, the time necessary to obtain a spectral estimate is small compared to the period of the heart cycle.

In disturbed, or turbulent, blood flow, oscillations may exist which are on the order of hundreds of cycles per second. The effects of these oscillations do appear in the Doppler spectrum because the periods of the oscillation are much shorter than the integration time of a spectrum analyzer. This is an important fact to note because there is evidence that the character of blood flow may be altered from laminar to disturbed, or turbulent, by abnormal valve leaflets,[37] or other cardiovascular disorders associated with mechanical defects. In fact, pulsed Doppler techniques are already being used qualitatively to detect disturbed flow in, and near, the heart associated with cardiovascular disorders.[38]

Recent work by Forster, et al.[39] has demonstrated that a quantitative measure of the intensity of turbulence can be made based on spectral broadening of the Doppler signal. This fact was initially noted by Goldstein and Hagan[40] and Pike, et al.[41] working with the laser Doppler anemometer, and later rigorously proved by George and Lumley[42] for CW systems. In the absence of velocity gradients, the turbulent intensity can be expressed as

$$\left(\frac{\overline{v'^2}}{v^2}\right)^{\frac{1}{2}} = \sqrt{\left(\frac{\sigma}{\mu}\right)_T^2 - \left(\frac{\sigma}{\mu}\right)_B^2}\ \cos\theta \qquad (34)$$

where θ is the angle between the beam direction and mean velocity, v, and $\overline{v'^2}$ is the mean square value of velocity fluctuations in the direction of the acoustic beam. μ and σ^2 are the spectral mean and variance respectively. B refers to the baseline ambiguity, and T the spectrum in the presence of turbulence.

In vitro experiments have been performed which demonstrate the feasibility of making turbulence measurements ultrasonically. In addition, the results indicate an advantage over thermal anemometry may exist, aside from the noninvasiveness of the Doppler method.

The flow system used was capable of producing either laminar or grid-generated turbulent flow having a flat velocity profile over most of a circular test section (see Figure 31). Centerline turbulence measurements were made at various distances from the grid. These measurements were calculated from spectral broadening of the Doppler spectrum according to Equation 34, and by direct measurements made with various hot-film probes. The results are shown in Figure 32. The Doppler measurements are higher than all the hot-film measurements. Note that the asymptotic value of the Doppler measurements of the turbulence intensity is approximately 3.5%, corresponding to the value for fully developed turbulent pipe flow.[43] Also note that the hot-film measurements decrease with increasing probe size, and all probes yield asymp-

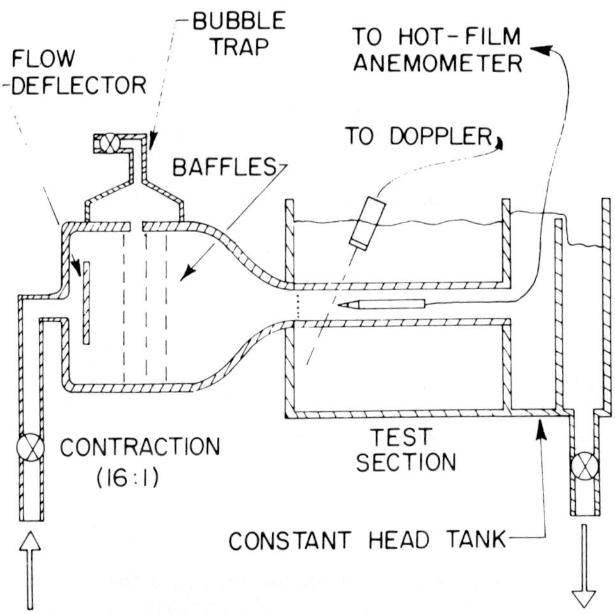

Fig. 31 Schematic representation of flow systems for producing controlled amounts of turbulence.

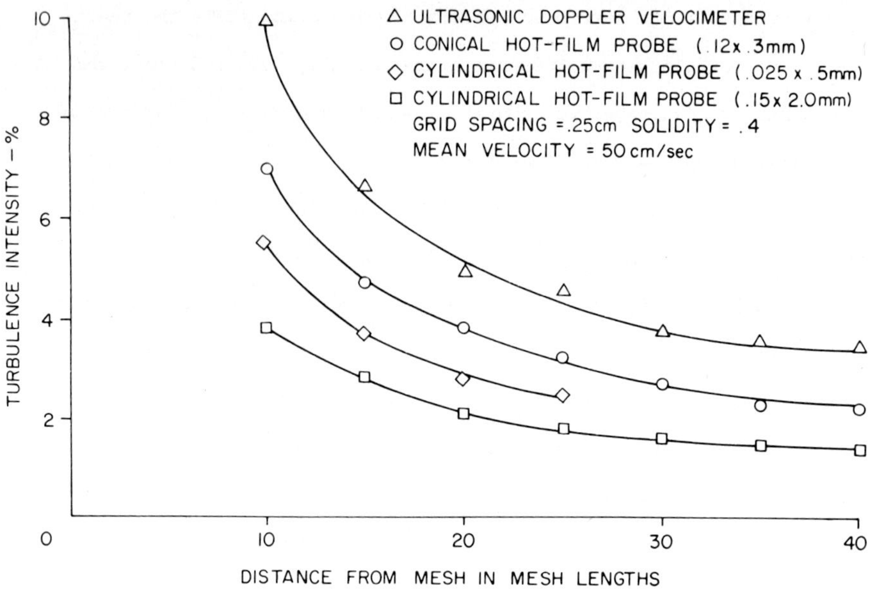

Fig. 32 Measurements of grid generated turbulence made with ultrasonic Doppler and thermal anemometer measurements.

totic values lower than expected. For accurate turbulence measurements with thermal probes, the active portion of a probe must be much smaller than the energy containing eddies. This is normally taken as the Lagrangian integral scale.[44] Near the grid, this scale is approximately the grid spacing of 2.54 mm. In fully developed pipe flow, it may be even smaller for the 2.54 mm pipe used.[43] Thus, both cylindrical probes are too large compared to the scale of turbulence to yield correct results (see Figure 30). It is possible that even the conical probe, which is one-eighth as large as the grid spacing, may be too large for accurate results. If the length scales of turbulence <u>in vivo</u> are on the order of those studied in the above experiments, thermal anemometer measurements must be taken with care using probes smaller than the smallest probe used in this study. Otherwise, appreciable attenuation in intensity measurements and turbulence spectral may result. A number of investigations have reported turbulence measurements obtained with larger than the conical probe used here.[45]

The advantage of using Doppler techniques for turbulence measurements is that spectral broadening due to turbulence is not affected by sample volume size as hot-film measurements are by probe size. However, the larger the sample volume, the larger the effect of velocity gradients if present. Thus, it is still important to use relatively small sample volumes for turbulence measurements.

4. METHODS OF QUANTITATION

4.1 The Problem of Quantitation

From the previous section, it should be clear that the process of making a blood velocity measurement with an ultrasonic Doppler device is indeed complex and subject to a number of errors. These errors can be divided into two categories depending on whether they are inherent to the system or technique being utilized (systematic errors) or stem from experimental difficulties in applying the technique (experimental errors). The choice of a particular measurement technique must be made so as to minimize the effect of systematic errors on the sought after parameter and then the technique must be implemented so as to minimize the effect of experimental errors. These types of considerations are the subject of this section.

The term "quantification," as applied to Doppler blood velocity sensing, generally refers to the accurate measurement of volume blood flow through a vessel or lumen, in units such as cc/sec. However, parameters other than volume blood flow, such as peak forward or reverse blood velocity, or blood velocity profiles, may be of primary interest to some researchers. In a more general sense, the term quantification should be used to describe two measurement processes - the accurate determination of blood velocity <u>and</u> the spatial distribution of the velocity field being measured.

Each process has its own set of measurement difficulties, but as pointed out previously, the two are related through a Doppler uncertainty principle which restricts the accuracy with which a target's position and velocity can be measured simultaneously. This Doppler uncertainty principle has several important consequences for the making of quantitative measurements. If accurate, regional blood velocity measurements are desired, a compromise must be sought between the size of the measurement region and the accuracy of the velocity determination. Depending on the frequency of ultrasound employed, this compromise may prohibit the measurement of such parameters as velocity profiles except in large vessels.

In general, the compromise between spatial resolution and velocity accuracy will be dictated by the parameter to be measured. For example, if volume blood flow is the sought after variable, there may be little point in making high resolution velocity profile measurements. It is important, therefore, to clearly define the parameter one desires to measure and then determine the appropriate measurement technique and method. The following sections will cover some of the important considerations in making accurate blood velocity measurements and the determination of the much sought after parameter of volume blood flow.

4.2 Blood Velocity Measurement

Ultrasonic Doppler devices for measuring blood flow should be termed velocimeters since this is the actual parameter measured. The relation between the Doppler shift, Δf, in energy scattered from red blood cells and their velocity, v, has been described previously by the Doppler equation and is restated here.

$$v = \frac{\Delta f \cdot c}{2f_o \cos\theta} \tag{35}$$

The simplicity of the above equation masks the complexity of a Doppler measurement. It describes the situation where there is only one moving particle moving through an infinite sound field. In actual blood flow measurements there are many cells which are distributed over a range of velocities and space, passing into and out of the sound beam. Under these actual conditions, the Doppler shift, Δf, is not a single frequency but is a distributed quantity, as described previously. Depending on the measurement conditions, the frequency distribution, P(f), of the Doppler signal may limit the accuracy of the velocity measurement.

4.2.1 *Measurement of* Δf. The measurement of Δf is perhaps the most difficult component of Doppler velocity measurements. In any real measurement, the detected Doppler shift frequencies form a frequency spectrum, whose shape depends on the conditions of the measurement and nature of the velocity field being interrogated. Generally, it is desired to extract the mean or first moment frequency of this spectrum since this frequency is the best approximation to the spatial average of blood velocity over the measurement region. For this extraction process to be meaningful, however, one must be aware of the many factors influencing the Doppler shift spectrum and the proper techniques for obtaining the first moment frequency.

4.2.2 *Instrument Effects*. Factors contributing to the broadening of the Doppler shift spectrum in the pulsed ultrasonic Doppler have been presented earlier and include transit time broadening, pulse width broadening, velocity gradients, and disturbed flow effects. While velocity gradient effects may not seem to belong in this class, they result from the spatial extent of the sample region, which depends on transducer characteristics, emission frequency, pulse width, sample gate time, etc. - all of which are parameters of the system. Spectral broadening in continuous wave Dopplers is primarily due to velocity gradient and transit time effects, both of which depend mainly on beam geometry.

Instrument effects on the Doppler shift spectrum can be best dealt with if the requirements are first defined, including the maximum range and spatial extent of the measurement region and the required velocity resolution. Next, the instrument parameters which are capable of modification should be optimized for the particular measurement requirements. This usually requires designing the transducer to achieve the desired beam characteristics, selecting the optimum ultrasound frequency for the range and resolution required, and selecting the proper pulse repetition frequency, pulse width, sample gate, etc. The next step is perhaps the most crucial - the measurement region must be experimentally determined. This can be accomplished using a thin jet of scatterers in a water tank to provide a line target which can be moved through the measurement region[46] or a moving sheet of scatterers, which provides a step velocity function response.[34] If the experimentally determined measurement region does not meet the resolution criterion, further instrument modification is in order, or the experimental goals need to be modified.

After achieving the spatial resolution requirements, the problem remains to extract meaningful information from the Doppler shift spectrum. The mean frequency has traditionally been extracted by zero-crossing detection, primarily because of the simplicity and ease of implementation of the technique. The method consists of averaging a train of impulses produced when the signal crosses a zero voltage threshold, Figure 33. Use of the zero-crossing frequency to voltage converter for Doppler shift signals has been described by Peronneau, et al.[14] and Flax, et al.[47] Use of these converters, however, can result in significant errors in the measured mean velocity.

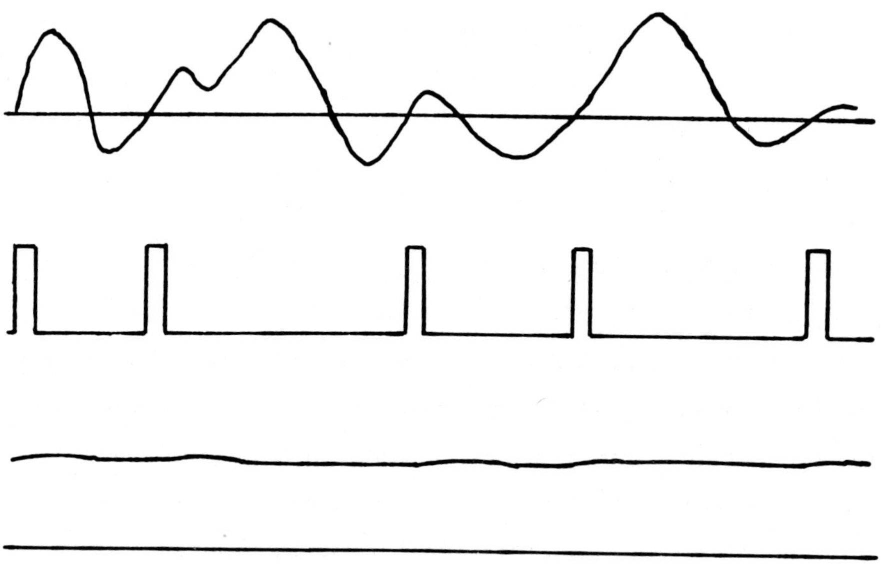

Fig. 33 Zero-crossing frequency detection. Illustrated from top to bottom are the input signal, the pulse train produced by the zero-crossing detector, and the analog waveform derived from averaging the pulse train.

The mean zero-crossing rate, $2f_m$, for a band-limited Gaussian random signal, such as a Doppler shift signal, is given by the following relationship.[48,49]

$$f_m^2 = \frac{\int f^2 P(f)df}{\int P(f)df} \qquad (36)$$

where f_m is the radius of gyration about the origin, f is the Doppler frequency, and P(f) is the power spectral density of the Doppler shift signal. The mean velocity in the sample region, however, is more closely approximated by the first moment frequency or

$$\Delta f = \frac{\int fP(f)df}{\int P(f)df} \qquad (37)$$

This has been shown by Brody and Meindl[26] for continuous wave systems. For narrow band signals, as normally encountered when dealing with small sample regions in laminar flow, the error introduced by taking the zero crosser output as an estimate of Δf is small. For broad band signals, however, the error can amount to as much as 15%. The conventional zero-crossing detector is also incapable of following rapid fluctuations in the mean frequency as may be encountered in flow disturbances.

The deficiencies in the zero-crossing detector may be overcome by more sophisticated signal processing techniques. A number of researchers have devised analog first moment processors for more accurate extraction of mean frequency.[11,21,50] These processors perform adequately in most flow sensing situations, but share the same deficiency of the zero-crossing detectors in their inability to follow rapid mean frequency fluctuations. A second technique involves shifting the Doppler spectrum up to a relatively high offset frequency, f_s (see Figure 34), typically between 50 and 100 kHz. The Doppler spectrum then becomes extremely narrow, relative to the offset frequency range, and zero-crossing detection closely approximates the mean frequency of the Doppler shift spectrum. The offset frequency corresponds to a DC level in the output which can be subtracted to yield a variation about zero volts corresponding to velocities toward or away from the transducer.

The offset frequency zero-crosser is only slightly more complex than

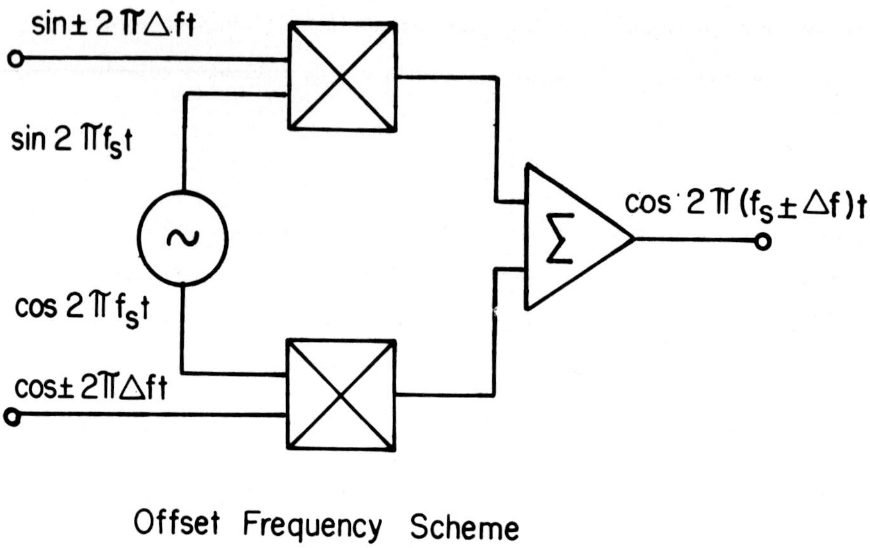

Fig. 34 Circuit for shifting Doppler spectrum to a higher offset frequency. Inputs are the two audio signals produced from quadrature detection of the Doppler signal.

the traditional zero-crossing detection scheme, but realizes several important advantages. Firstly, the output is more nearly proportional to the first moment of the Doppler shift spectrum, thus eliminating errors in measurement of mean velocity. Secondly, the offset frequency processor is capable of a higher frequency response, or the ability to follow rapid fluctuations in Δf, such as occur in disturbed flow conditions. This is due to the fact that the pulse train resulting from zero-crossing detection has a high frequency rate (about equal to the offset frequency), and less filtering is required to average the pulses to provide a voltage proportional to frequency. Finally, the offset frequency processor provides a simple means of displaying the frequency content of the Doppler shift spectrum in a form similar to a continuous power spectral density plot. This plot, more properly termed a time interval histogram, or TIH, is described below.

Often in the measurement of blood flow velocity with Doppler devices, the Doppler shift spectrum contains other relevant information in addition to the first moment frequency corresponding to the average velocity. Such situations occur frequently in cardiac measurements, where the flow may be in a "disturbed" or turbulent state. The identification of such states is necessary for diagnosis of such conditions as valvular insufficiency and septal defects. Generally, disturbed flow states result in "harsh" or "noiselike" Doppler shift audio signals, but such characterizations are of little value in documenting the magnitude and nature of the condition. A more quantitative approach involves evaluating the Doppler shift signal using spectral analysis techniques, carried out with a digital computer or with specialized hardware; however, these techniques are currently expensive and time consuming, and sometimes must be performed "off line."

The TIH represents a simple method for documenting the disturbed flow Doppler shift frequency fluctuations in real time. It is derived from the offset readout zero-crossing pulse train output as shown in Figure 35. Each

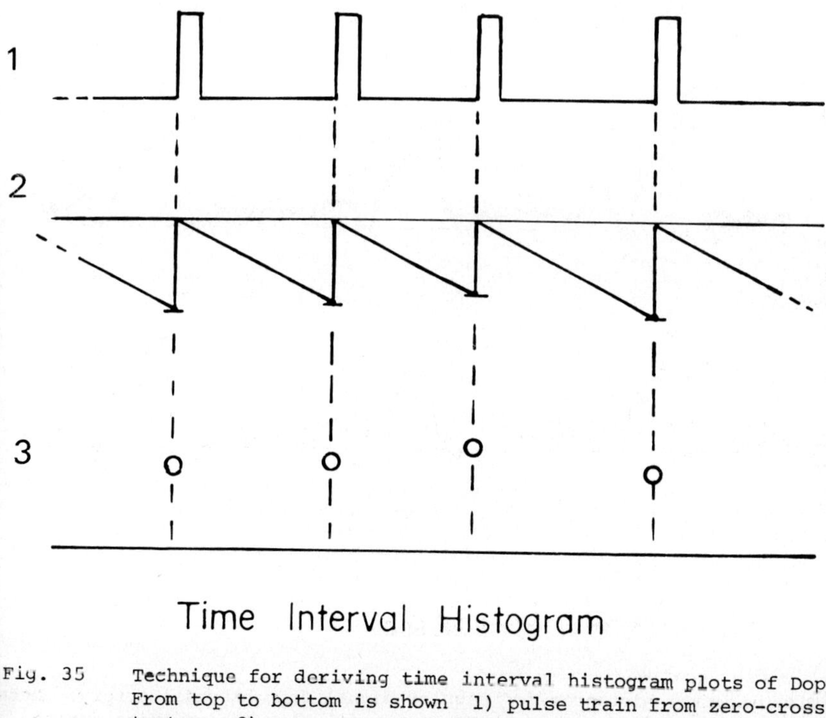

Fig. 35 Technique for deriving time interval histogram plots of Doppler signal. From top to bottom is shown 1) pulse train from zero-crossing detector, 2) ramp generated from pulse train, and 3) dot pattern produced from sampled values of ramp.

pulse starts a ramp which decreases linearly until the next zero-crossing pulse occurs. At this point the value of the ramp is sampled, stored, and immediately reset. The stored samples are amplified and plotted in real time using a fiber optic type recorder. Since the samples are produced at a high rate (offset frequency), the displayed pattern has an intensity or darkness equivalent to the number of samples at any given time interval. For a high offset frequency, f_s, the Doppler shift (Δf) represents only a small fraction of f_s. Under these conditions, a change in time interval between successive zero-crossings can be considered to be linearly related to a change in frequency. The TIH plot, therefore, has characteristics similar to a power spectral density plot, with each sample representing the nearly instantaneous value of the Doppler frequency.

Several TIH plots are illustrated in Figures 36 and 37, along with the first moment output from the offset frequency readout. Figure 36 shows blood

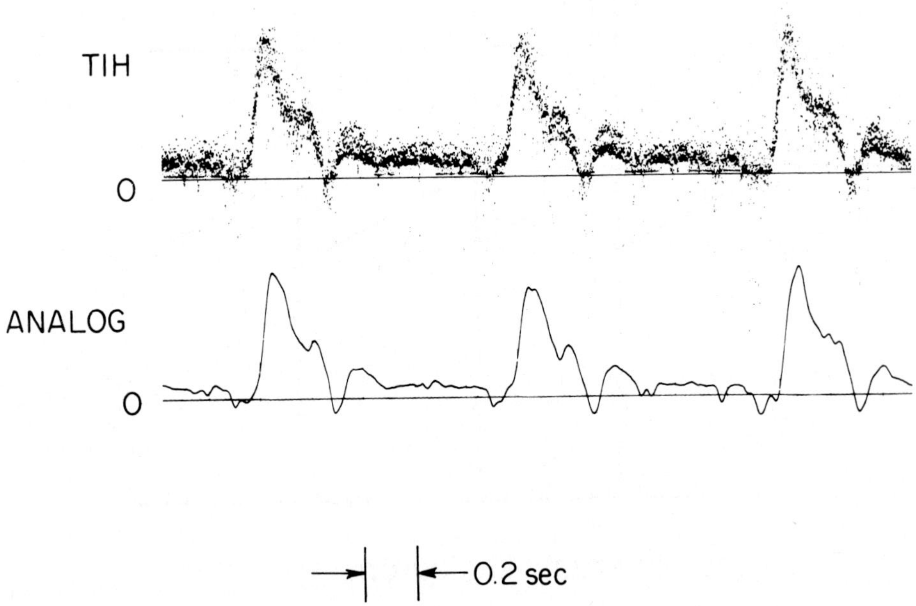

Fig. 36 Time interval histogram for Doppler signal from brachial artery. Note narrowness of dot pattern throughout, with slight broadening during deceleration phase. Analog waveform is shown below.

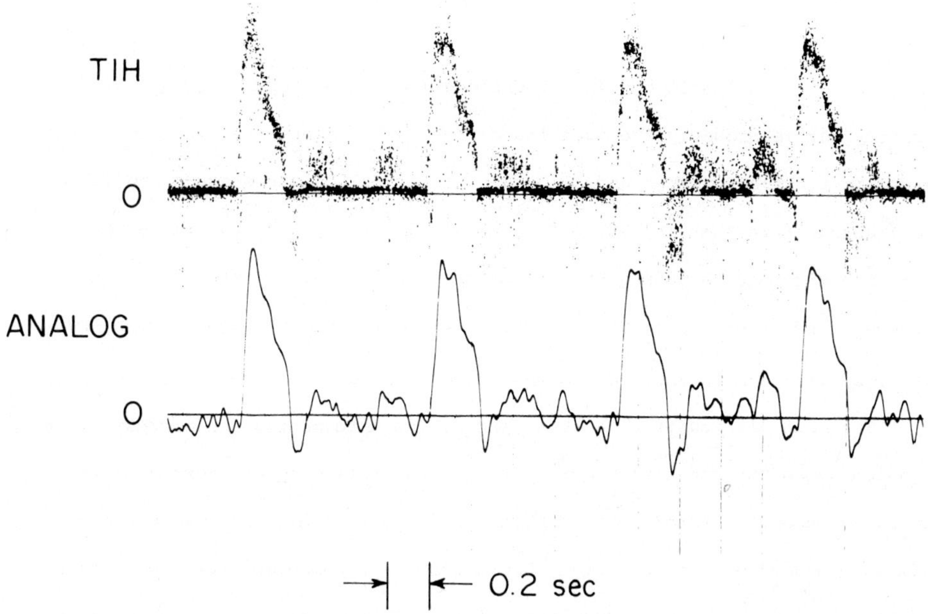

Fig. 37 Time interval histogram for flow in ascending aorta. Note broader dot pattern during forward flow phase with considerable spread immediately following valve closure.

velocity in the brachial artery while Figure 37 is a record of the ascending aortic flow recorded from the suprasternal notch. Note the additional information present in the TIH plots over the first moment output. In the brachial record, spectral broadening can be seen at the time of peak forward velocity and in the descending phase. The ascending aortic record shows considerably more spectral broadening, especially at the time of aortic valve closure and in the period immediately following. Records such as these provide valuable information on the characteristics of disturbed flow conditions.

a. *MEASUREMENT OF DOPPLER ANGLE*. In order to convert mean Doppler shift frequency to velocity, the angle between the velocity vector and the sound beam axis (θ in Equation 35) must be accurately known. The importance of accurate determination of θ can be seen in the fact that at a nominal Doppler angle of 60°, an error of only 3° in the measurement of θ results in an error of approximately 10% in velocity. Three accepted techniques for determining θ will be discussed.

1. *Velocity Null Method.* This technique is perhaps the most generally applied, but is probably the most inaccurate. The orientation of a subcutaneous vessel is first determined approximately by several measurements along its axis. The transducer is then angulated in a plane with respect to the vessel axis so as to minimize the returning Doppler shift frequencies. This orientation is assumed to be at $\theta = 90^{\circ}$. The transducer is then moved through a known angle, say α, in this same plane and the new Doppler angle is assumed to be $90^{\circ} - \alpha$.

The main difficulty with this approach is determining the correct plane in which to rotate the transducer. If the vessel is deep and tortuous, this obviously becomes a formidable problem. A further problem is that a broad null is produced due to beam width broadening of the Doppler spectrum. Also, the flow may not be smooth producing additional broadening. Another difficulty is that the transducer beam should be rotated about a point on the axis of the vessel to maintain the same measurement location. This usually requires a complicated transducer holder and arm. Considering these difficulties, the velocity null method may provide some information from shallow subcutaneous vessels with simple geometries.

2. *Triangulation Method.* The triangulation method involves measurement of the same velocity vector from different independent orientations and calculating the magnitude and direction of velocity from the various vector components, Figure 38. This triangulation can be carried out in two or three dimensions, depending on the complexity of the measurement situation and the accuracy desired. If the orientation of the vessel is desired, the measurement can be made during the peak forward velocity interval provided one assumes the vector to be coaxial with the vessel axis at that time.

Difficulties with this technique lie in the complexity of the calculations involved and problems with accurate probe positioning. For measurements at a known depth the situation can be simplified by mounting several transducers in a probe with a fixed and proper geometry. This assembly can then be positioned

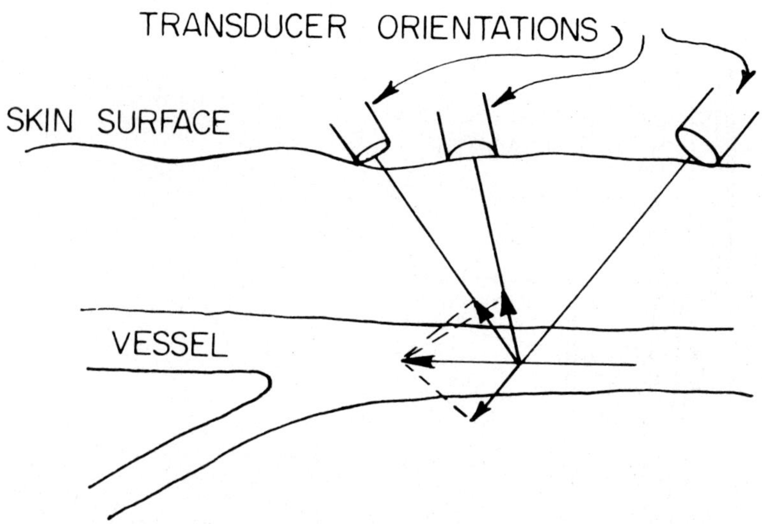

Fig. 38 Triangulation used to detect absolute orientation and magnitude of velocity vector. Vector is detected from three independent orientations and reconstructed from components.

over the vessel of interest. The geometry is thus easily maintained while measurements are made from each of the transducers. Figure 39 shows an example of the use of triangulation in an esophageal probe for obtaining quantitative velocity information from the descending aorta in dogs.[51]

3. *Imaging Methods*. For optimum determination of vessel orientation, the vessel can be imaged using ultrasound techniques. These scanning methods require precise knowledge of transducer position with respect to some fixed frame of reference. This can be accomplished using a scanning head with a readout of transducer orientation, a transducer arm with instrumented joints, or a noncontacting type of position readout utilizing acoustical or optical tracking techniques.[18]

Depending on the instrument used, the ultrasound image generated can show either the spatial pattern of velocity or anatomical structure of the vessel. Velocity or Doppler images are composed of recordings with suitable

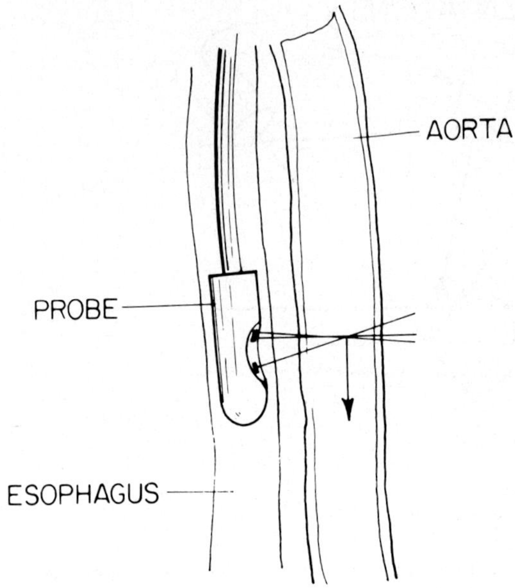

Fig. 39 Esophageal probe measurement of aortic blood velocity. Three transducers on probe are located at vertices of an equilateral triangle parallel to axis of probe. The sound beams intersect at a point near the center of the lumen of the aorta.

spatial resolution those regions from which a Doppler shift frequency is detected. This allows a two- or three-dimensional buildup of an image depicting the flow channel of a measured vessel. Such an image can provide information on flow obstructions and flow channel area as well as vessel geometry.[52,53] Vessel anatomy can also be imaged through detection of pulse echoes from interfaces representing changes in acoustical impedance. With two-dimensional scanning devices, the scan plane can be manipulated to include the vessel axis in the measurement region, providing a clear image of the vessel walls and lumen. If a Doppler probe is coupled to this scanning device, with a known relationship to the scan plane, the Doppler angle for velocity measurements can be observed directly provided the flow vector is parallel to the vessel axis. This technique is illustrated in Figure 40 which shows the block diagram and display of a duplex scanner, which combines echo and Doppler techniques.[17,54]

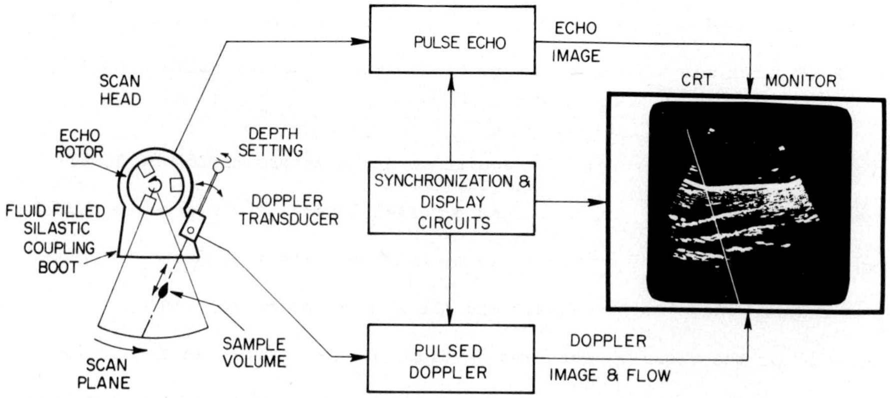

Fig. 40 A simple block diagram for a duplex scanner which combines real time, two-dimensional imaging with pulsed Doppler flow detection and vessel imaging. The display shows a long axis view of the common carotid artery with the simultaneous Doppler flow detection. The echo image is produced by a spinning rotor within the scan head which sweeps out three scans per revolution, for a total of 20 frames per second.

Clearly visualized are the walls of the carotid artery and its sonolucent lumen, along with the beam of the pulsed Doppler shown by the diagonal line intersecting the vessel axis. Since the Doppler beam is in the same plane as the echo scan, the Doppler angle can be read directly off the scanner display. While the implementation of this technique requires complex instrumentation, it is by far the simplest method with the greatest potential for velocity measurements in subcutaneous and even deep lying vessels. It has the added advantages of providing data on vessel geometry, including diameter and cross sectional area for volume flow measurement.

b. *VELOCITY PROFILE MEASUREMENT*. Velocity profile measurements provide useful information on the hemodynamics of flow through blood vessels. They are used to provide data on shear rates near the vessel walls and velocity disturbances across the vessel lumen. Velocity profile data is also used for computation of instantaneous volume flow rate via a process which will be covered in detail later. Measurement of velocity profiles requires a high degree of spatial resolution and can usually be accomplished with pulsed Doppler devices. As has been described, the accuracy of the velocity measurements with the pulsed Doppler is influenced by both systematic and experimental errors all of which may affect the users capability to obtain valid profile data.

Velocity profile measurements require accurate determination of local blood velocity at discrete points across a vessel lumen. The procedure usually followed with single gate instruments is as follows: The transducer is positioned at a known angle to the vessel and the sample volume located outside the near wall of the vessel. The sample volume is then moved electronically across the vessel in discrete steps, with several heart cycles of blood velocity information recorded at each point. The EKG must be recorded simultaneously so that velocity measurements can be synchronized in time. Generally the quantity of data is such that computer processing of the velocity waveforms is required.[55] The requirement that the Doppler angle be substantially less than 90° prohibits recording a velocity profile across a true diameter; however, for most measurements, the flow profiles can be assumed to be constant over the axial length in question. If a single gate pulsed Doppler is used, the flow characteristics are also assumed to remain constant over the time required to move the sample volume across the vessel. By using the EKG for synchronization, the velocity at a given instant in the cardiac cycle can be obtained for each of the measurement locations and then plotted versus distance to obtain a velocity profile.

Errors in velocity profile measurements can result from incorrect experimental application of the technique and from resolution errors in the measurement of local blood velocity. Misalignment of the sound beam across the vessel and inaccurate determination of Doppler angle are experimental difficulties which must be overcome. The transducer employed must produce a beam diameter which is small compared to the diameter of the vessel if the velocity measurements are to be accurate. A large diameter sound beam degrades resolution by increasing the radial projected length of the sample volume and introduces errors due to vessel curvature in measurements near the walls. Theoretical and experimental data indicate that errors due to sound beam width are negligible for beam diameters less than 1/4 the vessel radius. Beam diameter can be reduced by reducing the area of the transducer crystal or by focusing techniques.

Perhaps the most significant velocity profile measurement errors result from axial resolution limitations of the pulsed Doppler. Many variables influence the size of the measurement volume of the pulsed Doppler, as previously presented. While the lateral dimensions of the sample volume can be minimized by selection of the proper transducer characteristics, it is not always possible to obtain good axial resolution.

The axial length of the sample volume can be determined from a plot of Doppler signal intensity along the beam axis as a function of range using the underwater jet method described earlier. The resulting curve is known as the sample function since this function specifies the weighting of blood velocities along the sample volume length. The measured blood velocity is approximately the average of the velocities along the sample volume length, weighted by the sample function. Because of this property, the measured blood velocity profile when the sample volume is scanned across a blood vessel is described by the convolution of the sample function with the actual velocity profile. Obviously, the longer the length of the sample function, the greater the distortion of the actual velocity profile due to this convolution effect.

A second type of profile distortion occurs near the walls of a blood vessel.[56] As the sample volume is moved out of the flow stream, part of the sample volume enters a zero velocity region. Since a zero velocity region produces no Doppler shift signal, the device is unable to respond and the effective size of the sample volume is reduced. This effect introduces a boundary error in the velocity profile measurement which gives rise to higher velocities near the vessel walls than expected from simple convolution of the sample function with the actual profile.

For example, consider a rectangular velocity profile and a rectangular sample function, as shown in Figure 41. The convolution of these functions is trapezoidal in shape, as shown. If the boundary error effect is included,

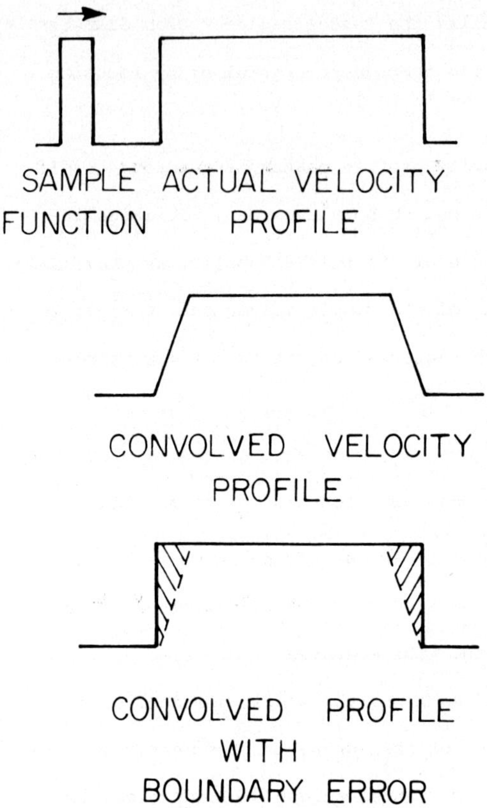

Fig. 41 Convolution and boundary error effects on a rectangular velocity profile.

however, the profile shape becomes rectangular, with an increase in width equal to the sample function width.

In actual velocity profile measurements, the boundary error effects may be diminished by reduced signal-to-noise ratios near the walls, and motion of the walls with the pressure pulse. The measured blood velocity profile generally falls somewhere between the convoluted velocity profile and the convoluted velocity profile with boundary error. This region is shown shaded in Figure 41.

To investigate the effects of sample function length on velocity profile measurement, a theoretical study was made using computer generated curves.[50] Velocity profiles were generated according to the formula,

$$y = V_p \ (1 - | x^n |) \quad (-1 \leq x \leq \ ) \tag{38}$$

This formula yields a family of curves as shown in Figure 42, with increasing values of n yielding profiles with increasing bluntness. For n = 2, the profile shape is parabolic. These theoretical profiles were convolved with rectangular sample functions of varying length. Nonlinearities introduced by the zero-crossing detection process were neglected. For long transmission pulse lengths, a rectangular shape is a good approximation to the actual sample function shape. For shorter pulse lengths, the rectangular shape is still a fair approximation if the width is taken at the half power points of the actual sample function curve. The length of the rectangular sample function is expressed as a fraction of the vessel radius, r. A second computation was also performed on the theoretical velocity profiles to include the effects of the boundary error. This convolution was carried out by reducing the length of the sample function at the walls to the length of the portion of the sample function within the vessel. These profiles are also calculated for various sample lengths for each n value.

Figure 43 shows both the convolution and boundary error effects for various sample function lengths on a velocity profile of order n = 2. Six sample function lengths are computed from 0.2r to 1.2r in 0.2r increments. Note the effect of increasing sample function length on the convolved velocity profile with boundary error can be considered as limits on the measured profile, which generally falls between the two extremes.

With more detailed knowledge of sample volume dimensions, frequency meter signal-to-noise characteristics, and returned signal intensity, it is possible to theoretically predict the effects of sample region extent on the velocity profile measurement and compensate for the errors introduced. A detailed mathematical description of this procedure has been given by

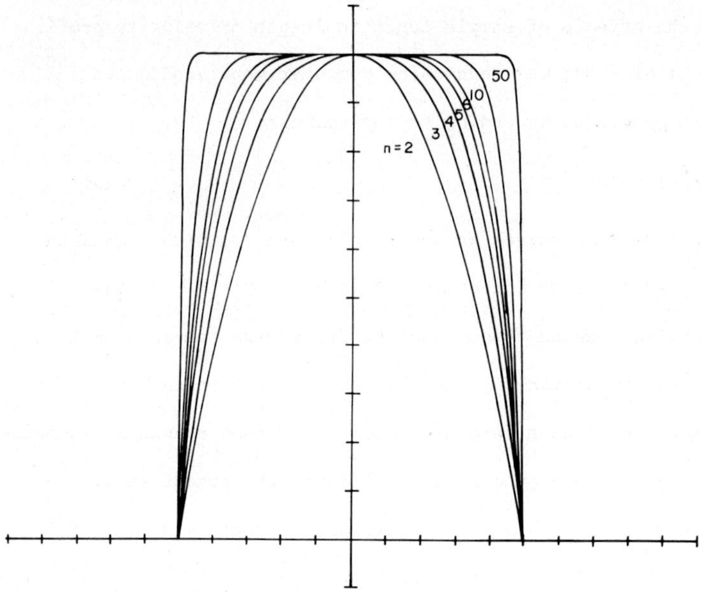

Fig. 42 Theoretical velocity profiles calculated from the equation 38.

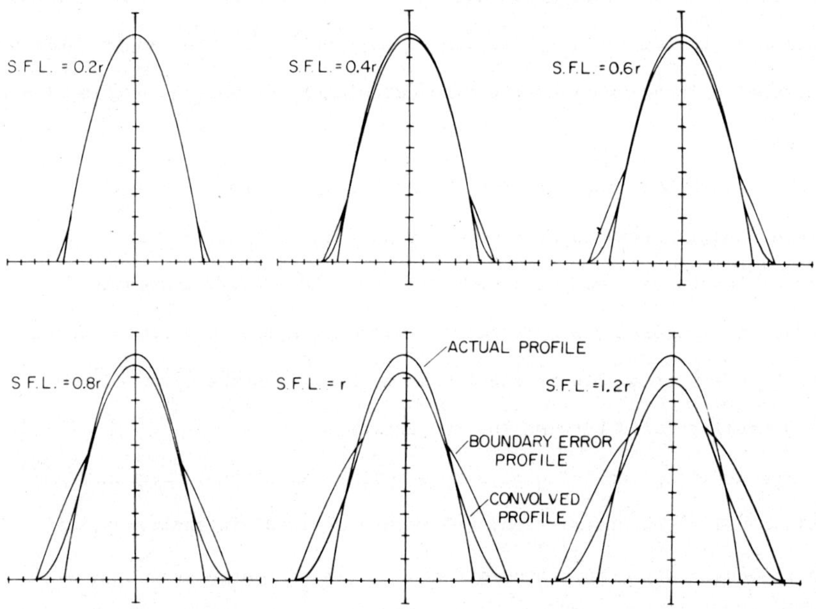

Fig. 43 Convolution and boundary error profiles for n = 2. The sample function length increases for each successive set of profiles by 0.2r.

Jorgensen, et al.[35] In practice, however, it is difficult, if not impossible, to predict the effect of all variables involved and it is reasonable to expect accurate profile measurement only in situations where the length of the sample function does not significantly alter the actual velocity profile. As seen in Figure 43, this situation applies when the sample function length is less than about 0.2 or 0.3 times the radius. For measurement situations outside this range, the experimenter must employ correction procedures with extreme care to achieve accurate velocity profile data.

c. *MEASUREMENT OF VOLUME FLOW RATE*. One of the most sought after blood flow parameters is volume flow rate. Volume flow rate measurements are necessary for determination of cardiac output, organ blood flows, limb and tissue perfusion, and a host of other physiologic parameters. The measurement of volume flow rate with ultrasonic Doppler devices involves deriving the necessary information from the appropriate velocity and dimension measurement.

The flow, Q, is defined as the mean velocity, v, over the entire cross sectional area of the vessel lumen, times this area, A.

$$Q = vA \tag{39}$$

If the velocity distribution over the cross section of the vessel is known, volume flow can be written as an integral.

$$Q = \int_A v(A)\, dA \tag{40}$$

where v(A) is the velocity at a point on the vessel cross section.

Unfortunately, in the case of ultrasonic blood velocity measurements, not all of the above parameters may be known. For continuous wave ultrasound measurements, provided the vessel is uniformly illuminated, the first moment of the Doppler shift power spectrum corresponds to v. The cross-sectional area, however, is not obtainable using CW techniques. With a pulsed system a narrow ultrasound beam through the center of the vessel can provide a measure of the diameter, thus allowing approximation of A, but the velocity distribution is then only known across this beam path. Therefore, CW Doppler measurements will require an independent measure of vessel geometry, while pulsed Doppler measurements involve assumptions concerning the velocity distribution over the area of the vessel cross section outside the beam axis.

Three basic techniques for obtaining volume flow from ultrasonic velocity measurements will be presented. All three can be implemented using a pulsed system with a variable receiver gate, which allows adjustment of the axial length of the sample volume. The first method employs uniform illumination of the vessel cross section similar to a CW ultrasonic technique. An independent measure of the vessel diameter must also be obtained. The second method uses the pulsed system to record blood velocity profiles which are then integrated to obtain volume flow. The third technique is a hybrid of the first and second and uses the pulsed device to measure the average velocity across a vessel along a narrow sound beam, thus eliminating the need for profile measurements. All three of these techniques can be carried out with single element transducers, introduced transcutaneously, to determine volume flow in underlying vessels.

1. *Uniform Illumination Method.* This technique involves separate determination of the mean blood velocity over the vessel lumen and vessel diameter. The determination of mean velocity is accomplished by exposing the entire vessel cross section to a uniform intensity sound field and then obtaining the first moment of the Doppler shift signal spectrum. The vessel diameter may be obtained by Doppler shift or echo imaging or by Doppler shift power scanning, to be discussed later.

Errors in applying the uniform illumination method are both experimental and systematic. Experimental errors are mostly due to improper selection of and positioning of the ultrasonic transducer. To provide a uniform intensity sound field, the transducer must have a larger beam cross-sectional area than the vessel. However, the beam area must not be so large as to include other adjacent blood vessels in its sound field. The sample volume of the pulsed Doppler can be set to include the entire vessel lumen by adjusting the duration of the receiver gate. This adjustment excludes unwanted signals from other vessels lying along the sound beam path. Since the sample volume of the pulsed

Doppler can be made somewhat larger than the vessel cross section, lateral and axial positioning errors are usually insignificant. The most important source of error due to probe positioning is the setting of the Doppler angle. This angle can be determined using one of the appropriate techniques described earlier.

Systematic errors are primarily due to signal processing and resolution. Since the sample region is quite large and can contain a wide range of velocities, the Doppler shift spectrum can also be quite broad. If zero-crossing detection is used to extract the mean Doppler shift frequency, significant errors can result. The error is profile dependent and is shown plotted in Figure 44 versus profile number, n, as defined in Equation 38. It can be seen from the figure that even at n = 12, corresponding to a very blunt profile, the volume flow error is approximately 4%. For parabolic flow, n = 2, the error is over 15%. These data demonstrate the need for a first moment processor for the Doppler shift signal.

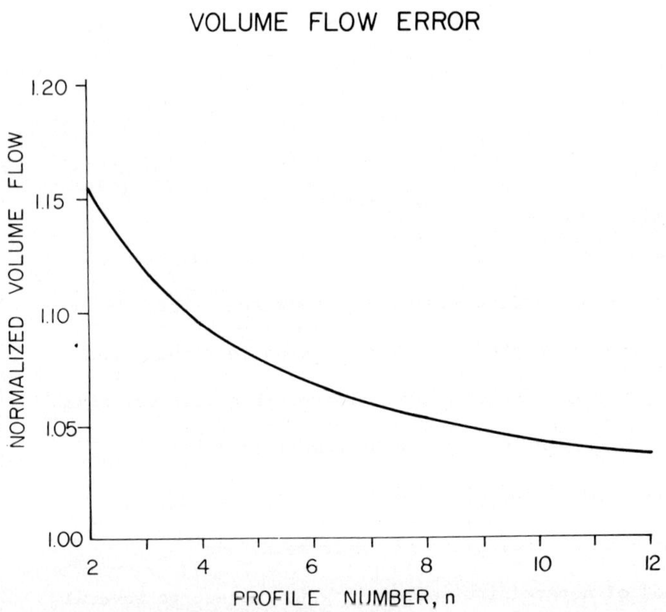

Fig. 44 Volume flow error introduced by zero-crossing detection for uniform illumination method. The profile number n is taken from Fig. 42.

The second source of systematic error in applying the uniform illumination technique derives from resolution errors in determining the geometry of the vessel cross section. The vessel diameter could be determined in a number of different ways, including Doppler signal power scanning, described later, echo scanning, and velocity profile measurement. Resolution errors associated with this latter method will also be discussed later.

Calculating the vessel cross-sectional area from the measured diameter assumes a cylindrical vessel shape. This assumption is reasonable for most subcutaneous vessels, however, care must be taken not to distort the vessel geometry by excessive pressure with the ultrasound transducer. Perhaps more significant is the error in diameter measurement, since the cross sectional area is proportional to the square of this measurement.

$$A = \frac{\pi}{4} d^2 \tag{41}$$

Let us assume that the actual diameter differs from the measured diameter by an amount Δd. The error in A is then given by

$$\begin{aligned} \Delta A &= \frac{\pi}{4} (d + \Delta d)^2 - \frac{\pi}{4} d^2 \\ &= \frac{\pi}{4} (2d\Delta d + \Delta d^2) \end{aligned} \tag{42}$$

The percent of error is

$$\frac{\Delta A}{A} = \frac{2\Delta d}{d} + \left(\frac{\Delta d}{d} \right)^2 \tag{43}$$

The error in calculated volume flow is also given by the above equation and is shown plotted as a function of $\Delta d/d$ in Figure 45. From the figure, we note that for a volume flow error of less than 10%, $\Delta d/d$ must be less than 0.05. For a 5 mm vessel, this amounts to a resolution of better than 0.25 mm, which is very demanding considering current diameter measurement techniques.

A list of the various types of errors associated with the uniform illumination technique along with an indication of their magnitude is shown in Table 1. While this method of determining volume flow is subject to several sources of error, it has the advantage of being one of the simplest and most

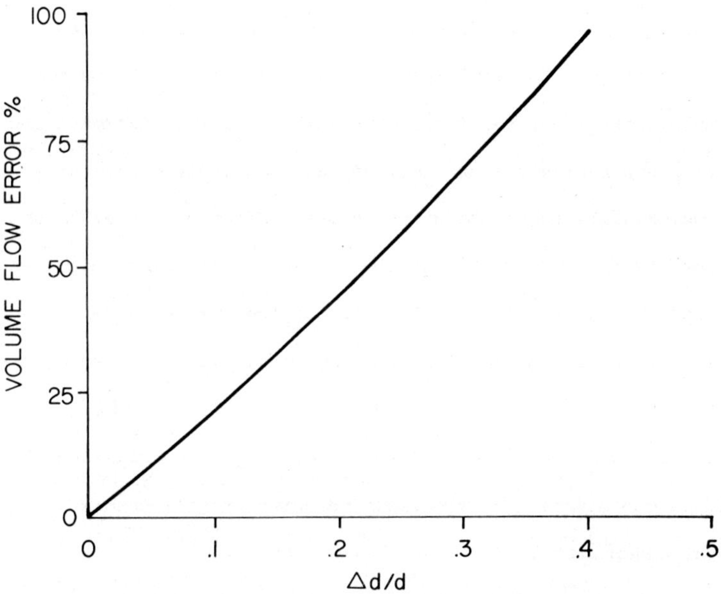

Fig. 45 Percent error in volume flow due to error in diameter measurement.

Table 1. Volume flow errors associated with the uniform illumination method.

Error Source	Magnitude
Doppler angle errors	$\sim \frac{\cos\sigma}{\cos(\sigma + \Delta\sigma)} - 1$
Zero-crossing detection	see Fig. 44
Noncylindrical vessel	~ distortion in diameter
Diameter measurement error	$\sim \frac{2\Delta d}{d} + (\frac{\Delta d}{d})^2$ (see Fig. 45)

direct techniques. Many of the error sources can also be eliminated or reduced with improved technology. The implementation of first moment processing can substantially improve the accuracy of the mean velocity determination. Improvements in resolving power of ultrasonic dimensional measurements are continually being made which will reduce the error in measurement of vessel cross-sectional error. Finally, the possibility exists of making real time measurements of a vessel cross section simultaneously with the mean velocity determination which, with appropriate signal processing, could yield a true ultrasonic volume blood flowmeter.

2. *Diameter Profile Integration Method.* A second technique for measurement of blood flow involves determination of the blood velocity profile across a diameter of the vessel under study.

Several instrument assumptions, including physiologic assumptions, are involved in extrapolating this data to volume flow rate. First of all, the vessel geometry is assumed to be cylindrical over the measurement region. Secondly, the blood velocity field should have a degree of radial symmetry about the vessel axis, specifically, the measured velocity profile should be valid for rotation about the vessel axis by $\pm 90^{\circ}$. Thirdly, the blood velocity vector is assumed to be parallel to the vessel axis at all times during the cardiac cycle, maintaining a constant Doppler angle to the sound beam. These latter two assumptions imply that the profile integration technique is not suitable for highly skewed velocity fields.

If the above conditions are met, the instantaneous volume flow can be found by dividing the vessel cross section into a series of concentric rings corresponding to the number of measurement locations (see Figure 46). The blood velocity at each location can be written as $\nu_i(t)$, where i ranges from 0 to n, the number of measurement locations. The volume flow is found by weighting each velocity by the area of the half-ring containing the measurement location and performing a summation. The area of the half-rings can be written in terms of the measurement increment, Δx. The center half-circle has an area (a) of

$$a = \frac{\pi}{2} \left(\frac{\Delta x}{2} \right)^2 \tag{44}$$

The area of the first half-ring is 8a, the second 16a, the third 24a, and so on. If we designate the center measurement location as the mth point, then the instantaneous volume flow can be written as

$$Q(t) = \sum_{i=0}^{n} 8\nu_i(t) \; |m-i| \; a + \nu_m(t)a \tag{45}$$

The average flow is found by integrating the above formula over a complete heart cycle, t_h.

$$\overline{Q} = \frac{\int_0^{t_h} \left[8\nu_i(t) \; |m-i| \; a + \nu_m(t)a \right] dt}{\int_0^{t_h} dt} \tag{46}$$

While the above calculations may seem quite tedious, they can be performed accurately and quickly on a digital computer, if the velocity waveforms are appropriately digitized.

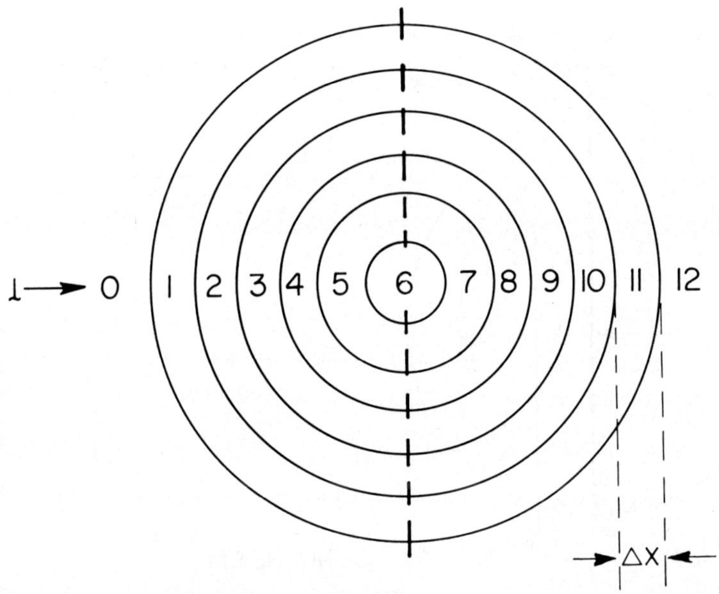

Fig. 46 Partitioning of vessel cross section into cylindrical shells for volume flow calculation. The diameter of the central element is Δx.

Experimental errors in applying the diameter profile integration technique include those errors previously mentioned in obtaining accurate velocity profile data. A further experimental error which affects volume flow calculation is misalignment of the sound beam across the vessel under study. If the sound beam is off the vessel axis, as shown in Figure 47, the calculation of volume flow is affected. This type of error is profile dependent and is at a minimum for uniform velocity field. Even for this case, however, the error in volume flow calculated goes up as the square of $\Delta r/r$.

$$\text{percent error in volume flow} = \left(\frac{\Delta r}{r}\right)^2 \times 100 \tag{47}$$

For a parabolic profile, the percent error increases to $2(\Delta r/r)^2 \cdot 100$ for small $\Delta r/r$. With these two cases as bounds on the error, we see that for a misalignment of only $\Delta r = 1/4r$, the random error in computed flow lies between 6.25 and 12.5 percent.

A second category of volume flow measurement errors are those introduced

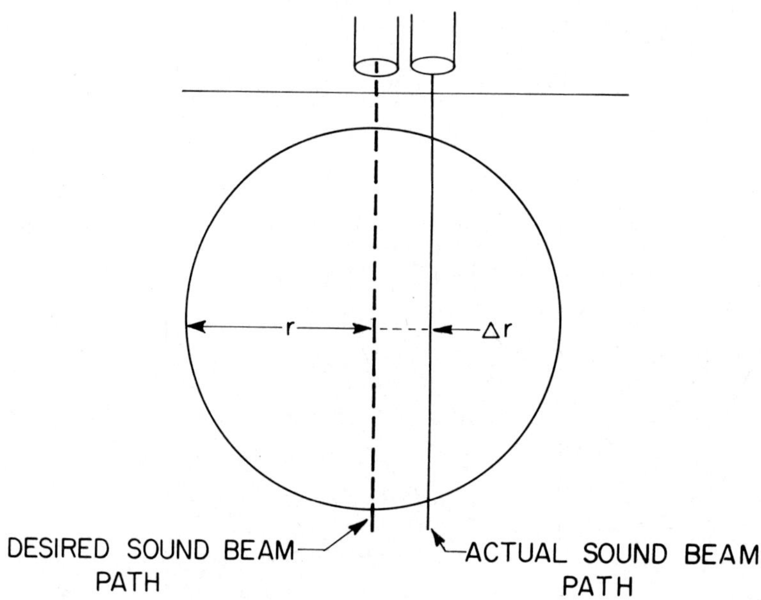

Fig. 47 Transducer beam offset, Δr, introduces error in velocity profile measurement and computed volume flow.

by resolution limitations of the pulsed Doppler velocity meter. As previously discussed, the finite extent of the sample volume leads to profile distortion effects which are quite significant for sample function lengths greater than 1/4 the vessel radius (refer to Figure 43). Next, we must compute the effects of these profile distortions on volume flow calculations.

A problem that immediately arises is where to specify the vessel wall locations. If the measured profile is the only data available, some method of wall location based on that profile must be constructed. One such method is to specify the wall location as the point at which the velocity goes to zero. This method is suitable for small sample function lengths, but obviously introduces rather large errors for lengths approaching the radius. An alternative method is to extrapolate the maximum and minimum profile slopes to zero velocity and take these points as the wall locations. This method reduces the error due to the convolution "tail" on the actual profile and allows wall determinations in noisy environments which disturb the baseline, but is also profile dependent and difficult to apply in irregular velocity fields. Occasionally, information on vessel diameter or wall locations is available from an independent source, such as echo imaging or ultrasound power scans. When such data is available, we can truncate our measured profiles to these new wall locations for purposes of calculating volume flow.

Taking into consideration the above method for determining the vessel wall locations, volume flow was computed from convolved profiles and convolved boundary error profiles using six different situations. For both profiles, flow was computed 1) over the entire width of the profile, 2) over the width of the in vessel profile, assuming an independent measure of the wall locations, and finally over a width determined by extrapolating the maximum and minimum profile slopes to the baseline.

Since the convolved velocity profile and the convolved boundary error velocity profile represent bounds on the measured velocity profile, the volume flow calculated from these curves should also represent the range of computed volume flow. Figures 48 through 50 are graphs of the volume flow error range

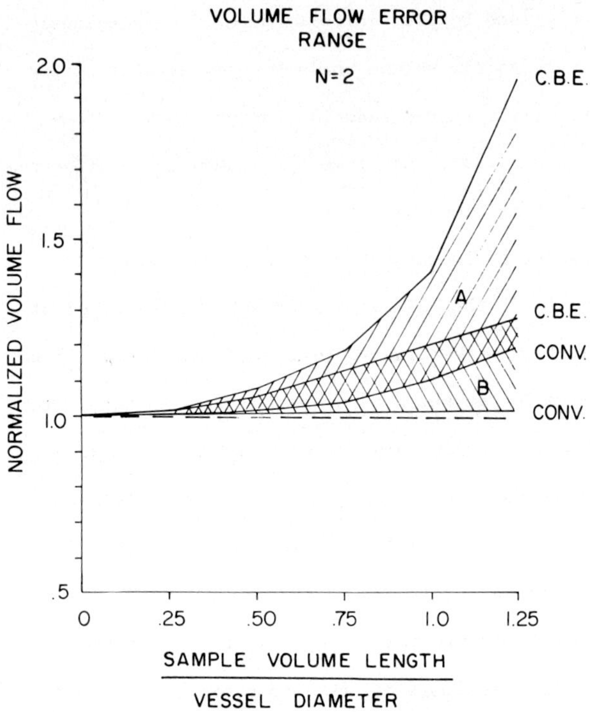

Fig. 48 Error range for computed volume flow for a parabolic velocity profile (n = 2). Error range B applies when the wall locations are known exactly, and error range A applies if the wall locations are computed from the maximum and minimum profile slopes.

as a function of sample function length for various n values. Only two error ranges are graphed - that from computing the vessel wall locations from the maximum and minimum profile slope, and that from calculations using independent measurements of the wall locations which are assumed to be exact.

As shown in Figure 48 for the parabolic profile case, the error ranges are both above the normalized actual volume flow line of 1.0. As the sample function length is increased, the error ranges increase dramatically. Figures 48 and 49 show the same error ranges for higher values of n. Note that as n increases, the error range for the computed wall location volume flow increases more rapidly with increasing sample function length. Somewhat unexpectedly, however, the error range for volume flow, as calculated from the profiles with an independent measurement of the wall location, tends to remain constant with increasing n, and even moves down to symmetrically enclose the actual volume flow lines.

Several conclusions can be made from the data in Figures 48 through 50. First of all, it is apparent that for sample function lengths less than 0.25r, the volume flow calculation from profile integration is quite accurate, regardless of the method of locating the vessel walls. Assuming a minimum practical sample function length of 1.5 mm, the technique should be valid, from the standpoint of resolution, in vessels 12 mm in diameter and larger. Since most vessels that are accessible transcutaneously are considerably smaller than this, we must consider the situation of larger relative sample function lengths. For these situations, it is apparent that the method of determining the vessel wall locations from the profile slope can lead to rather large errors in volume flow,

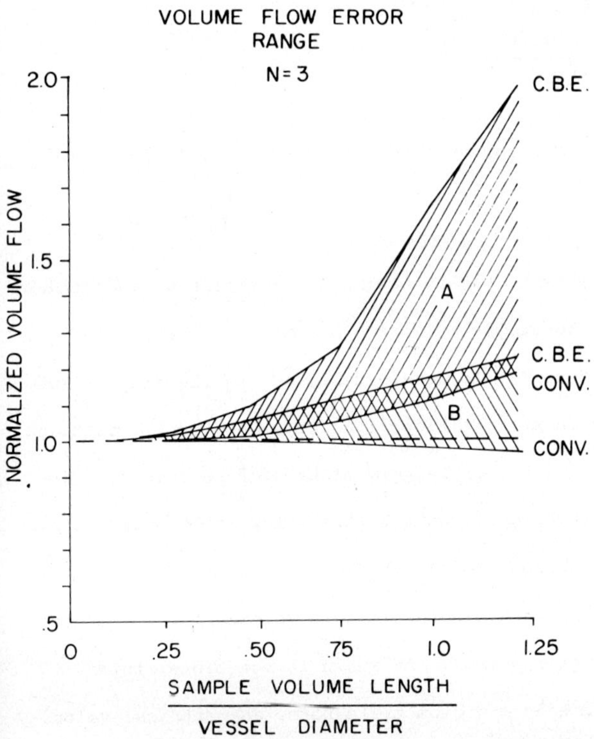

Fig. 49 Error range for computed volume flow from velocity profile of order n = 3. Error ranges are as defined in Figure 48.

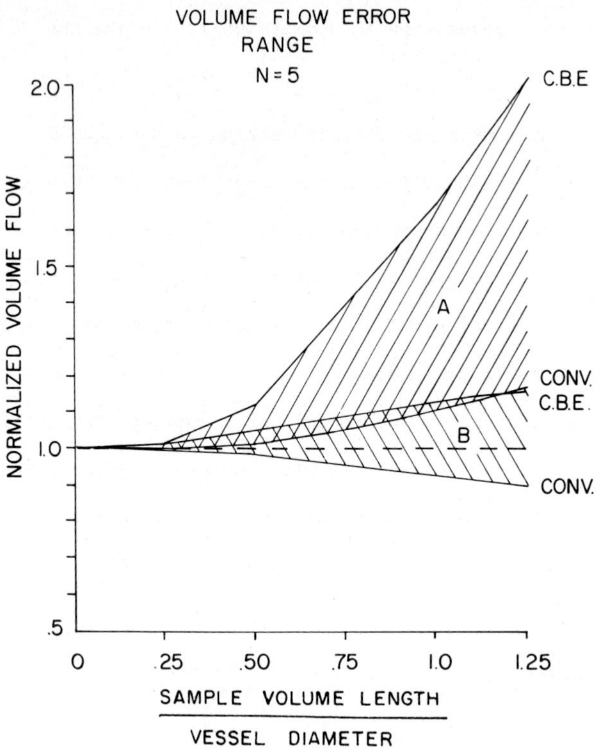

Fig. 50 Error range for computed volume flow from velocity profile of order n = 5.

even with ideal profiles. It appears highly desirable to obtain an independent measurement of the vessel wall locations by some other more accurate technique. Finally, we see that accuracy depends, to some extent, on profile shape. For volume flow calculations based on profile slope measurements of wall locations, the error is likely to increase with increasingly blunt profile shapes. For the independent wall location technique, however, the error in computed volume flow is likely to decrease with blunt profile shapes.

Table 2 summarizes the error sources inherent in volume flow measurement by the diameter profile integration method. As shown in the table, this method is subject to probe alignment and geometry errors, asymmetrical velocity field errors and resolution errors. The magnitude of the geometry and resolution errors indicate that this technique is best suited for large blood vessels

Table 2. Errors in volume flow computer by diameter profile integration method

Volume Flow Errors	
Error Source	Magnitude
Velocity profile not across true diameter	~ cos (Doppler angle, σ)
Physiological variation during scan*	~ time required for scan
Noncylindrical vessel	~ distortion in diameter
Nonsymmetrical velocity profile	~ degree of skewing
Blood velocity vector not parallel to vessel axis	~ function of off axis angle, radius of vector and duration of error during cardiac cycle
Beam diameter, d too large for vessel radius, r	~ function of $\frac{d}{r}$, small for $\frac{d}{r} < .25$
Sound beam not aligned with vessel axis	~ $(\frac{\Delta r}{r})^2$ for uniform velocity field
Doppler angle error	~ $\frac{\cos\sigma}{\cos(\sigma + \Delta\sigma)} - 1$
Resolution errors	~ function of $\frac{\text{sample function length}}{\text{radius}}$ and of wall location method

*single gate pulsed Doppler

on the order of 10 mm diameter and above. The small size of subcutaneous vessels accessible to the pulsed Doppler suggests that the profile integration technique is not the preferred technique for accurate transcutaneous volume flow measurements.

3. *Diameter Gate Average Velocity Method*. This technique for obtaining volume flow is in effect a hybrid of the uniform illumination method and the diameter profile integration method. It has the advantages of the uniform

illumination method in that it is simple and requires no computer processing. It shares with the diameter profile integration method the ability to measure vessel diameter with the same transducer used for velocity measurements. Unfortunately, the method also shares many of the error sources of both the other techniques.

The diameter gate average velocity technique utilizes a small transducer compared with the size of the vessel (again accuracy demands a transducer beam width $< 0.25r$). The transducer is oriented with respect to the vessel at a known Doppler angle so that the sound beam intersects the vessel axis. The gate of the instrument is then set to encompass the vessel lumen. If the flow characteristics are constant over the axial length of the vessel region under illumination, the average velocity along this thin sound beam represents the average velocity across the diameter of the vessel. If the gate is then narrowed and set to measure the peak velocity at the vessel centerline, the mean velocity for the entire vessel can be computed. A measurement of the diameter then allows calculation of total volume flow.

The mean velocity over the entire vessel cross section for profiles of the form

$$\nu(r) = \nu_p \, (1-r^n) \quad 0 \leq r \leq 1 \tag{48}$$

can be found by evaluating the surface integral over the region $r = 0$ to $r = 1$. The result is

$$\bar{\nu} = \nu_p \left(\frac{n}{n+2} \right) \tag{49}$$

The average velocity along a narrow sample beam across a diameter is given by the integral

$$\bar{\nu}_c = \nu_p \int_0^1 (1-r^n)\,dr \tag{50}$$

$$\bar{\nu}_c = \nu_p \left(\frac{n}{n+1} \right) \tag{51}$$

Solving for n in terms of $\bar{\nu}_c$ and ν_p .

$$n = \frac{\bar{\nu}_c}{\nu_p - \bar{\nu}_c} \tag{52}$$

Substituting the above equation for n in Equation 49, we obtain the mean velocity in terms of $\bar{\nu}_c$ and ν_p.

$$\nu = \frac{\nu_p \bar{\nu}_c}{2\nu_p - \bar{\nu}_c} \tag{53}$$

The above derivation is based on several assumptions. First of all, the velocity field is assumed to be axial symmetric, that is, the velocity profile is constant with respect to rotation about the vessel axis. Secondly, the profile shape must be describable by the mathematical form given in Equation 48. As for the diameter profile integration method, these assumptions preclude the application of the method to skewed velocity fields.

Experimental errors inherent in the diameter gate average velocity technique are much the same as those for the diameter profile integration method. The transducer must be oriented at a known Doppler angle, minimizing the offset of the sound beam from the vessel axis. The beam diameter must be equal to or smaller than 1/4 the radius. Once again, care must be taken not to distort the vessel shape by excessive pressure on the probe.

Systematic errors are also involved, as with the other techniques. Since the Doppler shift spectrum is broad, due to the range of velocities present, errors in converting frequency to voltage using zero-crossing detection result. The error in volume flow rate is essentially the same as that for the uniform illumination case, illustrated in Figure 44. Other systematic errors result from resolution errors in the determination of peak velocity and vessel diameter.

Table 3 lists the error sources associated with the diameter gate average velocity method. The large number of experimental and systematic error sources raise doubt as to the accuracy of this technique, especially in small vessels with skewed velocity fields. This method's greatest advantage lies with the ability to obtain volume flow estimates with a narrow beam width transducer without computer processing of profile data. Since the same small transducer

Table 3. Error source in volume flow determination by diameter gate average velocity method

Error Source	Magnitude
Measured velocity not across true diameter	~ velocity variation along vessel
Noncylindrical vessel	~ distortion in diameter
Nonsymmetrical velocity profile	~ degree of skewing
Blood velocity vector not parallel to vessel axis	~ function of off axis angle, radius of vector and duration of error
Beam diameter, d, too large for vessel radius, r	~ function of $\frac{d}{r}$; small for $\frac{d}{r} < .25$
Sound beam off vessel axis by Δr	~ $(\frac{\Delta r}{r})^2$ for uniform velocity field
Doppler angle error	~ $\frac{\cos\sigma}{\cos(\sigma + \Delta\sigma)} - 1$
Zero-crossing detection error	see Figure 44
Resolution error in diameter measurement	~ $\frac{2\Delta d}{d} + (\frac{\Delta d}{d})^2$

may be used to obtain a vessel diameter estimate, all essential data for computing volume flow can be acquired from a single transducer element, suggesting the possibility of multiplexing schemes. It is also possible that with better means of processing the Doppler shift spectrum, the peak frequencies, corresponding to peak velocity, could be extracted simultaneously with the mean, from which volume flow could be calculated using an analog or digital processor.

d. *VESSEL AREA MEASUREMENT*. In order to obtain accurate estimates of volume flow rate from average velocity measurements, the cross-sectional area of a vessel must be precisely determined. For flow measurements in the arterial circulation, the vessel cross section can be assumed to be circular, provided there are no external pressures applied. This reduces the area measurement to a diameter measurement, which is more easily accomplished.

Several techniques for determining vessel orientation and Doppler angle have been discussed, which included echo and Doppler imaging techniques. These same techniques may also be applied to the determination of vessel diameter. Echo measurement of vessel diameter has been successfully employed by Hokanson, et al.[57] and others.[58] The pulse widths employed in echo systems can be considerably shorter than those utilized in Doppler systems, allowing improved axial resolution. The technique involves simply determining the time differential between pulses returning from the near and far walls of a vessel. Pulsed echo diameter measurement can also be carried out using echo scanning devices.

The accuracy of echo imaging techniques for measurement of vessel diameter is difficult to define, due to a lack of knowledge concerning the precise characteristics of vessel wall reflections. It is generally assumed that the boundary between the endothelium and the fluid blood is the source of the strongest echoes, but the reflections off the outer layers of the arterial wall result in a smearing out of the total reflection. This uncertainty in echo source has led to development of other diameter measuring techniques involving Doppler detection.

The step response of pulsed Doppler velocity meters provides an accurate means of determining wall locations. This location is defined as the point at which the sample function is precisely centered on the wall and therefore half way into the flow stream. With the sample volume only half way into the vessel flow stream, the power scattered from the moving blood cells is one half that obtained when the volume is totally within the lumen. A power recording from the Doppler audio signal can be made while the sample region is scanned across the vessel diameter, and from the 1/2 power points on this curve, the diameter is determined.

Accuracy of the technique would seem to depend on the assumption that the sample volume is axially symmetric, but the error affects each wall location similarly, and not the difference measurement. Accuracy is also not resolution limited, but depends mainly on one's ability to locate the 1/2 power points of the power scan. The ability to locate these 1/2 power points is limited primarily by signal-to-noise ratio, wall motion, range attenuation, and vessel curvature. The quality of the power scan is affected by poor signal-to-noise ratios, but typically S/N ratios on the order of 20 to 50 db are obtainable in transcutaneous vessels. While motion tends to "blur" the wall location, a slow scan will provide a mean diameter. Range attenuation can be eliminated by range gain compensation, transducer design, or graphical correction of the power scan record. Finally, vessel curvature appears to be a problem only in vessels that are small compared to the beam diameter and for beam diameters up to 0.25 times the radius, the curvature is less than 1%. These limitations restrict the experimental accuracy of the wall location determination to about 1/4 the sample function length giving a diameter error on the order of $\sqrt{2/4}$ times the sample function length. For a 1 mm sample function length, the error in vessel diameter amounts to about $\pm$ .4 mm, or 10% of a 4 mm diameter vessel. While this error seems large, it represents possibly the best accuracy obtainable using ultrasonic techniques. With improved transducer and instrumentation design, it is possible this error will be further decreased.

5. CLINICAL APPLICATIONS OF DOPPLER TECHNIQUES

Most applications of Doppler devices for clinical diagnosis have been based on their use as qualitative instruments. Quantification, whether flow velocity acceleration or volume flow rate, has not yet found its way into clinical medicine. There are several reasons for this, the primary one being that practical quantitative techniques have not yet been developed and evaluated. Secondly, the potential role of quantitative blood flow velocity or volume rate measurements in clinical medicine is not fully appreciated for lack of familiarity with this parameter. Because of this, the demand or need for such developments, even though potentially useful, is somewhat uncertain.

Another important factor affecting the introduction of ultrasonic flow sensing as a clinical tool is the fact that blood flow measurement concepts and its role in medicine is not emphasized in clinical training, so that the physician does not think in terms of flow parameters. As a consequence, his demands for these devices and his potential use of them is low. As the diagnostic routines shift gradually from anatomical considerations toward physiologic considerations, we can anticipate an increasing use of flow detection and flow measurement ideas.

The principal applications of Doppler techniques to date are in the cardiac and peripheral vascular diagnostic specialties. Pulsed Doppler devices are used in both areas with continuous wave devices being of more usefulness for the peripheral vascular examination.

Current qualitative applications involve two aspects of blood flow evaluation. The first is the detection and localization of normal versus abnormal flow patterns. For example, this approach applied to the evaluation of peripheral circulation could mean a comparison of flow signals in the right femoral artery to those in the left femoral, or detection of velocity changes along the course of a vessel, as affected by stenosis or occlusion. Another growing peripheral vascular application of Doppler devices is the imaging of the flow lumen to show the location and alterations in the anatomy of a vessel. These latter applications are limited to a number of specialized laboratories, since the devices have only recently become available.

The cardiac applications of pulsed Doppler devices are primarily concerned with the detection and evaluation of blood flow disturbances resulting from valve or septal defects. Quantitation of intracardiac shunts, valvular insufficiencies, etc., is still to be evolved.

5.1 Arterial Occlusive Disease

This complicated disease process manifests itself in a variety of ways, most of which end up affecting blood flow behavior. The disease is one which goes through a lengthy development, beginning with small defects on or in the wall of the vessel. These may be in the form of fatty streaks or, later, ulcerated plaques. These early defects might not alter the flow patterns although they produce some type of flow disturbance. As the disease progresses, the formation of a soft plaque appears evident. These may intrude into the lumen of the vessel and begin to impair the flow rates. Eventually, these plaques begin to localize or form into some type of calcification, which ultimately can form a hard, well defined plaque intruding into the lumen of the vessel to totally block blood flow. This process may extend over an entire lifetime. It is important to detect these defects early. Blood flow analysis, in the terms of velocity or flow disturbances, appears to have potential to detect these changes and make a significant contribution to this difficult diagnostic problem.

Continuous wave Doppler devices are currently in routine clinical use in many peripheral vascular diagnostic laboratories for detecting the effect of these flow alterations.[59] Most of these applications depend upon auscultatory analysis of the Doppler signal. The medical conclusions are based on either the presence or absence of a flow signal or alterations of the signal in the form of high-pitched sounds due to acceleration or high velocities through narrow orifices. Other aspects of this signal analysis include the duration of the peak velocities, and the direction of flow at particular sites. With some experience, it is possible, using these techniques, to detect a partial occlusion by comparing the Doppler signal, for example, from the left carotid to that in the right carotid, or from the left femoral to the right femoral. Analog recording of these signals, while it may seem to be more logical, are generally not made. The clinical diagnosis is usually made at the time of the procedure by either the physician or sonographer, and is simply entered into the patient's record.

One important application of the continuous wave Doppler device is the assessment of cerebral vascular circulation as may be affected by carotid artery disease.[60] The arterial circulation to the brain is very complicated, with as many as six main input points. This disease is most frequently manifested by occlusions or plaques forming at the bifurcation or branch of the common and internal-external carotid arteries in the neck. When a blockage occurs in any of these vessels, often unpredictable but definite redistribution of blood flow usually occurs. Techniques have evolved using the continuous wave Doppler systems to assess the presence of carotid artery occlusions by evaluating this redistribution of flow.

A simple example can be used to demonstrate this procedure. In Figure 51 we see in schematic form the common carotid artery branching into the internal and external carotid arteries. The internal carotid blood flow moves primarily to the brain, a portion of this flow coupled with flow from the external carotid form, via small branches, a common supply to the eye. If these vessels have normal patency in the patient, their respective flow velocities will be toward the eye as shown in Figure 51a by the arrows. A continuous wave Doppler probe placed over the eye will detect this and indicate the proper flow direction. However, in Figure 51b, when the internal carotid may be totally or partially

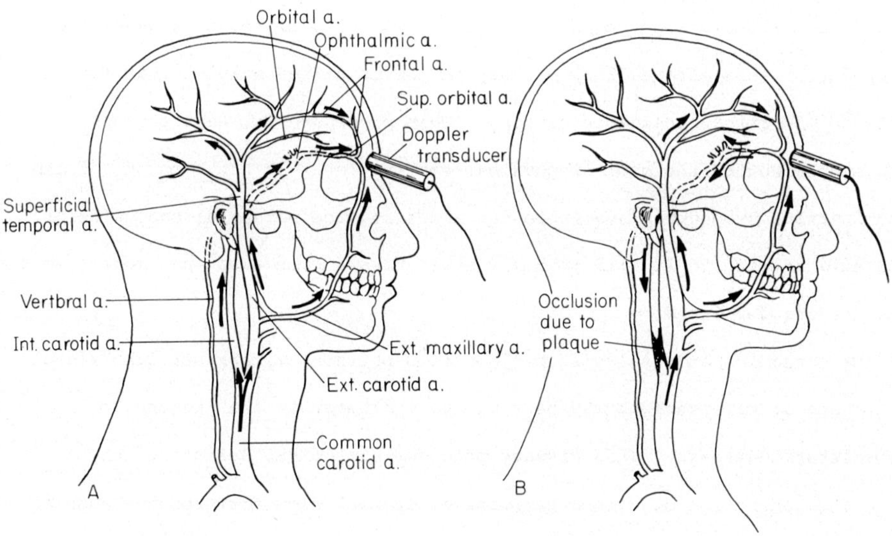

Fig. 51 A total or partial occlusion of the internal carotid artery, (B), can be detected by sensing the direction of blood flow through the supraorbital artery. The direction of flow ordinarily would be from the internal carotid anterior coming out over the eye. With an occlusion, this flow is reversed coming from either the external maxillary artery or frontal artery to flow in a reverse direction.

blocked, a number of things happen which can be evaluated using a simple Doppler.

In Figure 51b, if there is a partial occlusion of the internal carotid, the normal flow patterns will be altered such that the primary flow to the brain will come via the external carotid to the eye, and then back retrograde into the terminal branch of the ophthalmic and into the brain. If the external carotid artery is manually compressed, the flow signal detected in the ophthalmic artery will be diminished, or may be completely stopped. When the compression of the external carotid is relaxed there will be an augmented or high velocity flow signal detected in the ophthalmic artery. If the internal carotid is completely occluded, and it has been in this state for some time, a well developed retrograde flow will be moving from the eye through the ophthalmic artery towards the brain, in which case the flow signals of opposite polarity will be detected.

The test may not always be conclusive because of the very complicated intercommunication of the peripheral branches of these vessels. This pattern may occur normally in some people. In others who actually have disease, the development of this altered flow pathway may occur differently so that the results do not fit a repeatable pattern. This procedure is, nevertheless, a useful supplement to the overall physical examination. It has been one of the few noninvasive procedures available for assessing the potential occlusion of the carotid arteries. It will probably still have a role even when more specific techniques are developed.

The detection and analysis of blood flow patterns within straight vessel segments and at bifurcations may be a useful indicator of the presence of atherosclerotic disease. This disease produces anatomical defects which may lead to the generation of either temporal or spatial flow disturbances such as skewed velocity profiles or the generation of turbulent wakes. An example would be the case of an ulcerated plaque which tends to form a pit or alteration in the wall of the vessel. This type of plaque may not occlude the flow in any way, however, it may generate flow disturbances. The clinical problem resulting from this kind of defect will be to produce thrombi, which may pass downstream and be the source of transient ischemic attacks or potential strokes.

Pulsed Doppler devices show particular potential for the detection and evaluation of flow disturbances as indication of these wall defects. Felix, et al[61] have shown the use of pulsed Doppler for the detection of these disturbances. The spectral pattern, taken from Felix, of the Doppler shift signal resulting from normal or disturbed flow is shown in Figure 52. If the velocity gradients are small within the sample volume, this will produce a narrow band of Doppler shifts and show a characteristic dome shaped to the temporal flow pattern, Figure 52a. If the flow is disturbed so that there is a wide range of velocities present and they are moving in many different directions, then the dome shaped spectral pattern will be altered and filled in, Figure 52b. One would anticipate, then, that smooth flow should be proxi-

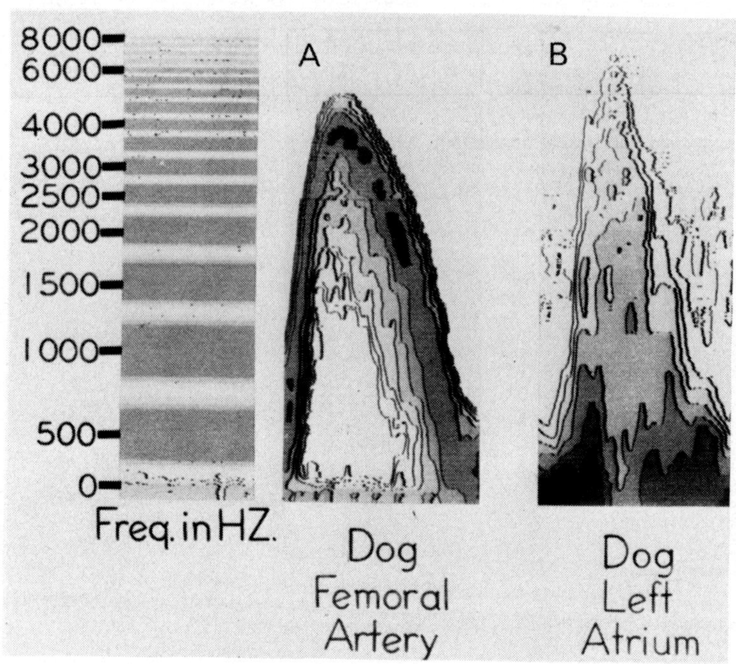

Fig. 52 The Doppler spectrum is shown as a narrow band pattern when there are small or few velocity gradients within the pulsed Doppler sample volume, dog femoral artery. Confused or disturbed flow shows as a filled-in pattern, dog left atrium. (From Felix, et al.[61])

mal to an occlusion and that if the degree of occlusion was sufficient, turbulence would be produced with a broad band spectral pattern, detected distal to the stenosis. Eventually this pattern should return to a normal or smooth flow pattern some distance downstream. Figure 53 demonstrates this measurement in a patient where the pulsed Doppler was used to evaluate the flow spectra in 2 cm increments, starting from the point of occlusion and stopping downstream. Notice that smooth or dome shaped flow spectral patterns just proximal to the partial occlusion become completely disturbed in the first 2 to 10 cm and eventually restores itself by the time it has reached 18 cm. The example used here is somewhat gross, but it serves to indicate the potential role of Doppler flow velocity spectra analysis as an early and sensitive indicator of small wall defects.

Doppler devices, both continuous wave and pulsed, can be used to generate images of the actual vessel flow lumen in a number of views at the bifurcation of these arterial vessels. The simplest form of Doppler imaging has been developed by Reid, et al.[53] This method utilizes continuous wave Doppler transducer and electronics. Figure 54a shows in schematic form the principles of their method. A Doppler transducer, having a focused beam with beam width

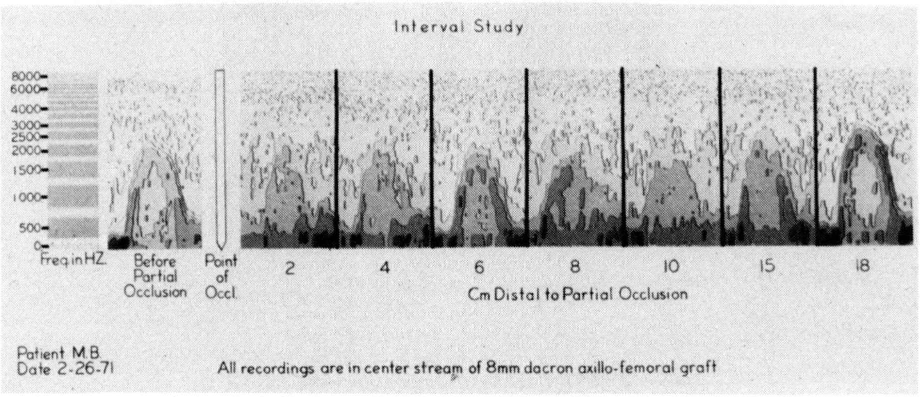

Fig. 53 The smooth dome-shaped spectral pattern analyzed from the Doppler signal taken just proximal to the partial occlusion becomes filled-in distal to the orifice. Finally, 18 cm downstream, the flow regains its smooth characteristics. (From Felix, et al.[61])

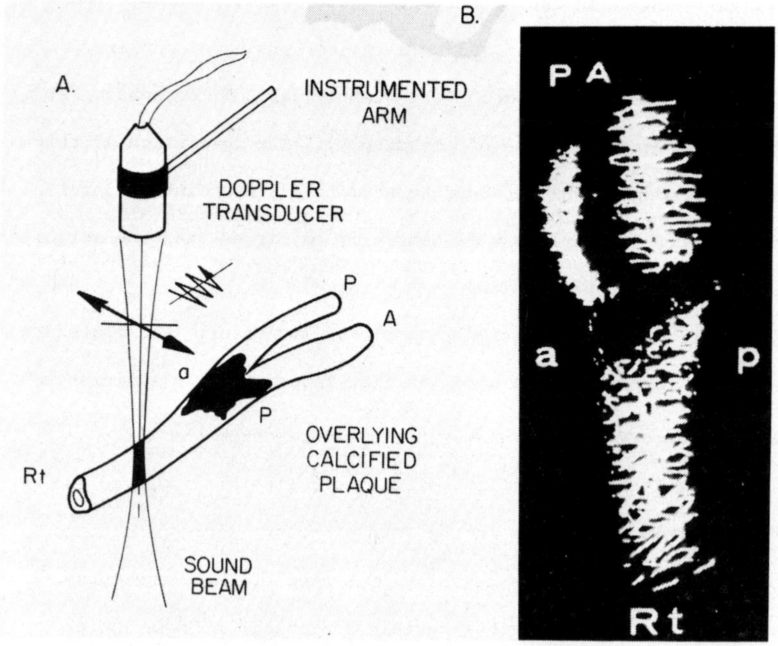

Fig. 54 A continuous wave Doppler can be used to develop images of a blood vessel arterial bifurcation by scanning the transducer transversely across the branch. At each point that flow is detected above a certain velocity threshold, a spot or line is stored on the oscilloscope display, (B). The resolution is dependent upon the beam width at the point of intersection with the vessel.

of a millimeter or so, is moved laterally across the vessel of interest. The position of the transducer is detected using an instrumented arm. The output from this arm controls the position of the spot on a cathode ray tube storage oscilloscope. As the transducer is moved back and forth laterally across the vessel of interest, an image will be produced of the type shown in Figure 54b. The image shows those points within the vessel where the blood flow velocity was detected above a certain threshold value. The somewhat cross-hatched appearance of the image is produced by each swing of the transducer through the vessel. In the example there is an overlying calcified plaque at the point of bifurcation. This plaque attenuates the sound beam to the point that Doppler signals cannot be detected beneath it. However, in the example, the vessel is still sufficiently patent to allow blood flow distal to the plaque. Therefore, the image in Figure 54 shows the two branches intact with a break across the image resulting from attenuation due to the intervening calcification. The images produced by this method are limited to a type of projected outline analogous to a conventional X-ray image. Consequently, all of the anatomical features will be mapped into the plane of the display. The resolution of this scanning system is dependent upon the sound beam width at the point of interception with the vessel. Such images will generally be larger than the actual vessel by at least 1/2 of the beam width.

This method is technically relatively simple to implement. The procedure takes 30 to 45 minutes to perform and provides flow information in addition to the image. The added information is simply not available from the X-ray arteriogram.[62] The magnitude of the flow velocity can also be displayed in the image by showing these variations in the form of colors corresponding to a range of velocities. This idea was recently presented by White and Curry.[63]

In summary, this technique provides certain anatomical information as well as gross flow data. The presence or absence of a stenosis or occlusion can be identified in most cases and the presence of flow disturbances that might result from an ulcerated plaque can be detected by auscultatory analysis.

A procedure similar to that used to evaluate the arterial flow is used with the venous system. For example, evaluation of the patency of valves in the veins of the legs can be accomplished using the continuous wave Doppler by noting whether or not reverse flow occurs below the valve following compression of the vein on the side nearer the heart.

A more complete evaluation of the status of the vessel bifurcation can be carried out using pulsed Doppler imaging, which has a number of advantages over the continuous wave approach.[64] In this case the image resolution is set by the dimensions of the sample volume which may be as small as 1 x 1-1/2 mm for a high resolution 5 MHz system. The additional advantage provided by the pulsed Doppler is the fact that it has range resolution. This permits mapping the vessel as either a cross-sectional or longitudinal image and with a variety of orientations. The continuous wave devices, since they lack range resolution, are unable to produce cross-sectional images of vessels, nor are they able to image bifurcations where the vessel may appear more than once along the beam axis.

The physical arrangement of the transducer scan arm for the pulsed Doppler scanner is very similar to the continuous wave system. Some of the differences are shown in Figure 55. In the pulsed Doppler case, the transducer can be moved both in an X and Y direction, as well as permitting rotation, angle θ, of the transducer about the axis of the arm. One of the primary purposes of the arm is to constrain the Doppler transducer so that it can only move within a plane. This plane must be oriented to intercept the vessel or bifurcation in the desired view. In the case of the pulsed Doppler, the range gate is set to intercept the vessel at the desired depth. During the course of the scan, a point is plotted on the oscilloscope face at each location the flow velocity exceeds the threshold level. Figure 56 shows an example of

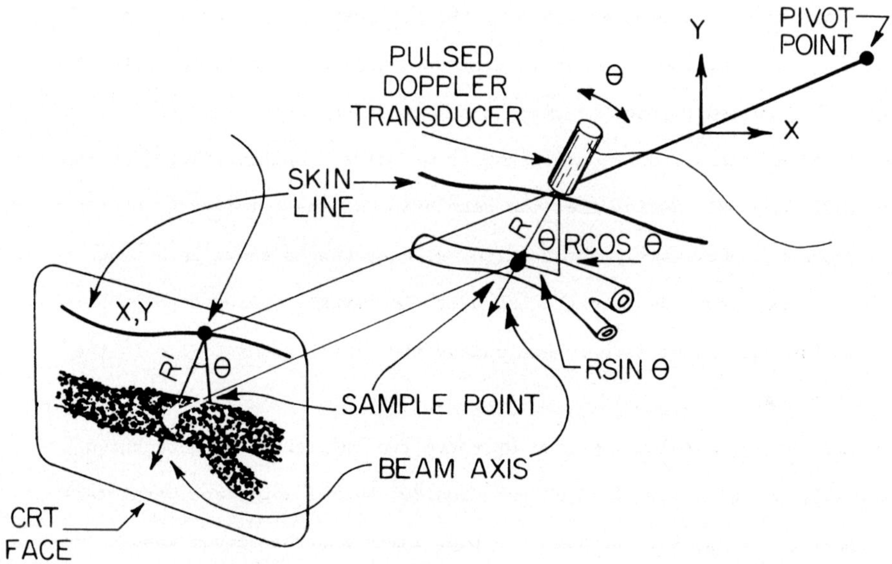

Fig. 55 A pulsed Doppler can also be used to generate flow images. The transducer scans in a plane as constrained by a mechanical arm. The position of the sample point is sensed by potentiometers attached to the arm and are used to position the beam on the cathode ray tube screen. At each point that flow exceeds a certain threshold value, a dot will be written, thus producing an image of the vessel lumen.

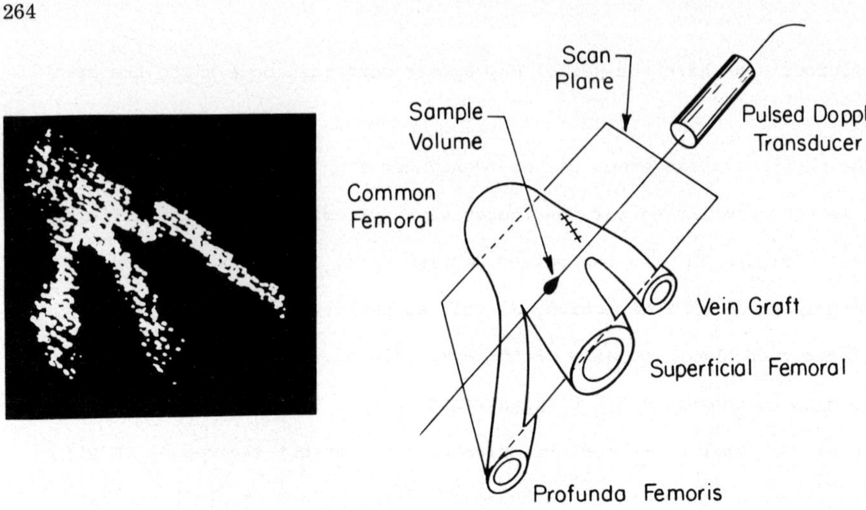

Fig. 56 An ultrasonic arteriogram of a common femoral, superficial femoral, perfunda bifurcation showing a partial occlusion of the vein graft attachment.

this type of image, taken from the common femoral bifurcation. In the example shown, a vein graft has been attached to the bifurcation. The graft is partially occluded at the point of attachment. This example demonstrates the ability of the pulsed Doppler to image a number of branches, all lying within the plane of the scan. Additional examples of pulsed Doppler imaging are shown in Figures 57 and 58. Notice that the resolution of these images is better than in the continuous wave method. Also, plaques can be shown in a long axis view as they protrude into the lumen of the vessel, which is not possible to do with the continuous wave scheme unless the view happens to lie in the plane of the scan.

While the information provided by these two Doppler imaging techniques is clinically useful, they still do not provide several important parameters. For example, a relationship between the flow lumen and the actual vessel lumen inside the walls is not shown. The possibility is slim for providing quantitative measurements of flow velocity or volume rate. Another disadvantage of these approaches is the fact that it takes some 10 minutes or more to generate the flow image, and this presupposes that the patient has remained perfectly

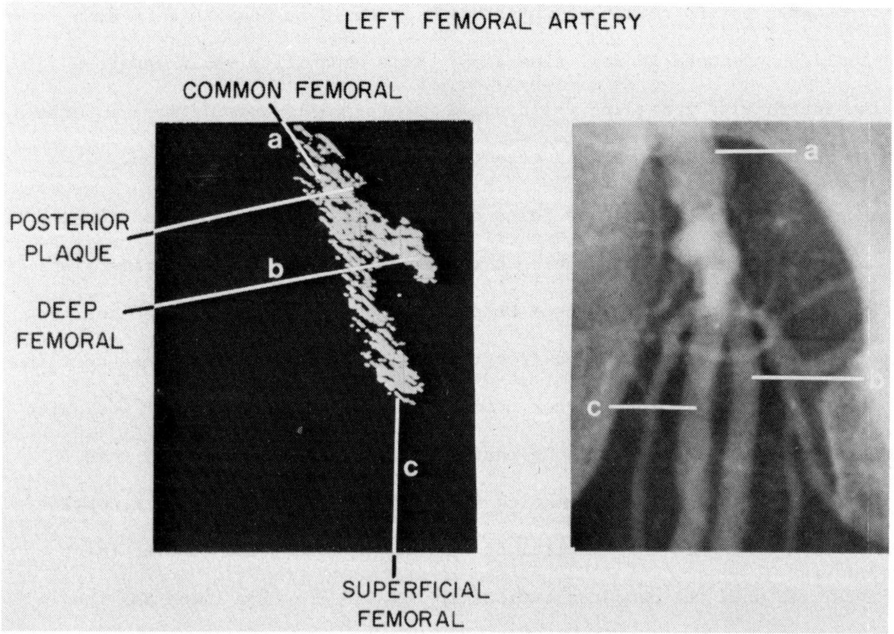

Fig. 57 A posterior plaque not readily visible on the X-ray arteriogram is visible in the long axis image of the femoral bifurcation.

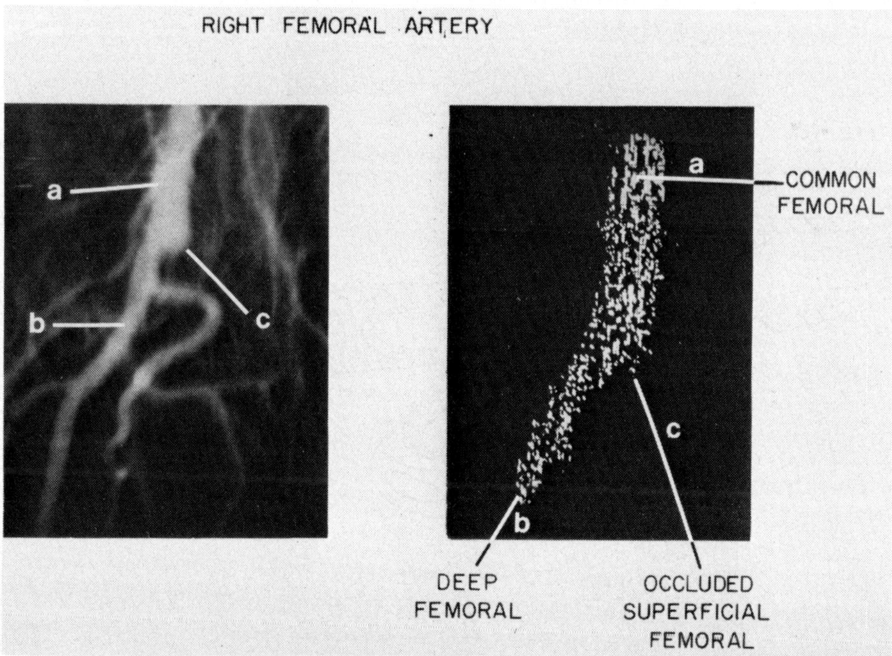

Fig. 58 The ultrasonic arteriogram shows a close correlation with the image produced by the X-ray arteriogram.

still during the entire course of the study. It would be desirable if such images could be produced in real time also. This capability would permit a real time interactive procedure which could theoretically provide more information in a shorter time and be less affected by possible patient motion.

A new instrument concept is being developed which has been given the name "Duplex" scanner.[17] This device combines the features of real time, two-dimensional echo imaging with pulsed Doppler blood flow velocity detection. The ideas for this instrument came from the realization that there were certain types of fatty plaques which did not produce images. As a result, it was quite possible that a bifurcation could be partially or totally occluded by some soft plaque material which echo imaging might not show. Conversely, a Doppler flow detector might show this partial occlusion. However, the Doppler image alone could not show the proper relationship between the flow lumen and the actual vessel walls. This situation is shown schematically in Figure 59.

The Duplex scanner device is shown in simplified block diagram form in Figure 40. This system integrates together in a time shared mode pulsed

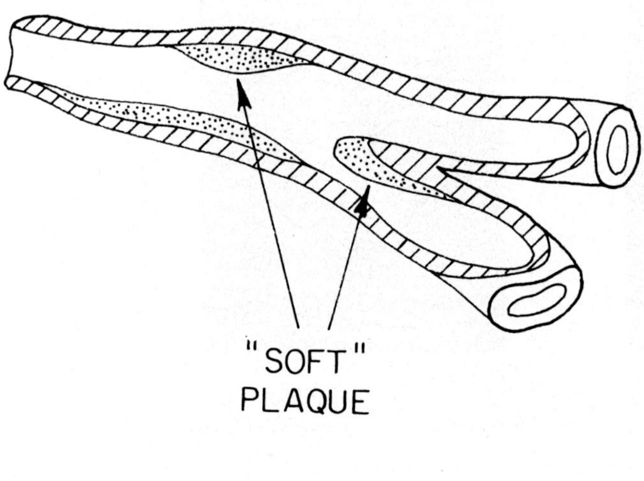

Fig. 59 Soft plaques can develop at bifurcations of peripheral arteries and may not produce echoes in the B-mode image.

echo and pulsed Doppler to form real time, two-dimensional images with simultaneous detection of blood flow velocity. The scan head is a rotor type in which the rotor contains three transducer element. As the rotor turns, three scans will be completed per revolution for a maximum of 30 frames or scans per second. The pulse repetition rate of the peripheral vascular version of this device is approximately 12 kHz, with 6 kHz devoted to the pulsed echo, and 6 kHz is devoted to the pulsed Doppler. Each system is activated alternately to avoid interaction. Their various outputs are multiplexed onto a high resolution cathode ray tube so that the real time, two-dimensional image of the vessel can be observed, while the Doppler transducer is steered into position to place the sample volume within the vessel of interest. This device permits rapid assessment of the anatomy of the carotid bifurcation where plaques are suspected. These plaques will produce shadows in the image[65] if calcified. If there are soft plaques or fatty streaks present which intrude on the flow lumen, these can be evaluated at the same time using the Doppler device to map out the actual geometry of the lumen at the bifurcation. The pulse Doppler flow image can be superimposed on the real time echo image to reveal the actual flow lumen in its proper relationship to the wall image. The initial evaluation of this device is currently underway and the results are extremely promising. Indications are that this device will be at least the forerunner of an advanced and more comprehensive ultrasonic device for peripheral vascular disease.

5.2 Cardiac Disorders

Clinically, the most important uses of pulsed Doppler echocardiography have been the detection, characterization, and display of normal or abnormal flow velocity patterns in the proximal great vessels and at each of the four cardiac valves.[38] These findings complement, and are additive to, the information obtained by standard echocardiography by being particularly sensitive to blood flow, and abnormalities of blood flow, not easily detected with standard echocardiographic devices.

There are two primary outputs available from the 3 MHz cardiac Doppler device. The first provides an audible signal in the range of 400 Hz to 5 kHz, which corresponds to the spectrum of Doppler shifts produced by normal or disturbed blood flow. The interpretation is based on evaluating the tonal quality of the signal. Smooth blood flow characteristically has a narrow bandwidth pattern which produces a tonal "musical-like" sound, whereas disturbed, or turbulent, flow has a wide bandwidth pattern which produces a harsh, scratchy sound. This method has been described in further detail in other publications.[66,67]

The second output available is an online, real time, graphical display of the Doppler shift frequencies printed on a fiberoptic strip chart recorder. A circuit has been built into the pulsed Doppler detector for performing spectral analysis of complex flow signals and for display on a cathode-ray tube. This circuit provides the time interval histogram (TIH) output, Figure 60.

When laminar flow exists, blood cells within the sample volume have fairly uniform direction and velocity. Therefore, backscattered sound waves

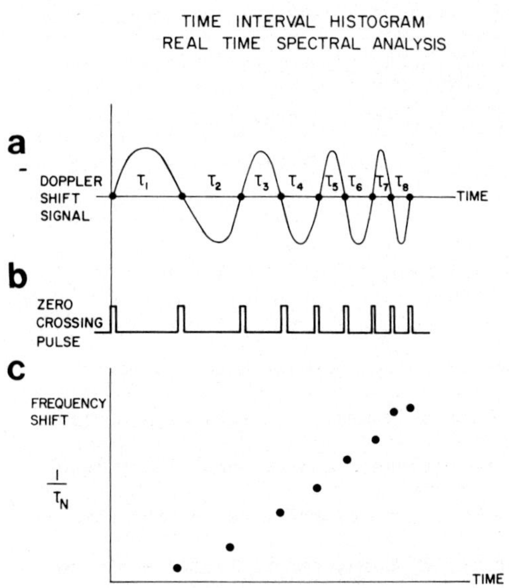

Fig. 60 The instantaneous Doppler shift frequency is determined using a time interval histogram analysis. The interval between each zero crossing point is determined, (A). A pulse is generated for each zero crossing event, (B), and a spot is plotted on the graph, (C), at a distance above the baseline which is inversely proportional to the distance between zero crossings.

have a fairly uniform Doppler shift. The time interval histogram produced by this flow has a narrowly clustered dot pattern (narrow band pattern), Figure 61a. When disturbed flow exists, the blood cells have varied direction and velocity which produce a time interval histogram with a widely dispersed dot pattern (wide band pattern), Figure 61b. Gradations between perfectly laminar and severely disturbed patterns are frequently seen.

Strip chart records contain simultaneous recordings of 1) the M-mode display of the received Doppler signal (in the form of a traditional M-mode echo display) with a superimposed line indicating the location of the sample volume at any particular time, 2) the time interval histogram of the Doppler

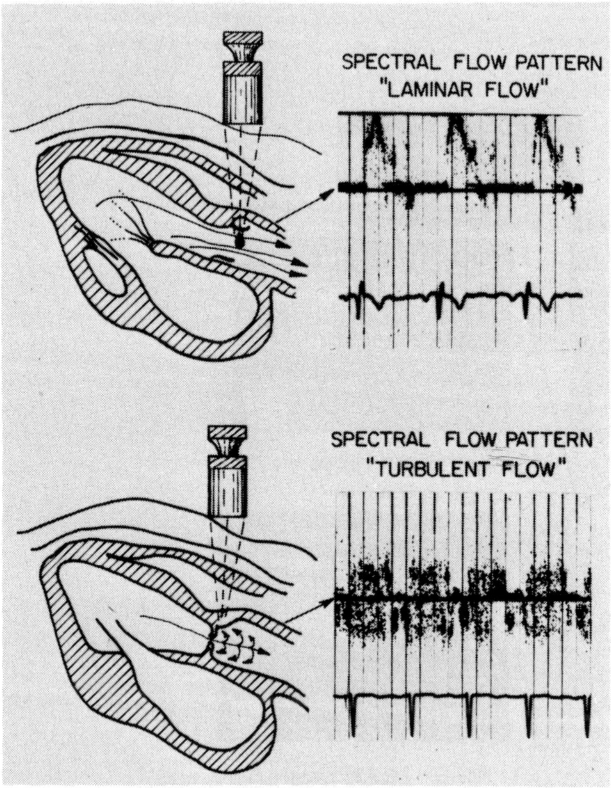

Fig. 61 The pulsed Doppler device can be used to detect the difference between smooth laminar flow, (A), versus disturbed flow, (B), which may result from a stenotic heart valve. The width of the dot pattern in the recording corresponds to the degree of disturbance present.

signal (also called a spectral display), and 3) a standard electrocardiogram, Figure 62. It is also possible to record an analog waveform of flow which corresponds to the mean value of the Doppler frequency shift; this can be shown as a separate channel on the recording.

There are many problems associated with flow quantitation within the heart by cardiac ultrasound.[68] The two parameters, mentioned previously, that determine flow quantitation include the instantaneous average blood flow velocity and the instantaneous area through which this velocity flows (usually valve orifice area).

Since both area and velocity are independent variables, it is necessary to measure them simultaneously. Although velocity profiles are obtainable with current equipment, they are tedious to measure. The variability in angulation between the ultrasonic beam and the flow across an intracardiac valve is a problem that can often be minimized by transducer angulation, allowing the

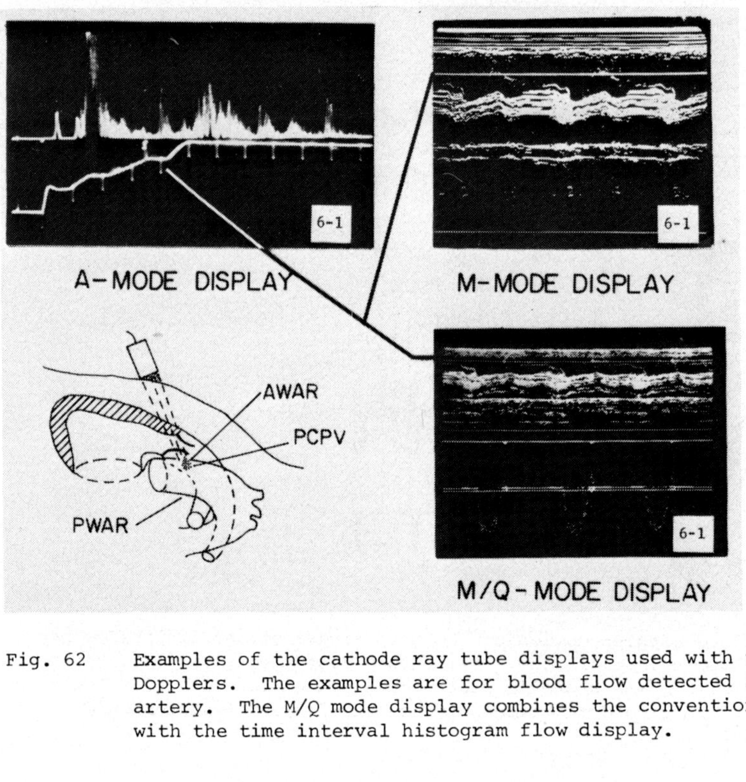

Fig. 62 Examples of the cathode ray tube displays used with the cardiac pulsed Dopplers. The examples are for blood flow detected in the pulmonary artery. The M/Q mode display combines the conventional M-mode display with the time interval histogram flow display.

assumption that the output represents a close measure of the maximum flow velocity.

Orifice size is also a variable. The mitral valve orifice, for example, may vary its area appreciably during a diastolic period. Fortunately, flow velocity profiles across the atrio-ventricular valves are fairly uniform except at the very edge of orifices (flat profile). It may be possible, in those circumstances, to determine the flow velocity at any one point in the orifice and then assume the flow velocity over the whole area of the mitral orifice.

Despite the inability to quantify beam angle, orifice size, and flow profile, considerable useful information is still obtainable from current systems. This data is basically of two types. The first is physiologic information involving flow velocity patterns as altered by heart rate, changes in filling characteristics, and by the ejection characteristics of the ventricles. The second type of information obtained involves the presence of disturbed flow superimposed on normal flow or disturbed flow in areas where flow does not normally occur, such as congenital intracardiac defects or regurgitant flows proximal to the atrio-ventricular or semilunar valves. Since all these calibration and measurement problems have not yet been solved to date, nor implemented, more practical applications have evolved.

The optimal Doppler flow signal is obtained when the ultrasonic beam is as parallel to blood flow as practical, in contrast to conventional pulsed echo techniques, where the maximum reflected signal is obtained when the sound beam is perpendicular to a cardiac structure.

The proximal great vessels are examined from the suprasternal notch with the transducer angulated posteriorly and inferiorly so that the ultrasonic beam intersects blood flow either in the ascending aorta or in the right pulmonary artery, which anatomically lies just posterior to the ascending aorta, Figure 63. Careful positioning of the sample volume allows flow velocity patterns to be recorded either in the ascending aorta, the right pulmonary artery,

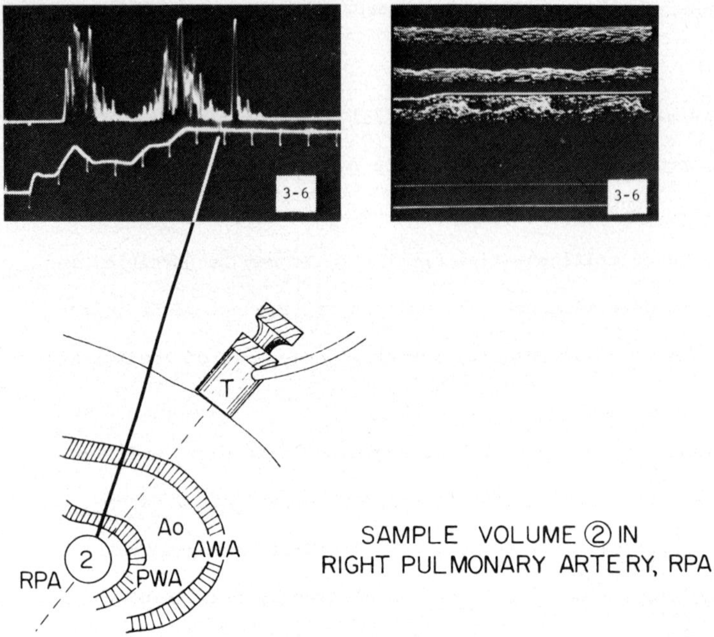

Fig. 63 Disturbed flow can be detected in the right pulmonary artery if the transducer is placed in the sternal notch with the sample volume located within the vessel. In this position, the sound beam passes directly through the aortic arch on its way to the pulmonary artery.

or with the transducer angulated more posteriorly, across the aortic arch and into the descending aorta.

Blood flow velocity patterns in the cardiac chambers, and at each of the four cardiac valves, is best detected from various precordial sites. Though flow velocity patterns can frequently be detected at the mitral and aortic valves from the usual third or fourth left intercostal space used for the pulsed echo examination, maximum flow velocities are more frequently obtained at positions somewhat offset from the usual echo locations. Maximum aortic valve flow velocities usually are obtained from the second left intercostal space with the transducer angulated inferiorly, while the maximum velocity patterns from the mitral valve in diastole are obtained from the left ventricular apex. This latter position also gives good flow velocity patterns from the left ventricular outflow tract in systole. The tricuspid valve, right ventricular outflow tract, and pulmonic valve are best examined by the pulsed Doppler technique from the third left intercostal space.

The flow velocities characteristics are displayed and recorded with an ease equal to that of the conventional pulsed echo examination, except for similar limitations involving ultrasound penetration in some patients.

5.2.1 *NORMAL FLOW PATTERNS.* Flow velocity patterns can be detected and recorded at each of the four intracardiac valves in the great majority of patients. Patterns are similar at the two semilunar valves and similar at the two atrio-ventricular valves. The aortic and mitral valves will, therefore, be used as examples.

As seen in Figure 64, where sampling is done at the aortic valve, flow

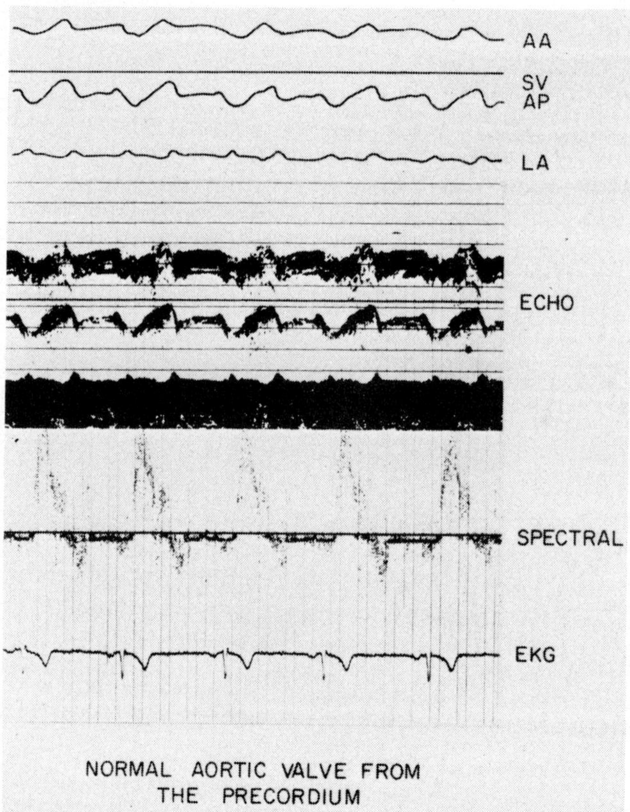

Fig. 64 The M/Q record shows the sample volume located in the aortic root. The time interval histogram spectral pattern illustrates normal, smooth flow through the valve orifice.

velocities increase rapidly with the opening of the aortic valve and velocities fall off at a somewhat slower rate in the latter half of the ejection phase.

Flow in the left ventricular outflow tract is essentially identical to that in the proximal aorta. Similarly, flow velocity patterns, in a direction away from the transducer, are detected in the right ventricular outflow tract, Figure 65. Normal flow patterns in the proximal great vessels are shown in Figure 66 with the aortic flow velocity waveform positive (above the baseline) for smooth "laminar" blood flow towards the transducer. The pulmonic blood

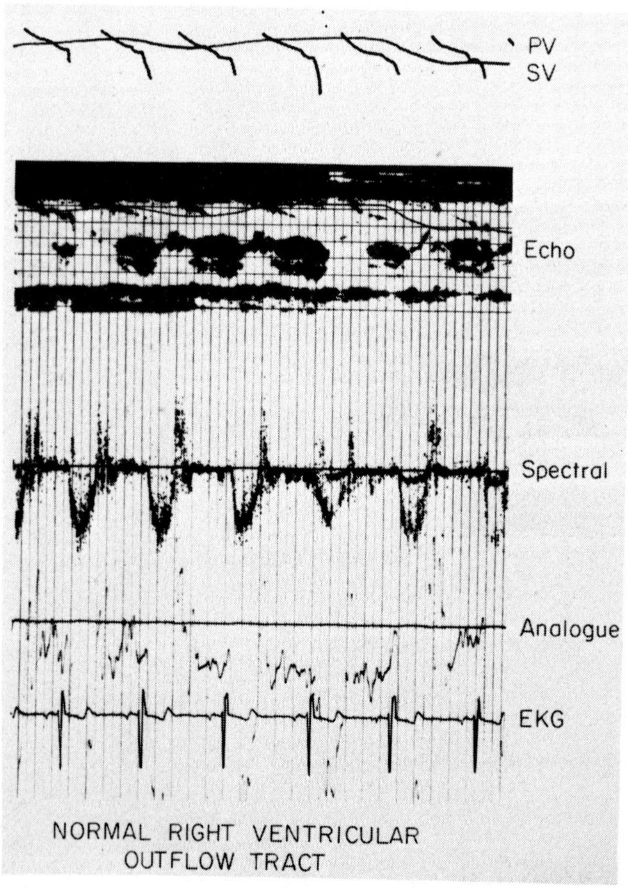

Fig. 65 The blood flow pattern through the pulmonary valve into the pulmonary artery is shown with a characteristically symmetrical dome-shaped flow velocity pattern. The deflection is below the zero baseline indicating that the flow velocity is away from the transducer.

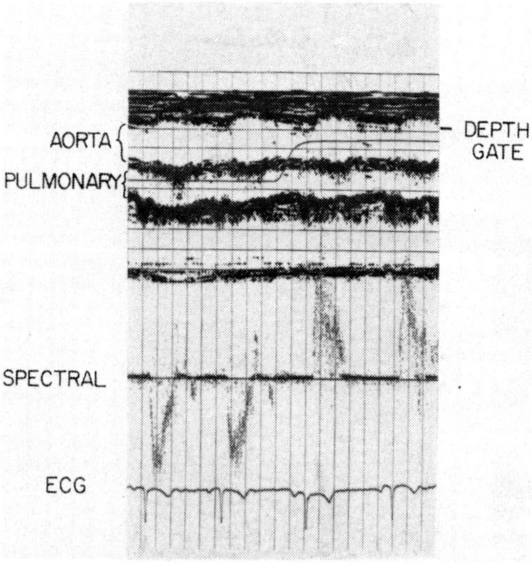

Fig. 66 The sample volume in the pulsed Doppler device scans smoothly from the pulmonary artery into the ascending aorta. The flow waveform shifts from one side of the baseline to the other as the direction of flow is different in these two vessels at this location. Taken from the suprasternal notch.

flow is moving away from the transducer and is shown as a smooth, or "laminar," pattern below the baseline. The horizontal line on the M-mode record represents the depth of the sample volume as it traverses from the proximal aorta into the right pulmonary artery. Sampling for mitral inflow is generally done in the area of the mitral annulus to avoid the velocities of the mitral valve leaflets passing through the sample volume. Two phases of flow are noted. The first occurring early in diastole commensurate with the timing of the E point of the echocardiogram and the rapid filling phase of the ventricle and then a variable period of diastasis, or low flow, is then noted followed by the second filling wave commensurate with atrial systole and the A point of the echocardiogram. These two periods of flow velocities have been arbitrarily labeled with small e (early diastolic filling) and small a (for the atrial systolic component of ventricle filling) to correlate with similar lettering of the standard echocardiogram, Figure 67.

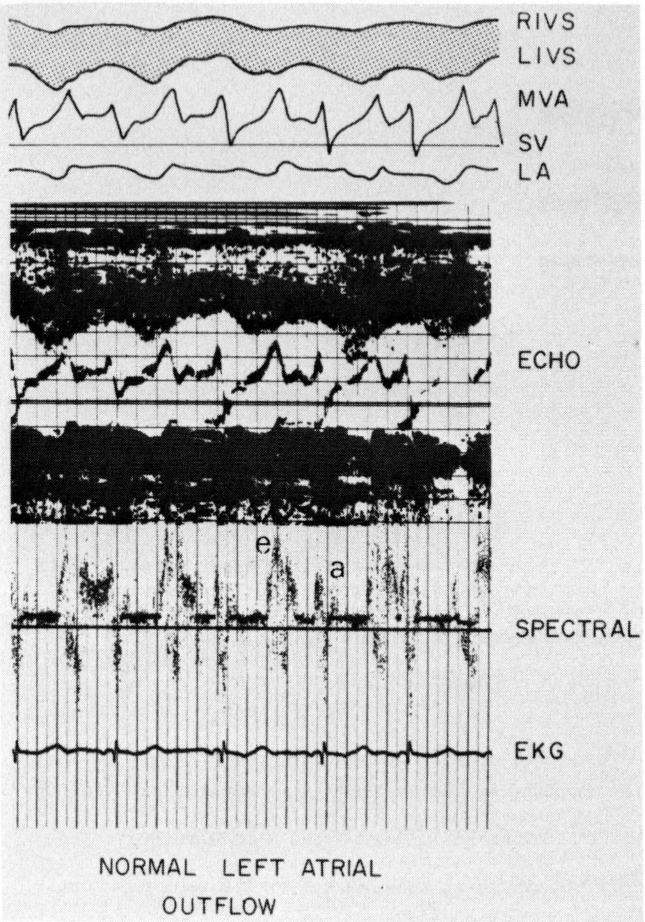

Fig. 67 Normal flow through the mitral valve is shown in the M/Q display. The sample volume is placed just posterior to the mitral valve annulus in the left atrium. The (e) and (a) components of the spectral pattern correspond to the initial ventricular filling and atrial contraction component.

5.2.2 *VALVE DEFECTS. PATHOLOGICAL FLOW PATTERNS.* Disturbances following the normal direction of flow has been best examined in the mitral valve in patients with mitral stenosis. As seen in Figure 68, two characteristic changes occur in the disturbed flow across the obstructed mitral valve. The first is the gradual decrement in flow velocities throughout the diastolic period in contrast to the rapid decrement in flow velocities that occurs in patients with

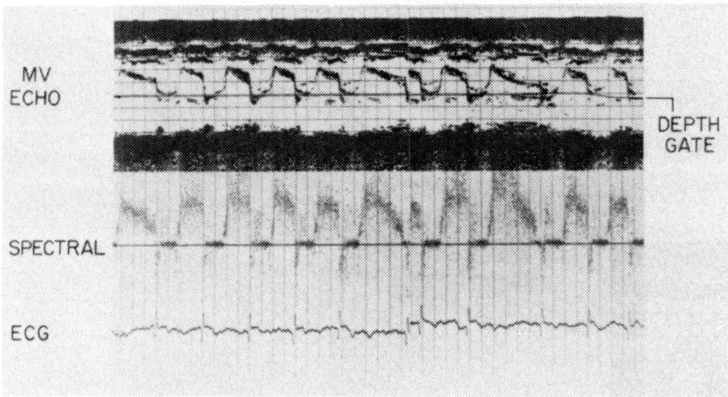

Fig. 68 The flow waveform through the stenotic mitral valve appears very similar to the stenotic mitral valve echo pattern on the M-mode display. Quantitation of flow at this site requires the simultaneous measurement of valve orifice area as well as velocity.

normal mitral valves. This, in many ways, correlates to the E/F slope of the mitral valve echogram and correlates with the known slowing of diastolic fill of the left ventricle in patients with obstructed mitral valves. The second characteristic change is the presence of turbulence superimposed upon the changes noted in the overall velocity. Correlation with the degrees of obstruction of the mitral valve has not been done with either the degree of disturbance or with the change in rate of fall-off of velocities throughout the diastolic period. The characteristics of the changes of the velocity patterns throughout the diastolic period may prove of benefit in the evaluation of prosthetic mitral valves when they cannot be adequately evaluated by standard echocardiographic techniques.

In aortic stenosis, a jet of systolic disturbance is detected within, and above, the aortic valve and is represented by a broad band Doppler signal, Figure 69. With idiopathic hypertrophic subaortic stenosis, or subaortic rings, a similar broad band systolic pattern is detected within the left ventricular outflow tract.

Pulsed Doppler echocardiography appears to be a more specific and sensitive technique for the detection of regurgitant jets at the aortic[69] and the mitral valve than standard echocardiography. These high velocity jets of

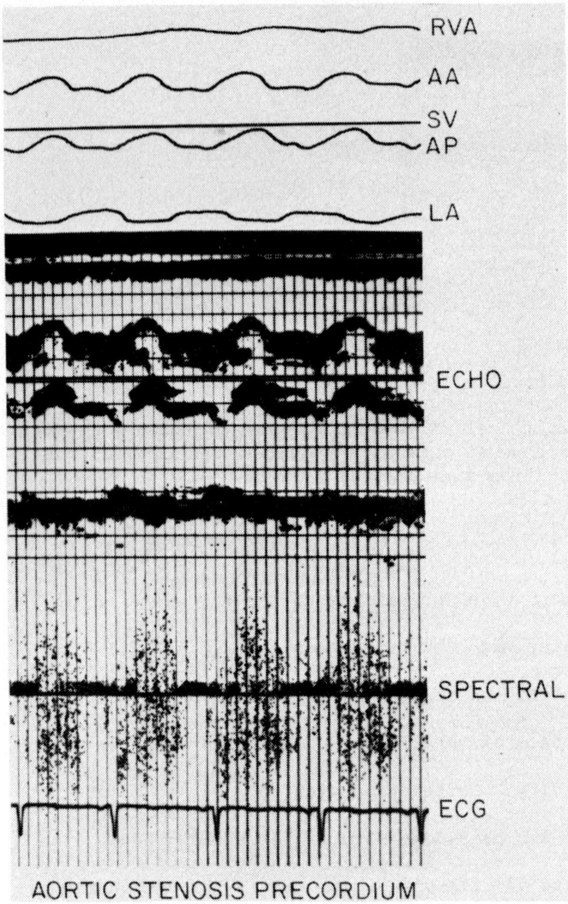

Fig. 69 The figure illustrates the M/Q display for a case of aortic stenosis. The sample volume is placed in the aortic valve orifice and the broad spectral pattern is one associated with a jet of flow at this location.

blood flow are, in most cases, easily detectable with the pulsed Doppler technique when the echocardiogram may either be normal, nonspecific, or only suggestive of areas of regurgitant flows. This detection of regurgitant flows, however, gives only limited information about regurgitant volumes of flow. Aortic insufficiency is detected by sampling for disturbed diastolic flow at, and just below, the aortic valve with the onset of flow occurring immediately after aortic valve closure, as seen in Figure 70. The flow disturbance associated with aortic insufficiency may last for a variable period of time through the diastolic period.

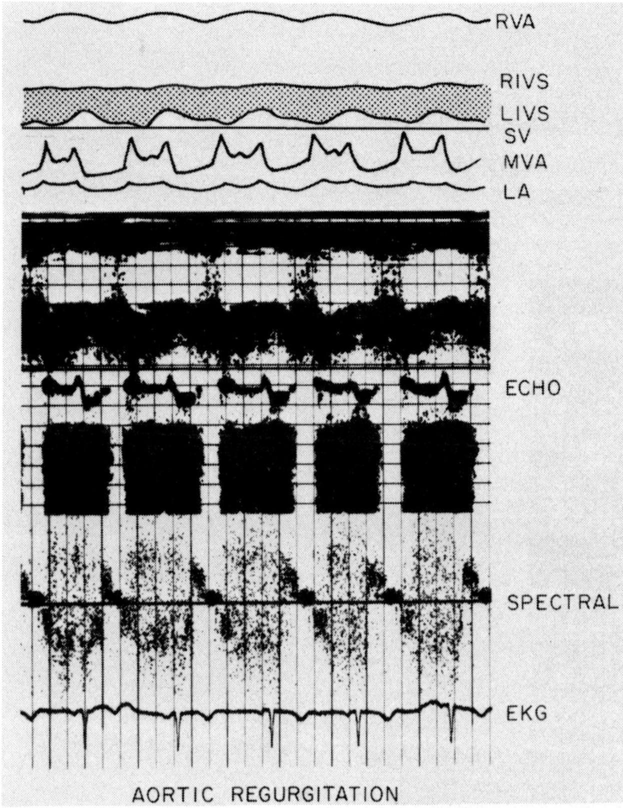

Fig. 70 Aortic insufficiency produces a jet of flow down the left ventricular outflow tract into the left ventricle. The broad dot pattern in diastole illustrates the turbulence present in this jet.

Mitral regurgitation may be well localized as seen most commonly in mitral valve prolapse, or it may be quite diffuse as seen with a ruptured mitral chordae and is detected on the left atrial side of the mitral valve and at times through the whole left atrium, Figure 71.

In both the normal and pathologic states, flow proximal and distal to the tricuspid valve is quite similar to that detected at the mitral valve with the Doppler detector, Figure 72.

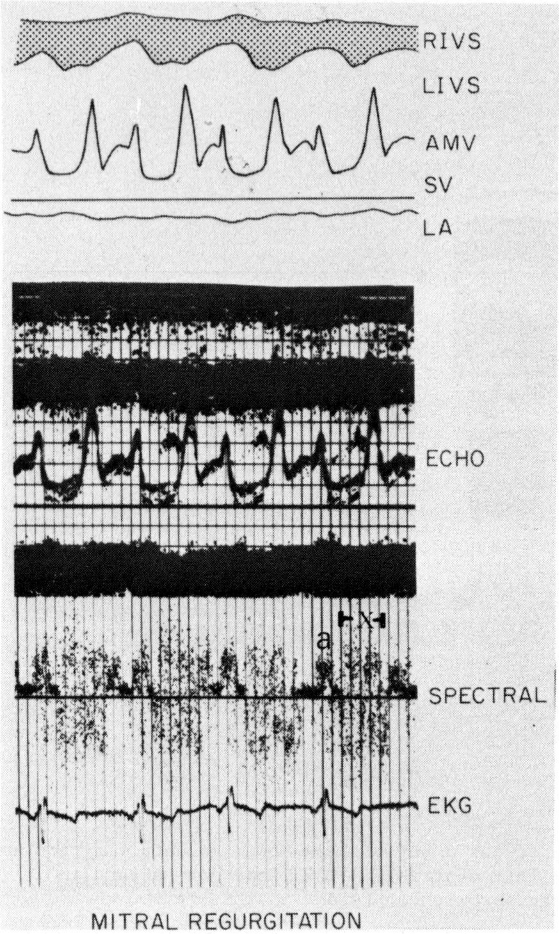

Fig. 71 Mitral regurgitation is detected by the presence of a small flow jet moving from the mitral valve orifice back into the left atrium. The wideband spectral pattern indicates the flow disturbance. The atrial contribution is labeled A, and the mitral regurgitant jet is shown in the interval X.

5.2.3 *INTRACARDIAC SHUNTS*. Flow through intracardiac shunts, both congenital and acquired (through disease process or surgical), is frequently detectable with pulsed Doppler ultrasound. Flows through atrial septal defects can be detected in approximately two-thirds the patients.[70] The flow patterns are characterized by three components of flow across the ASD that occur in ventri-

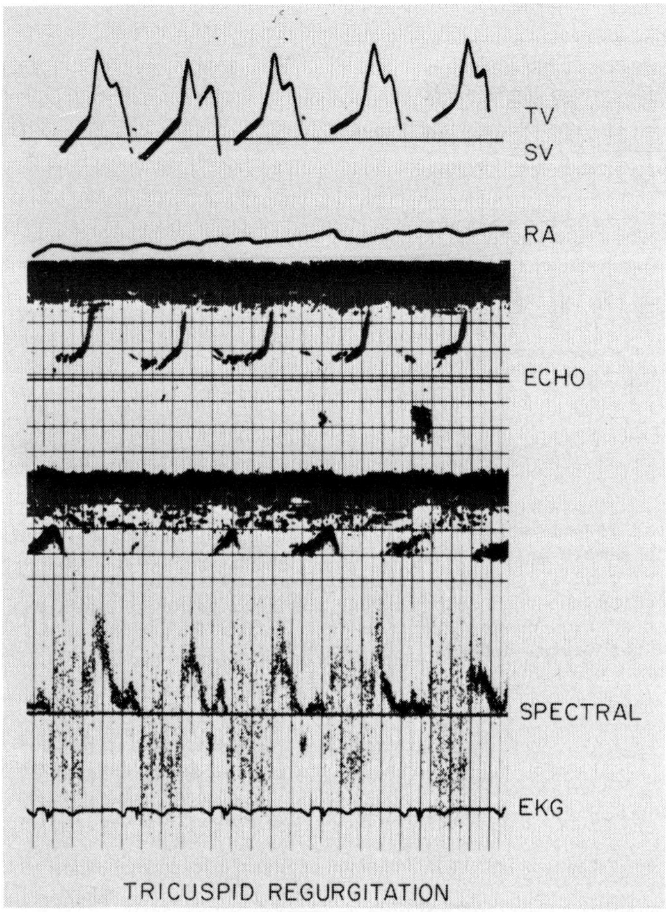

Fig. 72 Tricuspid regurgitation in the right atrium is detected in much the same fashion as mitral regurgitation. The flow jet is shown in ventricular systole by the broad spectral pattern.

cular diastole, atrial systole, and with ventricular systole. When an atrial septal defect is suspected, the transducer may be directed in such a manner that the sample volume is passed from the left atrial chamber to the right atrial chamber while flow is being assessed. If this maneuver is performed several times with the range gate set at slightly different distances, a disturbed flow can usually be detected from directly within, and on the outflow side of, the atrial septal defect, Figure 73.

Membranous ventricular septal defects which are frequently associated with normal or near normal echocardiograms, by the nature of the flow through

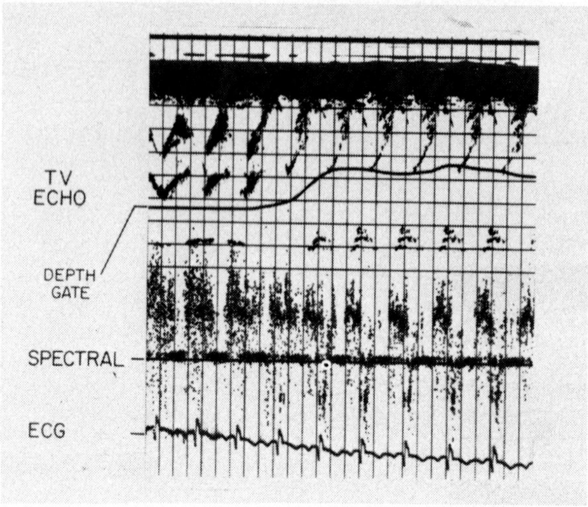

Fig. 73 An atrial septal defect can be readily detected with the pulsed Doppler by shifting the sample volume from the left atrium, directly behind the aortic root, gradually through the inter-atrial septum to the right atrium behind the tricuspid valve. The flow patterns are indicative of a jet of flow moving from the left atrium into the right atrium through the septal defect.

their orifices, give well localized, but extremely disturbed blood flow in the right ventricular outflow tract and are the easiest of the congenital cardiac shunts to detect by this technique. The localization of systolic disturbance in the right ventricular outflow tract and through a portion of the interventricular septum differentiates this lesion from infundibular pulmonic stenosis, Figure 74. Muscular ventricular septal defects are somewhat more difficult to detect, but generally give systolic turbulence in the body of the right ventricle.

Patent ductus arteriosis, with its biphasic systolic and diastolic disturbance noted in the pulmonary artery, is detectable even in cases that are "silent" by auscultation. Though shunt volume flow rates are not determined by this technique, the localization and presence or absence of the shunt, which may frequently occur in critically ill infants, is important clinical information. Surgically created shunts for the treatment of congenital heart disease are also assessable by the detection of flow through these conduits.

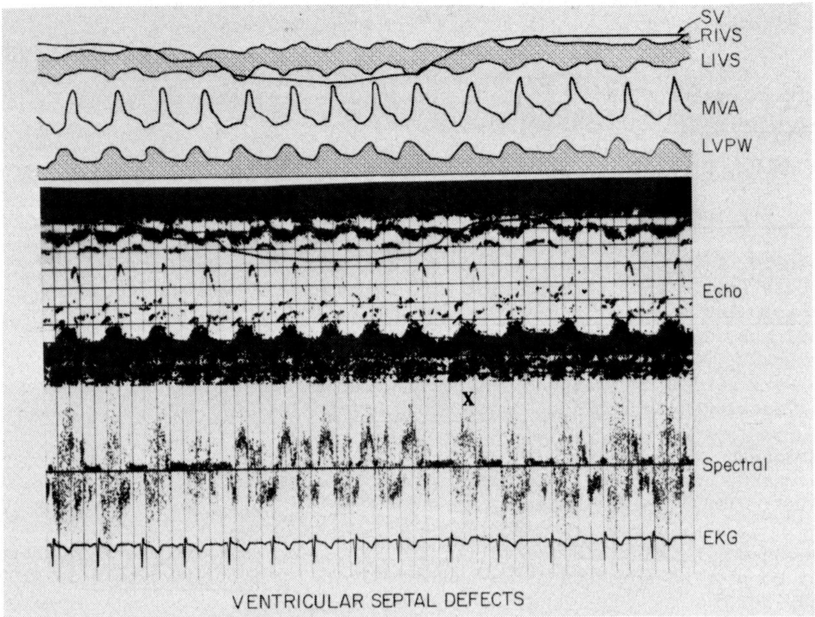

Fig. 74 A ventricular septal defect can be readily detected by the jet of flow moving through the membranous septum from the left ventricle into the right ventricle. The sample volume is shown near the septum in the right ventricle. The interval of the jet is shown as X.

REFERENCES

1. S. Satomura: A Study on Examining the Heart with Ultrasonics. I. Principles; II. Instrument. Jap. Circ. J. 20 (1956) p. 227.

2. S. Satomura and Z. Keneko: Ultrasonic Blood Rheography. Proc., 3rd Intl. Conf. on Med. Elect. (1960) pp. 254-258.

3. S. Satomura: Ultrasonic Doppler Method for the Inspection of Cardiac Functions. J. Acoust. Soc. Amer. 29 (1957) p. 1181.

4. S. Satomura: Study of the Flow Patterns in Peripheral Arteries by Ultrasonics. Nihon Onkyogakkai Shi (J. Acoust. Soc. Japan) 15 (1959) pp. 151-158.

5. D. L. Franklin, W. Schlegal and R. F. Rushmer: Blood Flow Measured by Doppler Frequency Shift of Backscattered Ultrasound. Science 134 (1961) pp. 564-565.

6. F. D. McLeod: A Doppler Ultrasonic Physiologic Flowmeter. Proc., 17th ACEMB 6 (1964) p. 81.

7. F. D. McLeod: A Directional Doppler Flowmeter. Digest, 7th ICMBE (1967) p. 271.

8. D. W. Baker, H. F. Stegall and W. A. Schlegal: A Sonic Transcutaneous Blood Flowmeter. Proc., 17th ACEMB 6 (1964) p. 76.

9. D. L. Franklin, N. W. Watson, R. L. VanCitters and K. E. Pierson: A Miniature Doppler Ultrasonic Blood Flow Meter -- Calorimetry System. Proc., 17th ACEMB 6 (1964) p. 78.

10. H. L. Stone, H. F. Stegall, V. S. Bishop and G. Laenger: Continuous Measurement of Blood Flow Velocity with an Intravascular Doppler Flowmeter. Digest, 7th ICMBE (1967) #13-6.

11. J. M. Reid, D. L. Davis, H. J. Ricketts and M. P. Spencer: A New Doppler Flowmeter System and its Operation with Catheter Mounted Transducers. In Cardiovascular Applications of Ultrasound, R. S. Reneman, ed., North Holland Pub. Co., Amsterdam (1974) pp. 183-192.

12. D. W. Baker and D. W. Watkins: A Phase Coherent Pulsed Doppler System for Cardiovascular Measurement. Proc., 20th ACEMB 27 (1967) p. 2.

13. P. Peronneau, A. Deloche, Bui-Mong-Hung and J. Hinglais: Débitmétrie ultrasonore. Developments et applications experimentales. Europ. Surg. Res. 1(2) (1969) pp. 147-156.

14. P. Peronneau, H. Hinglais, M. Pellet and F. Leger: Vélocimetre sanguin par effect Doppler a' émission ultra-sonore pulseé. L'onde Electrique 59 (1970) pp. 369-386.

15. D. W. Baker: Pulsed Ultrasonic Doppler Blood Flow Sensing. IEEE Trans. Sonics-Ultrasonics SU 17(3) (1970) pp. 170-185.

16. D. E. Hokanson, D. Mozersky, D. S. Sumner and D. E. Strandness, Jr.: Ultrasonic Arteriography: A new approach to arterial visualization. Biomed. Eng. 6 (1971) p. 420.

17. F. E. Barber, D. W. Baker, A. W. Nation, D. E. Strandness, Jr. and J. M. Reid: Ultrasonic Duplex Echo Doppler Scanner. IEEE Trans. Biomed. Eng. BME-21(2) (1974) pp. 109-133.

18. W. E. Moritz, P. L. Shreve and L. Mace: Analysis of an Ultrasonic Spatial Locating System. IEEE Trans. Inst. Meas. 25 (1) (1976) pp. 43-50.

19. J. M. Reid, R. A. Sigelmann, M. B. Nasser and D. W. Baker: The Scattering of Ultrasound by Human Blood. Proc., 8th ICMBE (1969) #10-7.

20. K. K. Shung, R. A. Sigelmann and J. M. Reid: The Scattering of Ultrasound by Red Blood Cells. Proc., Ultrasonic Tissue Characterization Seminar, NBS Special Pub. No. 453 (1976) pp. 207-212.

21. J. M. J. G. Roevros: Analogue Processing of CW Doppler Flowmeter Signals to Determine Average Frequency Shift Momentaneously Without the Use of a Wave Analyser. In Cardiovascular Applications of Ultrasound, R. S. Reneman, ed., North Holland Pub. Co., Amsterdam (1974) pp. 43-54.

22. F. Dunn: Temperature and Amplitude Dependence of Acoustic Absorption in Tissue. J. Acoust. Soc. Amer. 34 (1962) pp. 1545-1547.

23. F. D. McLeod: Multichannel Pulsed Doppler Techniques. In Cardiovascular Applications of Ultrasound, R. S. Reneman, ed., North Holland Pub. Co., Amsterdam (1974) pp. 85-107.

24. J. M. Reid and D. W. Baker: Principles of Doppler Ultrasound. In Fundamentals of Medical Ultrasonography, G. Baum, ed., J. P. Putnam and Sons, New York (1973) pp. 411-422.

25. R. W. Gill: An Implantable Pulsed Doppler Ultrasonic Blood Flowmeter Using Custom Integrated Circuits. Ph.D. Thesis (1975) Stanford Technical Report 4958-3.

26. W. R. Brody and J. D. Meindl: Theoretical Analysis of the CW Doppler Ultrasonic Flowmeter. IEEE Trans. Biomed. Eng. BME-21(3) (1974) pp. 183-192.

27. R. N. Bracewell: The Fourier Transform and Its Applications, McGraw-Hill Book Co., New York (1965).

28. A. W. Rihaczek: Principles of High-Resolution Radar, McGraw-Hill Book Co., New York (1969).

29. P. J. Bendick and V. L. Newhouse: Ultrasonic Random-Signal Flow Measurement System. J. Acoust. Soc. Amer. 56(3) (1974) pp. 860-865.

30. R. B. Chadwick and G. R. Cooper: Measurement of Distributed Targets with the Random Signal Radar. IEEE Trans. Aerosp. Elect. Systems AES-8(6) (1972) pp. 743-750.

31. C. P. Jethwa: Blood Flow Measurements Using Ultrasonic Pulsed Random Signal Doppler System. IEEE Trans. Sonics-Ultrasonics SU-22(1) (1975) pp. 1-11.

32. R. C. Waag, W. L. Rhoades and R. Gramiak: Instrumentation for Noninvasive Cardiac Chamber Flow Rate Measurement. Proc., IEEE Ultrasonic Symposium, Boston (1972).

33. K. McCarty and J. P. Woodcock: Frequency Modulated Ultrasonic Doppler Flowmeter. Med. Biol. Eng. 13 (1975) pp. 59-64.

34. R. L. Morris, M. B. Histand and C. W. Miller: The Resolution of the Ultrasonic Pulsed Doppler for Blood Velocity Measurements. J. Biomech. 6 (1973) pp. 701-710.

35. J. E. Jorgensen, D. N. Campau and D. W. Baker: Physical Characteristics and Mathematical Modelling of the Pulsed Ultrasonic Flowmeter. Med. Biol. Eng. 95 (1973) pp. 404-421.

36. J. E. Jorgensen and J. L. Garbini: An Analytical Procedure of Calibration for the Pulsed Ultrasonic Doppler Flowmeter. Trans. ASME J. Fluids Eng. 96 (1974) pp. 158-167.

37. D. L. Schultz, D. S. Tunstall-Pedoe, G. de J. Lee, A. J. Gunning and B. J. Bellhouse: Velocity Distribution and Transition in the Arterial System. In Circulatory and Respiratory Mass Transport (CIBA Foundation Symposium), G. E. W. Wolstenholme and J. Knight, eds., London, Churchill (1969) pp. 172-199.

38. S. L. Johnson, D. W. Baker, R. A. Lute and H. T. Dodge: Doppler Echocardiography: The localization of cardiac murmurs. Circulation 48 (1973) pp. 810-822.

39. F. K. Forster, J. L. Garbini and J. E. Jorgensen: Hemodynamic Turbulence Measurements Using Ultrasonic Techniques. Proc., 4th New Eng. Bioeng. Conf. S. Saha, ed., Pergamon Press, New York (1976) pp. 253-256.

40. R. W. Goldstein and W. F. Hagen: Turbulent Flow Measurements Utilizing the Doppler Shift of Scattered Laser Radiation. Physics of Fluids 10 (1967) pp. 1349-1352.

41. E. R. Pike, D. F. Jackson, P. J. Bourke and D. I. Page: Measurement of Turbulent Velocities from the Doppler Shift in Scattered Laser Light. J. Sci. Instrum. (J. Phys. E.) 1 Ser. 2 (1968) pp. 727-730.

42. W. K. George and J. L. Lumley: The Laser-Doppler Velocimeter and its Application to the Measurement of Turbulence. J. Fluid Mech. 60(2) (1973) pp. 321-362.

43. L. V. Baldwin and T. J. Walsh: Turbulent Diffusion in the Core of Fully Developed Pipe Flow. A.I.Ch.E.J. 7(1) (1961) pp. 53-61.

44. J. O. Hinze: Turbulence, McGraw-Hill Book Co., Inc., New York (1959).

45. K. M. Kiser, H. L. Falsetti, K. H. Yu, M. R. Resitarits, G. P. Francis and R. J. Carroll: Measurements of Velocity Waveforms in the Dog Aorta. Trans. ASME J. Fluids Eng. 98 (1976) pp. 297-304.

46. D. W. Baker and W. G. Yates: Technique for Studying the Sample Volume of Ultrasonic Doppler Devices. Med. Biol. Eng. 95 (1973) pp. 766-770.

47. S. W. Flax, J. G. Webster and S. J. Updike: Statistical Evaluation of the Ultrasonic Blood Flowmeter. In Biomedical Sciences Instrumentation, Plenum Press, New York (1970) pp. 201-222.

48. S. Q. Rice: Mathematical Analysis of Random Noise. Bell System Tech. J. 23 (1944) p. 282.

49. N. M. Bachman: Noise and Its Effects on Communications, McGraw-Hill,Inc., New York (1966) Chapter 4.

50. R. E. Daigle, F. D. McLeod, C. W. Miller, M. B. Histand and M. K. Wells: Transcutaneous Measurement of Volume Blood Flow. NASA Tech. Report NSG-7009 (1974).

51. R. E. Daigle, C. W. Miller, M. B. Histand, F. D. McLeod and D. E. Hokanson: Noninvasive Aortic Blood Flow Sensing Using an Ultrasonic Esophageal Probe. J. Appl. Physiol. 38(6) (1975) pp. 1153-1160.

52. D. E. Hokanson, D. J. Mozersky, D. S. Sumner and D. E. Strandness, Jr.: Ultrasonic Arteriography. Arch. Surg. 103 (1971) pp. 663-667.

53. J. M. Reid, D. L. Davis, J. B. Phillips and M. P. Spencer: Transcutaneous Flow Mapping with Continuous Wave Doppler. In Cardiovascular Applications of Ultrasound, R. S. Reneman, ed., North Holland Pub. Co., Amsterdam (1974) pp. 244-248.

54. R. E. Daigle, S. A. Rubenstein and D. W. Baker: A Duplex Scanning System for Pediatric Cardiology. In Ultrasound in Medicine, Vol. 3B, Plenum Press, New York (1977) p. 1209.

55. M. A. Brandestini: Application of the Phase Detection Principle in a Transcutaneous Velocity Profile Meter. In Proc., 2nd European Cong. on Ultrasonics in Med., Excerpta Medica International Congress Ser. No. 363 (1975) pp. 144-152.

56. F. D. McLeod, R. E. Daigle, C. W. Miller, R. J. Morgan and M. B. Histand: A Digital Doppler Velocity Profile Meter. Proc., 11th Annual Rocky Mountain Bioeng. Symp., ISA BM 73401 (1974) pp. 61-66.

57. D. E. Hokanson, D. J. Mozersky, D.S. Sumner and D. E. Strandness, Jr.: A Phase Locked Echo Tracking System for Recording Arterial Diameter Changes In Vivo. J. Appl. Physiol. 32 (1972) p. 728.

58. R. M. Olson and D. K. Shelton, Jr.: A Nondestructive Technique to Measure Wall Displacement in the Thoracic Aorta. J. Appl. Physiol. 32 (1972) pp. 147-151.

59. D. E. Strandness, Jr. and D. S. Sumner: Applications of Ultrasound to the Study of Arteriosclerosis Obliterans. Angiology 26 (1975) p. 187.

60. H. R. Müller: The Diagnosis of Internal Carotid Artery Occlusion by Directional Sonography of the Ophthalmic Artery. Neurology 22 (1972) p.816.

61. R. W. Felix, B. Sigel, M. S. Gibson, J. Williams, G. L. Popky, A. L. Edelstein and J. R. Justin: Pulsed Doppler Ultrasound Detection of Flow Disturbances in Atherosclerosis. J. Clin. Ultrasound 4 (1976) pp. 275-282.

62. M. P. Spencer, J. M. Reid and P. S. Paulson: Diagnosis of Carotid Artery Disease and Cerebral Vascular Insufficiency with Doppler Angiography and Ophthalmic Artery Sonography. In Cardiovascular Applications of Ultrasound, R. S. Reneman, ed., North Holland Pub. Co., Amsterdam (1974) pp. 249-265.

63. D. N. White and G. R. Curry: Clinical Results with a Colour Coded Differential Velocity Carotid Bifurcation Tomographic Scanner. In Ultrasound in Medicine, Vol. 3B, Plenum Press, New York (1977) p. 1389.

64. D. E. Strandness, Jr., T. W. Kornmesser and D. S. Sumner: A New Approach to Arterial Visualization. In Cardiovascular Applications of Ultrasound, R. S. Reneman, ed., North Holland Pub. Co., Amsterdam (1974) pp. 237-243.

65. C. J. Hartley and D. E. Strandness, Jr.: The Effects of Atherosclerosis on the Transmission of Ultrasound. J. Surg. Res. 9 (1969) pp. 575-582.

66. D. W. Baker and S. L. Johnson: Doppler Echocardiography. In Cardiac Ultrasound, R. Gramiak, ed., C. V. Mosby, St. Louis (1975) pp. 264-276.

67. D. W. Baker, R. A. Rubenstein and G. S. Lorch: Pulsed Doppler Echocardiography: Principles and applications. Amer. J. of Med. (1977) in press.

68. D. W. Baker and R. E. Daigle: Noninvasive Ultrasonic Flowmetry. In Cardiovascular Flow Dynamics and Measurements, N. H. C. Hwang and N. A. Norman, eds., University Park Press, Baltimore (1977) pp. 151-189.

69. J. M. Ward, D. W. Baker, S. A. Rubenstein and S. L. Johnson: Detection of Aortic Insufficiency by Pulsed Doppler Echocardiography. J. Clin. Ultrasound 5(1) (1977) pp. 5-10.

70. S. L. Johnson, S. A. Rubenstein, I. Kawabori, T. K. Dooley and D. W. Baker: The Detection of Atrial Septal Defect by Pulsed Doppler Flowmeter. Circulation 54 (Suppl. 2) (1976).

CHAPTER IV

TRANSDUCERS FOR ULTRASONIC SYSTEMS

Yoshimitsu Kikuchi

1. NOTES FOR TRANSDUCER USERS

1.1 Transducer Mechanisms Employable in Ultrasonics

In the application of ultrasonic waves to medicine and biology, the roles of ultrasonic transducers are: (1) The generation of ultrasonic waves, i.e., the active use, and (2) the reception of the waves, i.e., the passive use. In most cases involving this application, transducers are of electro-acoustic types, and the transduction mechanism is mainly based on either magnetostrictive, piezoelectric or electrostrictive phenomenon in solid substances. Mechanisms based on electromagnetic and electrodynamic phenomena are rarely employed and only in very special cases. Engineering reasons dictate this selection since the acoustic medium in medicine and biology through which the ultrasound propagates is mainly a liquid, solid or solid-liquid mixture with intentional gaseous transmission medium being rare. Sound sources in liquid or solid medium are required to have a relatively large force amplitude (vibro-motive force - v.m.f.) compared to vibrational amplitude. Sound sources in gaseous medium are, on the contrary, required to have relatively large vibration amplitude compared to v.m.f. Magnetostrictive, piezoelectric or electro-strictive phenomenon (hereafter collectively denoted as piezo-active phenomenon) are inherent properties of some solid substances, whereas electromagnetic or electrodynamic phenomenon are based on the mechanical deformation of structures produced by electric and/or magnetic fields. In principle, both types of phenomena actually connect mechanical vibration in an acoustic medium with electrical phenomenon in associated electric circuits, thus exhibiting the transducer action. The piezo-active mechanism, however, is convenient for use in generating large mechanical forces in comparison with the latter mechanisms which are useful for the generation of large mechanical amplitudes, but are inconvenient for the generation of large mechanical forces. Based on such a contrasting nature of the mechanisms described above, piezo-active phenomena are exclusively utilized in transmitters which are to generate ultrasonic waves in liquid or solid medium. The same conclusion is reached on technological grounds for the case of receivers to detect ultrasonic waves in any sound medium.

1.2 Specific Acoustic Impedance of Sound Medium

The character of the sound medium described above has been expressed by such technical terms as "acoustically hard" and "acoustically soft." Generally, liquids and solids are acoustically hard media, and gases are soft ones. When liquids and solids are compared with each other, solids are generally harder than liquids. And when solids are compared among themselves, resins, for example, are softer media than metals. It is not necessarily true that this classification always coincides with the common sensual description of the substances. Mercury, for example, is a hard acoustic medium, and some kinds of gas-solid mixtures happen to be acoustically softer than gases generally.

To describe this characteristic of the sound medium by means of a numerical expression, a concept called specific acoustic impedance, or surge impedance has been employed. The value for this impedance is defined by the complex ratio of the effective sound pressure at a point and the effective particle velocity of the same point. In the c.g.s. system this unit has the dimensions of dyne.sec/cm^3, (pressure, p, is in dyne/cm^2 and velocity, v, is in cm/sec.). In many cases, the unit has been written as absolute ohm/cm^2. The same definition has been extended for use with the MKS unit system.

In the sound medium in which absorption loss is negligible, the vibrational pressure amplitude, p, and the particle velocity, v, at a point in a plane progressive wave are in-phase (i.e., the phase difference is zero), and consequently the specific acoustic impedance defined by the ratio of both quantities is a real number and is called the characteristic impedance or specific acoustic resistance. In any electrical circuit in general, the real part of an impedance is the electrical resistance of the circuit. In this sense, the technical term, specific acoustic resistance, is used in an analogous way with the specific acoustic impedance. It is important to note that the term resistance in an electric circuit implies energy dissipation, whereas the term resistance here means that the sound energy passes the medium with no dissipation. In the mechanical parts of a transducer, however, the same terminology "resistance" expresses the energy dissipation of the corresponding parts.

1.3 Acoustical Load on Transducers

When viewed from the transducer a sound medium is a load on which the transducer transmitting performance is acting, and is called an acoustic load or mechanical load. In the case where the acoustic field near the transducer is approximately a plane progressive wave, the acoustic load per unit area of the transducer is nearly equal to the value of the specific acoustic resistance (characteristic impedance) of the sound medium.

1.4 Diffraction Due to Receiving Transducer

When the transducer is acting as a receiver, its performance is to withdraw ultrasonic energy from the sound transmitting medium, and consequently the specific acoustic resistance plays a role as the internal impedance of the energy source. The sound pressure in front of the receiving transducer varies depending on both the configuration of the transducer and the acoustical hardness of the receiving surface. This hardness is called input mechanical impedance. The pressure value generally differs considerably from the free-field pressure (i.e., the sound pressure before placing the receiver in the medium). When the input mechanical impedance is sufficiently high and the ultrasonic waves are incident perpendicularly on the receiving surface, the sound pressure on the surface is nearly doubled and the particle velocity of the wave is considerably smaller than that in the free field.

The degree of this diffraction is usually represented by a quantity called the diffraction coefficient of the receiving transducer and the internal impedance of the sound medium is treated as equal to its characteristic impedance (i.e., specific acoustic resistance of the medium times the area of the receiving surface).

1.5 Effects of Standing Waves

The foregoing descriptions are for the case where the acoustic field is a plane progressive wave. When reflective bodies are in the sound medium, the transmitted waves are interfered with by the reflected waves at every point in the field, and the relation between the sound pressure, p, and the particle velocity, v, is different from that in plane progressive waves. The extreme case involves standing wave phenomenon and the ratio of p to v is not equal to the specific acoustic resistance of the medium. Consequently, either the value of the acoustic load on the transmitting transducer, or the internal impedance of the medium for the receiving transducer, is not equal to the characteristic impedance of the medium. The performance of a transducer is thus very much dependent on the field patterns of reflected waves. In a standing wave field, the performance varies from one extreme to another when the transducer shifts its position by a quarter wave-length.

1.6 Radiated Ultrasonic Power

On the usual ultrasonic devices for medical or biological use, the available ultrasonic output power is indicated in watts (i.e., joules per second). This is the devices' nominal value when used in a medium of a specified specific acoustic impedance. When otherwise used, the ultrasonic radiation dose delivered to any biological sample may be very different depending on many factors including characteristics of the medium in which the sample is placed. Extreme examples occur when the sample is placed in a water vessel with anechoic walls compared with the sample placed directly on the transducer surface. When the sample size is smaller than the transducer surface, no acoustic load acts on the rest of the surface, and consequently the transducer performance is extremely modified so that the nominal indication of the output power shown on the device does not indicate its true operation.

1.7 Mechanical Fatigue and Break-Down of Transducers

When an ultrasonic apparatus designed for intense irradiation is used in other than the specified medium, the transducer may break down mechanically. This is because the apparatus as a whole has been designed based on the characteristic acoustic impedance of the specified medium as its expected load. For example, when designed for use in water, the transducer may break down if it is pulled out of the water since air is acoustically much softer than water, or if any distinct standing wave phenomenon occurs in front of the transducer surface even though the transducer is being used in water. Partial loading as described above may also yield a break-down.

When an ultrasonic transducer drives a negligible load the amplitude increases to an extreme, resulting in an excess stress amplitude in the transducer material, thus exceeding the mechanical strength.* The decrease of acoustic load also occurs in the case of strong cavitation resulting in an excess mechanical stress.

The phenomenon of mechanical fatigue generally occurs in any material when the material is mechanically forced to undergo a repeated stress which exceeds its fatigue limit. It has recently been confirmed that even in ceramics such as ferrites, barium titanate or PZT for piezo-active use the mechanical fatigue phenomenon exists in a similar manner to that which occurs in metals. Before any abrupt break-down is evident, permanent changes in various characteristics of the material may have been occurring which result in deterioration of the transducer performance.

1.8 Transducer Cable

In most ultrasonic devices, the transducer is connected to the electrical system by a cable. This cable, unlike the plug-in cord of the usual

---

* In such devices as ultrasonic cutters or surgical tools in which the mechanical (acoustical) load will vary, the vibrational amplitude of the transducer is stabilized by employing an electromechanical feed-back system.

electric home appliances, is a part of the high frequency electric circuit in the apparatus. It is not proper either to alter its length or to replace it with a different cable even though the new cable is for high frequency use. In the case of transducer transmitting or receiving use, alterations of the connecting cable may yield a considerable change in the overall sensitivity of the transducer. The ultrasonic frequency itself may change if the last stage of the electrical system has been designed to be a part of the high frequency driving oscillator. The cable length can be varied only in the case where the device has been designed to have an impedance-matched outlet for the transmission or inlet for reception together with an appropriately impedance-matched transducer and impedance-matched cables.

2. SOUND FIELD PATTERN OF TRANSDUCERS

The field pattern of a sound source placed in an infinite medium has long been widely studied theoretically as well as experimentally. Transducer design procedure for obtaining specific field patterns is well established and specifics are available in numerous texts. In this section, selected results of recent investigations will be described that are considered valuable from the standpoint of ultrasonic applications in medicine and biology.

2.1 Near-Field Patterns of Piston Membranes

For a sufficiently far distance from a piston membrane, the sound field pattern for a continuous wave has been described by a mathematical function called directivity[1,2*] which is a comparatively simple function since the wave propagates with a simple spherical wave front. In the vicinity of the mem-

* Directivity is defined by relative sound-pressure in relation to azimuth angle. Directivity expressions are generally those used when an infinite rigid baffle board is used with a piston of any shape (circular, rectangular, etc.). When used without the baffle, however, the directivity is not much different if the size of the piston is not much smaller than the wavelength, λ, in the sound medium. For approximate consideration, $a > \lambda$ can be the condition for a circular piston, a being the radius of the piston. The details for directivity index (i.e., beam gain) are recently summarized in: R. J. Bobber: Approximations to the directivity index, NRL Report 7750, National Technical Information Service, U.S. Dept. of Commerce, Springfield, Va., May 1, 1974.

brane, however, the sound field pattern is complex and the intensity contours show a pattern like a geographical map at a hill-group area.

The theory for the near field pattern is through to have been completed as far as its analytical treatment is concerned.[3] Numerical calculations for actual sound sources, however, is generally difficult. Those successfully obtained are the patterns for circular piston membranes, and the field distribution along the beam axis can be expressed by a comparatively simple analytical formula.[4] Denoting the sound intensity by I_z, the formula is expressed as

$$\frac{I_z}{I_o} = \sin^2 \left| \frac{\pi}{\lambda} \left(\sqrt{a^2 + z^2 - z}\right) \right| \tag{1}$$

where a is the radius of the circular piston and λ is the sound wavelength. Along with the axial distance z from the piston surface, I_z oscillates between a certain maximum I_o and zero until z reaches a distance z_o which is defined as

$$z_o = \frac{a^2}{\lambda} \tag{2}$$

This region is called Fresnel's interference zone. Any cross-section of the sound beam in this zone shows several concentric rings of intensity maximum and minimum, and the numbers of the rings increase with the decrease of z. An illustration is available in Wells'[5] employing the example of a 1.5 MHz transducer with an a of 1 cm. Investigation of interference in the immediate vicinity of the transducer surface is shown by Fry and Dunn.[6] Recently, Torikai et al.[7] published a convenient method for experimental observation of the near field pattern. A sheet of uniformly exposed photographic paper is immersed in a conventional developer so as to face the ultrasonic transducer under observation for its near field pattern. The ultrasonic waves promote the developing effect in accordance with the acoustic intensity pattern on the photographic paper. Torikai[8] also published a theory for simpler numerical calculations of near field patterns by employing acceptable approximations. Some actual examples will be shown later.

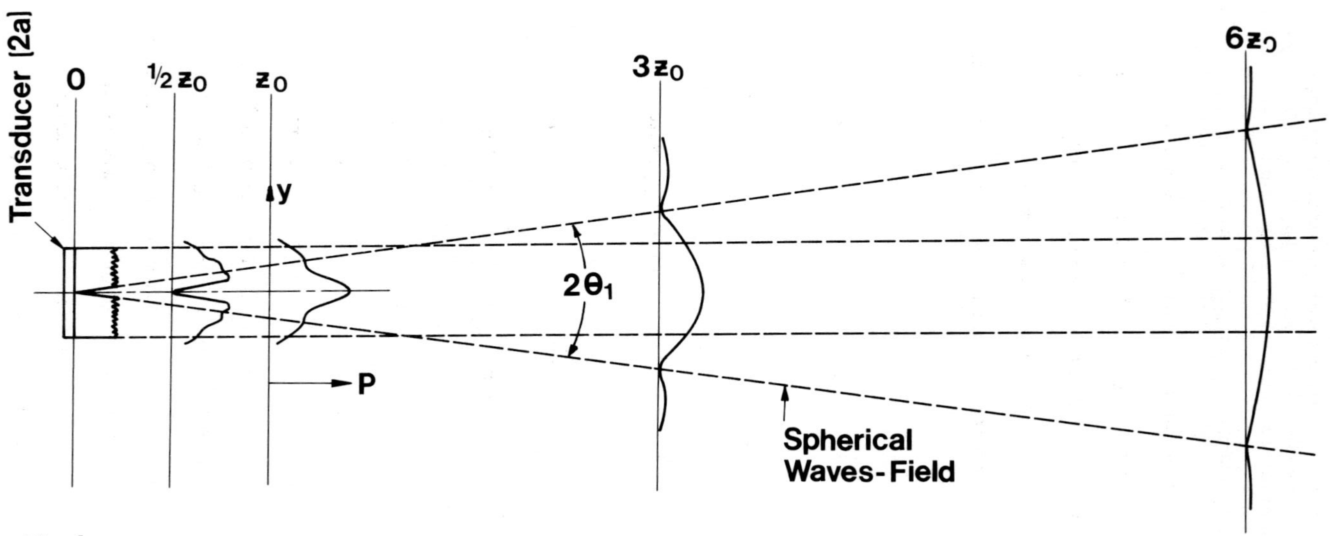

Fig.1

Fig. 1 Sound field of a circular piston.[9] $2a/\lambda = 16$. (Krautkrämer).

One practical way of considering the near field nature has been widely used. As shown in Figure 1, most of the transmitted ultrasonic energy propagates in the cylinder of 2a diameter up to the distance z_o, and then begins to gradually spread in the conical zone with 2θ as its apex angle, diverging with a spherical wavefront.[9] This is a very practical way for consideration, with θ_1 being equal to the first-zero angle of the far field directivity. In the case of a circular piston,

$$\theta_1 = \sin^{-1} \frac{0.61\lambda}{a} \qquad (3)*$$

As is understood from the foregoing statement, the width of the ultrasonic beam cannot be smaller than 2*a*. If one intends to obtain a sharper beam by decreasing 2*a*, z_o also decreases with the square of 2*a* as evident in Eq. (2). This fact must be remembered in every practical case of ultrasonic transducer application. In Table 1 are shown the numerical figures of z_o for some circular plane disks presently used in practice. Kossoff[10] has shown that the beam width becomes 1.35 x (2*a*) at the distance $2z_o$, and the intensity falls to 0.5.

Table 1 Axial length of Fresnel's zone, z_o, calculated by Eq. 2 with 1500 m/s as sound velocity

Frequency	z_o (cm)			
(MHz)	2a = 30	20	10	5 (mm)
1	15 cm	6.7 cm	1.7 cm	0.42 cm
2.25	34	15	3.7	0.93
5	75	33	8.3	2.1
10	150	67	17	4.2

* Directivity: $D = \frac{2J_1(ka \sin\theta)}{ka \sin\theta}$

where J_1: Bessel's function of the first kind, and $k = 2\pi/\lambda$. *D* becomes zero when $ka \sin\theta = 3.83$.

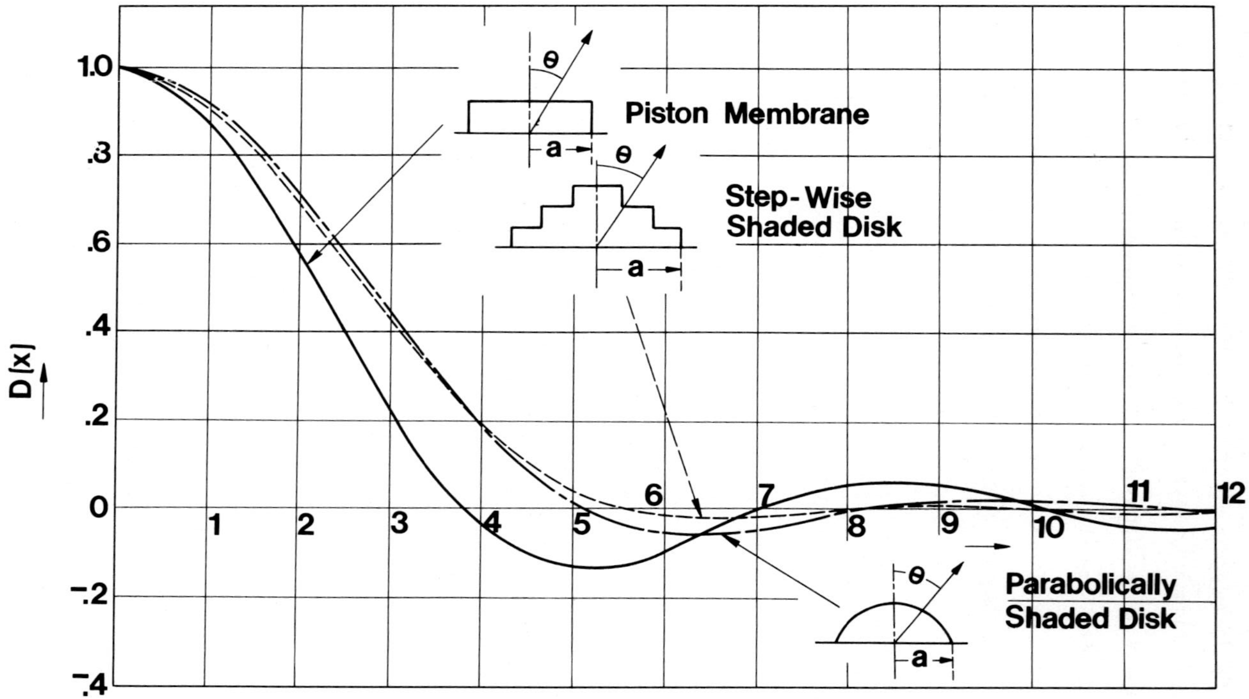

Fig. 2

Fig. 2 Directivity functions of circular disks with some shaded-amplitude distributions: Uniform, parabolic, and stepwise distributions. Ordinate D(x), directivity; Abscissa, Ka sin θ-

To realize as faithfully as possible the theoretical characteristics of a piston membrane in actual ultrasonic transducers, efforts have been made to obtain possibly uniform distribution of vibrational amplitude on the radiation surface when the transducers are designed and constructed of piezo-active materials. Longitudinal vibration of magnetostrictive materials, thickness mode vibration of piezo-electric disks, etc., represent some of the cases considered. For example, a comprehensive study on circular short columns has been published[11] to clarify the relation between the mode of vibration and the dimension ratio (i.e., the ratio of the radius and thickness). Various ways have been shown to suppress spurious modes of vibration.[12]

2.2 Shaded-Amplitude Membranes

Side lobes of directivity can be suppressed when the amplitude distribution of a radiating surface is intentionally designed for this suppression. Wolff,[13] Menges[14] and other research workers have contributed useful theories. Actual examples of calculated results[15] on circular disks are shown in Figure 2. A similar idea has also been applied to rectangular shapes.[15] Generally speaking, the suppression of side lobes can be achieved by a certain decrease of the amplitude distribution towards the disk edge. In this sense, the technique is called "shading." The present author proposed in 1954 some practical techniques which are easily applicable to circular disk transducers.[16] The techniques are based on such ideas as an asteroidal pattern (Figure 3) for

Fig. 3 Electrode pattern for amplitude-shading of a circular disk. (Kikuchi)

the electrode of the transducer disk to gradually decrease the sensitivity contribution towards the circumference, or, the idea of a concentric absorber to be attached to a uniform disk.

Near field patterns have been investigated with shaded-amplitude disks, and the considerable flattening of Fresnel's interference zone has been pointed out. Valuable suggestions are shown of a practical example in which a parabolic shading[17] is applied to a disk when $\underline{ka} = 10$, and also of a practice in which the amplitude of the ring-form part, $\frac{1}{7}\underline{a}$ in width, on a single circular disk is shaded[18] to be one-half the amplitude of the inner circular part. Haselberg and Krautkrämer[19] have published results for near field interference diminution* when shading with a Gaussian distribution is applied to a circular disk. Figure 4 shows one of their results. In the

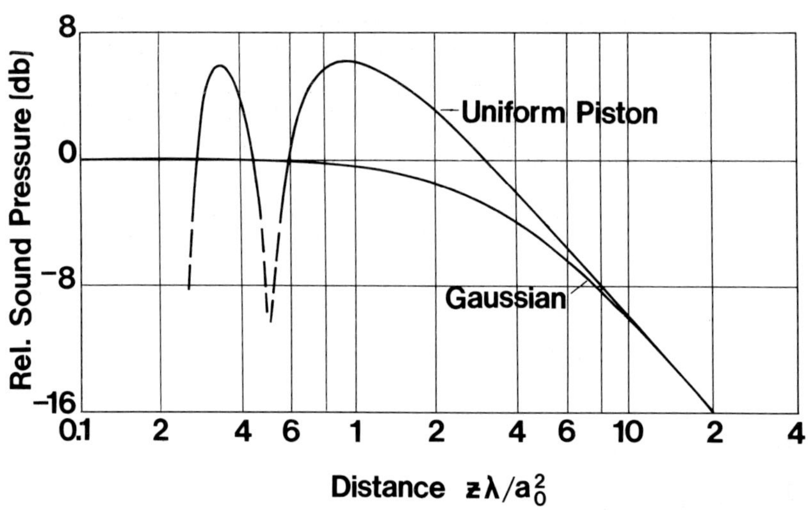

Fig.4

Fig. 4 Sound pressure distribution along the beam axis of a transducer with the Gaussian distribution of amplitude-shading. For comparison, that of an uniform piston is also shown. (Haselberg and Krautkrämer).

---

* $p(z) = p_o \left\{ \left(\frac{1}{\pi} \cdot \frac{z\lambda}{a_o^2}\right)^2 + 1 \right\}^{-\frac{1}{2}}$

far field, side lobes also disappear at the same time.* For actual cases, the tail of the Gaussian distribution can be cut off at an effective radius a_o at which the shading reaches the value 1/e of the central value. In this practice, the authors have proposed a pattern like a chrysanthemum flower for the electrode pattern to be applied on the disk. This pattern corresponds with the asteroidal pattern already shown in Figure 3, being modified by the Gaussian function.

2.3 Focusing Transducers

Presently, three principles are used for non-array type focusing transducers in ultrasonic applications: (1) Focusing by means of acoustic mirror systems, (2) focusing by acoustic lens systems, and (3) direct focusing by concave transducers. In actual transducers employing each principle, various ways and methods have been proposed and developed.

2.3.1 Acoustic Mirror System

In the methods of focusing by acoustic mirror systems there are many varieties ranging from the focusing of plane waves by a paraboloidal reflecting surface[20] as the simplest method up to the complex combination of an ellipsoidal mirror and a hyperboloidal mirror, proposed by Oloffson[21] in 1963, which aims at improvement of focus sharpness. The latter system can be characterized by its highly increased resolving power and its larger aperture in comparison with conventional mirror techniques.

Oloffson's system is shown in Figure 5. A conventional transmitter T sends an ultrasonic pulse (a wave-train) to illuminate a target near F_1. The part of the wave which enters into the aperture of the ellipsoidal concave mirror E converges toward F_2 but not up to F_2, as the wave is again reflected by the hyperboloidal convex mirror H placed on the way to F_2.

---

*

$$D(\theta) = e^{-\left(\frac{\pi a_o}{\lambda}\sin\theta\right)^2}$$

In these formulas, a_o is the radius at which the shading reaches the value 1/e of the central value, and is called the effective radius.

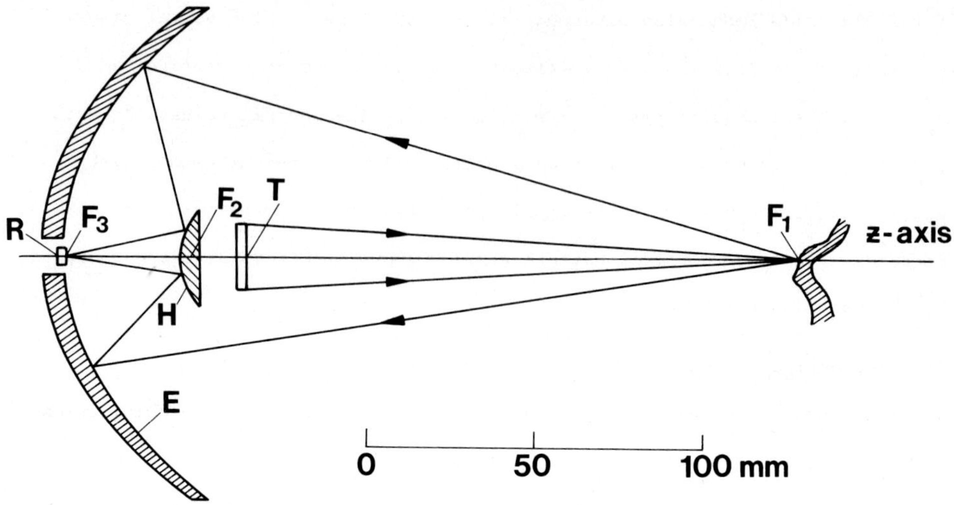

Fig.5

Fig. 5 Mirror system proposed by Oloffson.

Finally, the wave is focused at F_3. At the point F_3 there is placed a receiving transducer R of small size.

The lateral resolving power of this system is shown by formulas from which numerical calculations have been made. At 1 MHz ($\lambda = 1.5$ mm) the theoretical resolving power is 2.2 mm while a resolving power of 2.3 ± 0.2 mm has been determined experimentally when the radius of the receiving transducer is 1.5 mm. The resolving power for a target off F_1 by $\pm$ 10 mm is not much different than that for the on-focus values. (See Figure 5 of Ref. 21).

A mirror material made from tungsten-Araldite mixture was reported as superior to brass or wood. The accuracy of the surface finish was 5×10^{-2} mm. Later in 1967, Kelly, Okuyama and Fry[22] utilized this Oloffson's mirror system for the tomographical visualization of an excised liver.

Another mirror system has been reported[23] in which a paraboloidal and a prolate spheroidal mirror are used instead of the Oloffson ellipsoidal and hyperboloidal ones, and its advantages and limitations has been discussed from the standpoint of use for an ultrasonic imaging transducer.[24]

Yet another type of mirror system[25] has been proposed in which a convex circular cone with 90° apex angle is placed in front of a disk transducer so that the transmitted waves are once reflected at a 90° angle to a radially placed concave parabolic mirror where they are again reflected to focus at a region on the original axis of the transducer beam. This system was constructed[26] and used for experiments on biological tissue modification.[27]

2.3.2 Acoustic Lens System

Waves in general refract at an interface where the waves enter another medium of different propagation velocity, and the waves can be made to converge at a focus when the medium is a lens. Acoustic waves perform in this manner. Materials for acoustic lenses, however, are rather limited[28] since materials of different propagation velocity have generally different characteristic acoustic impedances, resulting in an impedance discontinuity. This discontinuity yields an energy reflection which depends on the impedance difference. When the sound medium is water, Lucite ($c = 2.8 \times 10^3$ m/s, $n = 1.9$) or Polystylene ($c = 2.4 \times 10^3$ m/s, $n = 1.6$), or other synthetic resins are often used for the lens material ($\underline{c}$ being the propagation velocity and $\underline{n}$ the acoustic refractive index). These materials are appropriate since the characteristic impedance of any of them is close to that of water and consequently the interface reflection loss is about 10%. The sound velocity, $\underline{c}$, in any of them is larger than that of water and, therefore, the shape for a sound converging lenses is concave which is opposite to the general situation for optical lenses.

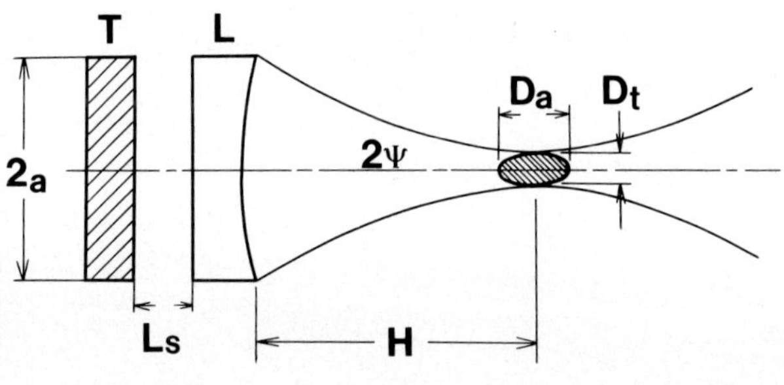

Fig. 6 Plano-concave lens L with transducer T, showing focal region D_a, D_t.

Plano-concave acoustic lenses[29] are often used in combination with piezoelectric disk transducers as shown in Figure 6. L is the lens and T the transducer. Generally for acoustic lenses, the diameter-to-wavelength ratio is much smaller than the ratio for optical lenses. This leads to a convergence which may be orders of magnitude larger than that common for optical cases. The dimensions of the focal region,* D_a and D_t are readily calculated.

The distance H shown in the above figure is called the working distance which must be quite large depending on the maximum size of the biological system to be irradiated. Thus, the availability of an appropriate lens aperture, 2Ψ, may be the limiting factor in designing an appropriate irradiating system. Fry and Dunn[30] have pointed out the following relation in a plano-concave lens:

$$H \simeq F\left(1 - \frac{F}{2R}\tan^2\Psi\right) \tag{4}$$

$$F \simeq \frac{R}{1-(1/n)}, \tag{5}$$

where F is the focal length, R the radius of curvature, and n the refractive index of the lens material.

The concave surface of a plano-concave lens can be spherical if the aperture is small. For larger apertures, the appropriate lens surface which is an ellipsoid of revolution must be used so as to focus the radiation into a common region. Formulas have been described[31] when $\lambda/4$ is given for the dimension of the common region.

The spacing L_s between the lens L and the transducer T in Figure 6 is best determined from the standpoint of impedance matching. Fry and Dunn[32] have

---

* For ultrasonic irradiation systems used to alter biologic tissues,[30] it is common to use values such that:

$$D_t \simeq k_t\,(F/2a)\lambda,\; D_2 \simeq k_a D_t,$$

where k_t and k_a are usually approximately unity, 2a being the lens diameter and F the focal length.

recommended that if liquid is in the interspace, which might also be used for cooling purposes, then the length $\underline{L}_s$ should be adjusted to be an odd multiple of a quarter wavelength when the characteristic impedance ($\rho_p c_p$), of the lens material is larger than ρc of the acoustic medium. This suggestion has come from the result of the authors'[33] theoretical analysis, developed for a composite transducer composed of parallel layers of different materials. This analysis was made from the standpoint of maximum electric power transmission to the transducer from a given voltage source by increasing the motional admittance of the transducer. This plane-wave transmission analysis, for the case of power gain involving a lens system must be modified. A method of approximate estimation for this case has been proposed by the same authors.[32]

In an actual design, it should be noted that it is desirable to choose the thickness $\underline{L}_s$ to be as thin as possible in order to reduce the characteristic changes of the composite system under conditions of temperature rise. A working system for 4 MHz use has a quarter wavelength spacing of castor oil or silicon oil.[34] Yoshioka and Oka,[35] however, used a duralmin plate of a half wavelength for this spacing (i.e., coupling medium) at 1 MHz. The transducer they used is a triple or quintuple layer of $\lambda/2$ resonant quartz disks cemented together with parallel connection of each electrode so that 3^2 or 4^2 times as much sonic output may be available from a given r.f. voltage source.

In recent years, such lower velocity liquids as carbon tetrachloride ($c = 938$ m/s, $n = 0.73$ in water),[28] ПЕРФТОРТРИЕЦИЛАМИН ($c = 564$ m/s)[36] etc. have appeared for making a focusing acoustic lens by putting the liquid in a capsule-type flexible bag which is appropriately pressurized to take the form of a convex lens.[37]

2.3.3 Concave Disk Transducers

Focused waves are also produced by fabricating a piezoelectric disk with a concave shape. This shape is called a shell, and when the concave surface is spherical as shown in Figure 7, it is called a spherical shell. When the shell is vibrated in the thickness mode, the sound wave radiated from the concave surface converges towards its focal region. An example[38]

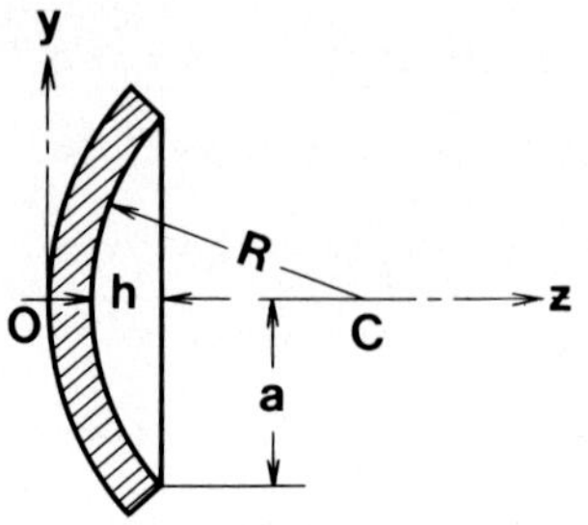

Fig. 7 Spherical shell-type focusing transducer.

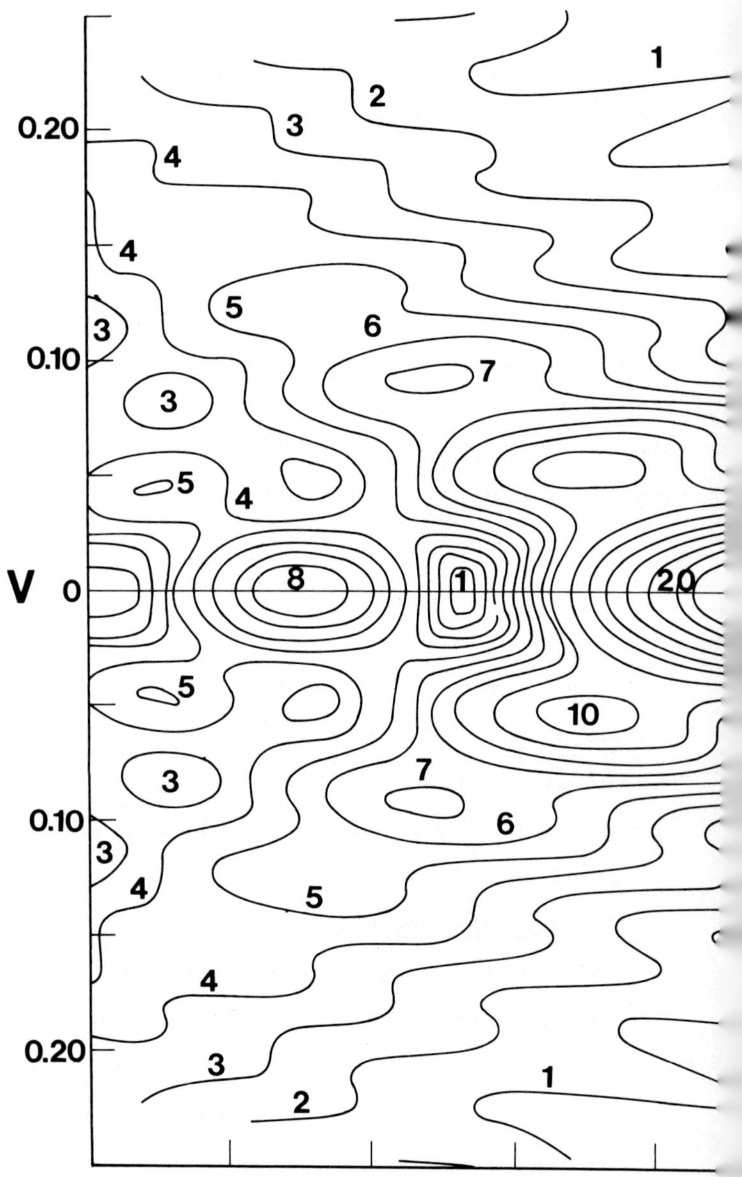

Fig. 8

Fig. 8 Near-field pressure pattern of a focusing transducer when the dimension parameter $\hat{D} = 10$, (Torikai). $\hat{D} = a^2/R$ (see Eq. 12), and Z and V are certain normalized distances.

of a near field pattern of such a shell is shown in Figure 8. The focal region is located near the center of curvature of the shell, C, in Figure 7, and the maximum point of the sound intensity comes nearer to the exact center with the increase of h/λ, _h_ being the depth of the shell. The intensity I_c at the center is:[39,40]

$$\frac{I_c}{I_i} \simeq \left(\frac{2\pi h}{\lambda} \right)^2 , \tag{6}$$

where I_i is the average intensity on the shell surface.

The lateral width, D_t, of the focal spot can be expressed as[41]

$$D_t \simeq 1.22 \frac{R\lambda}{a} , \tag{7}$$

where _a_ is the radius of the disk and _R_ is the radius of curvature (see Figure 7). This expression holds good provided that _a_ is sufficiently large compared with the sound wavelength λ and that the shell is shallow.

Kossoff[42] stated that 84% of radiated total power is in this main lobe with a diameter of D_t at the focal spot and that the beam width at any distance along the beam axis can be defined by the diameter D_t (z) of the cross-sectional circle through which 84% of the total power is involved. Denoting this axial intensity by I(z), we have

$$I(z)D_t^2(z) \simeq IcD_t^2 . \tag{8}$$

This approximate relation has been validated in the range where:

$$z = a^2/\lambda \text{ to } 2a^2/\lambda. \tag{9}$$

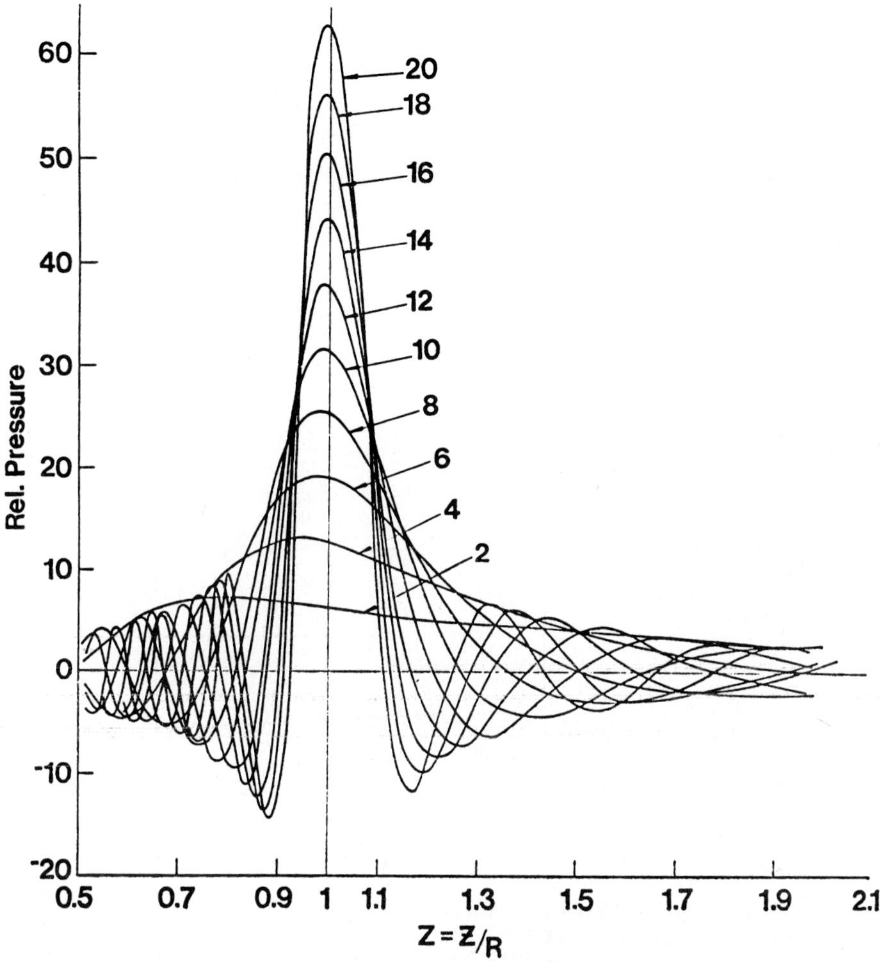

Fig. 9 Sound pressure distribution on the beam axis of concave disks, (Torikai The figures on each curve being the dimension parameter, $a^2/R\lambda$ (see Eq. 12), where a is the radius of the disk, and R the radius of curvature, and Z = z/R, the normalized distance.

The axial intensity $\hat{I}(z)$ is expressed as

$$\hat{I}(z) = \left(\frac{R}{R-z}\sin\frac{\pi a^2}{2\lambda}\frac{(R-z)}{Rz}\right)^2 \qquad (10)$$

when normalized to that at the focal plane, $\hat{I}(z)_{z=R}$.

Torikai[38] has recently offered a result shown in Figure 9, calculating the axial sound pressure for various concave disks using the approximations:

$$z/R \simeq 1, \quad a/R \ll 1. \qquad (11)$$

The parameter on each curve is the dimension parameter $\hat{D}$ of the disk normalized to λ as

$$\hat{D} = a^2/R\lambda. \tag{12}$$

The abscissa Z is the axial distance normalized to R, the radius of curvature of the disk.

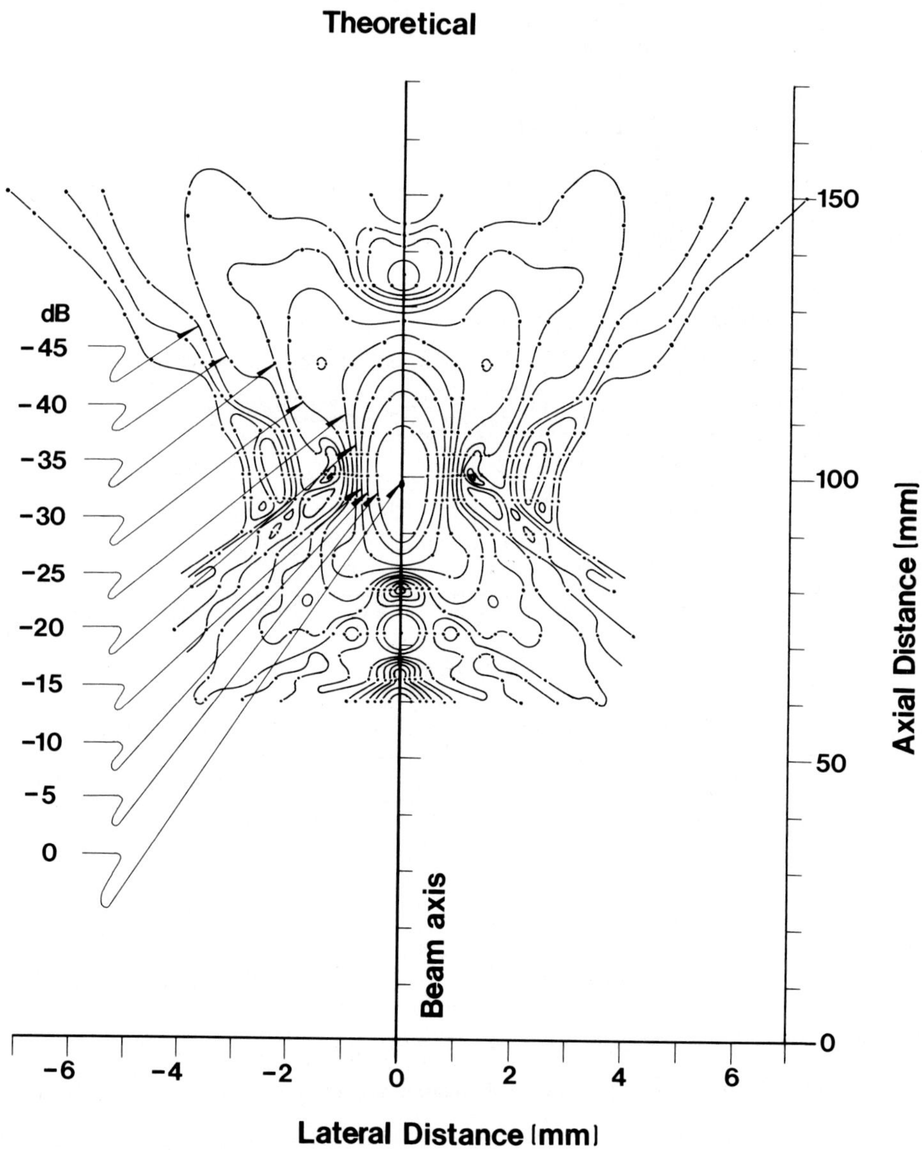

Fig. 10 Echo-intensity contour pattern of a concave spherical shell of $\hat{D}$ = 7.5.

(see also next page)

Based on this theoretical calculation, Kikuchi et al.[43,44] have recommended that $\hat{D}$ = 2 to 4 as an optimum dimension of a transducer for tissue visualization. This recommendation is a compromise between the usual requirement that the beam should be as narrow as possible in lateral width and longest in its

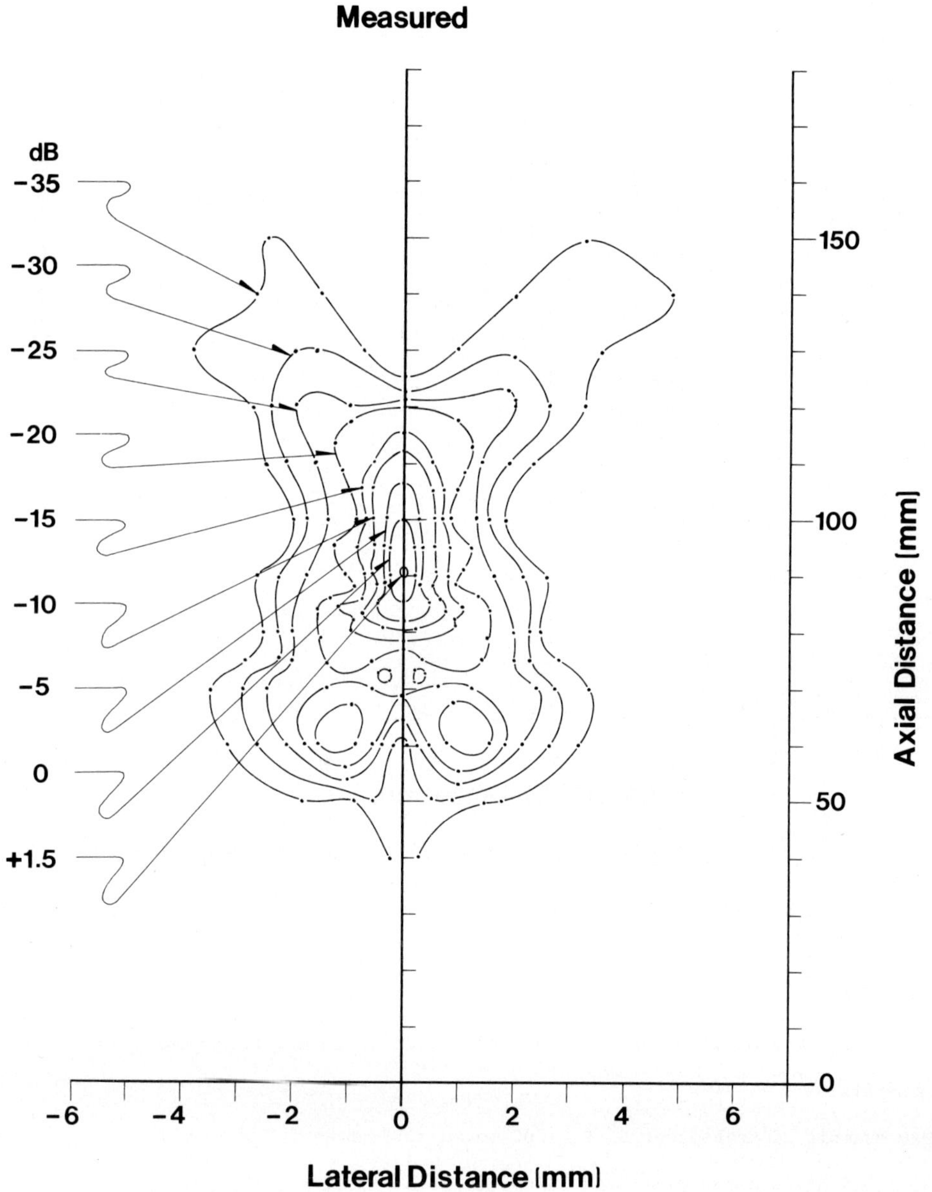

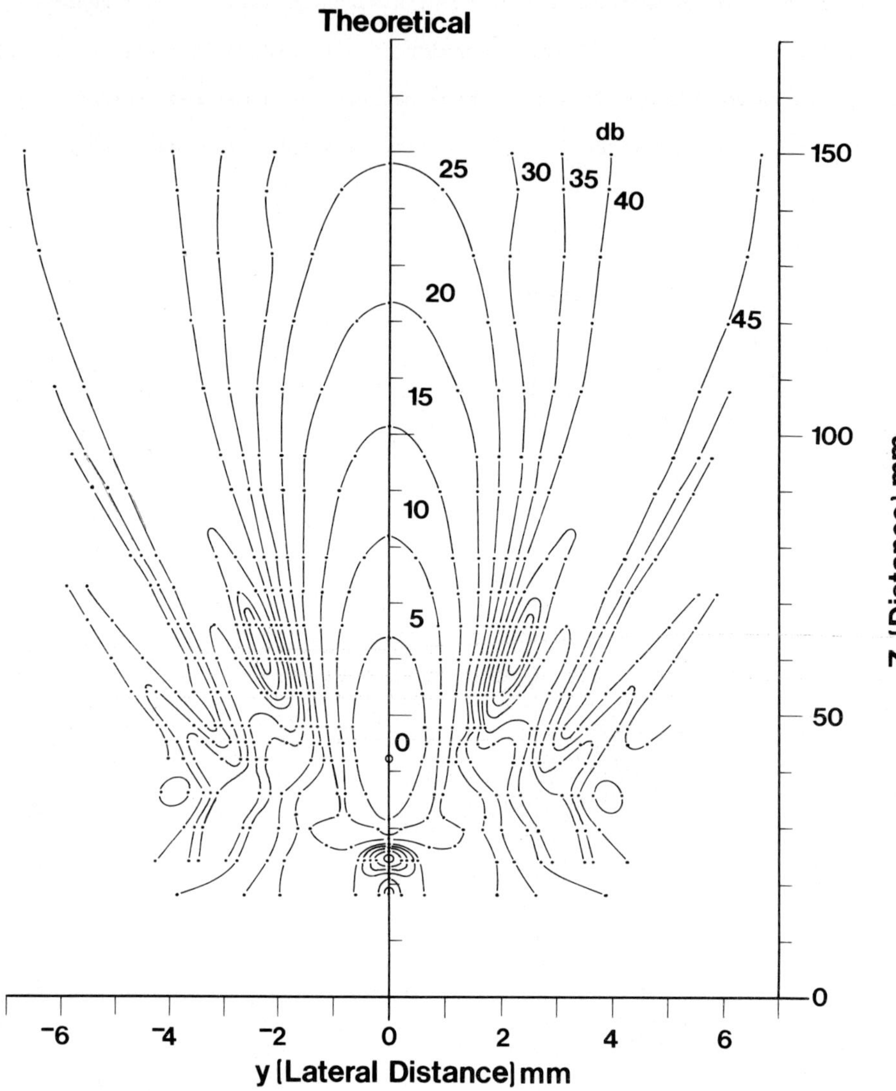

Fig. 11 Echo-intensity contour pattern when $\hat{D}$ = 14. (see also next page)

axial length and the other requirement that the number of the interference maxima and minima along the axis around the focus should be as small as possible.

For example, a transducer of R = 100 mm, a = 15 mm at 5 MHz in water (λ = 0.3 mm) has the dimension parameter of $\hat{D}$ = 7.5, and the pattern of

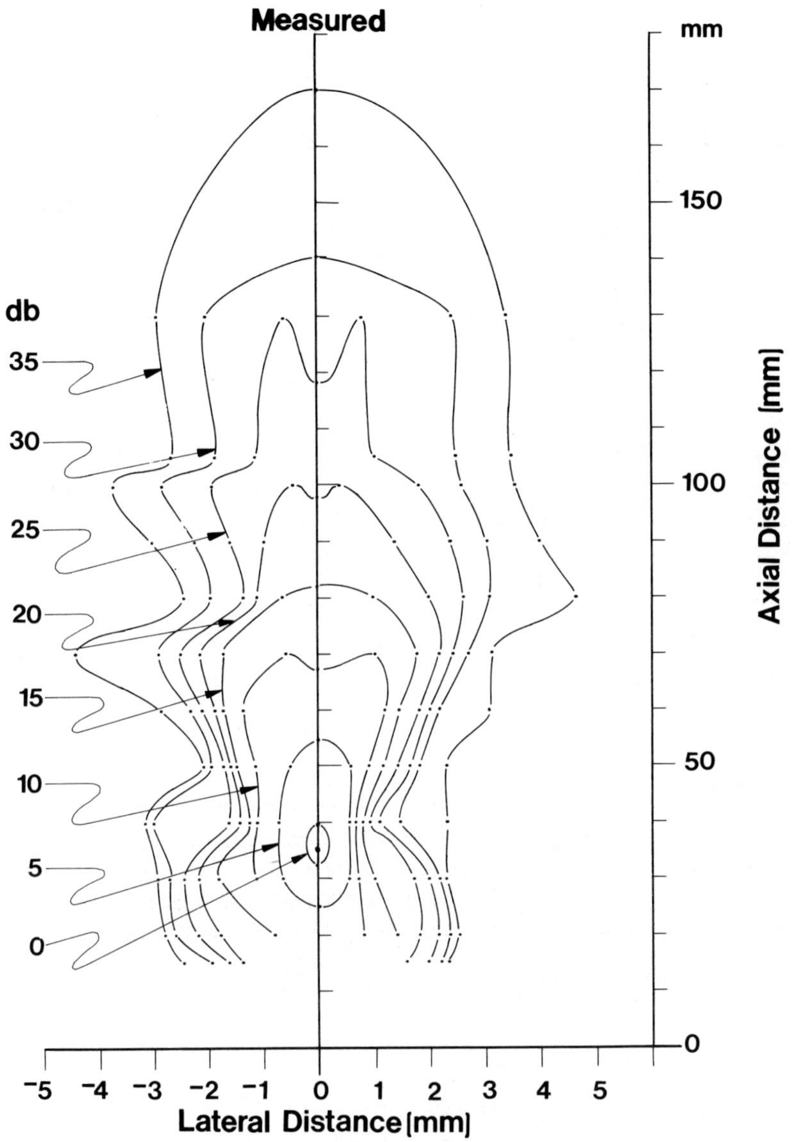

echo-intensity* around the focal region is as shown in Figure 10. For another transducer of R = 60 mm and a = 5 mm at 5 MHz, the dimension parameter is $\hat{D}$ = 1.4, and the pattern is shown in Figure 11. In each figure, the pattern to the left is the theoretical one** and that to the right is the measured one.

---

* For the experiment, the target is a small sponge ball placed in water; for theoretical patterns, the sound pressure value at every point is squared and the contour lines are drawn.

** Torikai (Private communication).

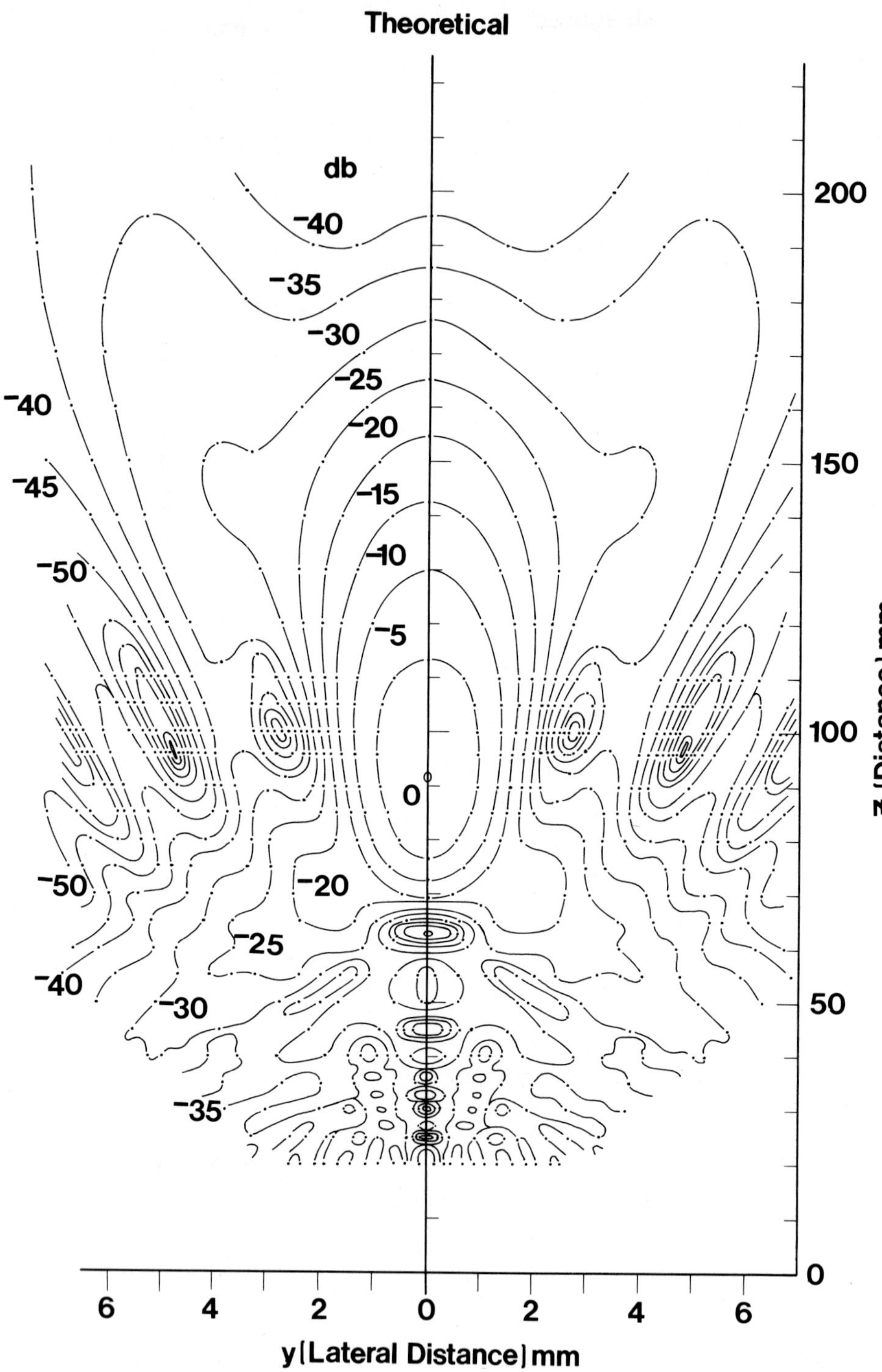

Fig. 12 Echo-intensity contour pattern when $\hat{D} = 3.4$. (see also next page)

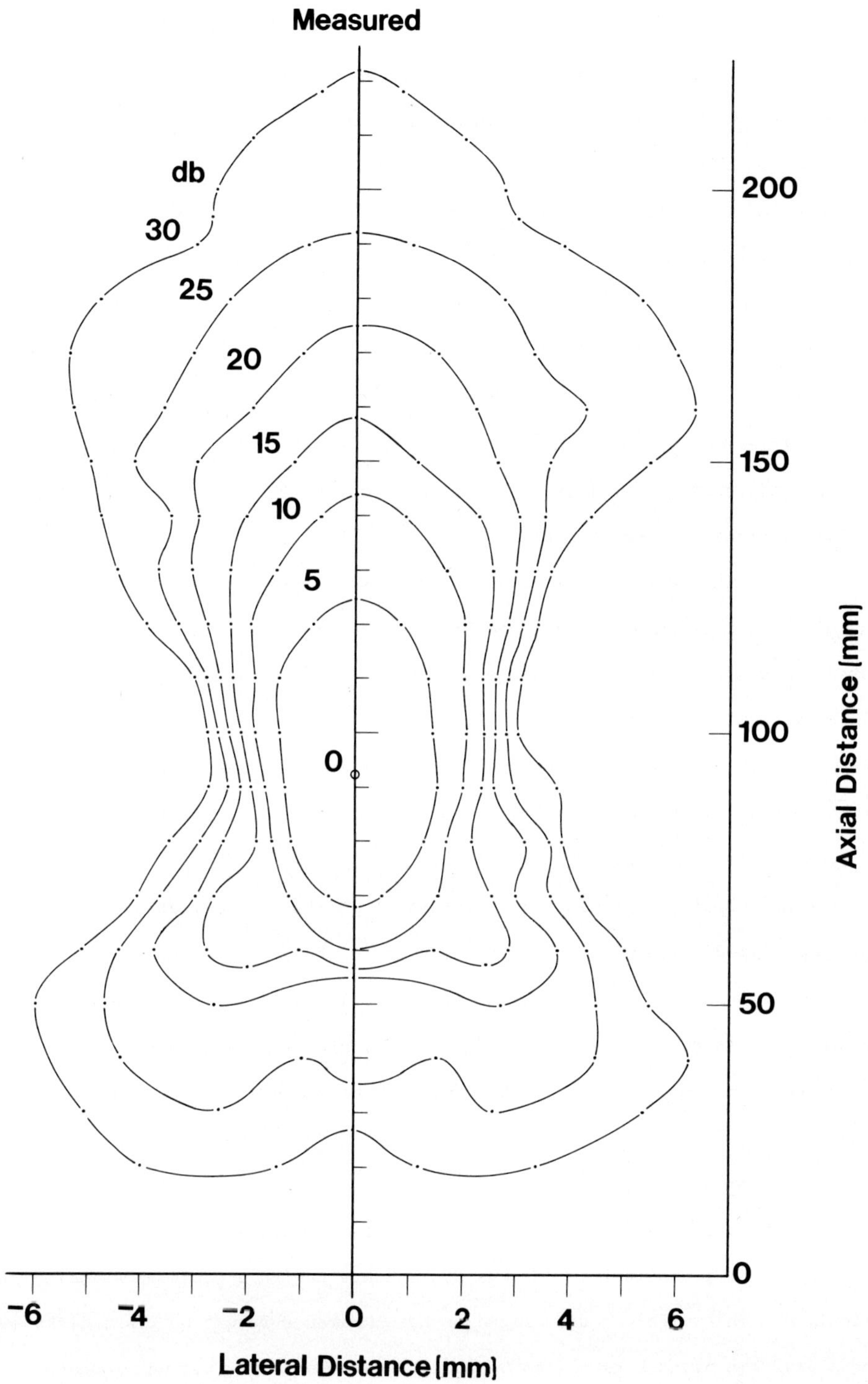
Measured
db
30
25
20
15
10
5
0
200
150
100
50
0
Axial Distance [mm]
−6
−4
−2
0
2
4
6
Lateral Distance [mm]

As stated above, a compromise can be found in any actual system design. For the case of $\hat{D} \simeq 3$, a transducer having R = 100 mm, a = 15 mm at 2.25 MHz (λ = 0.666 mm), the results are shown in Figure 12. The sought for compromise is attained by setting $\hat{D}$= 3.4, which fulfills the condition of the above recommended value of $\hat{D}$ = 2 to 4. The actual resolving power has proved to be sufficient for clinical use in cardiology[45,46] and in other clinical specialties.

Concave transducers can be made from quartz crystals of appropriate orientation by grinding them into spherical shells of uniform thickness. Electrostrictive ceramics such as Barium titanate or PZT are also used, obtaining larger sensitivity and easier matching to electrical circuits. Permanent polarization for the ceramics is given at the final stage of the fabrication so that the polarization is properly oriented with regard to the spherical surface.

Concave transducers are also used for intense irradiation of ultrasonic waves on biologic tissues. As an example,[47] a focusing irradiator of this type was used for stereotaxic encephalotomy at 3 MHz. The spherical shell is made of an electrostrictive ceramic, 50 mm in diameter and 65 mm in radius of curvature, and resonates at 3 MHz with fundamental thickness mode. It was reported that 1 kW/cm^2 in focal intensity was obtained by 7 W of input electric power, the focal diameter being observed to be 1.6 mm.

For ultrasonic processing on an industrial scale, large transducers are often required at lower frequencies such as 100 kHz, and curved mosaics have been proposed.[48] The radiating surface is built up as a mosaic of a large number of short columns made of barium titanate ceramics. When hexagonal columns are employed, the packing density of the mosaic can readily be increased.

2.3.4 Shaded Concave Disk

Filipczynski et al[49] conducted a theoretical analysis of the sound field produced by piezoelectric concave transducers having a Gaussian velocity distribution. This work was stimulated by Krautkrämer's result on plane disks having a Gaussian distribution (see Section 2.2). The theory was developed for the conditions of:

$$a \ll F \text{ and } a \gg \lambda,$$

i.e., the radius of the concave disk, a, is much smaller than the focal distance, F, and much larger than the wavelength. The final expression published for the sound pressure on the focal plane is expressed as

$$p = ik\rho c w_o \frac{G_o^2}{2F} e^{i\{\omega t - kF\left(1 + \frac{y^2}{2F^2}\right)\} - \frac{k^2 y^2 G_o^2}{F^2}} \quad (14)$$

where G_o is the characteristic parameter for the Gaussian distribution, w_o the

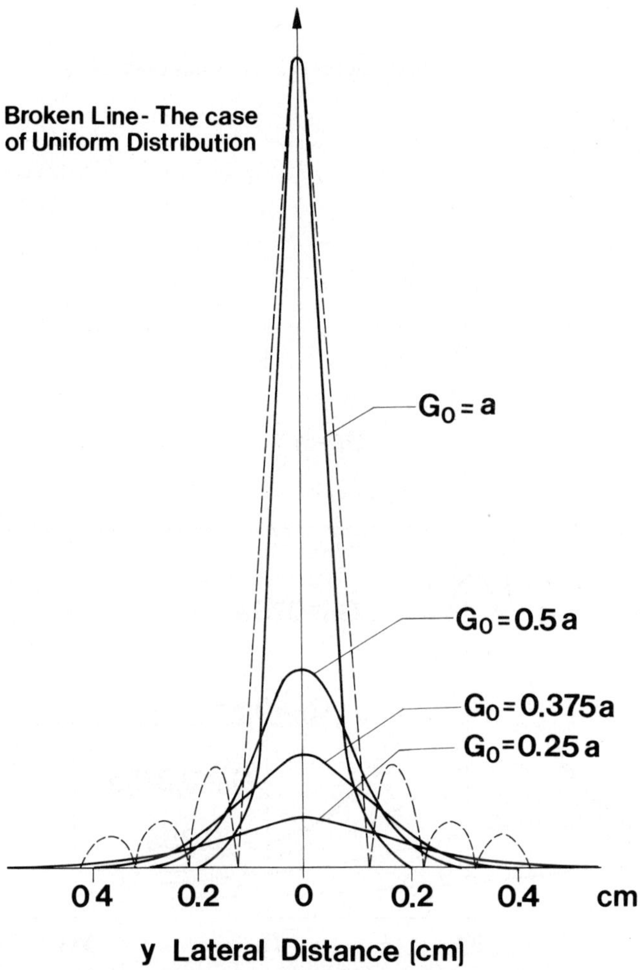

Fig. 13 Sound pressure on the focal plane produced by a concave spherical shell having Gaussian velocity distribution. Parameters being G, the Gaussian parameter (see Eq. 15). The curves being a few calculated examples with a transducer of: a = 1 cm, F = 10 cm, λ = 0.2 mm (Filipczynski).

maximal velocity, and y is a distance on the focal plane, measured from the beam axis. The velocity distribution $w(r)$ is given with reference to the radial coordinate, r, on the transducer surface as

$$w(r) = w_o e^{-\frac{r^2}{G_o^2}}. \qquad (15)$$

Figure 13 shows results for the pressure distribution on the focal plane, calculated by the above formulae. As expected from the idea of a Gaussian distribution, complex maxima and minima do not appear, in contrast to those appearing in Figure 8 of Section 2.3.3.

The pressure distribution along the beam axis was also derived as

$$p = \frac{ik\rho c w_o G_o e^{i(\omega t - kz)}}{2\{\frac{z}{G_o} + \frac{ikG_o}{2}(1 - \frac{z}{F})\}} \qquad (16)$$

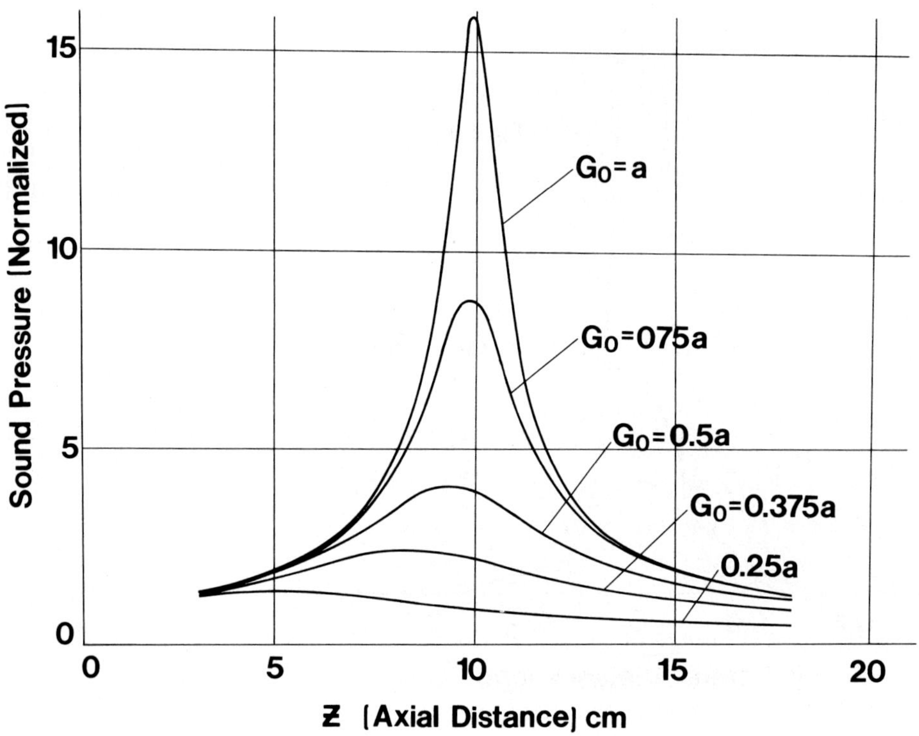

Fig. 14 Sound pressure along the beam axis of a concave spherical shell having Gaussian velocity distribution. Parameters being the same as Fig. 13.

Figure 14 shows results from this calculation, and the monotonic distribution also appears as expected.

Figures 13 and 14 above are for the case of a = 1 cm, F = 10 cm and λ = 0.2 mm (7.5 MHz in water).

2.3.5. Focusing Cylinder

Piezo-active cylindrical shells (i.e., a part of a hollow cylinder) were once used for ultrasonic processing. Presently, however, whole hollow cylinders are generally used. The ultrasonic field is the liquid contained in, or flowing in the cylinder. This circumstance involves a standing wave field with a pressure loop on the cylinder axis. The word "focusing" is used for this case, but the focusing is not that of progressive waves as stated in the foregoing sections.

For cavitation processing, ultrasonic waves of frequencies lower than 100 kHz are generally effective. In this frequency range, the radial vibrating mode of piezo-active hollow cylinders is more technically convenient than the thickness vibrating mode. The outer surface of the cylinder is non-loaded (i.e., air), and the resonant frequency of the cylinder-liquid system is designed with the considerations described below.[50]

The outer diameter, $\underline{d_o}$, and the thickness of the hollow cylinder, $\underline{t}$, are determined so that the radial mode resonant frequency of the composite cylindrical column, consisting of the hollow cylinder and the liquid column, is equal to the resonant frequency of a solid column of the same diameter, made of the same piezo-active material. A water-filled Barium titanate cylinder resonates in this mode when t/d_o = 0.225, provided that the length of the cylinder is sufficiently long. The formulas for this design have been given in the literature cited above.

2.4 Near-Field Patterns of Semi-Circular Transducers

Ohtsuki and Okujima[51] have developed, in recent years, a mathematical method to transform the surface integral of the usual Rayleigh's sound field function into a linear integral by introducing a sort of assisting function which they have named the ring function.[52] This method has facilitated com-

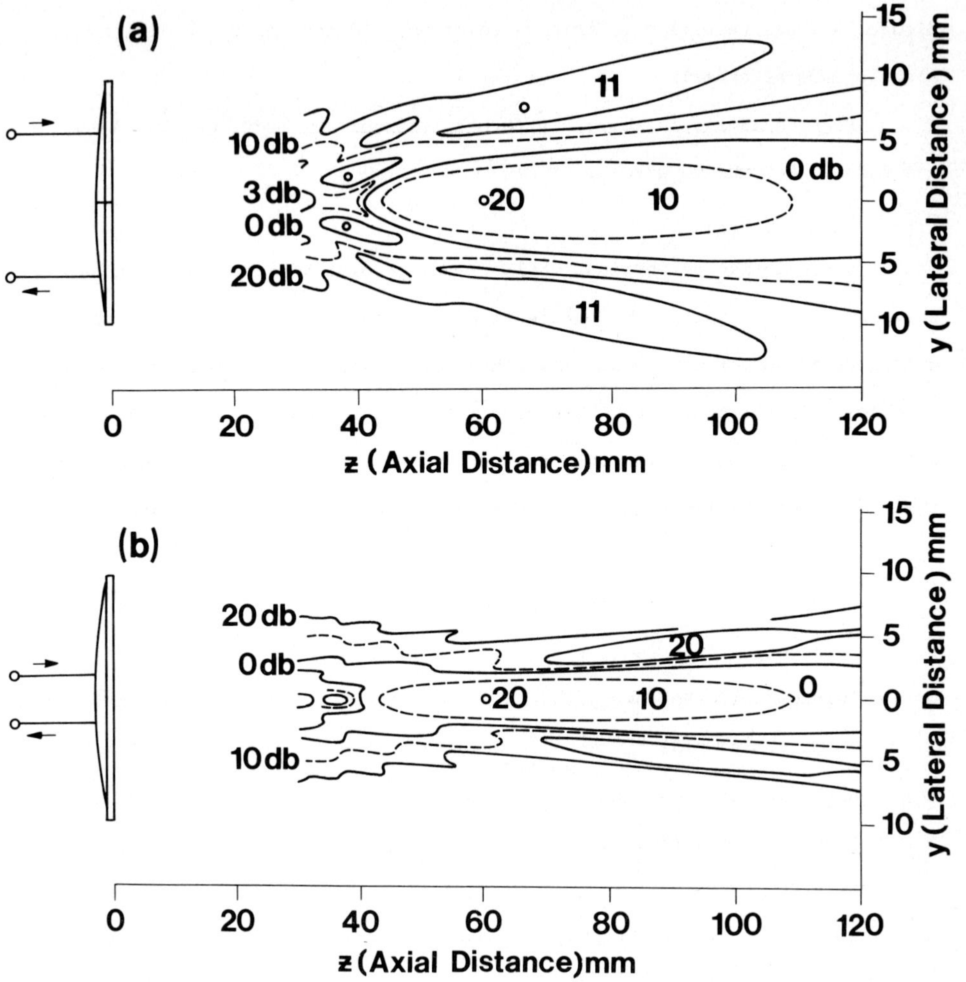

Fig. 15 Near-field gain pattern of a bisected circular concave shell when used for a pair of transmitting and receiving transducers. (a) In x-z plane. (b) In y-z plane. The coordinate axis y being parallel to the bisecting line. (Ohtsuki and Okujima).

puter calculation for the numerical values, and is considered useful for engineering-design procedures.

Using their method, they have published the near-field patterns of various semi-circular transducers of plane and concave shapes.[53] Figure 15 shows a numerical example, when a bi-sected circular concave shell is used as a pair of transmitting and receiving transducers of 2 MHz to be operated in a

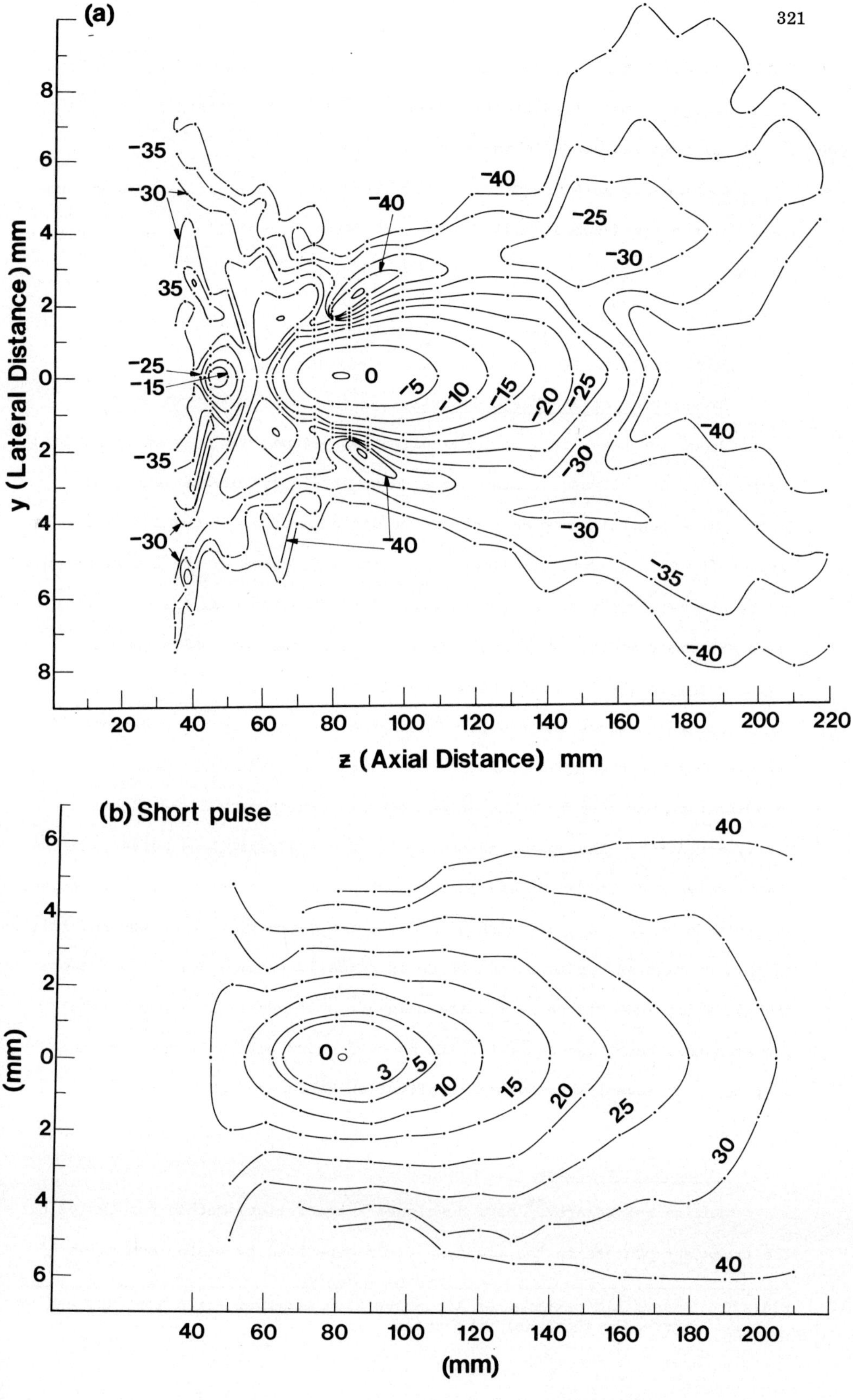
(a)
y (Lateral Distance) mm
z (Axial Distance) mm
-35
-30
35
-25
-15
-40
0
-5
-10
-15
-20
-25
-30
-35
(b) Short pulse
(mm)
40
0
3
5
10
15
20
25
30

Doppler velocity meter. The diameter and the radius of curvature of the shell are 20 mm and 80 mm, respectively. Figure 15(a) is the pattern in the X-Z plane, and (b) is that in the Y-Z plane.

A composite pair of quadri-sected circular concave shells is also proposed for the transducer, showing better patterns in both the planes of X-Z and Y-Z.

3. SOUND FIELD PATTERNS FOR EXTREMELY SHORT PULSES

3.1 Experiments on a Spherical Shell

Resolving power in tissue visualization depends not only on azimuthal sharpness of the ultrasonic beam but also on the length of the wave-train (i.e., pulse length) which determines the distance of range resolution of ultrasonic echoes. But, as azimuthal sharpness is based on the interference of the ultrasonic waves, the interference itself varies when the wave-train becomes extremely short. The short wave-train contains too small a number of waves to completely interfere with each other.

Figure 16 shows the result of measurements for a specific example.[54] The patterns are the equal intensity contours of an ultrasonic beam of the focusing type observed by means of the echo-intensity method (see Section 2.3.3, footnote). The upper pattern (a) is for the usual long pulse, and the lower pattern (b) is for an extremely short pulse, as short as one wavelength. The transducer is a concave disk with dimension parameter $\hat{D} = 3$ (see Eq. 12). In this particular configuration which is normally recommended for ultrasonic visualization (see Figure 12 and the relevant statement), it is found that the azimuthal resolving power for the case of short pulses is not very much degraded. Moreover, the favourable situation having no sub-maxima is attained.

3.2 Transient Processes

Kozina and Makarov[55] have developed a theoretical method for analyzing the transient process in the acoustic field generated by piston membranes of

---

Fig. 16 Near-field patterns of a concave disk observed for a long (a) and an extremely short (b) wave-train.

arbitrary shapes.* The method consists of using the Laplace transform of Schoch's linear integral while introducing the concept of a sound field divided into two regions of different nature in a transient process. One region is the space inside a cylinder whose base is the outer edge of the piston membrane of arbitrary shape, and the other region is the space outside the cylinder. The relating integrations are to be conducted for each region, taking into account the transient time of perturbations reaching the points of observation.

Later, the same authors[56] applied the method of analysis to circular, square and triangular piston membranes, and showed the transient waveforms in each region when the pistons are suddenly made to oscillate sinusoidally, and also displaced with a step function. Some interesting findings thus far obtained are as follows:

(i) The transient process of the field in the outside region generated by the circular piston undergoing a sinusoidal burst shows that the first half wave in the leading front is separable from the one which follows. In the trailing front of the transient, however, the sound field can either increase or decrease, depending on the direction-angle of the observation point; this sound field increases in the directions for which the stationary directivity has sub-maxima (i.e., side-lobes), and it decreases in the directions for the directivity valleys between the lobes.

(ii) When a circular piston is excited to make a sudden step displacement, the waveform of the pressure on the beam axis is a single square pulse, and the pulse-length becomes shorter with increasing distance.

(iii) In the case of a triangular piston, the transient acoustic field is dominant only at observation points which are facing perpendicularly to any of the three sides of the triangle.

Kasai et al.[57] have recently defined three such regions of different transient processes in the case of concave disks as shown in Figure 17, the point 0 bei

* It is assumed that the outer contour is fairly convex.

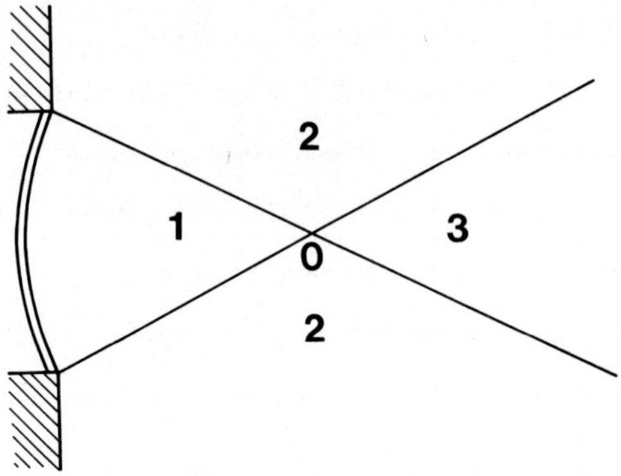

Fig. 17 Three regions of different transient processes in the acoustic field generated by an ultrasonic concave transducer.

the center of the radius of curvature. Region 1 corresponds with the inside region of Kozina-Makarov's, and Region 2 corresponds with the outer region. Echo waveforms were observed by using an extremely short ultrasonic pulse and the transient processes were confirmed as similar to Kozina-Makarov's descriptions. Region 3, however, is a new region where the transient process is otherwise discussed.[57]

Farn and Huang[58] have proposed a numerical method for computing transient processes in an acoustic field which is generated by any time-dependent motion of a body of arbitrary shape. The theoretical formulation has been designed by taking digital computation in mind, and no iteration is needed during the computing process. The method is based on source-density formulation and any transient pressure field can be predicted in general cases. A calculated example is shown for a sudden oscillation of a rigid sphere, and its error is discussed by comparing the result with the Kirchhoff's exact solution which has already been known. For a short lapse of time, the error is small. For a long lapse of time, however, the deviation becomes significant due to the advancing computation of the source density which is formulated with certain approximations. A slight deviation may be piled up every time when the time-step is advanced. The authors stated that the accuracy can be improved by "refining" the source elements together with reducing the intervals of the time-step accordingly.

4. GENERATION AND RECEPTION OF SHORT ULTRASONIC PULSES

Among factors affecting distance resolution of the pulse-echo method, the most important are the time length of the generated ultrasonic waves and the time length of signals when echoes from targets are received and transformed into electric signals. In case an extreme resolution is required, the transient response of the transmitting and receiving transducers is decisive.

For the theoretical analysis of the transient response it is possible, of course, to employ the Fourier transform method. However, as the method is based on frequency domain analysis, the handling is generally complex and the physical image of the phenomena is not easily conceived. Research on the analysis of the phenomena directly performable in the time domain has been developed by Ponomarev,[59] Redwood,[60,61] Eisenmenger,[62] Stuetzer[63,64] and other research workers. These results are applicable in cases of special boundary conditions. On the one hand, Cook,[65] Filipczynski,[66] Stephanishen[67,68] and Kasai[69] have been engaged in methods of approach using equivalent circuits for electro-acoustic systems. Especially, attacks have been made analyzing extremely short pulses, as short as one cycle or half-cycle excitation.[70]

As to practical methods for generation and reception of short pulses, several ideas have been in use:

4.1 Acoustic Backing

The rear surface of a transducer is directly coupled to a material having a specific acoustic impedance which is possibly close to that of the transducer material.[71,72,73] The nature of high absorption is simultaneously important. For a ceramic transducer, brass has been suggested by giving it a wedge shape and applying absorbing material on the tapered surfaces.[74] Extremely short pulses are successfully obtained. Acoustic characteristics of various resin-metal powder mixtures have been published.[73,75] An idea of using the same material as that of the transducer has recently been suggested[76] by giving a special shape to prevent multiple reflections in the backing.

4.2 Off-Resonance and End-Firing

As has been well known, a simple method still often used is one in which the frequency of the exciting electric wave-train (pulse) is chosen to be very different from the mechanical resonance of the transducer. Recently, an end-firing method has been proposed in which an interdigital electrode is used.[77]

4.3 Active Damping

Mechanical ringing in a transducer is quenched by impressing certain electrical waves on the transducer. Actual experimental results have been disclosed in which an electrical delay-line is used to obtain the quenching electrical voltage having the same waveform but opposite phase in relation to the original excitation voltage.[78] An extremely short ultrasonic pulse is successfully generated.

4.4 Intermediate Acoustic Layer

Some layers of different materials are inserted between a transducer and the acoustic medium in order to transform the latter impedance (resistance) into high mechanical impedance to be loaded on the transducer surface, so that the transducer is heavily damped. The simplest device is a quarter wavelength layer of a material to be applied directly on the transducer surface when the specific acoustic impedance (resistance) of the layer material has a value between those of the transducer and the acoustic medium. For example, a quarter wavelength layer is made of selected resin when a piezoelectric transducer is to be used in water. Literature covering analysis and design procedures involving this kind of problem have been published [79-83] for quite a long period of time. Some of the literature covers the case of acoustic delay-lines.

Kasai, Okuyama and Kikuchi[83,84] have recently shown that an extremely short ultrasonic pulse can be handled with high sensitivity both in its generation and reception in water without using any acoustic backing. The impedance of the layer material is designed as follows: Denoting the specific acoustic impedance of an acoustic medium by Z_F, that of a transducer by $\underline{Z}_o$, and that of the λ/4 layer material by $\underline{Z}_1$, then the technological conclusion is that the appropriate relation is expressed as

$$\frac{Z_1}{Z_F} = \left(\frac{Z_o}{Z_F}\right)^{\sigma} \tag{17}$$

At an optimum choice of the layer material, the ultrasonic wave-train, excited by a unit impulse voltage applied to the piezoelectric resonator, will have the shortest length obtainable for a given dynamic range. When the index (σ) in Eq. (17) is 0.27, there will be a 50 db-dynamic range, and for an index (σ) of 0.35 there will be a 20 db range. With the λ/4 layer device, a considerable gain in sensitivity generally applies, in comparison to transducers with heavy backing designed for the generation of the same pulse length. Because this λ/4 device needs no backing, there is no useless energy dissipation as compared to the lossy backing material case.

4.5 Use of Wide Frequency Band

As has been theoretically known, wide frequency band transducers can handle short ultrasonic pulses. Some proposals and concrete constructions have been published,[79] and the design procedure has also been suggested, taking into account the relation between transducer materials and driving electric circuits.[73] Some other ideas and experiments[85] have been reported, including a multi-element transducer having logarithmic taper for the thickness of each element, and also a folded element transducer. On the other hand, a transducer of non-uniform thickness, or of non-uniform material, has also been described. Transducers using the electrostatic principle have been suggested and experiments[86] have been shown for the ultrasonic frequency range. This report is concerned with polymeric foils to be used for a wide-band transducer, covering a wide frequency range from 10 kHz to 5 MHz.

5. REMARKS FOR MEDICAL ENGINEERS

As it is not the scope of this Chapter to describe engineering details of the medical arts, electronic or acoustic scientists who are engaged or will be engaged in medical engineering should consult many other papers and books which contain such details. Relevant publications which the present author considers to be valuable are listed later in this chapter, and some brief introductions of them are described in the following:

A. Interdisciplinary Problems

Interdisciplinary problems between medicine and ultrasonics are dealt with in the references already cited 5, 6 and in additional references A1 through A10. Many papers relevant to interdisciplinary problems have been published. Two of the present author's papers (Refs. A11 and A12) are helpful in understanding general features of the clinical arts prevailing in Japan; Ref. A12 is particularly relevant as it provides for a quantitative description in the clinical examination which is common in all fields of clinical medicine.

B. Ultrasonic Engineering in General

Books on general ultrasonic engineering include references (B1-B9) in addition to those already cited (Refs. 1,5,6 and 9). Among these, Ref. B2 deals mainly with ultrasonic transducers, covering magnetostrictive and piezo-electric transducers and their basic problems including the concept of piezo-active constants, design procedure, ultrasonic field problems relating to the transducers, etc. Ref. B3 describes mainly piezo-active materials in detail and covers their large amplitude use. Ref. B4 is a 1803 page handbook, though written in Japanese, comprehensively summarizes ultrasonics, ranging from physics to practical technology. Refs. B5, B7 and B8 include many monographs dealing with ultrasonic visualization including holographic and pulse echo methods. Ref. B6 is helpful in learning about ultrasonic energy applications, especially of ultrasonic tools. Ref. B9 describes ultrasonic methods for industrial inspection; so-called ultrasonic flaw detection, and is a good reference for medical and/or biological inspection, as this technique originated from the flaw detection technique.

C. Piezo-Active Materials and Mechanical Fatigue

In addition to the reference books cited above, Ref. C1 is a monograph discussing the limitation of ultrasonic power available from transducers. Its relation to cavitation phenomenon is included, taking into account mechanical fatigue of the transducer material. Ref. C3 discloses the fatigue phenomena experimentally observed in ferrite material.

As already illustrated in Section 1, ultrasonic cavitation, which accompanies intense vibration of a transducer in a liquid medium, reduces the acoustic radiation impedance acting on the transducer as its mechanical load, and results in mechanical fatigue of the transducer. Elaborate experimental investigations on cavitation have been carried out to describe the characteristics quantitatively. The decrease in the value of the impedance becomes dominant with the dissolved air content, decrease of ultrasonic frequency, increase of ultrasonic pulse length, decrease of the pulse rate, etc. Ref. C6 summarizes these data.

Ref. C2 is a review of magnetostrictive materials. Ref. C4 is also a review of ultrasonic transducers in general, covering the historical background up to present status of VHF and UHF transducers. High power operation is discussed together with the experimental method of observation. Two types of UHF transducers are shown in Ref. C7; one is a transmission-line-coupling type and the other is a resonant r.f. cavity type. This reference describes a system in which a piezoelectric plate is excited in a higher harmonic mode of vibration, (as high as the 487th harmonic) generating nearly 2 kMHz ultrasonic waves in liquids.

A comprehensive table and characteristic figures of piezoelectric ceramics are available in Ref. C5. For newer ceramics, the properties are described in Ref. B3.

D. Transducer Sensitivity Measurement

With regard to the measurement of transducer sensitivity and/or output power, various methods have been published such as radiation force methods in which a special reflecting plate with a spring balance, or a solid ball suspended in the acoustic field, is deflected by the radiation force. Thermal probes such as thermocouples or thermistors of special design have also been used for power measurement. In addition to these well-known methods applied in the continuous wave case, other developments have recently become available and include: pulsed-wave intensity measurements,[D1] the standard mutual calibration principle applicable in the MHz ranges[D2] utilizing a special microphone,[D3] and a uterine probe proposed to measure the ultrasound intensity of obstetric devices.[D5] A transient method of measuring $\underline{k}$ (electro-mechanical coupling factor) in circular piezoelectric plates is also available.[D4]

Other measurement methods include light diffraction methods and calorimetric methods. All of these methods are expanded on in Chapter V of this book.

E. High Speed Scanning Transducers

High speed scanning of biologic tissues is achieved by both mechanical and electronic means. Ref. E1 is concerned with mechanical scanning and describes the ultrasonic cinematography of the living heart. Ref. E2 proposes an electronic sector scanning system for diagnostic use and Ref. E7 describes its engineering details, in which the electronic steering of ultrasonic pulses and the real-time processing of phased array reception are dealt with. Ref. E10 describes clinical results when the system is applied to cerebral diagnostics.

Ref. E3 is concerned with mechanical scanning, in which a rotating ultrasonic beam is reflected by a mirror so that a linear scanning beam is obtained, and Ref. E4 describes clinical applications involving observation of fetal movements in the uterus.

Ref. E5 is a relatively new proposal involving electronic scanning, in which a linear array of 200 transducers are electronically controlled to obtain a linear scan. Each acoustic beam involves a phased array complex composed of any 20 adjacent transducers of the 200 total elements. Ref. E6 deals with the near field pattern of a phased array in the MHz range. In Ref. E8, phased array transducers of a focusing type are illustrated. Ref. E9 shows the application of electronic radial scanning to an intracardiac catheter, using the phased array principle. Ref. E12 is concerned with a sector-scanning electronic system, and Ref. E13 involves a discussion on the element arrangement of a phased-array system for sector scanning in diagnostic tomography.

Ref. E11 considers an electronic linear-scanning system in which 20 individual transducers of the directional beam type are simply switched from one to the next successively so that real-time observation of the living heart is possible.

Ref. E14 considers high speed mechanical sector-scanning system to be used for cardio-tomography, in which a directional transducer is rapidly swung so that a 40° - 90° sector scan of 30 frames per second is attained.

F. Signal Processing in Transducer Systems

Signal processing techniques have recently been introduced into ultrasonic transducer systems for medical use to improve the sharpness of the ultrasonic beam of a simple transducer arrangement,[F1] and/or to add such functions as variable focusing so that electronic rapid control of focal distance is attained.[F2,F3,F4] These techniques can be employed not only self-consistently but also additionally to high speed scanning techniques. Synchronized ultrasono-cardio-tomography[F5,F6] and cardiac kineto-ultrasonic tomography[F7,F6] have been proposed and used in clinical medicine. These methods are based on the electronic processing of the transducer motion and cathode-ray unblanking by synchronizing signals extracted from E.C.G. current. For the arrhythmia case, computer processing has been proposed in synchronized tomography.[F8,F6] Multiplex representation systems of ultrasonic tomograms have recently been disclosed in wich a simultaneous display of tomograms for different cardiac phases or different sensitivities[A12] of any single biologic section is attained by means of electronic switching in real-time mode.[F9,F10]

G. Trans-Ductal and Catheter Type Transducers

Special arrangements have been proposed for using ultrasonic transducers in the ductal organs of living human subjects. These are: Trans-vaginal,[G4,G5] Trans-esophagial,[G6,G7] and Trans-tracheal[G8,G9] uses, which will provide additional variations for clinical examinations other than the usual transrectal[G1,G2] transducers. Though not transductal, intracardiac catheter transducers have already been in use.[G10,G11]

H. Solid Horn

As is well-known, a solid bar of tapered cross-sectional area has transformer action for mechanical vibration when attached to a transducer surface, and has long been used for ultrasonic cutting and drilling. In recent years, the technique has been applied in medicine as a surgical tool. Design procedures are not much different from those for industrial use. The ultrasonic frequency is less than 100 kHz in most cases, and the solid horn is designed to resonate with the transducer resonance. For a simple conical horn, a description is available in Ref. H1 published in 1955. Later on, however, catenoidal and exponential horns have been suggested and their engineering characteristics discussed by many research workers.[H2]

For large amplitude use, the design procedure is not easy. As to the driving transducer, saturation of the vibro-motive force must be considered together with mechanical fatigue, both limit the mechanical power output and the electromechanical transformation efficiency. These latter two factors are dependent on the mechanical load which is, of course, the sending-end impedance of the solid horn itself.

As to the horn design, the selection of its configuration is one of the decisive factors for its power handling capability without producing fatigue breakdown even when the same transformation ratio of both end-velocities is given. In this connection, employment of a two- or three-step cascade horn has been introduced. Details of these problems have been discussed and concrete design procedures have been established. A comprehensive text is available.[H3] The suggested order of design calculations is as follows: (1) Mechanical power output at the horn tip (i.e., the small end), (2) the transformation ratio in relation to the driving transducer conditions, (3) choice of the horn shape, (4) selection of the horn material, (5) determination of the length of the horn, (6) mechanical power dissipation in the horn, and (7) the acting surface area of the transducer, which is determined by the required total power calculated through the above procedure from (1) to (6). For these design procedures, the most necessary properties are the large amplitude characteristics and fatigueless limits for alternating stress in both materials (transducers and horns).

REFERENCES

Section 2

1. H. Stenzel: Leitfaden zur Berechnung von Schallvorgagen, Julius Springer, Berlin, (1939).

2. T. F. Hueter and R. H. Bolt: Sonics, John Wiley & Sons, New York, (1955) p. 63.

3. H. Stenzel: l. c., pp. 51-80.

4. H. Backhaus and F. Trendelenburg: Z. tech. Phys., 7 (1926) p. 630.

5. P. N. T. Wells: Physical Principles of Ultrasonic Diagnosis, Academic Press, London (1969) p. 54.

6. W. J. Fry and F. Dunn: Ultrasound--Analysis and Experimental Methods in Biological Research. In Physical Techniques in Biological Research, W. L. Nastuk, ed., Academic Press, New York (1962), Chapt. 6, p. 296.

7. Y. Torikai and K. Negishi: On the Ultrasonic Field Near the Circular Piston-Like Source--(I) Visualization of the Ultrasonic Field. J. Acoust. Soc. Japan 13 (1957) p. 117.

8. Y. Torikai: On the Ultrasonic Field Near the Circular Piston-Like Source--(II) Theoretical Calculation of the Field. J. Acoust. Soc. Japan 14 (1958) p. 36.

9. J. u. H. Krautkrämer: Werkstoffprüfung mit Ultraschall, Springer-Verlag, Berlin (1961) pp. 59-60.

10. G. Kossoff, D. E. Robinson and W. J. Garrett: Ultrasonic Two-Dimensional Visualization Technique. IEEE Trans. Sonics-Ultrasonics, Vol. SU-12, No. 2, June (1965) pp. 31-37.

11. K. Shibayama and Y. Kikuchi: Studies on Vibration of Short Column. Sci. Rep., Res. Inst. Tohoku Univ., Vol. 8 (1956), Vol. 9 (1957), Vol. 11 (1960) and Vol. 13 (1961); also see Chapt. 9 in Ultrasonic Transducers, Y. Kikuchi, ed., Corona Pub. Co., Tokyo (1969) pp. 294-335.

12. For example see Figure 5 of Ref. 10.

13. I. Wolff: IRE, May (1937) p. 630.

14. K. Menges: Akust. Zeitschr., March (1941) p. 90.

15. K. Fukushima, J. Saneyoshi and Y. Kikuchi: Sound Field Characteristics in Relation to Ultrasonic Transducers. In Ultrasonic Transducers, Y. Kikuchi, ed., Corona Pub. Co., Tokyo (1969) pp. 336-382.

16. Y. Kikuchi: Japanese Patent No. 246569 and No. 247363. (Application in 1954, licensed in 1958).

17. See Figure 3.13 of Ref. 2, p. 71.

18. See Figure 11 of Ref. 6, p. 297.

19. K. von Haselberg and J. Krautkrämer: Ein Ultraschall-Strahler für die Werkstoffprüfung mit verbessertem Nahfeld. Acustica 9 (1959) pp. 359-364.

20. See, for example, Figure 7.13 of Ref. 2, p. 264 and/or Figure 3.2 of Ref. 5, p. 59.

21. S. Oloffson: An Ultrasonic Optical Mirror System. Acustica 33 (1963) pp. 361-367.

22. E. Kelly, D. Okuyama and W. J. Fry: Comparison of Ultrasonotomograms of the Liver of Swine. Record of the Spring Meeting of Japanese Soc. Ultrasonics in Medicine, No. 11-63, May (1967) p. 125.

23. F. L. Thurstone, N. I. Kjosnes and W. M. McKinney: Ultrasonic Scanning of Biologic Tissue by a New Technique. Science 139, No. 3681 (1965) p. 302.

24. F. L. Thurstone and W. M. McKinney: Focused Transducer Arrays in an Ultrasonic Scanning System for Biologic Tissue. In Diagnostic Ultrasound, C. C. Grossman, et al., eds., Plenum Press, New York (1969) p. 191.

25. See: W. J. Fry and F. Dunn, Figure 14(d) and Figure 48 of Ref. 6, p. 301 and p. 374.

26. M. Ide: Focusing by Means of a Parabolic Concentrator. (N.B.: the tentative translation of the Japanese title), J. Acoust. Soc. Japan, Meeting Record No. 1-1-12, October (1958) p. 23.

27. M. Ide, T. Wagai and T. Majima: Focused Ultrasonic Apparatus for Destruction of Biologic Tissues. (N.B.: The tentative translation of the Japanese title appeared in) J. Acoust. Soc. Japan, Meeting Record No. 2-1-9, May (1962) p. 119.

28. See, for example, Table 1 of Ref. 2, p. 265.

29. See: W. J. Fry and F. Dunn, Figure 17(c) of Ref. 6, p. 307.

30. W. J. Fry and F. Dunn: Ref. 6, p. 302-305.

31. Ibid, Eq. 46a, p. 304.

32. Ibid, p. 307-310.

33. W. J. Fry and F. Dunn: Ultrasonic Intensity Gain by Composite Transducers. J. Acoust. Soc. Amer., 34, No. 3 (1962) p. 188.

34. F. J. Fry: Recent Developments in Ultrasound at the Biological Research Laboratory and their Application to Basic Problems in Biology and Medicine. In Ultrasonic Energy, E. Kelly, ed., Univ. of Illinois Press, Urbana (1965) p. 204.

35. K. Yoshioka and M. Oka: Technical Developments of Focused Ultrasound and its Biological and Surgical Application in Japan. In Ultrasonic Energy, E. Kelly, ed., Univ. of Illinois Press, Urbana (1965) p. 190.

36. D. S. Grasynk, et al: Object Lens for Visualization of Ultrasonic Field. (N.B.: The tentative translation from the Russian title) in Abstract Paper (РЕФЕРАТЫ ДОКЛАДОВ) for the 8th All-Union Acoustic Conference of USSR, Moscow (1973) Vol. 2, Lecture No. U_{VB}-9, p. 271.

37. B. B. Brenden: Ultrasonic Holography - a Practical System. In Ultrasonic Imaging and Holography, Stroke, Kock, Kikuchi and Tsujuichi, eds., Plenum Press, New York (1974) p. 87.

38. Y. Torikai: Calculation of Acoustic Near-Field. (N.B.: the tentative translation of the Japanese title) Reports of the Professional Group on Ultrasonics, Inst. of Elect. Comm. Engrs. of Japan, October (1957) and November (1962); or see pp. 1399-1418 of Ref. B4.

39. H. T. O'Neil: Theory of Focusing Radiator. J. Acoust. Soc. Amer. 21, No. 3 (1949) pp. 516-526.

40. See P. N. T. Wells, p. 62 of Ref. 5.

41. See Eq. 7.7 of Ref. 2, p. 267.

42. G. Kossoff, D. E. Robinson and W. J. Garrett: See Ref. 10, p. 37.

43. Y. Kikuchi and M. Tanaka: Some Improvements in Ultrasonotomography for the Heart and Great Vessels -- Azimuthal Resolution of Concave Transducers. Reports of Professional Group on Ultrasonics, Inst. of Elect. Comm. Engrs. of Japan, October 26 (1966), in Japanese.

44. M. Tanaka and Y. Kikuchi: Favourable Effects of a Concave Transducer on the Ultrasono-Tomogram. Japanese J. Med. Elect. and Biolog. Engr. 5, No. 5 (1967) p. 18.

45. For example, Y. Kikuchi, D. Okuyama, M. Tanaka, T. Ebina and S. Oka: Ultrasonocardio-tomography and its Application to Morphological Measurement of the Heart. In Ultrasono Graphia Medica, J. Bock, et al., eds. Verlag der Wiener Medizinischen Akademie, Vienna (1971) Vol. III, p. 423.

46. Y. Kikuchi: Development and Present Aspects of Ultrasono-cardiotomography. In Ultrasonics in Medicine, M. deVlieger, D. N. White and V. R. McCready, eds., Excerpta Medica, Amsterdam (1974) p. 230.

47. M. Ide, T. Wagai, M. Ishii and T. Takeuchi: Instrumentation for Stereotaxic Encephalotomy for Animal Experiment with 3 Megacycle Focused Ultrasound. Digest of 6th Intl. Conference on Medical Electronics and Biological Engineering, Tokyo (1965) p. 259.

48. See Figure 7.16 of Ref. 2, p. 270.

49. L. Filipczynski and J. Etienne: Ultrasonic Focusing Piezoelectric System with Gaussian Velocity Distribution. Proc. 7th Intl. Cong. on Acoustics, Budapest (1971), No. 35-U-5, p. 637-640; also see Acustica 28, No. 2 (1973) p. 121.

50. T. F. Hueter and R. H. Bolt: Focusing Cylinder. In Sonics, John Wiley & Sons, Inc., New York (1955) p. 272.

51. S. Ohtsuki and M. Okujima: Calculation of Fresnel Zone of Slightly Confronted Semicircular Transducers. Med. Ultrasonics 9, No. 1 (1971) p. 95.

52. S. Ohtsuki: Calculation Method for the Near-Field of a Sound Source With Ring Function. J. Acoust. Soc. Japan 30, No. 2, February (1974) p. 76; or S. Ohtsuki: Ring Function Method for Calculating Near-Field of sound source. Bull. Tokyo Inst. Techn., No. 123, March (1974) p. 23, in English.

53. S. Ohtsuki and M. Okujima: Calculation of Near-Field Gain Patterns of Composite Transducers. Med. Ultrasonics 10, No. 1 (1972) p. 69.

Section 3

54. See the text and its references in: Y. Kikuchi: Development and Present Aspects of Ultrasono-cardio-tomography, Ref. 46, p. 235.

55. O. G. Kozina and G. I. Makarov: Transient Process in the Acoustic Fields Generated by a Piston Membrane of Arbitrary Shape. Soviet Physics-Acoustics 7, No. 1, July-September (1961) p. 39.

56. O. G. Kozina and G. I. Makarov: Transient Processes in the Acoustic Fields of Special Piston Membranes. Soviet Physics-Acoustics 8, No. 1, July-September (1962) p. 49.

57. C. Kasai: Doctoral Thesis, Tohoku University, March (1974); also see: Y. Kikuchi, D. Okuyama and C. Kasai: On the Sound Field of an Extremely Short Ultrasonic Pulse. Meeting Record of Acoustical Soc. of Japan, No. 1-1-22, October (1973) p. 43.

58. C. L. S. Farn and H. Huang: Transient Acoustic Fields Generated by a Body of Arbitrary Shape. J. Acoust. Soc. Amer. 43, No. 2 (1968) p. 252.

Section 4

59. P. V. Ponomarev: Transient in Piezoelectric Resonators. Soviet Physics-Acoustics 3 (1957) p. 260.

60. M. Redwood: Transient Performance of a Piezoelectric Transducer. J. Acoust. Soc. Amer. 33(1961) p. 527.

61. M. Redwood: Piezoelectric Generation of an Electrical Impulse. J. Acoust. Soc. Amer. 33 (1961) p. 1386.

62. W. Eisenmenger: Eine Kontrollschallquelle für breitbandige Mikrofone zur Messung von Druckimpulsen in Flüssigkeiten. Acustica 12 (1962) p. 165.

63. O. M. Stuetzer: Multiple Reflections in a Free Piezoelectric Plate. J. Acoust. Soc. Amer. 42 (1967) p. 502.

64. O. M. Stuetzer: Secondary Stresses in a Stress-Pulse-Activated Piezo-electric element. J. Appl. Phys. 38 (1967) p. 3901.

65. E. G. Cook: Transient and Steady-State Response of Ultrasonic Piezo-electric Transducers. IRE Conv. Record, Part 9 (1956).

66. L. Filipczynski: Transient and the Equivalent Electrical Circuit of the Piezoelectric Transducer. Acustica 10 (1960) p. 149.

67. P. R. Stephanishen: Transient Analysis of Lumped and Distributed Parameter Systems Using an Approximate Z-Transform Technique. J. Acoust. Soc. Amer., 52 (1972) p. 270.

68. P. R. Stephanishen: The Time-Dependent Force and Impedance on a Piston in a Rigid Infinite Planar Baffle. J. Acoust. Soc. Amer., 49 (1971) p. 841.

69. C. Kasai, D. Okuyama and Y. Kikuchi: Transient Response Analysis of Piezoelectric Transducers Using Reflection and Transmission Coefficient at their Surface. Elect. and Comm. in Japan (in English, published by Scripta Pub. Corp., Washington, D.C.) (1973) Vol. 56-A, No. 3, p. 19.

70. M. Redwood: A Study of Waveforms in the Generation and Detection of Short Ultrasonic Pulses. In Applied Material Research, Heywood & Co., Ltd. London (1963) Vol. 2, No. 2, p. 76.

71. J. M. Reid and C. R. Joyner: The Use of Ultrasound to Record the Motion of Heart Structure. In Ultrasonic Energy, E. Kelly, ed., University of Illinois Press, Urbana (1965) pp. 278-293, Fig. 16.

72. See: P. N. T. Wells: Physical Principles of Ultrasonic Diagnosis, Academic Press, London (1969) p. 38.

73. G. Kossoff: The Effect of Backing and Matching on the Performance of Piezoelectric Ceramic Transducers. IEEE Trans. Sonics-Ultrasonics, Vol. SU-13, No. 1 (1966) p. 20-30.

74. Y. Kikuchi, D. Okuyama, C. Kasai and K. Tatebayashi: Generation of Extremely Short Ultrasonic Pulses into Liquid by Acoustic Backing Transducer. Record of Spring Meeting, Japan Soc. Ultrasonics in Med., May (1970) p. 53; also see Fig. 14 of Ref. B7, p. 253.

75. A. Lutch: Solid Mixtures with Specified Impedances and High Attenuation for Ultrasonic Waves. J. Acoust. Soc. Amer., 34 (1962) p. 131-132.

76. A. F. Brown and J. P. Weight: See Figure 1 of Ref. 85, p. 162.

77. L. J. van der Pauw: The Planar Transducer--A New Type of Transducer for Exciting Longitudinal Acoustic Waves. Appl. Phys. Letters, Vol. 9, No. 3, (1966) p. 129.

78. C. Kasai, D. Okuyama, Y. Kikuchi and G. H. Leichner: Active Damping of Piezoelectric Transducers to Generate Extremely Short Ultrasonic Pulse. Med. Ultrasonics 8, No. 1 (1970) (in English, published by Japan Soc. of Ultrasonics in Med., Juntendo Univ., Hongo, Tokyo) p. 40.

79. H. J. McSkimin: Transducer Design for Ultrasonic Delay Lines. J. Acoust. Soc. Amer. 27, No. 2 (1955) p. 302; also see: H. J. McSkimin: Performance of High Frequency Barium Titanate Transducers Generating Ultrasonic Waves in Liquids. J. Acoust. Soc. Amer., 31, No. 11 (1959) p. 1519.

80. K. S. Aleksandrov, et al.: Effect of an Intermediate Layer on the Frequency Characteristics of Ultrasonic Delay Lines. Soviet Physics-Acoustics 6, No. 2 (1960) p. 170.

81. G. Kossoff: see Ref. 73.

82. R. J. Kazys and V. Pomarkas: The Frequency and Transient Response of Piezotransducers with Intermediate Layers and Electrical Matching Circuits. Proc. 7th Intl. Cong. Acoustics, Budapest (1971) No. 25U3, p. 629.

83. C. Kasai, D. Okuyama and Y. Kikuchi: Generation Detection of Short Ultrasonic Pulses via Piezoelectric Transducers with an Intermediate Layer of Quarter Wavelength. Elect. and Comm. in Japan (1973) (In English, Scripta Pub. Corp., Washington, D.C.), Vol. 56-6A, No. 4, p. 43.

84. Y. Kikuchi: Development and Present Aspects of Ultrasono-Cardio-Tomography. In Ultrasonics in Medicine, M. deVlieger, et al., eds., Excerpta Medica, Amsterdam (1974) p. 230-238.

85. A. F. Brown and J. P. Weight: Generation and Reception of Wideband Ultrasonics. Ultrasonics 12, July (1974) pp. 161-167.

86. G. Curtis: A Broadband Polymeric Foil Transducer. Ultrasonics 12, July (1974) p. 148-154.

SPECIAL REFERENCES FOR ENGINEERS

Interdisciplinary Problems

A1. E. Kelly: Ultrasound in Biology and Medicine, American Institute of Biological Sciences, Washington, D.C. (1957).

A2. E. Kelly: Ultrasonic Energy--Biological Investigation and Medical Applications, University of Illinois Press, Urbana (1965).

A3. L. C. Gross, et al.: Diagnostic Ultrasound, Plenum Press, New York (1966) (The Proc. 1st Intl. Cong., Univ. Pittsburgh, 1965).

A4. J. Bock, et al: Ultrasono Graphia Medica, Verlag der Wiener Medizinischen Akademie (1971), Vol. I, II and III; (Proc. 1st World Congress on Ultrasonics in Medicine and SIDUO III, Vienna, June 2-7, 1969).

A5. D. Gordon: Ultrasound as a Diagnostic and Surgical Tool, E & S Livingston, Ltd., London (1964).

A6. K. Tanaka: Diagnosis of Brain Disease by Ultrasound, Shindan-To-Chiryo Co., Ltd., Marunouchi, Tokyo, English Ed. (1969).

A7. D. N. White: Ultrasonic Encephalography, Hanson & Edgar, Ltd., Kingston, Ont., Canada (1970).

A8. JSUM Publication Committee: Ultrasonics in Medicine--Principle, Research and Practice (in Japanese) Y. Kikuchi, ed.-in-chief, Igaku Shoin, Ltd., Tokyo (1973) Second Ed.

A9. See Ref. B7, appearing in a later page.

A10. M. deVlieger, D. N. White and V. R. McCready: Ultrasonics in Medicine, Excerpta Medica, Amsterdam/American Elsevier Pub. Co., Inc., New York (1974); Proc. 2nd World Cong. on Ultrasonics in Medicine, Rotterdam, June 4-8, 1973.

A11. Y. Kikuchi: Ultrasonic Approach to Internal Medicine--Application to Diagnostic Examination. Jap. J. Med. (English edition by Jap. Soc. of Intl. Med., Tokyo), Vol. 9, No. 2, April (1970) p. 70-80.

A12. Y. Kikuchi: Way to Quantitative Examination in Ultrasonic Diagnosis. Med. Ultrasonics, Vol. 6, No. 1 (1968) p. 1-8.

Ultrasonic Engineering in General

B1. T. F. Hueter and R. H. Bolt: Sonics, John Wiley & Sons, Inc., New York (1955).

B2. Y. Kikuchi: Ultrasonic Transducers, Corona Pub. Co., Ltd., Tokyo (1969).

B3. O. E. Mattiat: Ultrasonic Transducer Materials, Plenum Press, New York (1971).

B4. J. Saneyoshi, Y. Kikuchi and O. Nomoto: Cho-onpa Gijutsu Benran (Handbook of Ultrasonic Technology, in Japanese), Nikkan Koygo Press, Tokyo, (1971) 4th ed.

B5. G. Wade: Acoustical Holography, Plenum Press, New York (1972) Vol. 4.

B6. L. D. Rosenberg: Physical Principles of Ultrasonic Technology, Plenum Press, New York (1973) Vol. 2.

B7. G. W. Stroke, W. E. Kock, Y. Kikuchi and J. Tsujuichi: Ultrasonic Imaging and Holography--Medical, Sonar and Optical Applications, Plenum Press, New York (1974).

B8. P. S. Green: Acoustical Holography, Plenum Press, New York (1974) Vol. 5.

B9. J. u. H. Krautkrämer: Werkstoffprüfung mit Ultraschall, Springer-Verlag, Berlin (1961).

Piezo-Active Materials and Mechanical Fatigue

C1. Y. Kikuchi: The Theoretical Limit of Mechanical Output Power Available from Megnetostriction Vibrators. IEEE Trans. Sonics-Ultrasonics, Vol. SU-15, No. 1, January (1968) p. 1-12.

C2. Y. Kikuchi: Magnetostrictive Materials and Applications. IEEE Trans. Magnetics, Vol. MAG-4, No. 2, June (1968) p. 107-117.

C3. Y. Kikuchi, D. Okuyama and T. Sannomiya: On the Mechanical Fatigue in Magnetostrictive Ferrite for Ultrasonic Vibrators. Summaries of International Conference on Ferrites, Jap. Soc. of Powder & Powder Metallurgy, eds., Inst. Phys. and Chem. Research, Asaka-Saitama (1970) p. 9B3-6.

C4. Y. Kikuchi: Ultrasonic Transducers, Proc. 7th Intl. Cong. on Acoustics, Budapest (1971) Vol. 1, No. 23-G-1, p. 75.

C5. D. A. Berlincourt, D. R. Curran and H. Jaffe: Piezoelectric and Piezomagnetic Materials. In Physical Acoustics, W. P. Mason, ed., Academic Press, New York (1964), Vol. 1, Part A, p. 202.

C6. Y. Kikuchi: Some Aspects of Ultrasonic Cavitation. In Ultrasonic Energy, E. Kelly, ed., Univ. of Illinois Press, Urbana (1965) p. 23-50.

C7. F. Dunn and J. E. Breyer: Generation and Detection of UHF Sound in Liquids. J. Acoust. Soc. Amer. 34 (1962) p. 775-778.

Transducer Sensitivity Measurement

D1. G. Kossoff: The Measurement of Peak Acoustic Intensity Generated by Pulsed Ultrasonic Equipment. Ultrasonics, October (1969) p. 249-251.

D2. M. Okujima and K. Kaburagi: Near-Field Sensitivity Calibration of Transducers for Ultrasonic Diagnosis. (In Japanese) Record of Meeting of Jap. Soc. Ultrasonics in Med., November (1971), No. 20-42, p. 77.

D3. M. Okujima: Microphone for Measurement of Sound Pressure Radiated from Ultrasono-Diagnostic Equipment. Proc. 8th Intl. Cong. Acoustics, London (1974) p. 353.

D4. C. Kasai, D. Okuyama, Y. Kikuchi and K. Miyata: Method for Measuring the Electromechanical Coupling Factor of Piezoelectric Materials by Means of Transient Process Analysis. Reports of 1973 Autumn Meeting of Acoustical Soc. of Japan, (1973) No. 2-1-8, p. 255 (In Japanese); see also: Y.

Kikuchi and C. Kasai: (same title), Proc. 8th Intl. Cong. Acoustics, London (1974) p. 437; also see: (same authors and title), Ultrasonics, September (1975) p. 208.

D5. Y. Kikuchi, D. Okuyama, C. Kasai and Y. Yoshida: Bougie Type Probe for the Measurement of Ultrasound Intensity in the Human Uterus. (In Japanese) Record of Meeting of Jap. Soc. Ultrasonic Med., October (1974) No. 26-62, p. 115.

High Speed and/or Electronic Scannings

E1. A. Asberg: Ultrasonic Cinematography of the living heart. Ultrasonics, Vol. 5 (1967) p. 113.

E2. J. C. Somer: Electronic Sector Scanning for Ultrasonic Diagnosis. Ultrasonics 6 (1968) p. 153-159.

E3. J. Pätzold, W. Kranse, H. Kresse and R. Soldner: Present State of an Ultrasonic Cross-Section Procedure with Rapid Image Rate. IEEE Trans. Bio-Medical Eng. (1970) p. 263.

E4. E. Reinold: The Observation of Foetal Movements by Means of Ultrasonics. Electromedica, No. 2 (1971).

E5. R. Uchida, Y. Hagiwara and T. Irie: Electro-Scanning Ultrasonic Diagnostic Equipment. Japan Med. Engrg. (English edition of JEE, Dempa Pub. Tokyo) (1971) 71-72, p. 58.

E6. Y. Kikuchi, D. Okuyama and H. Kamata: On the Near Field Pattern of an Acoustic Phased Array Transducer for Use in MHz Range. Record of Spring Meeting of Acoustical Soc. Japan (in Japanese), Tokyo (1972) p. 343; see also: Y. Kikuchi: Ref. 46, p. 230.

E7. J. C. Somer: Some Results Obtained with Electronic Sector Scanning in Patients with Cerebral Disease. Progress Report of Inst. Med. Phys. TNO, Utrecht, Netherlands, August (1972) No. PR-3, p. 105-119.

E8. G. Kossoff: Ultrasonic Research in Medicine in Australia--A Review. Ultrasonic Intl. Conference Proc. (1972), Session 6 (Med. & Biolog. Ultrasonics) p. 199-205.

E9. N. Bom, C. T. Lancée and F. C. van Egmond: An Ultrasonic Intracardiac Scanner. Ultrasound 10, No. 2 (1972) p. 72-76.

E10. J. C. Somer: Electronic Sector Scanning in Cerebral Diagnostics. See Ref. A10, p. 304.

E11. N. Bom, et al.: Multi-Element System and Its Application to Cardiology. See Ref. A10, p. 297.

E12. F. L. Thurstone and O. T. vonRamm: Electronic Beam Scanning for Ultrasonic Imaging. See Ref. A10, p. 43.

E13. N. Endo and M. Okujima: Element Arrangement of Array Transducer for Electronic Sector Scanning in Ultrasonodiagnostic Tomograph. J. Acoust. Soc. Japan 30, No. 6 (1974) p. 319-324; for the principle, see Med. Ultrasonics (In English) Vol. 9 No. 1 (1971) p. 118-120, and Vol. 10 No. 1 (1972) p. 71-74.

E14. Y. Takemura, K. Nakagawa and S. Sato: High Speed Ultrasonocardiotomograph. Toshiba Review (Japanese ed.) 20, No. 10 (1974) p. 889.

Signal Processing in Transducer Systems

F1. M. Okujima, K. Katakura and K. Nakano: Ring Transducer with Very Sharp Receiving Directivity for Ultrasonic Tomograph. Med. Ultrasonics, Vol. 6, No. 2 (1968) p. 135-137; see also: (same title), Proc. 6th Intl. Cong. on Acoustics, Tokyo (1968) No. M-1-4, p. M-9/M-12.

F2. For example, see G. Kossoff: Ref. E8.

F3. K. Iinuma and M. Hashiguchi: Moving Focus Transducers. (I) and (II), (in Japanese). Record of Meeting of Jap. Soc. Ultrasonics in Med., May (1972) No. 21-11, p. 19; also May (1973) No. 22-46, p. 91.

F4. M. Ueda, T. Sato, Y. Maeda and K. Yamamoto: Dynamic Focusing Ultrasonic Transducer by Using Analog Switches. (In Japanese) Record of Professional Group on Ultrasonics, Inst. EEE, Japan, November (1973) No. US73-38, p. 1-14.

F5. Y. Kikuchi, et al.: Ultrasonocardio-tomography and Its Application to Morphological Measurement of the Heart. See Ref. A4 (1971), Vol. III, p. 423

F6. Y. Kikuchi, D. Okuyama and T. Tanaka: Ultrasono-cardio-tomography. See Ref. B7, p. 425.

F7. Y. Kikuchi, D. Okuyama, T. Ebina, S. Oka and M. Tanaka: Cardiac "Kineto-Ultrasonotomography". Reports of the 6th Intl. Cong. Acoustics, Tokyo (1968) Vol. VI, No. M-1-8, p. M-25.

F8. Y. Kikuchi, et al.: Multi-Information Recording and Reproduction in the Ultrasono-cardio-tomography. See Ref. B5 (1972) p. 113.

F9. Y. Kikuchi, D. Okuyama, T. Sannomiya, Y. Yoshida and M. Tanaka: Kinematographic Presentation of Ultrasono-cardio-tomograms by Means of Multiplex Presentation System with a Scanning Convertor of Storage Type. (In Japanese) Record of Meeting of Jap. Soc. Ultrasonics Med. November (1973) No. 24-21, p. 37.

F10. Y. Kikuchi, D. Okuyama and T. Sannomiya: A Multiplex Presentation System for the Sensitivity Graded Tomogram Pairs in Ultrasonotomography. (In Japanese) Record of Meeting of Jap. Soc. Ultrasonics Med., May (1973) No. 23-51, p. 101.

Trans-Ductal and Catheter Type Transducers

Trans-Rectal Transducers

G1. J. J. Wild and J. M. Reid: Progress in the Technique of Soft Tissue Examination by 15 MC Pulsed Ultrasound. In Ultrasound in Biology and Medicine, E. Kelly, ed., American Institute of Biological Sciences, Washington, D.C. (1957) Fig. 18, p. 30-48.

G2. H. Watanabe: Development and Application of New Equipment for Trans-rectal Ultrasonotomography. In International Congress Series, No. 277, Excerpta Medica (1973) Proc. 2nd World Cong. on Ultrasonics in Medicine, Rotterdam, p. 33.

Laparoscopic Transducers

G3. S. Hayashi, et al.: Ultrasonic Diagnosis of Breast Tumor and Cholelithiasis. Western J. Surgery, Obstet. & Gynecol. 70, January-February (1962) Fig. 11 p. 34-40; also see Fig. 4 of Y. Kikuchi: Recent Results of Research and Development in the Field of Ultrasonics in Japan. Proc. 3rd Intl. Cong.

Acoustics, Stuttgart, L. Cremer, ed., Elsevier Pub. Co., Amsterdam (1959) Vol. 2, p. 1193-1206.

Trans-Vaginal Transducers

G4. T. Wagai, et al.: Recent Advance of Medical Application of Ultrasound Developed by the Ultrasonic Research Group, Juntendo University, Tokyo (1963) Figs. 11 and 12, p. 9.

G5. Y. Kikuchi, et al.: (Uterus Probe), see Ref. D5.

Trans-Esophageal Transducer

G6. T. Ebina, et al.: The Diagnostic Application of Ultrasound to the Disease in Mediastinal Organs. Sci. Rep. Res. Inst. Tohoku Univ., Series C, Vol. 12, No. 3-4, December (1965) p. 199-212.

G7. F. A. Duck, C. J. Hodson and P. J. Tomlin: An Esophageal Doppler Probe for Aortic Flow Velocity Monitoring. Ultrasound in Med. Biol., Vol. 1, No. 3, August (1974) p. 233-241.

Trans-Tracheal Transducer

G8. Y. Haneda, et al.: Ultrasonic Visualization of Mediastinal Tumor--Especially Regarding the Transtracheal Method. Med. Ultrasonics, Vol. 6, No. 1 (1968) p. 54.

G9. Y. Kikuchi, M. Tanaka, M. Oshiba and Y. Haneda: Ultrasonic Diagnosis of Mediastinal Tumors. In Ultrasono Graphia Medica, Verlag der Wiener Med. Akademie (1969), Vol. III, p. 107-114.

Catheter Transducers

G10. S. Kimoto, et al: Ultrasonic Tomography of the Liver and Detection of Heart Atrial Septal Defect with the Aid of Ultrasonic Intravenous Probes. Ultrasonics, Vol. 2 (1964) p. 86.

G11. N. Bom: New Concepts in Electrocardiography, H. E. Stenfert Kroese N.V., Leiden (1972) p. 54.

Solid Horn

H1. T. F. Hueter and R. H. Bolt: Sonics, John Wiley & Sons, Inc., New York (1955) p. 274.

H2. For example, N. Clark: Trans. IRE, Ultrason. Eng., 11 (1954) p. 19; W. P. Mason: J. Acoust. Soc. Amer. 28 (1956) p. 1207; G. Merkulov: Soviet Physics-Acoustics 3 (1957) p. 246; E. Mori: Bull. Tokyo Inst. Techn., Ser. B, No. 43 (1961); C. Sugi: J. Acoust. Soc. of Japan 18, No. 2 (1962) p. 54; E. A. Neppiras: Acustica 13 (1963) p. 369, etc.

H3. For example, E. Mori: Design Procedures of the Vibration System for Mechanical Fabrication Use. (In Japanese). See Ref. B4, Chapt. 25, p. 1634-1650.

CHAPTER V

ULTRASONIC DOSIMETRY

W. D. O'Brien, Jr.

SECTION I

1. INTRODUCTION

The objective of this chapter is to describe the general concepts of dosimetry with specific emphasis on ultrasonic dosimetry and relate its applicability to ultrasound in medicine and biology.

Since the emphasis of this volume is on the biological and medical applications of ultrasound, the role of ultrasonic dosimetry will be presented with the view towards radiation protection. This is natural since the purpose of ultrasonic dosimetry is to relate magnitudes of specific ultrasound variables to the likelihood or extent of occurrence of identifiable biological action in the living system, usually located at the same region. This approach is in no way meant to imply that the use of ultrasound in the clinical practice of medicine represents a hazard. But it is commonly known that at sufficient levels ultrasound can produce irreversible biological damage and, as such, the potential does exist to produce damage so it is imperative to develop a common nomenclature to express meaningful radiation protection information.

With the increasing use of ultrasound comes the obligation to properly assess the risk associated with such human exposure. While no statistically based survey has been conducted to document the extent to which ultrasound is being used, a number of indicators do, however, strongly support the view that its use is increasing and that a large fraction of the human population will eventually be exposed.

A market analysis in 1969 predicted that the dollar value of the ultrasonic market would increase 300 percent during the period between 1968 and 1973.[1] This represents an annual increase of 75 percent. More recently, based upon discussions with clinical manufacturers, the U. S.

Food and Drug Administration's Bureau of Radiological Health predicted that in 1976 the industry will grow at an annual rate of 50 percent and that annual dollar sales will be around $40 million.[2]

In 1971, the Bureau of Radiological Health surveyed 301 out of 6306 short-term general hospitals in the United States and found that 12 percent of the hospitals used diagnostic ultrasound.[3] In an editorial in the Journal of Clinical Ultrasound, the editor[4] doubted the 12 percent figure since he believed Doppler ultrasound to be used by over 50 percent of the obstetricians in the United States. In 1976 the same government agency reported that a hospital survey showed 35 percent used ultrasound.[5] While questions may be raised as to the validity of the 1971 and 1976 survey estimates, they do show, however, an approximate annual increase in use of 24 percent.

An international mail survey was conducted under the sponsorship of the IEEE Group on Engineering in Medicine and Biology's Subcommittee on Ultrasound Safety and Standards and with the help of the American Institute of Ultrasound in Medicine, the Biomedical Engineering Society, the Bureau of Radiological Health, the United States Public Health Service and the United Kingdom Medical Research Council.[6] In part, the survey was to determine the extent to which ultrasound is used. On the average, between 1963 and 1971, there was an annual increase in use of clinical ultrasound of approximately 10 percent.

It was estimated that, in the United Kingdom, the number of ultrasonic diagnostic examinations is doubling every three years.[8] This represents an annual 26 percent increase.

The United States National Science Foundation, through its Office of Experimental Research and Development Incentives, conducted an international state-of-the-art survey of diagnostic ultrasound in March, 1973. One of the conclusions of the survey team was that between 1971 and 1973, the number of ultrasonic diagnostic instruments sold in the United States increased by 300 percent, which is an annual increase of 73 percent. Another conclusion of the NSF survey team was an estimate that the sales of clinical ultrasonic devices will match those of x-ray devices by 1983.[9]

Use and trends of use information for therapeutic ultrasound is as difficult to obtain as with diagnostic ultrasound, even though it has been recognized as a potential hazard since the initial applications in the 1930's.[10,11,12] It is suggested that there is presently little change in the extent of use but information is quite limited. Three surveys have been conducted, all within the last few years, which may provide some guidance, but it is difficult to extract precise information since the surveys were limited in design.[13,14,15] The 1970 Pinellas County, Florida survey[13] indicated approximately 45,000 ultrasonic treatments per month to 6000 patients in a county population of 500,000. Assuming this is typical of the United States, and there is no supportive information for this assumption, in a population of 200 million, this would represent on an annual basis approximately 216 million treatments to 2.4 million patients.

Additionally, although there have been only a few reports of the output parameters, an approximation can be rendered as to the general range of ultrasonic power and intensity outputs from commercial diagnostic instrumentation. The measurements of pulse devices show that the spatial average ultrasonic power ranges from 60 μW to 21 mW, the spatial average, temporal peak intensity ranges from 1 to 95 W/cm^2 and the spatial peak, temporal peak intensity from 2 to 177 W/cm^2. The nominal ultrasonic frequency range is from 1 to 10 MHz with a variation of 400 to 2000 Hz in the pulse repetition frequency.[16,17,18,19] One of the pulse devices is a pulsed Doppler and exhibits a pulse repetition frequency of 48 kHz and a spatial average, temporal average intensity of 296 mW/cm^2.[19] For continuous wave, Doppler devices, the spatial average ultrasonic power ranges from 1 to 37 mW and the spatial average intensity ranges from 1 to 305 mW/cm^2.[16,18,19,20,21]

SECTION II

2. DOSIMETRIC CONCEPTS

Dosimetry is concerned with the quantitative determination of energy interaction with matter, or, in other words, defining the quantitative relationship between some physical agent and the biological effect it produces. In one sense dosimetry is the determination of a dose, or similar type of physical parameter, which characterizes the physical agent as to its potential or actual interaction with the biological material of interest.

In the case of ultrasonic dosimetry, the object is to relate magnitudes of specific parameters - such as intensity, acoustic pressure, particle displacement, etc., or perhaps some parameter yet to be developed - to the likelihood of occurrence of a biological alteration. To accomplish this, it is necessary (1) to quantify the output parameter or parameters of the source, (2) to determine the effect of the material on the propagating energy, viz., reflections, refraction, scattering, absorption, etc., and (3) to relate the first two items to quantitative parameter determination at the site of interest. To this end, Section 4 of this Chapter describes techniques by which the source output can be quantified and Chapters VI, VIII and XIII describe ultrasonic propagation properties of tissue such as absorption, velocity and impedance along with Chapter VII on scattering properties of tissue. Chapters IX, II, X and XIV provide the kinds of biophysical interaction data from which it is potentially feasible to begin developing dose-effective responses.

Thus dosimetry has two important objectives. The first is to define physical quantities which properly reflect an interaction at some site in biological material which may be expressed in units such as joules/kgm, joules/m^3, etc. The second is to develop a concept or concepts of the quantity that is applicable for radiation protection purposes.

At this point it is appropriate to briefly describe two terms which have resulted in some confusion in ionizing radiation dosimetry with the hope that such confusion will not arise or, at least, be minimal in ultrasonic dosimetry, viz., quantity and unit. Unit represents the amount and quantity represents the thing for which the unit was devised. For example, length, time and mass are quantities for which the meter, second and kilogram are their units, respectively. This distinction is also valid for special units such as hertz (cycle per second) for which frequency is the quantity.

Typically, dose connotes something that is given or imparted in a quantitative manner. The history of other forms of radiation has documented that defining dose, or dose-like concepts, is difficult, especially when the object is to include all possible physical and biological variables. Otherwise, and the more common, special quantities are developed for the specific case or biological action under consideration. In ionizing radiation, for example, dose generally refers to the quantity absorbed dose which has been specifically defined as the energy imparted to matter by ionizing radiation per unit mass of irradiated material at the site of interest.[22] But other dose quantities have been defined for specific purposes such as genetically significant dose which is the gonad dose from medical exposure, or cumulative dose, dose equivalent, threshold dose, etc.[23] In photobiology, dose sometimes refers to the quantity dose of ultraviolet radiation which has been defined as the energy per unit surface area applied to an object.[24,25] Other quantities which have been used to characterize ultraviolet radiation were chosen to quantify a specific bioeffect. These included minimal erythema dose, minimal perceptible erythema, subvesicular dose, minimal color dose, etc.[26] There is currently much discussion regarding microwave dosimetry. Terms such as specific absorption rate,[27,28] absorbed power density and specific absorption density[29] and energy dose-rate[28] have all been either used or suggested as a basic quantity to describe absorbed electromagnetic energy. Appropriateness of units has also not been agreed upon.[30]

It is useful, for purposes of illustration, to examine in some detail the history of another form of radiation with a view towards its dosimetry development. For this purpose, ionizing radiation is chosen because it represents a well developed history.[31,32,33,34,35,36] However, this examination is not meant to indicate that ultrasonic dosimetry should take this route. The mechanisms of interaction between ultrasonic and ionizing radiation are quite different. Rather, the rationale for ionizing radiation dosimetry, as with other radiations, is one of developing an acceptable and reasonable nomenclature by which researchers in various fields can communicate and, if necessary, by which radiation protection guidelines can be developed. Knowledge of the energy disposition of the tissue site of interest is one of the critical elements in understanding the interaction between the radiation form and matter.

One of the earliest dosimetric concepts for ionizing radiation was the _skin unit dose_, more commonly known as the _skin erythema dose_ or _threshold erythema dose_. One skin unit dose was the amount of radiation which just produced reddening, or erythema, of the skin. The reddening followed the exposure within a period of about one week. The detector, human skin, was very imprecise but the skin unit dose was, nevertheless, used as a basis for the first radiation protection guideline in the mid-1920's. The _tolerance dose_ was suggested to be a small fraction, around one percent, of a skin unit dose, averaged over a one month period.

Not until sensitive and reproducible ionizing radiation measurement devices were developed was there a physical measurement of ionizing radiation. The concept and value of the unit _roentgen_ was established in 1928 and defined in terms of the ionization, or interaction, of x-ray radiation in air.[37] It was a special unit of exposure but no specific quantity was defined at that time for which the roentgen was its unit.

In an effort to relate the _tolerance dose_ to a physical parameter, radiotherapists were polled as to the number of roentgens required to produce one skin unit dose. Thus, based upon a rough value of 600 R for one skin unit dose, the tolerance dose worked out to be 6 R on a monthly basis, or 0.2 R/day. In the early to mid-1930's national and international organizations endorsed a tolerance dose of 0.2 R/day but neither gave a description of their concept of a tolerance dose. Later this value was reduced to 0.1 R/day and remained at this level for 12 years. The term tolerance dose created many problems because it was impossible to predict just what level was tolerable over a long period of time. With the realization in the mid-1930's that ionizing radiation effects may not be threshold type reactions, the term maximum permissible dose was substituted for tolerance dose.

As a result of biophysical and biological effect studies with various types (qualities) of ionizing radiation, it was recognized that broader dosimetric concepts were required to define and describe quantitatively ionizing radiation fields. This was especially important when applying dosimetry to radiation protection in that the roentgen was inadequate because of its limitation to x- and γ-rays and because it was not a measure of absorbed energy. In the early 1930's, it was shown that the biological effect of ionizing radiation depended not only upon the exposure intensity and time but also upon the quality of radiation since differences between x-rays and γ-rays were observed in growth reduction and mortality studies.[38] This was termed the relative biological effectiveness (RBE) concept and became even more important in the 1940's with the production and discovery of other ionizing radiation particles. In terms of the absorbed dose unit rad, which will be discussed shortly, this meant that the same number of rads of neutrons, for example, would produce a greater biological effect as compared to x- or γ-rays.

In the late 1930's a unit, termed the energy unit, was suggested as a dose of γ-rays delivered to tissue in terms of absorbed energy per gram of tissue.[39] Also, around this time, another unit, the gram-roentgen, was suggested[40] as the amount of energy absorbed by one gram of air when irradiated by one roentgen. In the late 1940's another unit was suggested to describe energy absorption, viz., the rep for roentgen-equivalent-physical.[41] Originally, one rep was defined as that dose of ionizing radiation which produced an energy absorption of 84 $ergs/cm^3$ in tissue. It was based upon the roentgen in that this was meant to be the energy absorbed by tissue when exposed to one roentgen. That meant that the definition depended upon a calculation of energy absorption and upon other tissue parameters which were subject, in some cases, to wide uncertainty. These difficulties were reflected in redefining one rep from the original value to other values in order to reflect actual tissue absorption properties and to remove dependence of tissue density.

The rep concept led to what is currently the quantity absorbed dose with the unit rad for radiation absorbed dose. The advantage of the unit rad over the rep is that the one rad has been arbitrarily defined as 100 ergs/gm and thus is independent of material properties.[42]

At the same time the rep was being suggested as a unit to describe dose, the unit rem, for roentgen-equivalent-man, was also being suggested for radiation protection purposes. The rem was defined as the product of energy absorption (in reps) and the relative biological effectiveness (dimensionless) of the energy under consideration. If there were energies of different RBE, the rem was then the sum of each respective product.[41]

In the mid-1950's, the rem concept was adopted, using the rad instead of the rep. The quantity RBE dose in rems was equal to the product of absorbed dose in rads and the RBE and, in the case of multiple RBE's, the sum of each product.[43]

A few years later, RBE was changed to quality factor, QF, and assigned fixed values which were closely representative of actual RBE's for specific conditions and energies. This was done because RBE itself was dependent upon a large number of variables and for radiation protection purposes, the quality factor values chosen were representative of RBE's. Thus, the quantity dose equivalent was adopted, its unit the rem, and was equal to the product of absorbed dose, quality factor and other dose modifying factors to account for spatial and temporal dose distribution.[44]

In the history of ionizing radiation dosimetry is reflected the rationale for which national and international commissions have labored to develop concepts and define units and quantities. Initially, the threshold dose was defined as a monthly fraction of the skin unit dose and later, following proper instrumentation development, in terms of an exposure in roentgens. The term threshold dose was later called maximum permissible dose because of the realization that risk from ionizing radiation may not be represented by a threshold. Because of the desire to express the effect of ionizing radiation in terms of the interaction with or absorption by tissue at the site of interest, the absorbed dose concept was defined. Eventually the dose equivalent quantity evolved to embody both physical and biological parameters.

By comparison, the field of ultrasonic dosimetry has not developed to the extent of ionizing radiation dosimetry. The most widely used dosimetric parameter in ultrasonic bioeffect and biophysical studies is intensity in the mixed unit of W/cm^2. The principal reason for the use of intensity is, perhaps, convenience since it is an easily measured parameter as will be discussed in Section IV. As a dosimetric quantity, intensity represents many of the same problems as does the ionizing radiation quantity "exposure" in that it is not a measure of dose, or the like. Yet the majority of bioeffect and biophysical reports use intensity as the measured physical parameter of the ultrasonic field. This extensive literature documents the actions of ultrasound but, in most cases, lacks the necessary characterization of the field at the site of interest. An ideal situation would be to know the instantaneous particle velocity, the instantaneous acoustic pressure and the phase between these two field parameters at the site or sites of interest.[45]

Through both calculations[46] and experimentation[46,47,48,49] attempts have been made to determine *in utero* ultrasonic intensity in both the gravid and nongravid human uterus. A model of the tissue layers between the skin surface and fetal sac have yielded a total attenuation of 23 dB at a frequency of 2.5 MHz.[46] One of the earliest *in vivo* experiments[47] showed an average loss between the skin and uterus to be around 2.5 dB at

2.25 MHz but more recent studies have shown this to be higher, in the range from 9 to 20 dB[46,48] or from 6 to 14 dB.[49]

There have been ultrasonic dosimetric quantities which are noteworthy of comment in that they represent, in concept, the basic approach to dosimetry. The cataract-producing unit, CPU, was a quantity defined as the length of exposure necessary to produce a grossly observable cataract and expressed in units of seconds.[50] The dosimetric concept damage ability index with the unit $second^{-1}$ is a quantity intended to describe the effect of ultrasound on spinal cord hemorrhage.[51]

Quite recently it has been suggested[52] that a universal dosimetric response to ultrasonic exposure may exist for different tissues but the response has only been demonstrated, in a limited manner, in mammalian brain tissue. The response is in terms of the "energy absorbed per unit volume" (J/mm^3) for histologically observable lesions at superthreshold levels[53] as a function of the "delivered intensity." It is shown that at two different ultrasonic frequencies, 3 and 4 MHz, identical constant volume curves result even though there are two different "threshold levels."[53]

A step or two removed from knowledge of the *in situ* ultrasonic field parameters, but yet an essential part of field characterization, is a thorough characterization of the ultrasonic field at the site where the specimen is placed but without the specimen in place. This provides a basis from which *in situ* field parameters can be calculated and a basis whereby duplication of the experimental arrangement is made possible. As will be discussed, there are techniques available to measure many of the ultrasonic field parameters when the medium of interest is macroscopic, homogeneous, isotropic and low-loss, such as water. But this does not hold for *in vivo* measurements and, as a result, both instrumentation and dosimetric concepts need to be developed.

To properly develop ultrasonic dosimetric concepts, a much greater understanding of the processes by which ultrasonic energy interacts with biological material is required. This point can be illustrated by a number of examples. Whenever ultrasonic energy is absorbed by any biological material, heat results. In fact, biological tissues absorb ultrasound at a relatively high rate (see Chapter VI). If temperature rise, or for that matter the rate of temperature rise, is the response under study, then the amount and rate of ultrasonic energy being absorbed per unit volume or mass will be important. To determine this, the spatial and temporal intensity, the exposure time and the material absorption properties will be required to calculate the absorbed energy. Additionally, geometric beam considerations affect the temperature response since the initial rate of rise of temperature has been shown to be quite similar for both plane and focused ultrasonic beams but diffusion properties of the tissue markedly affect the temporal temperature development.[54,55] The importance of thermal diffusion has also been shown to explain the apparent frequency independence of histological lesions in mammalian central nervous system tissue under focused ultrasonic conditions.[56] If intensity is to be the ultrasonic quantity at the site of interest, then the specific heat will have to be considered in order to calculate the temperature response.[57,58] Thus, to know the intensity _in situ_ and, in most cases where the irradiation would be in the near field, would require detailed information of the acoustic pressure and particle velocity.

If cavitation, as another example, is the mechanism of action, then the acoustic pressure may be the primary physical parameter which should be specified, or, at least, the parameter from which a dose-related term is derived. An approach to develop a dosimetric concept for considering cavitation as the causative mechanism could begin by defining the parameters responsible for the onset of bubble activity in tissue (while this is an extremely difficult and, perhaps, not yet an achievable goal, the approach is what is being emphasized). These parameters would include, but are not necessarily limited to, the ultrasonic field quantities, such as acoustic pressure frequency and amplitude, the environmental conditions such as ambient pressure and the tissue properties such as nuclei sites, degree and state of gas, etc. Following the formation of the bubble(s), the rate of growth could be considered. If the ultrasonic forces were insufficient to cause the bubble to grow to the "dynamic threshold radius,"[59,60] the bubble would remain stable and radially oscillate. This has been termed stable cavitation.[61] Under this condition, a relatively small amount of energy is extracted from the ultrasonic wave but relatively high and localized energies are established in the bubble vicinity as it functions as a secondary source. Examples of the types of resultant forces from such oscillatory bubbles can be found in the literature[62,63,64,65] and in this Volume. Should the bubble grow greater than the dynamic threshold radius, transient or collapse type cavitation would result, causing extremely high energy densities within its immediate vicinity with the certainty of adversely affecting biological material. The temporal exposure conditions would be an essential parameter to consider in view of the fact that an exposure time appears to be necessary to elicit a biological effect.[66,67] This, along with other biological observations[53,68,69] resulting from cavitation-type phenomenon have all been reported in terms of ultrasonic intensity.

Other phenomena which need to be at least considered in any ultrasonic dosimetric approach, especially when aimed at radiation protection, includes the question of whether or not ultrasonic biological actions are cumulative, the role of synergism, frequency dependence of an effect, critical organ or tissue concept and, perhaps, others. Although ultrasonic energy does not have an analogy to ionizing radiation's "quality of radiation," the relative biological effectiveness represents an important radiology concept and thus should be kept in mind.

There appears to be, at least, a reasonable doubt whether or not cumulative effects occur from exposure to ultrasound. Summation of subparalytic doses of ultrasound, with sufficient time for temperature equilibrium to be re-established between pulses, produced paralysis in frog hind limbs.[70] It has also been demonstrated that, under pulsed ultrasonic exposure conditions, by varying only the duty cycle with a constant pulse width, spinal cord hemorrhage occurred only when the total sum of on-time of pulses reaches the same value.[51]

Synergism between ultrasound and other agents should properly be considered for any ultrasonic dosimetric concept. There have been both positive[71] and negative[72] synergistic findings with ionizing radiation and positive findings with both hypoxia[51] and one chemotherapy drug.[73]

Structural lesions in mammalian adult brain were initially thought to be frequency independent over the range from 1 to 9 MHz[53,56,69] but after further examination of the data, a slight but oscillatory frequency dependence was shown.[74] The mechanism responsible for this dependence has been identified as a capsular layer surrounding the brain.[75] It is interesting to observe that most organs possess some type of capsular layer and thus its dependence upon ultrasonic energy transmission should be considered.[76] By compensating for the capsular layer effect, the intensity threshold responsible for lesion production appears to be frequency independent.

Other types of frequency dependent examples, all based upon intensity as the ultrasonic parameter, include greater damage to liver at lower frequencies over the range 0.5 to 6 MHz,[51,77] greater change in the electrophoretic mobility of irradiated cells at lower frequencies over the range 0.5 to 3.2 MHz[78] and greater susceptibility to the production of cataracts at higher frequencies over the range 5 to 15 MHz.[79]

SECTION III

3. MEASUREMENT CONSIDERATIONS

3.1 The Non-Idealized Field

The typical ultrasonic use of diagnostic ultrasound in the clinical practice of medicine suggests that the source is circular in shape and exhibits a uniform velocity distribution of simple harmonic motion.[80,81,82,83,84,85] Most standard texts[86,87,88] describe the resultant ultrasonic field from the plane piston source. Generally, it is divided into two regions, the near or Fresnel region and the far or Fraunhofer region. This is an idealized model, one which is highly convenient for instructional purposes but one which may not in fact describe the field of real transducer sources.

It is, at least, prudent to make a few observations regarding the ultrasonic field from a flat, circular piston source operating with a uniform velocity distribution. It is known that as the mechanical back loading of a ceramic crystal and quartz plate is increased, and back loading is widely used for diagnostic transducers, the surface displacement becomes more uniform and, hence more piston-like.[89,90,91] The dividing point between the near and far field is commonly taken at the last axial maximum which occurs at a distance of a^2/λ from the transducer surface (a is the transducer radius and λ is the acoustic wavelength).[82,88] Other distances have also been used to designate this dividing point such as $2a^2/\lambda$ when the axial intensity just begins to behave as r^{-2} (r is the distance from the transducer surface)[87] or $0.75\ a^2/\lambda$ where the beam width is the narrowest.[92]

Even with the idealized piston source, the near field is extremely complex. Numerical procedures have been employed to construct detailed perspective images of the near field and demonstrate quite a complicated field parameter distribution.[92] Also, it has been demonstrated that a reactive component of axial intensity exists near the transducer surface.[93] Thus, even with the idealized piston source, plane wave assumptions are not valid in the near field.

Non-uniform velocity distribution on the transducer surface markedly affects the propagated field distribution pattern.[94,95,96] For example, a Gaussian velocity distribution on the transducer surface results in an axial intensity distribution free of oscillations and a directivity pattern Gaussian in shape and thus free of side lobes.[94] Clamped and simply supported boundary conditions of a circular transducer result in axial intensity distributions more typical of the Gaussian case but with small oscillations, the clamped case exhibiting oscillations of a lesser amplitude than the simply supported case.[96] Also, the clamped and simply supported disks have a greater divergence than that of the piston source.[96]

At the spatial peak, temporal peak intensity levels at which some of the ultrasonic diagnostic instrumentation operates, the effect of finite amplitude effects upon the measurement process must be considered. Finite amplitude effects simply mean that a fraction of the propagated ultrasonic energy is transferred to its higher harmonics as a function of distance.[97,98,99] For example, at an ultrasonic frequency of 2.5 MHz and with an initial intensity of 3 W/cm^2 at the source, as the energy propagates away from the source in water 4 percent of the energy has been transferred into the second harmonic, 5 MHz, at 10 cm and 12 percent at 20 cm. Also about 4 percent has been transferred into the third harmonic, 7.5 MHz, at 20 cm. With an initial intensity of approximately 12 W/cm^2, 12 percent of the energy is transferred into the second harmonic at 10 cm from the source.[97]

On the other hand, in Chapter IX of this Volume, at extremely high intensities, it is argued that finite wave effects are negligible under certain experimental conditions.

3.2 Absolute and Relative Measurement Process

There does not appear to be a standard definition as to what is an absolute measurement process for ultrasonic field parameter determination. So a working definition which will be employed here is one in which the measurement process does not need to be calibrated in a known field in order to yield quantitative information about the ultrasonic field parameter which is being measured. Therefore, a relative measurement process must be one which requires calibration in a known field. It should be noted that the total measurement process is either absolute or relative, not the measurement itself.

The piezoelectric technique developed at the National Bureau of Standards[100,101] and described in the next section under Piezoelectric Techniques qualifies as a National Bureau of Standards reference standard for total ultrasonic power. The National Bureau of Standards is also developing a calorimetric method for determining total ultrasonic power[102] but no description is currently available. This may also be available as an NBS reference standard.

3.3 Accuracy and Precision

The process of taking and analyzing measurements requires both human skills and judgment along with instrumentation which is subject to uncertainties. As such the reporting of analyzed measurements are commonly represented by numerical qualifiers which are intended to convey a degree of confidence. One of the more typical statements is "the accuracy of... is ___ percent." It has been suggested[103] that such a quasi-absolute statement represents poor practice and should be avoided. An understanding of this statement is embraced in the terms error, true value, systematic error, precision and accuracy.

Error is the difference between the value obtained and the true value. In order to quantify the error of a measurement, it is necessary to first quantify the true value. In concept, true value seems to be one of those terms which is implicitly understood but in practice it may never be exactly determined. For example, in measuring, say, ultrasonic intensity, the system will yield a number. But to what is it compared for the determination of the error. If the United States National Bureau of Standards (or some other standards organization) had an ultrasonic intensity standard, the measured value would be compared to it, and this would be explicitly stated in reporting the measurement, but there would be no guarantee that the NBS standard truly represented the true value.

Whereas error can be defined in terms of a measurement, precision and accuracy are defined in terms of a measurement process, that is a method of measurement which can be described in statistical terms or, as Eisenhart[104] states, has "attained a state of statistical control." Precision has been defined as "... the typical closeness together of successive independent measurements of a single magnitude generated by repeated applications of the process under specified conditions."[103] High precision is indicative of a measurement process which exhibits small experimental or random errors.[105] Accuracy, on the other hand, has been defined as "... the closeness to the true value characteristic of such measurements,"[103] and a highly accurate measurement process is indicative of a small systematic error, or bias.[105] In other words, precision is concerned with how close together the measurements are to each other, irrespective of how close the measurements are to the true value, and accuracy is concerned with how close the measurements are to the true value.

Now it is, perhaps, clearer why specifying the accuracy of something in terms of some percentage represents poor practice. How, in fact, do you specify the accuracy unless you know the true value. It is, of course, possible to indicate all possible sources of systematic errors, and make an informed judgment as to the degree of bias. Systematic errors are those which introduce a constant bias and are affected by such items as an error in instrumentation calibration, a flaw in the fundamental theory being imposed, improper use of an instrument, observer error resulting from habit, etc. If the measurement process is calibrated against a reference standard, such as maintained by the National Bureau of Standards, then it is recommended that the systematic error be considered negligible and that an explicit statement to this effect be indicated when reporting the results.[103]

Precision, or more properly, imprecision, can be numerically qualified through statistical tests, that is the precision of a measurement process can be known and measured. This is accomplished by computing and reporting quantities such as the standard deviation.

A much more comprehensive discussion of those concepts can be found in the literature.[103,104,105] Specific recommendations are provided by Eisenhart[103] for reporting experimental uncertainties for each of the four possible situations of whether or not systematic error or imprecision are negligible.

SECTION IV

4. MEASUREMENT TECHNIQUES

4.1 Capacitance Probe

When an ultrasonic wave is normally incident upon a boundary and the characteristic acoustic impedance mismatch is sufficient to assume total reflection such as with a solid-air interface, the boundary displacement is twice that of the wave's particle displacement. This is the principle behind the capacitance probe wherein one surface of a capacitor is free and the other fixed such that when an ultrasonic wave is incident at the free surface boundary, the capacitor thickness is modulated. One of the first applications of the capacitance probe was the absolute determination of the fundamental and second harmonics in distorted ultrasonic waves to study third order elastic constants.[106,107,108] This principle has more recently been applied to the absolute determination of the instantaneous particle displacement of plane[17,109,110,111] and focused[112] ultrasonic pulses in liquids. The capacitance probe is very desirable for measuring the ultrasonic output from clinical pulse-echo diagnostic equipment in that it represents one of two techniques (the other is the electrodynamic probe) which is absolute, operates over a wide frequency range and yields the response of pulses. The major disadvantage is that frequency compensation is required to convert the instantaneous particle displacement to intensity. The maximum width of the pulse is a function of the distance the pulse traverses in a buffer rod which provides the free capacitor surface. The buffer rod also introduces an impedance mismatch with the liquid, and thus must be acoustically characterized to determine the fraction of signal transmitted into it.

Filipczynski[110] used a gap thickness of 49 μm, an aluminum buffer rod and a polarization voltage across the gap of 208 volts and obtained a free surface displacement of 6.4 Å for a temporal peak intensity in the liquid of 76 mW/cm^2.

4.2 Chemical

The ultrasonic vibration potential results by virtue of the interaction between the ultrasound and ions in solution. As a compressional or vibrational wave is propagated in an ionic fluid, the differing mobilities between the cations and anions cause a local electrical deviation from neutrality. Debye[113] developed the first theory which predicted and described the phenomenon. Later[114] a more exact form of the Debye effect equation was developed. Some sixteen years after Debye first predicted the effect, the suggestion was put forth that it may prove useful and practical as a means for determining absolute ultrasonic intensity.[115] Around that same time the Debye effect was first experimentally observed.[116] Later, vibration potentials were reported in pure water[117] and in pure organic liquids[118] but these findings, which contradicted theory, were attributed to the probe design.[119]

While the mechanism responsible for the velocity potential has not been precisely defined (so that this technique can be used as a primary method), there appears to be no disagreement that the potential is directly proportional to the amplitude particle velocity.[117,118,119,120,121,122,123] Specifically, in a moderately dilute solution where the diffusion forces are negligible, the ultrasonic vibration potential Φ_o of the $C^{Z+}A^{Z-}$ electrolytic is given by[122]

$$\Phi_o = 0.155\, a_o \left[\frac{W_+}{Z_+} t_+ - \frac{W_-}{Z_-} t_- \right]$$

where Φ_o is the vibration potential in microvolts, a_o is the amplitude particle velocity in cm/sec and W, Z, and t are, respectively, the apparent molar mass, the number of unit charges and the transport number for cations (+) and anions (-). In a 0.1 M solution of NaCl, for example, at an ultrasonic frequency of 220 kHz, an amplitude particle velocity of 1 cm/sec yields a potential of 1 μV.[122] Assuming idealized plane wave propagation, this corresponds to an ultrasonic intensity around 86 mW/cm^2.

4.3 Electret Transducer

An electret is a material, or dielectric body, which possesses separate electric poles of opposite sign of a permanent or semi-permanent nature. At the macroscopic level, it can be considered the electrostatic analogue to a permanent magnet.[124] Electret transducers, and specifically the foil-electret transducer, have been employed over the extensive frequency range from 10^{-3} to 2×10^8 Hz.[125,126,127]

The applicability of foil-electret transducers to megahertz ultrasound in liquids has been demonstrated,[128,129] wherein an N x N array of electret microphones sampled the ultrasonic field. Two different two-dimensional arrays of 16 x 16 were designed, one which consisted of a foil electret and a 16 x 16 element backplate and the other which consisted of the foil electret and backplate each with 16 stripes in an overlapping matrix fashion. In the former design 16^2, or 256, switches are required and in the latter 2 x 16, or 32, switches are required to sample each electret receiver. When the ultrasonic signal is incident upon the array, each element yields a voltage proportional to the instantaneous acoustic pressure. The useable frequency range of these arrays is from 0.3 to 2 MHz. The extension to arrays of 200 x 200 are suggested as feasible.

4.4 Electrodynamic Probe

The electrodynamic probe is similar in concept to the air loud-speaker wherein the incident sound wave on the diaphram moves the coil positioned in a permanent magnetic field and thus generates a voltage. For this probe wires are wound around an insulator, and the surface which the ultrasound is incident upon is positioned in a permanent magnetic field with the wires normal to the flux lines. The ultrasound vibrates the windings yielding a voltage directly proportional to the instantaneous particle velocity.

Filipczynski[130] was the first, and perhaps still the only one, to apply this absolute method to the measurement of ultrasonic pulses from clinical diagnostic equipment. The electrodynamic probe has been developed for the measurement of both plane[17,110,130] and focused[112] ultrasonic pulses. For plane waves, as an example, Filipczynski[110] used an effective length of 2 μm aluminum wire of 8.0 cm in a magnetic field of 4200 Gauss and yielded a particle velocity of 3 cm/sec which is equivalent to a pulse intensity of 58 mW/cm^2.

4.5 Optical Techniques

The principle advantage of the acousto-optic interaction techniques for the measurement of acoustic pressure is the avoidance of placing any measuring device in the sound field. It is assumed that the interaction between light and sound perturbs the sound field negligibly. Brillouin[131,132] first predicted that elastic waves in liquids and solids diffracted light. The phenomenon was experimentally verified by Debye and Sears[133] and by Lucas and Biquard.[134] Bär[135] conducted rather extensive experimentation which, in part, led to a fundamental theory for describing the distribution of light in the different diffracted orders.[136,137,138,139,140,141] Thus Brillouin scattering, Debye-Sears effect, Raman-Nath diffraction or scattering and Bragg diffraction have all been terms used to describe the general phenomenon. Further discussion of the early work can be found in references 142,143,144.

The diffraction problem, as it relates more directly to the interaction phenomenon between laser light and megahertz ultrasound has been reduced to a set of coupled difference-differential equations.[147] Subsequently, Klein and Cook[148] numerically examined the two limiting cases termed the Raman-Nath and Bragg regions and also the intermediate region with the view towards establishing criteria for quantitative measurements of acoustic pressure amplitude. The three regions, in fact, are basically determined by the parameter Q which is governed completely by the geometry of the experiment. Mathematically,

$$Q = \frac{k^* L}{\mu_o k}$$

where k^* and k are the respective wave numbers for sound and light, μ_o is the undisturbed optical index of refraction for the material in which the sound is propagating and L is the distance over which the light interacts with the sound, the interaction length.

In the Raman-Nath region, where $Q \ll 1$, the ultrasound beam is typically narrow and the frequency is low. Under this condition, when the light beam enters normal to the sound beam, the parallel acoustic wavefronts function equivalently to an optical grating by virtue of the sound field modulating the material's index of refraction. The resulting Fraunhofer diffraction pattern is symmetrical and equally spaced about the primary (zeroth order) light beam. The normalized nth order light intensity, I_n, is described by

$$I_n = J_n^2 \quad (v)$$

where J_n is the nth-order cylindrical Bessel function and v is the Raman-Nath parameter, which can be determined uniquely by measuring three adjacent diffraction orders.[149,150,151] The acoustic pressure amplitude can be determined from

$$p = \frac{v}{kL\left(\frac{\partial \mu}{\partial p}\right)_s}$$

where $\left(\frac{\partial \mu}{\partial p}\right)_s$ is the adiabatic piezo-optic coefficient. Raman and Ventakaraman[152] and Riley and Klein[153] have determined values for the adiabatic piezo-optic coefficient for different materials, including water.

This procedure has been suggested in order to obtain the acoustic pressure amplitude.[151,154,155,156] However, in order for this technique to be absolute, the interaction length must be known. Breazeale and Hiedemann[154] used a schlieren technique.[157,158,159] Cook[160,161] has shown, however, that transducer orientation is critical in a non-uniform field. A principle assumption is plane wave propagation. In the Fresnel region of the ultrasonic beam, the interaction length may be assumed to be equivalent to the active surface of the transducer. But, theoretical analyses between the Raman-Nath parameter and the strength of the effective optical phase grating have been reported for circular[162] and square[162,163,164] transducers and indicate problems associated with determining the Raman-Nath parameter by assuming a transducer configuration.

Haran and colleagues[151] have described an experimental procedure by which the total acoustic power can be obtained without the need to make assumptions about the interaction length.

The intermediate and Bragg diffraction regions[148] have not been utilized for acoustic field measurement owing to the fact that the predominant clinical ultrasonic instrumentation operates at frequencies and possesses beam diameters which result in low values of Q.

An imaging broadening method for the determination of acoustic pressure amplitude was suggested by Hueter and Pohlman[165] based upon work by Lucas.[166] Also, a decrease in the zeroth order light intensity method was developed by Loeber and Hiedemann.[167] Both of these techniques, along with the above Raman-Nath region technique were used a number of years ago but little has been done with these two procedures since that time.[154,155]

The above acousto-optic methods are applicable in determining the acoustic pressure amplitude and continuous wave conditions. Newman[159] describes a schlieren technique by which an image of individual acoustic wavefronts can be obtained. Cook and Berlinghiere[160,161,168,169] have described a method by which a cross-sectional map of the acoustic pressure amplitude and phase can be obtained from a progressive sound wave. The procedure involves measuring the optical phase retardation of a light beam which traverses the sound beam laterally as the sound source, in effect, is rotated about its axis. They have shown[169] that the data requirements are not too excessive to yield an appropriate cross-sectional map. Cook[170,171] further describes a method by which the optical retardation, both amplitude and relative phase, can be determined and thus yield absolutely the Raman-Nath parameter. This latter procedure is applicable for both continuous wave and pulsed ultrasonic waves and is capable of yielding average and peak acoustic pressure.

An acousto-optic technique which requires a "probe" in the field and yields the absolute spatial particle displacement amplitude is based upon the Michelson interferometer whereby a thin, flexible membrane (metalized plastic film) is placed in the sound field and exhibits a particle displacement of the sound wave.[172,173] The interferometric arrangement is designed to detect the amplitude displacement of the membrane as a function of position on the membrane, hence yielding the spatial distribution of the particle displacement amplitude of ultrasonic pulses. The reported sensitivity of this technique is approximately 0.1 Å which is comparable to an intensity of 1 μW/cm^2 at 1.5 MHz in water.

4.6 Piezoelectric Techniques

Piezoelectric techniques are generally applied to measure the instantaneous acoustic pressure. They are, perhaps, the oldest techniques employed to measure underwater sound, following the discovery of the piezoelectric effect in 1880.[174,175] A most interesting and readable history of its discovery and development has been prepared by Cady[176] and Hunt.[177] In this section, the term piezoelectric techniques refers to all the associated phenomena which embody the transduction of the instantaneous acoustic pressure to an electrical signal in the megahertz frequency range. This includes not only piezoelectricity but also magnetostriction, ferroelectricity, etc. There are a number of excellent books[86,176,177,178,179] and monographs[180,181,182] which aptly describe these phenomena and their association with transduction.

One class of piezoelectric materials not included are the polymer films which have yet to be successfully applied to megahertz ultrasound. Yet they appear to have a potential application and thus are mentioned for completeness.[183,184,185,186]

Brendel[187] gives a brief summary of the basic requirements for hydrophones used to measure clinical ultrasonic instrumentation. These include being (1) temporally stable, (2) small relative to the wavelength, (3) adequately sensitive, and (4) broad-banded. While no one hydrophone satisfies all these criteria,[188] reasonable attempts have been made to construct hydrophones suitable for diagnostic ultrasound measurements.[187,189,190,191,192,193] Most were designed so that calibration could be accomplished with reciprocity. Only three have quantified the hydrophone response with sensitivities of 0.004 - 0.007 μV/bar (0.04 - 0.07 x 10^{-6} μV/Pa),[189] 0.2 mV/Wcm^{-2} (1.2 x 10^{-3} μV/Pa at 1 W cm^{-2})[192] and -109 db re 1 V/Pa (3.6 μV/Pa)[193] over the approximate bandwidth of 1-10 MHz with approximate transducer surface areas, respectively, of 0.0003, 0.17 and 38 mm^2.

The application of piezoelectric techniques are well suited for mapping the acoustic pressure distribution. In one report[194] a PZT-5H ceramic cylinder, 1.6 mm by 1.6 mm by 0.25 mm, positioned and attached coaxially on the end of a 1.6 mm diameter stainless steel tube, was used in an automated measuring system to obtain field plots of the acoustic pressure. The hydrophone exhibited an approximate sensitivity of 2.8 μV/Pa at 1 MHz.[195] Another piezoelectric probe, constructed with the same type of ceramic cylinder in basically the same fashion exhibited a sensitivity of approximately 10 μV/Pa.[190]

Two other experimental procedures are described to map out the ultrasonic field distribution with piezoelectric probes but little detail is provided regarding the probe calibration.[196,197] However, both do yield detailed and highly descriptive perspective maps of the ultrasonic field distribution.

The Underwater Sound Reference Division (USRD) of the Naval Research Laboratory USRD type E8 transducer is a calibrated ultrasonic transducer over the frequency range from 150 kHz to 2 MHz.[198] It is used as a projector, hydrophone and reciprocal transducer and is available on a rental basis.[199] It has a 2 cm diameter piezoelectric disk with a coupling media of castor oil between the disk and rubber window. It is designed to operate over the temperature range from 5 to 35°C which may represent some problems. The beam pattern closely follows the theoretical shape in the megahertz frequency range.

The National Bureau of Standards has developed a standard method to determine ultrasonic power[101] and is now offering a calibration service.[100] The technique employs a quartz transducer, operating at its fundamental frequency as a plane piston source. The total ultrasonic power is determined by the product of the applied voltage squared and the resonant radiation conductance, viz., $P = V^2 G_r$. The resonant radiation conduction is determined by measuring the input admittance of the transducer at resonance both in water (loaded resonant condition) and in air (unloaded resonant condition) and at twice the resonant frequency in air (clamped condition), thus yielding, respectively, G, G' and G" from which

$$G_r = \frac{(G' - G)\ (G - G'')}{(G' - G'')}$$

Because of the difficulty in determining G and G' directly, an experimental procedure has been established, using sophisticated NBS equipment, such that G_r can be determined within an uncertainty of about $\pm$ 2%. Typical value of the resonant radiation conductance for an air-backed, 2 MHz, 6.35 mm radius active surface quartz transducer radiating into water is 4.92 microsiemens. Taking into account additional uncertainties, including about a $\pm$ 1% in the applied voltage, the total radiated power can be known to within $\pm$ 5%.[101]

4.7 Radiation Force

Radiation force techniques are very desirable methods for determining ultrasonic power and spatially averaged intensity under both continuous wave and pulse conditions in that they are absolute, capable of sensitivities into the microwatt range, relatively simple in design, and generally do not require frequency compensation. When an object is placed in an ultrasonic beam, a force is exerted on the object by virtue of the time-independent pressure, called radiation pressure. This force is a direct result of the transport of energy by the ultrasonic wave and is equal to the time rate of change of momentum per unit area and, thus, is directly related to ultrasonic intensity and power. Chapter I provides mathematical and theoretical details of the phenomenon.

There have been two basic types of objects, or targets, used to measure radiation force, one which intercepts the entire beam and one which intercepts only part of the beam, thus yielding, respectively, the total ultrasonic power and spatially averaged intensity. In the case where the target intercepts the entire beam, is normal to the beam and is a perfect absorber (that is, all the energy incident upon the target is dissipated into it), a one milliwatt ultrasonic beam exerts a force equivalent to a weight of 67 micrograms upon the target, should the target be in a medium such as water with a speed of sound of 1500 m/s. Rooney[20] utilized a perfect absorber target at normal incidence in conjunction with a sensitive electrobalance to measure total power. The sensitivity of the balance was 0.1 micrograms, which is equivalent to 1.5 microwatts, but the system's reported minimum capability was 30 microwatts with a 5 microwatt standard deviation. Kossoff[200] earlier used a modified analytic balance with a sensitivity of 10 micrograms (0.15 milliwatts) and a totally absorbing target at normal incidence to the beam and achieved a minimum detectability of 80 microwatts. Kossoff[201] cautioned that an absorbing target is subject to Archimedes upthrust force because target temperature could increase due to energy absorption. At normal incidence, it should also be noted that if the target is not a "perfect" absorber, a standing wave develops between target and source which would affect the results.

To minimize thermal problems, a perfect reflecting target can be used. The target surface is typically positioned at an angle of 45° with respect to the beam which redirects the energy perpendicular to the beam and into an absorbing and/or scattering material. Under these conditions in water, the force exerted by a one milliwatt ultrasonic beam is equivalent to a weight of 134 $\cos^2 \theta$ micrograms where θ is the angle of incidence between the beam direction and the normal of the target surface, which is 67 microgram at 45°. Kossoff[200] used the same balance for a totally reflecting target as with absorbing target reported above and achieved a minimum detectability of 40 microwatts. Hill[190] described the construction of a balance system which was a modification of Kossoff's[200] design. This design exhibited a precision (variance) of $\pm$ 300 microwatts with a sensitivity of 300 microwatts. A less sensitive balance system was also described for application at therapeutic power levels. Wells and coworkers[202] described two radiation force balances and compared them at the one watt level to a calorimeter. Agreement was reported to be within 6 percent. The torsion balance principle has been utilized to measure radiation force incident upon a reflecting target.[203,204,205] The minimum detectability reported has been 30 microwatts. Zeidonis[206] also reported on the use of an analytic balance to measure ultrasonic power and obtained a sensitivity of 10 microwatts.

Another experimental set-up to measure ultrasonic power with a totally reflecting target has not yet shown itself sensitive enough to measure diagnostic levels. The bouyant float technique, first described by Oberst and Richmann,[207] is used to measure the radiation force of a downwardly directed ultrasonic beam by measuring the displacement of a float bouyant between two liquids, say water and a denser material such as carbon tetrachloride. The float target is an inverted conically shaped surface which permits the float to be self centering when the ultrasonic power is incident upon the target. The sensitivity of the technique depends predominately upon the diameter of the float stem which traverses both liquids and density difference between the two liquids. This technique has mainly been used to measure ultrasonic powers in the high milliwatt to tens of watts range[15,194,208] although Reid[209] has suggested that this technique may serve as a field standard and indicated that a sensitivity of 1 milliwatt per millimeter of float displacement may be achievable.

The horizontal deflection of a target has been used to measure both power and intensity. Wells and colleagues[210] suspended a totally reflecting target at an angle of 45° with respect to the direction of the sound beam and measured its deflection. In this arrangement, a power of one milliwatt caused a horizontal deflection of 200 micrometers. The minimum detectability of this arrangement was 10 microwatts. Using the deflection of a spherical target mounted on a bifilar suspension to measure radiation force was suggested as early as 1949 by Fox and Griffing.[211] The contemporary techniques generally consist of a steel ball connected to a filament (either with glue or by threading the filament through the finely drilled hole in the ball) and suspended vertically into a liquid such as water. The horizontal displacement of the suspended ball is determined for small angular deflections and yields the radiation force

$$F_r = \frac{dm_b g}{l}$$

where m_b is the bouyant mass of the sphere, g is the gravitational constant and l is the suspension length.[194,212,213] The relationship between the radiation force and the ultrasonic intensity is given by

$$I = \frac{F_r c}{\pi a^2 Y}$$

where c is the speed of sound in the fluid, a is the ball radius and Y is the "acoustic radiation force function"[214,215,216] which depends upon the elastic properties of the ball, the density of the fluid, ball size, and acoustic frequency. The suspended ball technique has not been applied to measuring clinical instrumentation because it typically is not sensitive enough. For example, if a stainless steel 440C ball, radius of 0.794 mm, has a suspension length of 11 cm, a 1 mm deflection is equivalent to 111 mW/cm^2.[194] A comparison of this technique with that of two other measurement techniques, optical and thermal, in 1958 showed agreement to within $\pm$ 10 percent.[217] Also, a recent evaluation of the suspended ball technique using elastic spheres showed that by comparing the acoustic radiation force function to experimental data, the ultrasonic intensity can be determined with an accuracy of about $\pm$ 3 percent.[218]

Zieniuk and Chivers[219] in a reasonably complete review have compared the radiation force and thermal techniques for the measurement of ultrasonic power and intensity.

Experimentally, both acoustic streaming and radiation force effects occur simultaneously.[220,221] As a result, Sokollu[221] argues that since the two phenomena cannot be differentiated, the radiation force techniques have a built-in bias or systematic error. There is no effective procedure to separate out the acoustic streaming except with the insertion of a membrane which, ideally, would serve as a barrier to the flow and permit the sound wave to be transmitted undisturbed. Rooney,[20] for example, used a 25 μm thick stretched membrane to shield the target from acoustic streaming.

The phenomenon of acoustic streaming, on the other hand, can be utilized to measure the absolute ultrasonic energy density.[220,222] Here an experimental arrangement permits the measurements of acoustic streaming flow rate and with known vessel dimensions and acoustical properties of fluid the energy density can be evaluated.

4.8 Reciprocity

On their own MacLean[223] and Cook[224] applied the principle of reciprocity[225] to the development of a method for absolute calibration of electroacoustic transducers. The complete range of reciprocity techniques are extensively covered by Hunt,[177] Ackerman and Holak[226] and Bobber[198] as they predominately apply to transducer calibration at frequencies typically lower than those employed in medicine; however, the principles are the same. To apply conventional reciprocity to a single absolute calibration, three transducers are required and three sets of voltage measurements must be made. The three transducers are a projector (source, loudspeaker, etc.), a hydrophone (receiver, microphone, etc.) and a linear, passive, reversible electro-acoustic transducer. The latter transducer is reciprocal, viz., the ratio of its receiving sensitivity to its transmitting response must be equal to a constant, the reciprocity parameter. This parameter is a function of the transmission medium, the acoustic frequency, and boundary conditions. Bobber[227] derived a unifying concept for all reciprocity parameters in terms of a general reciprocity parameter. The only requirement was that the medium satisfy the acoustical theorem. The reciprocity parameter for conventional (spherical-wave) reciprocity is[87]

$$J = \frac{2d\lambda}{\rho c}$$

where d is the distance between transducers, λ is the acoustic wavelength and ρc is the characteristic acoustic impedance of the medium.

The two-transducer reciprocity techniques, a special case of conventional reciprocity, require that both the hydrophone and reciprocal transducer have the same sensitivity which, without a third transducer, cannot be verified. Carstensen[228] devised a self reciprocity technique in which both the hydrophone and reciprocal transducer exhibited the same sensitivity. This was accomplished by reflecting the transmitted signal off a perfect reflector and back into the same transducer. It is possible to obtain an absolute calibration on a single transducer without the need for additional transducers. Koppelmann and colleagues[229] and Reid[230] applied the self-reciprocity technique to calibrate transducers. In the former, a two transducer procedure was used wherein one was calibrated _via_ the self-reciprocity technique and then the unknown transducer was calibrated in the known sound field. Reid[230] developed the theory for and described the use of the self-reciprocity technique for calibrating medical pulse-echo transducers. He developed theoretical expressions for both near and far field conditions under both continuous wave and pulsed regimes.

4.9 Thermal Techniques

Both transient and steady state thermal techniques have been employed as primary methods for the measurement of ultrasonic intensity and power. It has been pointed out[221] that thermal techniques have been employed since early in the century. Two reasonably comprehensive articles dealing with early work have been reported by Richards.[231,232]

Palmer[233] was among the first to demonstrate the feasibility of a thermocouple instrument to measure the minute intensity variations in an ultrasonic field. For this device, the thermocouple junction was coated with various absorbing materials. It was necessary to calibrate this device.

Fry and Fry,[57,58] on the other hand, demonstrated how the thermocouple probe was an absolute method for the determining ultrasonic intensity. The transient thermoelectric technique consists of a thermocouple junction embedded in a liquid of known ultrasonic absorption.[57,58,213,234,235,236] The absorbing liquid is contained in a holder with polyethylene, as an acoustic window, providing minimal reflection. A transient ultrasonic pulse, usually of one second in duration, is incident upon the junction. The time rate of temperature change, dT/dt, is related to the ultrasonic intensity by the expression

$$I = \frac{\rho C J}{2\alpha} \left(\frac{dT}{dt}\right)_o$$

where ρ, C, J and α are, respectively, the density, the heat capacity per unit volume, the mechanical equivalent of heat and the amplitude ultrasonic absorption coefficient, of the absorbing liquid in which the junction is embedded. A comparison with optical and radiation force techniques in 1958 showed agreement to within approximately $\pm$ 10 percent.[217] This transient thermoelectric technique also provides the opportunity to determine the absorption coefficient in tissue (see Chapter III in this Volume).

Zienuik[237] employed the principal used by Palmer[233] whereby the difference in voltage was determined between sets of thermocouples in and out (shielded) of the sound field.

Thermistors have also been employed to measure the fine details of an ultrasonic field.[238] In this technique, the thermistor probe senses the intensity of a continuous wave ultrasonic field but must be calibrated.

The principal attraction of calorimetric techniques lies in the fact that when ultrasonic energy is completely absorbed, a precise determination of the temperature change assures a precise determination of ultrasonic power. This, of course, requires that the thermal losses to the environment be at a minimum or, at least, determinable and under control. Also, calorimetric techniques are preferred for determining ultrasonic power in those cases where nonlinear effects, including cavitation, exists since these methods are ideally suited to transfer energy into heat, regardless of its spectral distribution.[221,239]

Three types of calorimetric techniques have been reviewed[240] and include a steady flow system, a transient system and a substitution system, all of which require the conversion of ultrasonic energy into thermal energy *via* a suitable high absorbing material. Thus each yields an absolute measure of ultrasonic power. In the steady flow system the change in temperature between the input and output of a flowing absorbing liquid is determined. In the transient system, the time change of temperature of the absorbing medium is determined and, in principle, this is similar to the transient thermoelectric probe with the exception that the total power is the determinable parameter. The substitute system duplicates the thermal history in the absorbing medium between the absorbed ultrasound and an embedded electrical heater and thus equates the electrical power to the ultrasonic power.

Sokollu[241] described a miniaturized calorimeter designed with the purpose of measuring total ultrasonic power from a focused source. The system's sensitivity restricted its application to levels above one watt. Its accuracy was highly dependent upon the accuracy of the electrical power measurement comparison. van den Ende[242] also described a similar calorimeter but it was not necessarily designed for a focused signal. Comparison with a radiation force technique over the power range from about 0.3 to 1.2 watts showed agreement to within ± 10 percent. Comparison of a spherical calorimeter within the same power range to a radiation force technique agreed to within ± 6 percent.[202]

Zienuik[243,244] developed the theory for and argues that the non-adiabatic, non-isothermal calorimeter does not represent two drawbacks of quasi-adiabatic calorimeters, viz., corrections of heat exchange are extremely difficult and high thermal inertia.

Mikhailov[245] developed a calorimeter which made use of the thermal expansion of the sound absorbing liquid. The device is calibrated by applying a known electrical power through the heating coil, embedded in the absorbing liquid. The thermal expansion of the absorber is determined by measuring the rise of the liquid in a small capillary tube.

A calorimetric technique has been employed to yield ultrasonic intensity by providing a finite size aperture, diameter of 5 wavelengths, for the ultrasonic energy to enter and hence to be absorbed.[246] This device has been used to measure the peak ultrasonic intensity (over limited spatial extent).

Szilard[247] reported on an instrument which could be used to measure the ultrasonic energy delivered to patients during therapeutic treatment. A plate, which is affixed to the therapeutic transducer surface during treatment, has a series of thermistors embedded to monitor the temperature change and, hopefully, the ultrasonic energy delivered to the patient.

REFERENCES

1. J. Y. Harris: Electronic Product Investory Survey. Bureau of Radiological Health Report. BRH/DEP 70-29 (1970) Rockville, Maryland.

2. S. W. Smith: Diagnostic Equipment and Its Use. Eighth Annual National Conference on Radiation Control, May 2-7 (1976) Springfield, Illinois.

3. E. Landau: Are There Ultrasonic Dangers for the Unborn? Practical Radiology, 1 (1973) pp. 27-31.

4. J. H. Holmes: Editorial in Journal of Clinical Ultrasound. 2 (1974) pp. 87-90.

5. P. Roney: Hospital Survey of Obstetrical Ultrasound. Ad Hoc Review Panel on Ultrasound Bioeffects and Measurements, Bureau of Radiological Health, April 9-10 (1976).

6. M. C. Ziskin: Survey of Patient Exposure to Diagnostic Ultrasound. In <u>Interaction of Ultrasound and Biological Tissues Workshop Proceedings.</u> DHEW Publication (FDA) 73-8008, BRH/DBE 73-1, U.S. Government Printing Office, Washington, D.C. (1972) pp. 203-205.

7. J. M. Reid and M. R. Sikov, eds.: <u>Interaction of Ultrasound and Biological Tissues Workshop Proceedings.</u> DHEW Publication (FDA) 73-8008, BRH/DBE 73-1, U.S. Government Printing Office, Washington, D.C. (1972).

8. P. N. T. Wells: What Future for Ultrasonics? - Biomedical Ultrasonics. Ultrasonics 11 (1973) p. 16.

9. J. L. Marx: Diagnostic Medicine: The Coming Ultrasonic Boom. Science 186 (1974) pp. 247-250 (in Research News).

10. R. E. DeForest: Present Status of Use of Ultrasonic Energy in Physical Medicine. J. Amer. Med. Assoc. 148 (1952) pp. 646-651.

11. J. F. Herrick and F. H. Krusen: Ultrasound and Medicine. A Survey of Experimental Studies. J. Acoust. Soc. Amer., 26 (1954) pp. 236-240.

12. S. Licht: History of Therapeutic Heat. In Therapeutic Heat and Cold, ed. S. Licht (1965) pp. 198-231, Waverly Press, Inc., Baltimore, Maryland.

13. D. G. Remark: Survey of Diathermy Equipment Use in Pinellas County, Florida. DHEW Publication No. BRH/NERHL 71-1, (1971) U. S. Government Printing Office, Washington, D.C.

14. H. F. Stewart, G. R. Harris, B. A. Herman, R. A. Robinson and M. E. Haran: Survey of Use and Performance of Ultrasonic Therapy Equipment in Pinellas County, Florida. DHEW Publication No. (FDA) 73-8039, (1973) U. S. Government Printing Office, Washington, D. C.

15. H. R. Stewart, G. Harris, R. Robinson and J. Garry: Survey of Use and Performance of Ultrasonic Therapy Units in the Washington, D.C. area. In Health Physics in the Healing Arts, (1973) pp. 467-472, DHEW Publication No. (FDA) 73-8029, U. S. Government Printing Office, Washington, D.C.

16. C. R. Hill: Acoustic Intensity Measurements on Ultrasonic Diagnostic Devices. In Proc. 1st World Congress on Ultrasonic Diagnostics in Medicine and SIDUO III, eds. J. Bock and K. Ossoinig (1969) pp. 21-27, Verlag der Wiener Medizinischen Akademie.

17. L. Filipczynski: Ultrasonic Measurement Methods and Instrumentation Developed in Poland for Medical Diagnostics. In Proc. Conference Biol. Med., UBIOMED-70, ed. L. Filipczynski (1970) pp. 75-80, Polish Scientific Publisher, Warsaw, Poland.

18. P. L. Carson, W. R. Hendee and S. Leung: Performance Evaluation and Dosimetry in Diagnostic Ultrasound. Quarterly Bulletin, AAPM, December, (1973) pp. 200-205.

19. P. L. Carson and T. F. Oughton: Survey of Acoustic Power and Intensities Produced by Diagnostic Ultrasound. In Proc. Intern. Conf. Med. Physics (1976).

20. J. A. Rooney: Determination of Acoustic Power Outputs in the Microwatt-Milliwatt Range. Ultrasound Med. Biol. 1 (1973) pp. 13-16.

21. M. C. Ziskin, N. Romayananda and K. Harris: Ophthalmologic Effect of Ultrasound at Diagnostic Intensities. J. Clin. Ultrasound 2 (1974) pp. 119-122.

22. ICRU: Radiation Quantities and Units. International Commission on Radiation Units and Measurements, Report No. 19 (1971) Washington, D.C.

23. BEIR: The Effects on Populations of Exposure to Low Levels of Ionizing Radiation. National Academy of Sciences - National Research Council (1972) Washington, D.C.

24. ICP: Proposals Concerning Definitions and Units for the Biological and Medical Use of UV Radiation. In Proc. First Int. Photobiological Congress, (1954) p. 440, Amsterdam.

25. C. S. Rupert: Dosimetric Concepts in Photobiology. Photochem. Photobiol. 20 (1974) pp. 203-212.

26. S. M. Michaelson: Human Exposure to Nonionizing Radiant Energy - Potential Hazards and Safety Standards. Proc. IEEE 60 (1972) pp. 389-421.

27. C. C. Johnson: Recommendations for Specifying EM Wave Irradiation Conditions in Bioeffects Research. J. Microwave Power, 10 (1975) pp. 249-250.

28. C. Susskind: Correspondence on D. R. Justesen's "Prescriptive Grammar for the Radiobiology of Nonionizing Radiation." J. Microwave Power, 10 (1975) p. 357.

29. A. W. Guy: Correspondence on D. R. Justesen's "Prescriptive Grammar for the Radiobiology of Nonionizing Radiation." J. Microwave Power, 10 (1975) pp. 358-359.

30. D. R. Justesen: Toward a Prescriptive Grammar for the Radiobiology of Nonionizing Radiation: Quantities, Definitions, and Units of Absorbed Electromagnetic Energy - An Essay. J. Microwave Power, 10 (1975) pp. 343-356.

31. F. W. Spiers: Radiation Units and Theory of Ionizing Dosimetry. In Radiation Dosimetry, eds. G. J. Hine and G. L. Brownell (1956) pp. 1-47, Academic Press, New York, N.Y.

32. C. C. Wang and L. L. Robbins: Biological and Medical Effects of Radiation. In Radiation Dosimetry, eds. G. J. Hine and G. L. Brownell, (1956) pp. 125-152, Academic Press, New York, N.Y.

33. F. W. Spiers: The Historical Development of Radiation Dosimetry. In Proc. International School of Physics, Course XXX, Radiation Dosimetry, ed. G. W. Reid (1964) pp. 1-6, Academic Press, New York, N.Y.

34. W. C. Roesch and F. H. Attix: Basic Concepts of Dosimetry. In Radiation Dosimetry, eds. F. H. Attix and W. C. Roesch, Vol. 1, (1968) pp. 1-41, Academic Press, New York, N.Y.

35. L. S. Taylor: Radiation Protection Standards. CRC Press, (1971) Cleveland, Ohio.

36. S. C. Bushong: The Development of Radiation Protection in Diagnostic Radiology. CRC Press (1973) Cleveland, Ohio.

37. ICRU: International X-ray Unit of Intensity. International Commission on Radiological Units and Measurements, Report No. 2, Brit. J. Radiol., 1 (1928) p. 363.

38. G. Failla and P. S. Henshaw: The Relative Biological Effectiveness of X-rays and Gamma Rays. Radiology 17 (1931) pp. 1-43.

39. L. H. Gray and J. Read: Measurement of Neutron Dose in Biological Experiments. Nature 144 (1939) pp. 439-440.

40. W. V. Mayneord: Energy Absorption. Brit. J. Radiol., 13 (1940) pp. 235-247.

41. H. M. Parker: Health Physics, Instrumentation and Radiation Protection. Adv. Biol. Med. Phys., 1 (1948) pp. 223-285.

42. ICRU: Recommendations of the International Committee for Radiological Units. International Commission on Radiological Units and Measurements, Report No. 7, Radiology 62 (1954) pp. 106-109.

43. ICRU: Report of the ICRU. International Committee for Radiological Units and Measurements, Report No. 8 (1956) National Bureau of Standards Handbook 62.

44. ICRU: Radiation Quantities and Units. International Committee for Radiological Units and Measurements, Report No. 10a, (1962) National Bureau of Standards Handbook 84.

45. W. D. O'Brien, Jr., M. L. Shore, R. K. Fred and W. M. Leach: On the Assessment of Risk to Ultrasound. In 1972 Ultrasonic Symposium Proceedings, ed. J. de Klerk, pp. 486-490, IEEE Catalog No. 72 CHO 708-8SU (1972) New York, N.Y.

46. A. J. Hall: Ultrasonic Dosimetry in Obstetrics. Presented at 2nd World Congress on Ultrasonics in Medicine (1973) Rotterdam, June 4-8 (Abstract).

47. J. Bang: The Intensity of Ultrasound in the Uterus During Examination for Diagnostic Purposes. Acta Path. Microbiol. Scan., Section A, 80 (1972) pp. 341-344.

48. A. J. Hall and H. P. Robinson: Transmitted Ultrasonic Energy Levels in the Nongravid Uterus. Colloquium on Ultrasound Bioeffects and Dosimetry, (1974) London, July 22-23 (Abstract).

49. J. Etienne, L. Filipczynski, A. Ficek, J. Groniowski, J. Kretowicz, G. Lypacewicz and J. Salkowski: Intensity Determination of Ultrasonic Focused Beams Used in Ultrasonography in the Case of Gravid Uterus. Ultrasound Med. Biol., 2 (1976) pp. 119-122.

50. E. W. Purnell, A. Sokollu, R. Torchia and N. Taner: Focal Chorioretinitis Produced by Ultrasound. Invest. Ophth., 3 (1964) pp. 657-664.

51. K. J. W. Taylor and J. B. Pond: A Study of the Production of Hemorrhagic Injury and Paraplegia in Rat Spinal Cord by Pulsed Ultrasound of Low Megahertz Frequencies in the Context of the Safety for Clinical Usage. Brit. J. Radiol. 45 (1972) pp. 343-353.

52. R. L. Johnston and F. Dunn: Ultrasonic Absorbed Dose, Dose Rate, and Produced Lesion Volume. Ultrasonics 14 (1976) pp. 153-155.

53. F. Dunn and F. J. Fry: Ultrasonic Threshold Dosages for the Mammalian Central Nervous System. IEEE Trans. Biomed. Eng., BME-18 (1971) pp. 253-256.

54. J. B. Pond: A Study of the Biological Action of Focussed Mechanical Waves (Focussed Ultrasound). Ph.D. Thesis (1968) London University.

55. J. B. Pond: The Role of Heat in the Production of Ultrasonic Focal Lesions. J. Acoust. Soc. Amer., 47 (1970) pp. 1607-1611.

56. R. M. Lerner, E. L. Carstensen and F. Dunn: Frequency Dependence of Thresholds for Ultrasonic Production of Thermal Lesions in Tissue. J. Acoust. Soc. Amer., 54 (1973) pp. 504-506.

57. W. J. Fry and R. B. Fry: Determination of Absolute Sound Levels and Acoustic Absorption Coefficients by Thermocouple Probes - Theory. J. Acoust. Soc. Amer., 26 (1954) pp. 294-310.

58. W. J. Fry and R. B. Fry: Determination of Absolute Sound Levels and Acoustic Absorption Coefficients by Thermocouple Probes - Experiment. J. Acoust. Soc. Amer., 26 (1954) pp. 311-317.

59. H. G. Flynn: Cavitation Dynamics. I. A Mathematical Formulation. J. Acoust. Soc. Amer., 57 (1975) pp. 1379-1396.

60. H. G. Flynn: Cavitation Dynamics. II. Free Pulsations and Models for Cavitation Bubbles. J. Acoust. Soc. Amer., 58 (1975) pp. 1160-1170.

61. H. G. Flynn: Physics of Acoustic Cavitation. In Physical Acoustics, ed. W. P. Mason, Vol. I, Part B, (1964) pp. 57-172, Academic Press, New York, N.Y.

62. W. L. Nyborg: Acoustic Streaming. In Physical Acoustics, ed. W. P. Mason, Vol. II, Part B, (1965) pp. 265-331, Academic Press, New York, N.Y.

63. W. L. Nyborg: Mechanisms for Nonthermal Effects of Sound. J. Acoust. Soc. Amer., 44 (1968) pp. 1302-1309.

64. J. A. Rooney: Shear as a Mechanism for Sonically Induced Biological Effects. J. Acoust. Soc. Amer., 52 (1972) pp. 1718-1724.

65. W. L. Nyborg: Intermediate Biophysical Mechanics. (1976) Cummings Publishing Co., Menlo Park, California.

66. P. R. Clarke and C. R. Hill: Physical and Chemical Aspects of Ultrasonic Disruption of Cells. J. Acoust. Soc. Amer., 47 (1969) pp. 649-653.

67. S. Z. Child, E. L. Carstensen and M. W. Miller: Growth of Pea Roots Exposed to Pulsed Ultrasound. J. Acoust. Soc. Amer., 58 (1975) pp. 1109-1110.

68. R. Esche: Investigation of Vibratory Cavitation in Liquids (in German). Akust. Beih., 4 (1952) pp. 208-218. Partial translation in Ultrasonic Biophysics, eds. F. Dunn and W. D. O'Brien, Jr., Dowden, Hutchinson and Ross, Inc., Stroudsburg, Pennsylvania, 1976.

69. F. J. Fry, G. Kossoff, R. C. Eggleton and F. Dunn: Threshold Ultrasonic Dosages for Structural Changes in the Mammalian Brain. J. Acoust. Soc. Amer., 48 (1970) pp. 1413-1417.

70. W. J. Fry, V. J. Wulff, D. Tucker and F. J. Fry: Physical Factors Involved in Ultrasonically Induced Changes in Living Systems: I. Identification of Non-Temperature Effects. J. Acoust. Soc. Amer., 22 (1950) pp. 867-876.

71. K. Woeber: The Effect of Ultrasound in the Treatment of Cancer. In Ultrasonic Energy, ed. E. Kelly (1965) pp. 137-149, University of Illinois Press, Urbana, Illinois.

72. P. R. Clarke, C. R. Hill and K. Adams: Synergism between Ultrasound and X-Rays in Tumour Therapy. Brit. J. Radiol., 43 (1970) pp. 97-99.

73. F. W. Kremkau, J. S. Kaufmann, M. M. Walker, P. G. Burch and C. L. Spurr: Ultrasonic Enhancement of Nitrogen Mustard Cytotoxicity in Mouse Leukemia. Cancer 37 (1976) pp. 1643-1647.

74. F. Dunn, J. E. Lohnes and F. J. Fry: Frequency Dependence of Threshold Ultrasonic Dosages for Irreversible Structural Changes in Mammalian Brain. J. Acoust. Soc. Amer., 58 (1975) pp. 512-514.

75. R. L. Johnston and F. Dunn: Influence of Subarachnoid Structures on Transmenninges Ultrasonic Propagation. J. Acoust. Soc. Amer., 59 (1976) p. S76 (Abstract).

76. S. Fields and F. Dunn: Correlation of Echographic Visualizability of Tissue with Biological Composition and Physiological State. J. Acoust. Soc. Amer., 54 (1973) pp. 809-812.

77. K. J. W. Taylor and J. Pond: The Effects of Ultrasound on Varying Frequencies on Rat Liver. J. Path., 100 (1970) pp. 287-293.

78. K. J. W. Taylor and D. L. Newman: Effects on Electrophoretic Mobility of Ehrlich Cells Exposed to Ultrasound of Variable Parameters. Phys. Med. Biol., 17 (1972) pp. 270-276.

79. A. Sokollu: Destructive Effect of Ultrasound on Ocular Tissues. In Interaction of Ultrasound and Biological Tissues Workshop Proceedings. eds. J. M. Reid and M. R. Sikov, (1972) DHEW Publication (FDA) 73-8008 BRH/DBE 73-1, U.S. Government Printing Office, Washington, D.C.

80. J. Blitz: Fundamentals of Ultrasonics. (1963) Butterworth, London.

81. G. L. Gooberman: Ultrasonics, Theory and Application. (1968) The English Universities Press, London.

82. P. N. T. Wells: Physical Principles of Ultrasonic Diagnosis. (1969) Academic Press, New York, N.Y.

83. P. N. T. Wells: Basic Principles and Diagnostic Methods. In Ultrasonics in Clinical Diagnosis, ed. P. N. T. Wells (1972) pp. 3-38, Churchill Livingston, London.

84. E. N. Carlsen: Ultrasound Physics for the Physician - A Brief Review. J. Clinical Ultrasound, 3 (1975) pp. 69-75.

85. M. Hussey: Diagnostic Ultrasound. (1975) John Wiley and Sons, New York, N.Y.

86. T. F. Hueter and R. H. Bolt: Sonics (1956) John Wiley & Sons, Inc., New York, N.Y.

87. L. E. Kinsler and A. R. Frey: Fundamentals of Acoustics, (1962) Wiley, New York, N.Y.

88. R. T. Beyer and S. V. Letcher: Physical Ultrasonics, (1969) Academic Press, New York, N.Y.

89. E. A. G. Shaw and R. J. Sujir: Vibration Patterns of Loaded Barium Titanate and Quartz Disks. J. Acoust. Soc. Amer., 32 (1960) pp. 1463-1467.

90. L. Filipczynski and G. Lypacewicz: Vibration Patterns and Properties of Piezoelectric Ceramic Transducers for Diagnostical Application in Medicine. In Ultrasonics in Biology and Medicine, ed. by L. Filipczynski, (1970) pp. 81-89, Polish Scientific Publishers, Warsaw, Poland.

91. G. Lypacewicz and L. Filipczynski: Measurement Method and Experimental Study of Ceramic Transducer Vibrations. Acustica 24 (1971) pp. 216-221.

92. J. Zemanek: Beam Behavior Within the Nearfield of a Vibrating Piston. J. Acoust. Soc. Amer., 49 (1971A) pp. 181-191.

93. L. Filipczynski: The Near Field Distribution on the Axis of a Vibrating Piston. The 6th International Congress on Acoustics, (1968) paper L-3-2, pp. L-69 - L-72.

94. K. von Hasselberg and J. Krautkrämer: An Ultrasonic Radiator for Flaw Detection with Improved Near Field (In German). Acustica 9 (1959) pp. 359-364.

95. G. W. Miner, II and P. A. Laura: Calculation of the Nearfield Pressure Induced by Vibrating Circular Plates. J. Acoust. Soc. Amer., 42 (1967) pp. 1025-1030.

96. D. L. Dekker, R. L. Piziali and E. Dong, Jr.: Effect of Boundary Conditions on the Ultrasonic Beam Characteristics of Circular Disks. J. Acoust. Soc. Amer., 56 (1974) pp. 87-93.

97. R. P. Ryan, A. Lutsch and R. T. Beyer: Measurement of the Distortion of Finite Ultrasonic Waves in Liquids by a Pulsed Method. J. Acoust. Soc. Amer., 34 (1962) pp. 31-35.

98. R. K. Gould, C. W. Smith, A. O. Williams, Jr., and R. P. Ryan: Measured Structure of Harmonics Self-Generated in an Acoustic Beam. J. Acoust. Soc. Amer., 40 (1966) pp. 421-427.

99. F. Ingenito and A. O. Williams, Jr.: Calculation of Second-Harmonic Generation in a Piston Beam. J. Acoust. Soc. Amer., 49 (1971) pp. 319-328.

100. NBS: NBS Offers Ultrasonic Power Calibration Service. In United States Department of Commerce News, Technical News from the National Bureau of Standards, April (1974) TN-4581.

101. T. L. Zapf: Calibration of Quartz Transducers as Ultrasonic Power Standards by an Electrical Method. In 1974 Ultrasonics Symposium Proceedings, ed. J. de Klerk, pp. 45-50, IEEE Cat. No. 74-CHO 896-ISU (1974) New York, N.Y.

102. T. L. Zapf: Personal Communication (1976).

103. C. Eisenhart: Expression of the Uncertainties of Final Results. Science 160 (1968) pp. 1201-1204.

104. C. Eisenhart: Realistic Evaluation of the Precision and Accuracy of Instrument Calibration Systems. J. Res. Nat. Bureau Stds., 67c (1963) pp. 161-187

105. Y. Beers: Introduction to the Theory of Error. (1953) Addison-Wesley Publishing Co., Cambridge, Mass.

106. D. O. Thompson, M. A. Breazeale and W. B. Gauster: A Capacitive Microphone for Ultrasonic Measurements at Megacycle Frequencies. In Proc. Conf. on Acoustics of Solid Media, (1964) p. 223, Warsaw, Poland.

107. W. B. Gauster and M. A. Breazeale: Detector for Measurement of Ultrasonic Strain Amplitudes in Solids. Rev. Sci. Instr., 37 (1966) pp. 1544-1548.

108. R. D. Peters, M. A. Breazeale and V. K. Pare: Variable Gap Capacitive Detector for the Measurement of Ultrasonic Displacement Amplitude in Solids. Rev. Sci. Instrum., 39 (1968) pp. 1505-1506.

109. J. Blitz and D. G. Warren: Absolute Measurement of the Intensity of Pulsed Ultrasonic Waves in Solids and Liquids with a Capacitor Microphone at Megahertz Frequencies. Ultrasonics 6 (1968) pp. 235-239.

110. L. Filipczynski: Absolute Measurements of Particle Velocity, Displacement or Intensity of Ultrasonic Pulses in Liquids and Solids. Acustica 21 (1969) pp. 173-180.

111. L. Filipczynski: Intensity Determination of Ultrasonic Pulses by Means of a Capacitance Transducer. In Interaction of Ultrasound and Biological Tissues Workshop Proceedings, eds. J. M. Reid and M. R. Sikov. DHEW Publication (FDA) 72-8008 BRH/DBE 73-1, U.S. Government Printing Office, Washington, D. C. (1972) pp. 171-172.

112. L. Filipczynski, G. Lypacewicz and J. Salkowski: Intensity Determination of Focused Ultrasonic Beams by Means of Electrodynamic and Capacitance Methods. Proc. Vibration Problems 15 (1974) pp. 297-305.

113. P. Debye: A Method for the Determination of the Mass of Electrolytic Ions. J. Chem. Phys. 1 (1933) pp. 13-16.

114. J. Bugosh, E. Yeager and F. Hovorka: The Application of Ultrasonic Waves to the Study of Electrolytic Solutions I. A Modification of Debye's Equation for the Determination of the Masses of Electrolytic Ions by Means of Ultrasonic Waves. J. Chem. Phys. 15 (1947) pp. 592-597.

115. E. Yeager and F. Hovorka: The Application of Ultrasonics to the Study of Electrolytic Solutions III. The Effect of Acoustical Waves on the Hydrogen Electrode. J. Chem. Phys. 17 (1949) pp. 416-417.

116. E. Yeager, J. Bugosh, F. Hovorka and J. McCarthy: The Application of Ultrasonic Waves to the Study of Electrolytic Solutions II. The Detection of the Debye Effect. J. Chem. Phys. 17 (1949) pp. 411-415.

117. A. N. Hunter: Some Observations on the Debye Effect in Electrolytes. Proc. Phys. Soc., 71 (1958) pp. 847-851.

118. A. J. Rutgers and W. Rigole: Ultrasonic Vibration Potentials in Colloidal Solutions, in Solutions of Electrolytes and in Pure Liquids. Trans. Faraday Soc., 54 (1958) pp. 139-143.

119. E. Yeager, J. Booker and F. Hovorka: Ultrasonic Vibration Potentials in Non-Ionic Liquids. Proc. Phys. Soc., 73 (1959) pp. 690-692.

120. E. Yeager, H. Dietrick and F. Hovorka: Ultrasonic Waves and Electrochemistry. II. Colloidal and Ionic Vibration Potentials. J. Acoust. Soc. Amer., 25 (1953) pp. 456-460.

121. A. Weinmann: Theory of Ultrasonic Vibration Potentials in Pure Liquids. Proc. Phys. Soc., 75 (1960) pp. 102-108.

122. R. Zana and E. Yeager: Ultrasonic Vibration Potentials and Their Use in the Determination of Ionic Partial Molal Volumes. J. Phys. Chem., 71 (1967) pp. 521-536.

123. R. Zana and E. Yeager: Quantitative Studies of Ultrasonic Vibration Potentials in Polyelectrolyte Solutions. J. Phys. Chem., 71 (1967) pp. 3502-3516.

124. H. J. Wintle: Introduction to Electrets. J. Acoust. Soc. Amer., 53 (1973) pp. 1578-1588.

125. D. Legros, J. Lewiner and P. Biquard: Generation of Ultrasound by a Dielectric Transducer. J. Acoust. Soc. Amer., 52 (1972) pp. 196-198.

126. D. Legros and J. Lewiner: Electrostatic Ultrasonic Transducers and Their Utilization with Foil Electrets. J. Acoust. Soc. Amer., 53 (1973) pp. 1663-1672.

127. G. M. Sessler and J. E. West: Electret Transducers: A Review. J. Acoust. Soc. Amer., 53 (1973) pp. 1589-1600.

128. A. K. Nigam and G. M. Sessler: Condenser-Transducer Array for Acoustical Holography. Appl. Phys. Letters, 21 (1972) pp. 229-231.

129. A. K. Nigam, K. J. Taylor and G. M. Sessler: Foil-Electret Transducer Arrays for Real-Time Acoustical Holography. In Acoustical Holography, 4, ed. G. Wach, Plenum Press, (1972) pp. 173-194, New York, N.Y.

130. L. Filipczynski: The Absolute Method for Intensity Measurements of Liquid-Borne Ultrasonic Pulses with the Electrodynamic Transducer. Proc. Vibration Problems, Warsaw 8 (1967) pp. 21-26.

131. L. Brillouin: Diffusion de la Lumiere par un Corps Transparent Homogene. Comptes Rendus, 158 (1914) pp. 1331-1334. Translation in Physical Acoustics Benchmark Papers in Acoustics, ed. by R. B. Lindsay, (1974) pp. 416-418, Dowden, Hutchinson and Ross, Inc. Stroudsburg, Pennsylvania.

132. L. Brillouin: Diffusion de la Lumiere et des Rayons X par un Corps Transparent Homogene. Ann. Phys. (France), Series 9, 17 (1922) pp. 88-122.

133. P. Debye and F. W. Sears: On the Scattering of Light by Supersonic Waves. Proc. National Acad. Sci. (U.S.A.) 18 (1932) pp. 409-414.

134. R. Lucas and P. Biquard: Optical Properties of Solids and Liquids under Ultrasonic Vibrations. J. Phys. Radium, Series 7, 3 (1932) p. 464.

135. R. Bär: Uber einige Demonstrationsversuche zur Beugung des Lichtes an Ultraschallwellen. Helv. Physica Acta, 6 (1933) pp. 570-580.

136. N. S. N. Nath: The Diffraction of Light by High Frequency Sound Waves: Generalized Theory. The Asymmetry of the Diffraction Phenomena of Oblique Incidence. Proc. Indian Acad. Sci., Section A, 4 (1936) pp. 222-242.

137. C. V. Raman and N. S. N. Nath: The Diffraction of Light by High Frequency Sound Waves: Part I. Proc. Indian Acad. Sci., Section A, 2 (1935) pp. 406-412.

138. C. V. Raman and N. S. N. Nath: The Diffraction of Light by Sound Waves of High Frequency: Part II. Proc. Indian Acad. Sci., Section A, 2 (1935) pp. 413-420.

139. C. V. Raman and N. S. N. Nath: The Diffraction of Light by High Frequency Sound Waves: Part III. Doppler Effect and Coherence Phenomena. Proc. Indian Acad. Sci., Section A, 3 (1936) pp. 75-84.

140. C. V. Raman and N. S. N. Nath: The Diffraction of Light by High Frequency Sound Waves: Part IV. Generalized Theory. Proc. Indian Acad. Sci., Section A, 3 (1936) pp. 119-125.

141. C. V. Raman and N. S. N. Nath: The Diffraction of Light by High Frequency Sound Waves: Part V. General Considerations Oblique Incidence and Amplitude Changes. Proc. Indian Acad. Sci., Section A, 3 (1936) pp. 459-465.

142. L. Bergman: Der Ultraschall and seine Anwendung in Wissenschraft und Techik. (1954) Hirzel Verlag, Stuttgart.

143. J. C. Slater: Interaction of Waves in Crystals. Rev. Mod. Phys. 30 (1958) p. 197.

144. M. Born and E. Wolf: Principles of Optics. Second (Revised) Edition (1964) pp. 594-610, Macmillan Company, New York.

145. C. F. Quate, C. D. W. Wilkinson and D. K. Winslow: Interaction of Light and Microwave Sound. Proc. IEEE, 53 (1965) pp. 1604-1623.

146. E. I. Gordon: A Review of Acousto-Optical Deflection and Modulation Devices. Proc. IEEE, 54 (1966) pp. 1391-1401.

147. W. R. Klein, B. D. Cook and W. G. Mayer: Light Diffraction by Ultrasonic Gratings. Acustica 15 (1965) pp. 67-74.

148. W. R. Klein and B. D. Cook: Unified Approach to Ultrasonic Light Diffraction. IEEE Trans. Sonics-Ultrasonics, SU-14 (1967) pp. 123-134.

149. B. D. Cook: Determination of Ultrasonic Waveforms by Optical Methods. IEEE Trans. Sonics-Ultrasonics, SU-11 (1964) pp. 89-94.

150. B. D. Cook: Optical Method for Ultrasonic Waveform Analysis, Using a Recursion Relation. J. Acoust. Soc. Amer., 37 (1965) pp. 172-173.

151. M. E. Haran, B. D. Cook and H. F. Stewart: Comparison of an Acousto-Optic and a Radiation Force Method of Measuring Ultrasonic Power. J. Acoust. Soc. Amer., 57 (1975) pp. 1436-1440.

152. V. Raman and K. S. Venkataraman: Determination of the Adiabatic Piezo-Optic Coefficient in Liquids. Proc. Royal Soc. London, Series A, 171 (1939) pp. 137-147.

153. W. A. Riley and W. R. Klein: Piezo-Optic Coefficients of Liquids. J. Acoust. Soc. Amer., 42 (1967) pp. 1258-1261.

154. M. A. Breazeale and E. A. Hiedermann: Optical Methods for the Measurement of the Sound Pressure in Liquids. J. Acoust. Soc. Amer., 31 (1959) pp. 24-28.

155. M. A. Breazeale, L. E. Hargrove and E. A. Hiedemann: Measurement of Sound Pressure Amplitude by Optical Methods. U.S. Navy J. Underwater Acoustics 10 (1960) pp. 381-387.

156. K. R. Erikson: Calibration of Standard Ultrasonic Probe Transducers Using Light Diffraction. In Interaction of Ultrasound and Biological Tissues Workshop Proceedings. (1972) DHEW Publication (FDA) 73-8008 BRH/DBE 73-1, U.S. Government Printing Office, Washington, D.C. pp. 193-197.

157. V. M. Berry: The Diffraction of Light by Ultrasound. Academic Press, (1966) New York, N.Y.

158. G. Kossoff: Visualization Systems. In Interaction of Ultrasound and Biological Tissues Workshop Proceedings. (1972) DHEW Publication (FDA) 73-8008 BRH/DBE 73-1, U. S. Government Printing Office, Washington, D. C. pp. 177-179.

159. D. R. Newman: Ultrasonic Schlieren System Using a Pulsed Gas Laser. IEEE Trans. Sonics-Ultrasonics, SU-20 (1973) pp. 282-285.

160. B. D. Cook: Mapping of Ultrasonic Fields with Conventional Light Diffraction. J. Acoust. Soc. Amer., 53 (1972) p. 340 (A).

161. B. D. Cook: Proposed Mapping of Ultrasonic Fields with Conventional Light Diffraction. J. Opt. Soc. Amer., 65 (1975) pp. 682-684.

162. F. Ingenito and B. D. Cook: Theoretical Investigation of the Integrated Optical Effect Produced by Sound Fields Radiated from Plane Piston Transducers. J. Acoust. Soc. Amer., 45 (1969) pp. 572-577.

163. B. D. Cook and F. Ingenito: Ultrasonic Transducer Configuration for Producing a Phase Grating of Nearly Uniform Strength. Proc. IEEE, 56 (1968) pp. 871-872.

164. W. T. Maloney, G. Meltz and R. L. Gravel: Optical Probing of the Fresnel and Fraunhofer Regions of a Rectangular Acoustic Transducer. IEEE Trans. Sonics-Ultrasonics, SU-15 (1968) pp. 167-172.

165. T. Hueter and R. Pohlman: Eine Optische Methode zur Bestimmung der Ultraschallabsorption in Undurchsichtigen, Schallweichen Medien. Zeit. Agnew.Physik, 1 (1949) pp. 405-411.

166. R. Lucas: Sur les Phenomenes de Mirage Optique dus aux ondes Elastiques. Comptes Rendus, 199 (1934) pp. 1107-1108.

167. A. P. Loeber and E. A. Hiedemann: Investigation of Stationary Ultrasonic Waves by Light Refraction. J. Acoust. Soc. Amer., 28 (1956) pp. 27-35.

168. B. D. Cook and J. C. Berlinghieri: Experimental Mapping of Ultrasonic Fields by Light Diffraction. In 1975 Ultrasonics Symposium Proceedings, ed. J. de Klerk, pp. 133-135, IEEE Catelog No. 75 CHO 994-4SU, (1973) New York, N.Y.

169. J. C. Berlinghieri and B. D. Cook: Data Requirements for the Mapping of Ultrasonic Fields with Conventional Light Diffraction. J. Acoust. Soc. Amer., 58 (1975) pp. 823-827.

170. B. D. Cook: Measurement from the Nearfield of an Ultrasonically Produced Phase Grating. J. Acoust. Soc. Amer., 58 (1975) p. S128.

171. B. D. Cook: Measurement from the Optical Nearfield of an Ultrasonically Produced Phase Grating. J. Acoust. Soc. Amer., 60 (1976) pp. 95-99.

172. R. S. Mezrich, K. F. Etzold and D. H. R. Vilkomerson: Ultrasonovision. In 1974 Ultrasonics Symposium Proceedings, ed. J. de Klerk, pp. 1-4, IEEE Cat. No. 74-CHO 896-1SU, (1974) New York, N.Y.

173. R. S. Mezrich, K. F. Etzold and D. H. R. Vilkomerson: System for Visualizing and Measuring Ultrasonic Wavefronts. In Acoustical Holography, Vol. 6, ed. N. Booth (1975) pp. 165-191, Plenum Press, New York, N.Y.

174. J. Curie and P. Curie: Développement par Compression de l'électricité Polaire dans les Cristaux Hémièdres à Faces Inclinées. Bulletin de la Société Minéralogique de France, 3 (1880) pp. 90-93.

175. P. J. Curie and P. Curie: Crystal Physics-Development by Pressure of Polar Electricity in Hemihedral Crystals with Inclined Faces. Compt. Rend. 91 (1880) p. 294. Translation in Acoustics: Historical and Philosophical Development, ed. R. B. Lindsay (1972) pp. 373-374, Dowden, Hutchinson and Ross, Inc., Stroudsburg, Pennsylvania.

176. W. G. Cady: Piezoelectricity. McGraw Hill, New York, N.Y. Revised edition by Dover Publications, Inc. New York, N.Y. (1946).

177. F. V. Hunt: Electroacoustics. An Analysis of Transduction, and its Historical Background. (1954) Harvard University Press, Cambridge.

178. W. P. Mason: Piezoelectric Crystals and Their Application to Ultrasonics. (1950) Van Nostrand, New York, N.Y.

179. B. Jaffe, W. R. Cook, Jr. and H. Jaffe: Piezoelectric Ceramics. (1971) Academic Press, New York, N.Y.

180. D. A. Berlincourt, D. R. Curran and H. Jaffe: Piezoelectric and Piezomagnetic Materials and Their Function in Transducers. In Physical Acoustics, ed. W. P. Mason, Vol. I, Part A, (1964) pp. 169-270, Academic Press, New York, N.Y.

181. D. Berlincourt: Piezoelectric Crystals and Ceramics. In Ultrasonic Transducer Materials, ed. O. E. Mattiat, (1971) pp. 63-124, Plenum Press, New York, N.Y.

182. Y. Kikuchi: Magnetostrictive Metals and Piezomagnetic Ceramics as Transducer Materials. In Ultrasonic Transducer Materials, ed. O. E. Mattiat, (1971) pp. 1-61, Plenum Press, New York, N.Y.

183. S. Edelman, L. R. Grisham, S. C. Roth and J. Cohen: Improved Piezoelectric Effect in Polymers. J. Acoust. Soc. Amer., 48 (1970) pp. 1040-1043.

184. J. Cohen and S. Edelman: Piezoelectric Effect in Oriented Polyvinylchloride and Polyvinylfluoride. J. Appl. Physics 42 (1971) pp. 3072-3074.

185. J. Cohen and S. Edelman: Direct Piezoelectric Effect in Polyvinylchloride Films. J. Appl. Physics, 42 (1971) pp. 893-894.

186. N. Murayama, K. Nakamura, H. Obara and M. Segawa: The Strong Piezoelectricity in Polyvinylidene Fluoride (PVDF). Ultrasonics 14 (1976) pp. 15-23.

187. K. Brendel: Hydrophones. In Interaction of Ultrasound and Biological Tissues Workshop Proceedings. DHEW Publication (FDA) 72-8008 BRH/DBE 73-1, U. S. Government Printing Office, Washington, D.C. (1972) pp. 181-182.

188. C. R. Hill: Ultrasound Dosimetry - Chairman's Summary. In Interaction of Ultrasound and Biological Tissues Workshop Proceedings. DHEW Publication (FDA) 73-8008 BRH/DBE 73-1, U. S. Government Printing Office, Washington, D.C. (1972) pp. 153-158.

189. E. V. Romanenko: Miniature Piezoelectric Ultrasonic Receivers. Soviet Physics - Acoustics, 3 (1957) pp. 364-370.

190. C. R. Hill: Calibration of Ultrasonic Beams for Bio-Medical Applications. Phys. Med. Biol., 15 (1970) pp. 241-248.

191. D. R. Newman: Measurement of Diagnostic Level Ultrasound Using Small Piezoelectric Transducers. In Interaction of Ultrasound and Biological Tissues Workshop Proceedings. DHEW Publication (FDA) 73-8008 BRH/DBE 73-1, U. S. Government Printing Office, Washington, D.C. (1972) pp. 183-186.

192. J. R. Colbert, R. C. Eggleton and A. J. Weidner: Intensity Calibration of Pulsed Ultrasonic Beams. In Interaction of Ultrasound and Biological Tissues Workshop Proceedings, DHEW Publication (FDA) 73-8008 BRH/DBE 73-1, U.S. Government Printing Office, Washington, D.C. (1972) pp. 187-192.

193. M. Okujima: Microphone for Measurement of Sound Pressure Radiated from Ultrasono-Diagnostic Equipment. In Proc. Eighth International Congress on Acoustics (1974) p. 353, Goldcrest Press, Trowbridge, Wilts, England.

194. W. D. O'Brien, Jr., C. L. Christman and S. Yarrow: Ultrasonic Biological Effect Exposure System. In 1974 Ultrasonics Symposium Proceedings, ed. J. de Klerk, pp. 57-64, IEEE Catalog No. 74 CHO 896-1 SU, (1974) New York, N.Y.

195. R. K. Gould: Personal Communication. (See Reference 194).

196. A. J. Cousin, K. C. Smith and I. H. Rowe: A Calibrated Ultrasonic Field Measuring System for General Use. IEEE Trans. Instrum. Meas., IM-24 (1975) pp. 317-321.

197. P. C. Pedersen and D. A. Christensen: Power Measurement Techniques Applied to Imaging Systems. In Acoustical Holography, Vol. 6, ed., N. Booth, (1975) pp. 711-739, Plenum Press, New York, N.Y.

198. R. J. Bobber: Underwater Electroacoustic Measurement. Superintendent of Documents, U. S. Government Printing Office, Washington, D.C. (1970).

199. Naval Research Laboratory, Underwater Sound Reference Division, Orlando, Florida 32806.

200. G. Kossoff: Balance Technique for the Measurement of Very Low Ultrasonic Power Outputs. J. Acoust. Soc. Amer., 38 (1965) pp. 880-881.

201. G. Kossoff: Radiation Force. In Interaction of Ultrasound and Biological Tissues Workshop Proceedings. DHEW Publication (FDA) 73-8008 BRH/DBE 73-1, U.S. Government Printing Office, Washington, D.C. (1972) pp. 159-161.

202. P. N. T. Wells, M. A. Bullen, D. H. Follett, H. T. Freundlich and T. Angell James: The Dosimetry of Small Ultrasonic Beams. Ultrasonics 1 (1963) pp. 106-110.

203. I. Gabrielli and G. Iernetti: Torsion Balance for Radiation Pressure Measurements. Acustica 13 (1963) pp. 175-180.

204. A. Wemlen: A Milliwatt Ultrasonic Servo-Controlled Balance. Med. Biol. Engng., 6 (1968) pp. 159-165.

205. L. E. Dunbar: Torque Feedback Radiometer for Ultrasonic Power Measurements. J. Clinical Ultrasound 4 (1976) pp. 293-295.

206. J. G. Ziedonis: Technique for Measuring Ultrasonic Power. J. Acoust. Soc. Amer., 53 (1973) pp. 341-342 (A).

207. H. Oberst and P. Riechmann: Methods of Measurement Used by the Federal Institute of Physical Technology for the Evaluation and Certification of Medical Ultrasonic Instrumentation (in German). Physikalisch-Technischen. Bundesanstalt, 2 (1953) p. S35.

208. G. Kossoff: Calibration of Ultrasonic Therapeutic Equipment. Acustica 12 (1962) pp. 84-90.

209. J. M. Reid: A Proposed Interim Field Standard for Measuring Ultrasonic Power Emitted by Medical Diagnostic Apparatus. In Interaction of Ultrasound and Biological Tissues Workshop Proceedings, eds. J. M. Reid and M. R. Sikov, DHEW Publication (FDA) 73-8008 BRH/DBE 73-1, U. S. Government Printing Office, Washington, D. C. (1972) pp. 163-164.

210. P. N. T. Wells, M. A. Bullen and H. F. Freundlich: Milliwatt Ultrasonic Radiometer. Ultrasonics 2 (1964) pp. 124-128.

211. F. E. Fox and V. Griffing: Experimental Investigation of Ultrasonic Intensity Gain in Water Due to Concave Reflectors. J. Acoust. Soc. Amer., 21 (1949) pp. 352-359.

212. W. J. Fry and F. Dunn: Ultrasound: Analysis and Experimental Methods in Biological Research. In Phys. Tech. Biol. Res. 4 (1962) pp. 261-394.

213. F. Dunn and F. J. Fry: Ultrasonic Field Measurement Using the Suspended Ball Radiometer and Thermocouple Probe. In Interaction of Ultrasound and Biological Tissues Workshop Proceedings. eds. J. M. Reid and M. R. Sikov, DHEW Publication (FDA) 73-8008 BRH/DBE 73-1, U.S. Government Printing Office, Washington, D.C. (1972) pp. 173-176.

214. T. Hasegawa and K. Yosioka: Acoustic-Radiation Force on a Solid Elastic Sphere. J. Acoust. Soc. Amer., 46 (1969) pp. 1139-1143.

215. K. Yosioka, T. Hasewaga and A. Omura: Comparison of Ultrasonic Intensity from the Radiation Force on Steel Spheres with that on Liquid Spheres. Acustica 22 (1969/70) pp. 145-152.

216. T. Hasegawa and K. Yosioka: Acoustic Radiation Force on Fused Silica Spheres, and Intensity Determination. J. Acoust. Soc. Amer., 58 (1975) pp. 581-585.

217. M. A. Breazeale and F. Dunn: Comparison of Methods for Absolute Calibration of Ultrasonic Fields. J. Acoust. Soc. Amer., 55 (1974) pp. 671-672.

218. F. Dunn, A. J. Averbuch and W. D. O'Brien, Jr.: A Primary Method for the Determination of Ultrasonic Intensity with the Elastic Sphere Radiometer. Acustica (in press, 1976).

219. J. Zienuik and R. C. Chivers: Measurement of Ultrasonic Exposure with Tadiation Force and Thermal Methods. Ultrasonics 14 (1976) pp. 161-172.

220. J. E. Piercy: Ecoulement Acoustique et Pression de Radiation. J. Physique Radium, 17 (1956) pp. 405-410.

221. A. Sokollu: On the Measurement of Ultrasound Intensity in Tissue: Reservations on Radiation Force and the Case for Calorimetry. In Interaction of Ultrasound and Biological Tissues Workshop Proceedings, eds. J. M. Reid and M. R. Sikov, DHEW Publication (FDA) 73-8008 BRH/DBE 73-1, U. S. Government Printing Office, Washington, D. C. (1972) pp. 165-170.

222. J. E. Piercy and J. Lamb: Acoustic Streaming in Liquids. Proc. Royal Soc., A, 226 (1954) pp. 43-50.

223. W. R. MacLean: Absolute Measurement of Sound Without a Primary Standard. J. Acoust. Soc. Amer., 12 (1940) pp. 140-146.

224. R. K. Cook: Absolute Pressure Calibration of Microphones. J. Acoust. Soc. Amer., 12 (1941) pp. 415-420.

225. S. Ballantine: Reciprocity in Electromagnetic, Mechanical, Acoustical, and Interconnected Systems. Proc. Inst. Radio. Engrs. 17 (1929) p. 929.

226. E. Ackerman and W. Holak: Reciprocity Calibrations. Amer. J. Physics, 25 (1957) pp. 454-460.

227. R. J. Bobber: General Reciprocity Parameter. J. Acoust. Soc. Amer., 39 (1966) pp. 680-687.

228. E. L. Carstensen: Self-Reciprocity Calibration of Electroacoustic Transducers. J. Acoust. Soc. Amer., 19 (1947) pp. 961-965.

229. J. Koppelmann, K. Brendel and J. Wolf: Calibration of Underwater Sound Transducers in the Frequency Range 75 kHz to 2 MHz (in German). Acustica 25 (1971) pp. 73-80.

230. J. M. Reid: Self-Reciprocity Calibration of Echo-Ranging Transducers. J. Acoust. Soc. Amer., 55 (1974) pp. 862-868.

231. W. T. Richards: An Intensity Gauge for "Supersonic" Radiation in Liquids. Proc. National Acad. Sci., 15 (1929) pp. 310-314.

232. W. T. Richards: The Heating of Liquids by the Absorption of Sound, and its Relation to the Energy of Intense High Frequency Sound Waves. Proc. National Acad. Sci., 17 (1931) pp. 611-616.

233. R. B. J. Palmer: A Thermoelectric Method of Comparing Intensities of Ultrasonic Fields in Liquids. J. Sci. Instr. 30 (1953) pp. 177-179.

234. F. Dunn and W. J. Fry: Precision Calibration of Ultrasonic Fields by Thermoelectric Probes. Paper 12.5 presented at the IRE National Convention (1957) March, New York, N.Y.

235. K. Yosioka, Y. Kawasima and H. Hirano: Absolute Determination of High Ultrasonic Intensities by Thermoelectric Probes. Memoirs Institute Sci. Indust. Res., XVIII (1961) pp. 15-22.

236. S. A. Hawley, J. E. Breyer and F. Dunn: Fabrication of Miniature Thermocouples for uhf Acoustic Detectors. Rev. Sci. Instr., 33 (1962) pp. 1118-1119.

237. J. K. Zienuik: Ultrasonic Thermoprobe. Ultrasonics 4 (1966) pp. 136-137.

238. J. B. Pond, B. Woodward and M. Dyson: A Microscope Viewing Ultrasonic Irradiation Chamber. Phys. Med. Biol. 16 (1971) pp. 521-524.

239. E. A. Neppiras: Measurements in Liquids at Medium and High Ultrasonic Intensities. Ultrasonics 3 (1965) pp. 9-17.

240. E. A. Lloyd: Energy Measurements. In <u>Ultrasonic Techniques in Biology and Medicine</u>, ed. by B. Brown and D. Gordon, (1967) pp. 46-86, Charles C. Thomas, Springfield, Illinois.

241. A. Sokollu: Absolute Measurement of Total Radiated Power of Ultrasonic Transducers for Biomedical Use. Extrait du Bulletin du Laboratoire d'Electroacoustique, Universite de Liege, No. 9 (1966) pp. 23-27.

242. H. van den Ende: A Radiation Power Calorimeter and a Radiation Force Meter for Small Ultrasonic Beams. Med. Biol. Engr., 7 (1969) pp. 411-417.

243. J. K. Zienuik: On a Non-Adiabatic Non-Isothermal Calorimeter for Measurements of Ultrasonic Wave Power in Liquids. Proc. Vibration Problems Warsaw, 6 (1965) pp. 367-378.

244. J. K. Zieniuk and B. Evans: The Influence of Thermal Parameters on Making Ultrasonic Power Measurements by a Calorimetric Method. Acustica, 25 (1971) pp. 47-52.

245. I. G. Mikhailov: Methods of Measuring the Absolute Intensity of Ultrasonic Waves in Liquids and Solids. Ultrasonics 2 (1964) pp. 129-133.

246. C. A. Wiederhielm: Ultrasonic Intensity Meter. Rev. Sci. Instr., 27 (1956) pp. 540-541.

247. J. Szilard: A New Device for Monitoring Ultrasound Dosage. In Proc. Eighth International Congress on Acoustics, (1974) p. 352, Goldcrest Press, Trowbridge, Wilts, England.

CHAPTER VI

ULTRASONIC ABSORPTION AND DISPERSION

F. Dunn and W. D. O'Brien, Jr.

1. INTRODUCTION

Complete specification of wave propagation properties of a medium require knowledge of inertial, restoring and loss parameters as functions of the state, wave, and media variables. For an infinitesimal ultrasonic wave propagating in an isotropic unbounded liquid-like medium, these become, respectively, the density, an elastic parameter such as the adiabatic compressibility, and the attenuation. The inertial and restoring parameters embody the more easily measured impedance and velocity functions. The attenuation parameter includes absorption of the wave energy, in the body of the medium and degraded to heat, and all other mechanisms by which energy is extracted from the wave process or redirected at media inhomogeneities. Thus knowledge of the speed of sound and loss terms as functions of acoustic and state variables, position, tissue and state of tissue, are important for interpretations of, for example, clinical diagnostic ultrasonic data.

Biological macromolecules in solution constitute a simpler medium but one which interrogation by ultrasound, again involving observations of the wave velocity and loss functions, can yield details of reactions occurring in the time range 10^{-3} to 10^{-10} sec.

This chapter includes reviews of both of these kinds of media. As scattering phenomena are treated in Chapter VII, no further mention is made here. The relatively simpler media of solutions are considered first. As it has been established that proteins are largely responsible for absorption in tissues, studies to elucidate the underlying mechanisms have involved solutions of this chemical species, of its components and of somewhat related chemical species. These are each treated separately in a comparative fashion. Tissues are then treated and the dependence of the absorption and velocity on the principal parameters frequency, temperature, and wave amplitude are exhibited. Remarks are made of tissues requiring special attention, viz., bone, lung, refractive media of the eye and those with high collagen content.

The attempt here has been to be timely rather than complete and in this regard the reader may wish to consult Dunn et al.[1] and Reid and Sikov[2] for additional details.

2. SOLUTIONS

2.1 Amino Acids

Amino acids will be considered first since they are elementary organic solute molecules of biological significance. When dissolved in water without additional ionic constituents to influence the state of charge of the amino acid, the solution exhibits a magnitude of the absorption parameter, α/f^2, which varies only slightly with frequency in the megahertz range, and differs only slightly from that of the solvent water.[3-8]

Amino acids in water solution at neutral pH may be considered according to their action of structure making or breaking in the solvent. In neutral solution the amino acids are present as doubly charged molecules (zwitterions) and are susceptible to dissociation and recombination reactions on change of the environment. Hydrogen bonding sites are located on both the amino and carboxyl groups, while the side chain may be acidic, nonpolar, or basic. The potentialities for breaking or making water structure in the vicinity of the solute molecules are therefore considerable. Hammes and Pace[3] suggested that the ultrasonic absorption mechanism responsible for aqueous solutions of glycine, diglycine and triglycine is that which involves solute-solvent (water) interaction.

Breaking the water structure will result in an increase in the concentration of less-bonded water molecules which form a closer-packed and less compressible structure, recognized by an increase in the velocity of sound propagation. Such an effect is observed for all the amino acids studies by Barrett (see Dunn, Edmonds and Fry[10]) namely, glycine, alanine, serine, valine, glycine ethyl ester hydrochloride, lycine hydrochloride, and glutamic acid. Other measurements of the velocity of propagation and, hence, of the compressibility of aqueous solutions of amino acids have been made by Passynsky,[11] Gucker and Haag,[12] and Goto and Isemura.[13] The presence of predominantly hydrogen-bonding solute species, for example, glycine, may increase the concentration of relatively unbonded water molecules in secondary hydration regions, and such molecules would become available to participate in the transition between ice-like (bonded) and unbonded water:

$$(H_2O)_{\text{bonded}} \rightleftarrows (H_2O)_{\text{unbonded}} \tag{1}$$

An alternative approach is to consider that the transition probabilities and hence the rate constants of the reaction (Eq. 1) are affected by the presence of solutes.[14] If the transition probabilities were substantially influenced and reduced, then the relaxation frequency for the transition between bonded and unbonded water would be increased with a consequent reduction in the absorption measured at much lower frequencies. Depression of the absorption coefficient below the value for water has been observed in aqueous solutions of potassium and sodium acetates, bromides, and iodides[9] in urea[5] and in guanidine hydrochloride.[16] The effect is evidently produced by the presence of anions of low charge density which have a small electrostrictive action on the surrounding water molecules.

These results for solutions that depress the value of the absorption coefficient at higher frequencies disclose the possibility of significant errors arising in the interpretation of ultrasonic absorption data for other solutions. It is not justified to assume without qualification that the absorption attributable to the solvent phase of a solution is the same as that of the pure solvent, nor to derive under such assumption the "absorption coefficient" of the solute from a simple mixing formula. The potential error in ascribing the difference, $(\alpha\lambda)_{\text{solution}} - (\alpha\lambda)_{\text{solvent}}$, to absorption by the solute becomes progressively greater as the frequency increases. The preferred procedure, though more exacting experimentally, is to make measurements at sufficiently high frequencies so that $(\alpha f^2)_\infty$ for the solution may be determined. Then the effective solute absorption at any lower frequency will be given by $\alpha/f^2 - (\alpha/f^2)_\infty$.

When the pH of an amino acid aqueous solution is within the range of 2 to 4 or 11 to 13, it has been experimentally observed that the relaxational behavior can be described by a single relaxation frequency. A number of amino acids have been investigated as a function of pH, viz., glycine,[5,17-20] serine and threonine,[4] glutamic acid, aspartic acid and alanine[18] and arginine and lysine.[19,6] Absorption maxima have been observed within the pH ranges 2 to 4 and 11 to 13, with such praking being described quantitatively by assuming that the proton-transfer reaction dominates the absorption.

Hussey and Edmonds[5] have conducted the most extensive study of glycine in aqueous solution as a function of pH and compared their results to those of Maas[17] and Applegate, Slutsky and Parker.[18] The acidic absorption peak involves the proton-transfer reaction of the α-carboxyl group and the alkaline absorption peak involves the α-amino group. The theory described by Eigen and deMaeyer[21] quantitatively describes the amplitude and shape of the two absorption peaks. Likewise, the separate reactions involving the α-amino and α-carboxyl groups in serine and threonine are responsible for absorption peaks.[4] However, for the case of lysine and arginine, the ε-amino and guanidine groups, respectively, interact in the proton-transfer reaction with the α-amino group such that one can not assume that the α-amino group and side-chain amino group proceed independently.

2.2 Polypeptides

Four possible mechanisms of ultrasonic absorption in aqueous solutions of polypeptides have been examined, viz., proton-transfer reaction, helix-coil transition, solvent-polymer interaction and relaxation of the shear viscosity. Major interests have tended to be focused on the proton-transfer reaction and helix-coil transition.

Burke, Hammes and Lewis[22] measured ultrasonic attenuation in polyglutamic acid solutions over the frequency range from 6 to 175 MHz and the pH range from 5.4 to 9.0 for a mixed solvent of aqueous sodium chloride and dioxane. Their data refer to 25°C and exhibit a distribution of relaxation times under all conditions, with a tendency to approach a single relaxation time at the highest value of pH. The molecular weight of their sample was 75,000 and the concentrations employed were 3 to 1.5 g/100 ml, corresponding to modalities of the monomer of 0.24 and 0.12, respectively (3.53 and 1.77 g monomer/100 ml). The absorption values were smaller than the classical absorption calculated from shear viscosity at zero frequency, indicating that shear-viscosity relaxation had occurred at lower frequencies. A single value of the sound velocity, 1.59×10^5 cm/s, was reported for all values of pH and both concentrations. In this regard it should be noted that resolution of the very small velocity dispersion would not be expected.

The mechanism preferred by Burke, Hammes, and Lewis[22] to account for the distribution of relaxation times is the interaction of solvent and polymer involving a breakdown of the water structure around the polypeptide and solvation of the latter by the water. Subsidiary experiments indicated that the solvating species was more probably water than dioxane. The alternative mechanisms of interest in this system, namely, helix-coil transition and protolytic reactions, were regarded as contributing only to a minor extent in the response to ultrasonic perturbation.

According to Schwarz,[23] the mean relaxation frequency for the helix-coil transition in polyglutamic acid might be expected to be around 2 MHz, and so a broad spectrum of relaxation times centering around the value 8×10^{-8} would be anticipated. However, a single relaxation time will approximate the behavior as long as the period of the perturbation is small compared to that mean relaxation time, e.g., for frequencies higher than about 20 MHz. The helix-coil transition is known to occur in polyglutamic acid solutions around pH 6.[24] Wada, et al.[25] have measured the complex shear viscosity, at 50 kHz, of polyglutamic acid in the same solvent as

Burke, et al.[22] Their results indicate a substantial variation of the imaginary part of the shear viscosity near pH 6, which was attributed to the helix-coil transition. Their estimate of the maximum relaxation time was about 10^{-6}s, in rough agreement with a result calculated from Schwarz's[23] theory. These authors consider compressional wave experiments in the megahertz frequency range to be inappropriate for investigation of the helix-coil transition mechanism, since they have observed solutions of polyglutamic acid and glutamic acid to behave similarly in such experiments.

Saksena, Michels and Zana[26] examined polyglutamic acid solutions of molecular weight 94,000 in 0.2 M NaCl at concentrations of 0.15 and 0.18 mole-residue/liter at the three ultrasonic frequencies of 5.08, 26.2 and 99.6 MHz. A peak in the absorption versus pH curve occurred at the pH which corresponds to the helix-coil transition point. Assuming the helix-coil transition was the mechanism responsible for the absorption maximum, mean relaxation time and standard volume change, as determined from the experimental data, calculated to be 1.1×10^{-8} x and 0.3 cm^2/mole.

More recently Barksdale and Stuehr[27] examined very dilute aqueous solutions of polyglutamic acid of molecular weight 58,000 over the extended frequency range from 40 kHz to 35 MHz which required three different techniques to measure the attenuation, viz., resonant (40-250 kHz), statistical reverberation (250-1000 kHz) and pulse transmission (15-35 MHz). Their two concentrations were 0.0027 and 0.0011 mole-residue/liter in 0.03 M NaCl which were, respectively, 55 and 136 times less concentrated as those used by Saksena, Michels and Zana.[26] A relaxational absorption was observed over the pH range 4.8 to 5.6, at 37°C. No excess absorption was observed outside this pH range. The absorption amplitude near the helix-coil transition midpoint of pH 5.21 was around 100×10^{-17} sec^2/cm and the relaxation time was around 10^{-6} s or a relaxation frequency of 160 kHz. They demonstrated the following four important facts with the polyglutamic acid data which quantitatively favors the helix-coil transition: that the excess absorption occurred at the helix-coil transition pH, a maximum was observed in both the relaxation amplitude and relaxation time at the transition pH, the excess absorption was a linear function of concentration, and the relaxation time was independent of concentration.

Parker, Slutsky and Applegate[28,29] have obtained compressional wave data which they attribute to the helix-coil transition. Their measurements were made on solutions of poly-L-lysine of weight-average molecular weight 86,300 in 0.6 M sodium chloride at 35.8°C and at a polymer concentration of 2.0 g/100 ml. The data were fitted satisfactorily to curves described by single relaxation times. Each possible mechanism was examined for its ability to account for the observed absorption, viz., proton-transfer, solvation, viscous relaxation and conformational change. The principal difference between solutions of polylysine and polyglutamic acid is the greater likelihood that proton-transfer reactions would be significant in the former. The volume change for protonation of an amino group is considerably larger than that for a carboxyl group. Nevertheless, the authors rejected this mechanism because an unreasonably large volume change, associated with too small a backward rate constant, would have been required. The solvation mechanism could not be so conclusively eliminated, but evidence was presented to show that it was less satisfactory than the helix-coil mechanism.

Zana and Tondre[30] examined the helix-coil mechanism of ultrasonic absorption in aqueous solutions of poly-L-lysine. They induced the helix-coil transition by the addition of small electrolytes while the polypeptide concentration was kept between 0.10 and 0.11 M/l. It was concluded that the absorption peaks did not result from the equilibrium between the helix and coil forms of the molecule.

Zana and Tondre[31] then studied the ultrasonic absorption behavior as a function of pH of poly-L-lysine, which undergoes a helix-coil transition at pH 10.3, and of poly-DL-lysine, which does not undergo a helix-coil transition over the entire pH range. The ultrasonic absorption of the polymer solution at concentrations of 0.06 M in 0.3 M NaBr was measured over the pH range from 5 to 13 at a temperature of 25°C. Both polymer solutions yielded an absorption maximum at pH 11.5. The position of the absorption maximum was in excellent agreement for the proton-transfer reaction on the α-amino groups and, thus, not attributed to the helix-coil transition.

Measurements of absorption of poly-L-ornithine were carried out by Hammes and Roberts[32] at 25°C in water (0.2 M NaCl) - methanol (85:15 v/v) solvent at a polymer concentration around 2 g/100 ml. Under these conditions, poly-L-ornithine undergoes conversion from coiled to 90% helical conformation, with increasing pH, with the transition midpoint at pH 11. An absorption maximum and a relaxation time maximum were both observed around pH 11.2 - 11.3 and the authors interpreted these observations in terms of a helix-coil equilibrium.

Zana and Tondre[31] measured poly-L-ornithine as a function of pH over the range 5 to 13 in two solvents, viz., water (0.3 M NaBr) and water (0.3 NaBr) - methanol (85:15 v/v) in which the maximum helical content at high pH is approximately 25% and 100%, respectively. It was observed that the absorption maximum, which occurred at pH 11.5, was slightly greater for poly-L-ornithine in the water solvent than in the methanol solvent. If the helix-coil equilibrium were responsible for the absorption maximum, the polymer in the methanol solvent would have yielded a greater amplitude than that in the water solvent. In addition, at a frequency of 6.49 MHz, the absorption versus pH curves for poly-L-lysine and poly-L-ornithine in the water (0.3 M NaBr) solvent are practically coincident leading the authors to suggest strongly that the absorption maxima at pH 11.5 are due to the proton-transfer reaction among the side chain amino groups.

Zana and Tondre[31] have compared results of poly-L-glutamic acid with those of poly-L-lysine and poly-L-ornithine and the absorption mechanisms responsible for the amplitude peaks, viz., proton-transfer reaction versus helix-coil equilibrium. Schwarz[23] has shown that the absorption maximum at the midpoint of the helix-coil transition would be most pronounced at the relaxation frequency.

$$f_R = \frac{2\sigma k_F}{\pi} \tag{2}$$

where σ is the nucleation parameter which describes the difficulty of introducing the first helical segment into a region of coiled residues and k_F is the forward rate constant for helical growth. For aqueous solutions of polyglutamic acid, f_R has been found to be 160 kHz.[27,33] And, since the nucleation parameter is smaller for poly-L-lysine and poly-L-ornithine, f_R would be lower. Therefore, in the megahertz frequency range, the excess absorption due to the helix-coil transition would be only slightly detectable. In the case of poly-L-lysine and poly-L-ornithine, the midpoint pH for the helix-coil equilibrium is only 1 pH unit less than the absorption peak from the proton-transfer reaction. For poly-L-glutamic acid the midpoint pH for the helix-coil equilibrium is at least 2 pH units greater than the absorption peak from the proton-transfer reaction of the side chain carboxyl groups. Thus Zana and Tondre[31] suggest that experimentally, the helix-coil equilibrium can be observed in the megahertz frequency range when the nucleation parameter is high and the pH value characterizing the proton-transfer reaction is separated by at least 2 pH units from the midpoint pH for the helix-coil equilibrium.

2.3 Proteins

Measurements of the acoustic properties of aqueous hemoglobin solutions have been more comprehensive than those on any other solution of biological significance. The earliest work of importance is that of Carstensen, Li and Schwan,[34] who investigated the ultrasonic absorption and velocity in blood, plasma and solutions of albumin and hemoglobin, and concluded that the acoustical properties of blood are largely determined by the protein concentration. In addition, they showed that the absorption coefficient of hemoglobin is approximately the same as that of serum albumin within the frequency range 0.8 - 3 MHz in the temperature range 10 - 40°C. Some of the most accurate data on aqueous solutions of hemoglobin are those obtained by Carstensen and Schwan[35] in the frequency range of 0.3 to 10 MHz. At lower frequencies, data have been obtained by Gramberg[36] at 35 and 82 kHz and at higher frequencies by Edmonds[37] in the range of 32 to 232 MHz, by Schneider,

Muller-Landau and Mayer[38] in the ranges of 0.5 to 5, 10 to 30 and 300 to 1000 MHz and by Edmonds, et al.[39] at 270 and 470 MHz. In the case of hemoglobin solutions there is, therefore, some hope that the major portion of the frequency range of relaxational absorption at neutral pH may have been covered by experimental measurement. However, ambiguities in interpreting the data still exist.

Carstensen and Schwan[35] reported data obtained by the use of their fixed-path pulse transmission apparatus. Their measurements of the absorption and velocity of sound were performed under a variety of conditions which enabled them to study the effects of temperature, concentration, pH, and the presence of neutral salts. The data relate to solutions of oxyhemoglobin with a possible impurity amounting to no more than 20 percent methemoglobin. A specific experiment designed to show any difference in the absorption by these two forms of the protein exhibited none. Schneider, Muller-Landau and Mayer,[38] however, determined the absorption of oxygenated and deoxygenated bovine hemoglobin aqueous solutions at 1, 10 and 500 MHz (25°C, 15 g/100 ml). They found the absorption of oxygenated hemoglobin solutions greater at 500 MHz, smaller at 10 MHz and equal at 1 MHz to the absorption of deoxygenated hemoglobin solutions.

Hemoglobin samples of Carstensen and Schwan,[35] were obtained originally from blood specimens of horse, cattle, sheep, and humans, from which the hemoglobin was liberated by treatment with toluene followed by centrifugation. Slight differences in the absolute magnitude of the absorption and its temperature dependence were observed for hemoglobins derived from the different mammalian species. Where specific data are given in the following text, they refer to bovine hemoglobin unless otherwise stated.

Figures 1 and 2 show respectively the absorption per wavelength and the velocity of propagation as a function of frequency at 25°C as measured by Carstensen and Schwan.[35] The concentration of hemoglobin in the solution is shown as the parameter on the curves. A correction has been applied to the absorption data by assuming a simple mixing formula and subtracting from

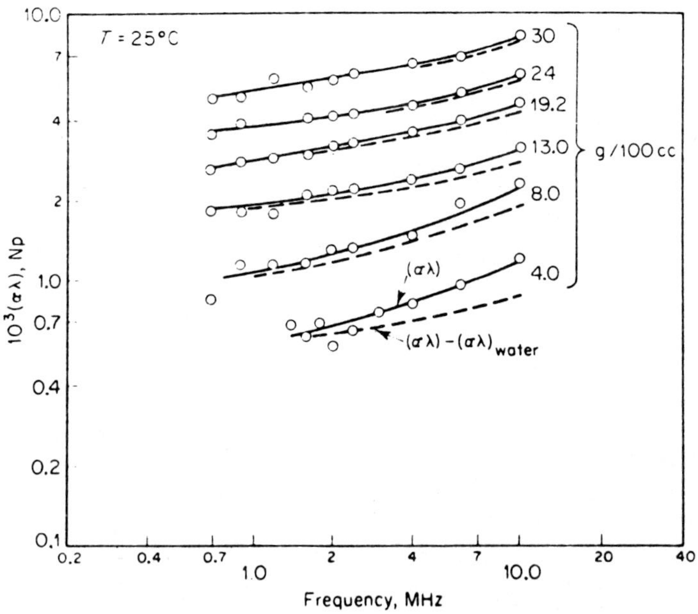

Fig. 1 Absorption per wavelength for bovine hemoglobin solutions in water at $25^{o}C$ as a function of frequency and concentration. Dashed curves show the residual absorption after correcting for the presence of the solvent by a simple mixing formula.[35]

the measured data a contribution that is attributed to the water alone. The mixture is presumed to consist of the species "simple water" and the species "hydrated macromolecules." The result of applying this correction is to indicate that the frequency dependence of the absorption attributable to the hydrated macromolecule is the same at all concentrations. Furthermore, a reduced form of the data shows absorption to be proportional to concentration of hemoglobin up to 15 g/100 ml. At higher concentrations the absorption increases faster than linearly with concentration, showing a 20 percent excess at a concentration of 30 g Hb/100 ml solution. Therefore, they indicate that to the limit of the linear proportionality, it is reasonable to discuss the absorption coefficient of the hydrated macromolecular species as though one were dealing with a dilute solution of the species in water. The dispersion of the velocity of propagation is consistent with the hypothesis that the excess absorption of the hydrated macromolecule may be described in terms of relaxation processes.

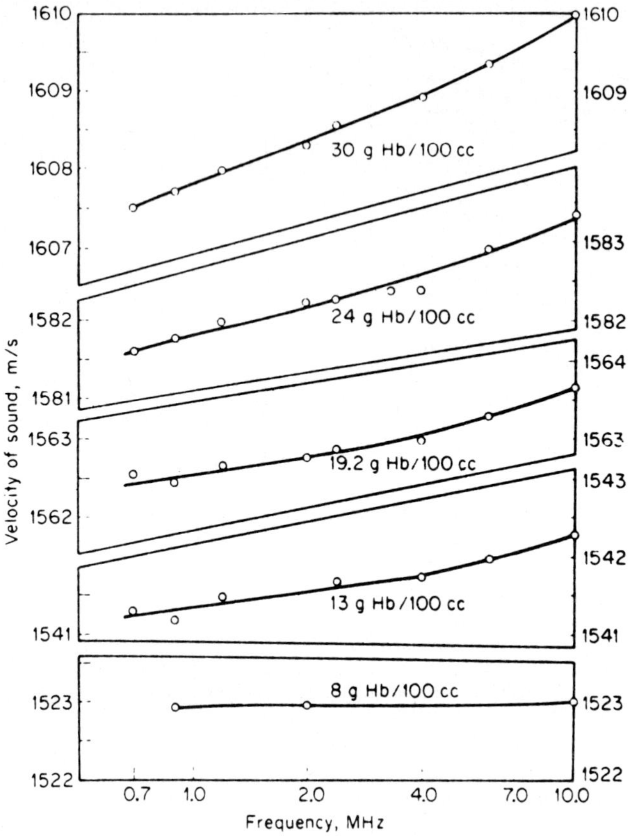

Fig. 2 Dispersion of the velocity in solutions of bovine hemoglobin in water at 25°C.[35]

As indicated by Carstensen and Schwan,[35] the concentration dependence of ultrasonic absorption in hemoglobin solutions is linear to approximately 15 g/100 ml. The same linear dependence has been suggested by Kessler and Dunn[40] in bovine serum albumin solutions, by Kessler, O'Brien and Dunn[41] in polyethylene glycol solutions, by O'Brien[42] in ovalbumin solutions and by O'Brien and Dunn[43,44] in both uncrystallized and twice crystallized bovine hemoglobin solutions, although the concentration at which the absorption deviated from linearity varied among polymer type and polymer preparation. In the case of "dilute" solutions, where the absorption is considered to vary linearly with polymer concentration, the mechanism of absorption is attributed to processes involving the interaction between the solvent (usually water) and the solute (the polymer). On the other hand, in other than dilute solutions the non-linear increase in absorption, it is suggested, arises from intermolecular interaction processes.[40,43,44] It has recently been suggested,[45] following absorption measurements of bovine serum albumin to concentrations around 40 g/100 ml, over the frequency range 3.4 to 15 MHz, at a temperature of 20°C, that the absorption dependence upon centration does not possess a linear region. In fact, it was experimentally shown that the excess ultrasonic absorption can be adequately described, over the entire concentration range up to 40 g/100 ml, by a concentration dependence to the 1.2 power. It thus appears that one must at least consider intermolecular interaction over the entire range. Along similar lines, Kremkau and Carstensen[7] and Kremkau, Carstensen and Aldridge[46] investigated the ultrasonic absorption as a function of hemoglobin concentration up to 60 g/100 ml. At concentrations greater than about 40 g/100 ml, bovine erythrocytes were acrolein-fixed which meant that the hemoglobin was intracellular. Thus, at a solution hemoglobin concentration of 60 g/100 ml, there is a four-fold increase in the specific absorption (these authors defined specific absorption as the measured absorption minus solvent absorption divided by hemoglobin concentration) over that of concentrations less than 10 g/100 ml.

Carstensen and Schwan[35] obtained absorption data on human hemoglobin at a concentration of 16.5 g/100 ml where the temperature was varied from 7 to 35°C. It was shown that the level of the plateau in absorption is shifted slightly to higher frequencies and lowered slightly as temperature was increased. Such a variation with temperature is consistent with the concept of a redistribution of relaxational processes, each exhibiting a temperature dependence determined by its energy of activation.

Two alternative procedures (instead of the application of a simple mixing formula) may be adopted in an attempt to isolate the relaxational contributions to the total measured absorption. One may either calculate the classical absorption coefficient from the static viscosity and the measured velocity of propagation and consider the classical absorption to be identical with the high-frequency absorption; or one may actually attempt to measure the absorption at sufficiently high frequencies to identify an asymptotic value. The former procedure assumes that relaxation of the viscous mechanism does not occur in the frequency range of interest. The latter procedure was adopted by Edmonds,[37] Schneider, Muller-Landau and Mayer,[38] and Edmonds, et al.[39] Edmonds obtained an asymptotic value of $\alpha/f^2 = 38 \times 10^{-17}$ sec^2/cm, applicable to a solution of bovine hemoglobin (predominately, 80 to 90%, oxygenated form) in water at a concentration of 15 g/100 ml at a temperature of 25°C (see Figure 3). Schneider, Muller-Landau and Mayer, for the same temperature and concentration conditions, obtained asymptotic values of $\alpha/f^2 = 33 \times 10^{-17}$ sec^2/cm for oxygenated hemoglobin and $\alpha/f^2 = 30 \times 10^{-17}$ sec^2/cm for deoxygenated hemoglobin. These results were from data up to 100 MHz whereas the Edmonds et al upper frequency was 470 MHz. Use of the asymptotic value (by subtraction from measured values of α/f^2 at lower frequencies) was made to determine the alleged relaxational contribution to the measured absorption. Edmonds et al[39] fitted the absorption spectrum of hemoglobin solutions with a continuous ramp distribution of relaxation frequencies (shown in Figure 3) and predicted the upper-bound relaxation frequency of 260 MHz and the lower-bound relaxation frequency of 22 kHz, which only reflects the lowest data point at 35 kHz. Schneider, Muller-Landau and Mayer,[38] with a discrete distribution, assigned four distinct relaxations which adequately described the ultrasonic absorption spectrum. Their upper relaxation frequency limit is at 80 MHz and suggest that the absorption processes are attributed in either the protein or the hydrated moiety.

The absorption coefficient of a 12.5 g/100 ml solution of human serum albumin in water has been found to be indistinguishable from that of human hemoglobin in the same concentration.[34] These conclusions apply within the frequency range 0.8 to 3 MHz and the temperature range 10° to 40°C. The

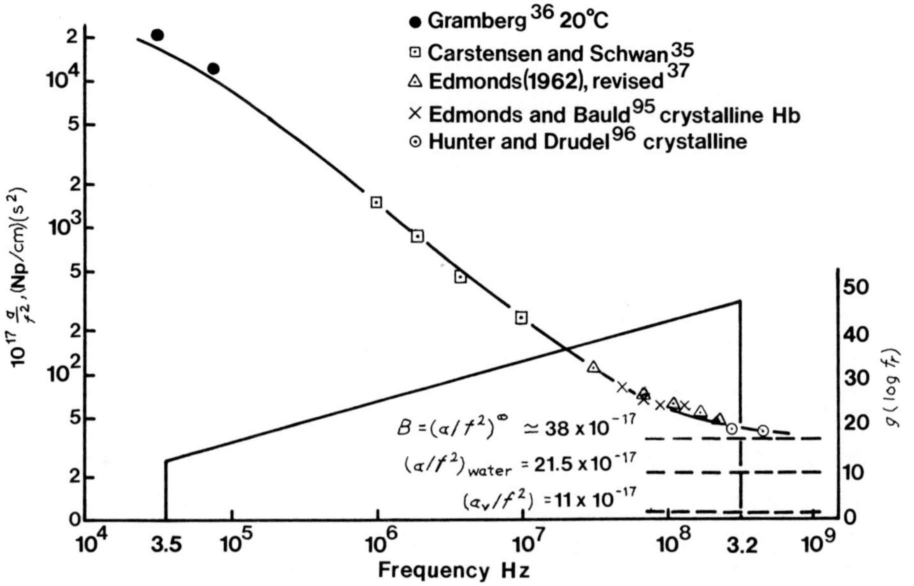

Fig. 3 Absorption parameter α/f^2 for an aqueous solution of bovine hemoglobin at concentration 12 g Hb/100 ml and 25°C. The superscript ∞ refers to the asymptotic measured values at high frequencies and subscript ν to the classical viscous contribution.[39]

absorption data in these ranges were also coincident with the absorption values of red blood cells, when compared on the basis of reduced concentration, that is, the absorption coefficient divided by the concentration.

The role of the conformation of proteins in the absorption of ultrasonic energy in biological materials has received some attention. The ultrasonic absorption spectrum of gelatin, a protein which assumes a random coil conformation in aqueous solution, exhibits a magnitude considerably less than that of hemoglobin, a globular protein which exists in a compact rigid structure, as shown in Figure 4.[47] Hemoglobin is composed of four polypeptide chains and has a quaternary structure while the albumins, serum albumin and ovalbumin, each consist of a single polypeptide chain and have tertiary structure in aqueous solution. Hemoglobin, bovine serum albumin and ovalbumin all possess a similar absorption spectra.[42-44]

Lang, Tondre and Zana[48] examined bovine serum albumin, β-lacto-

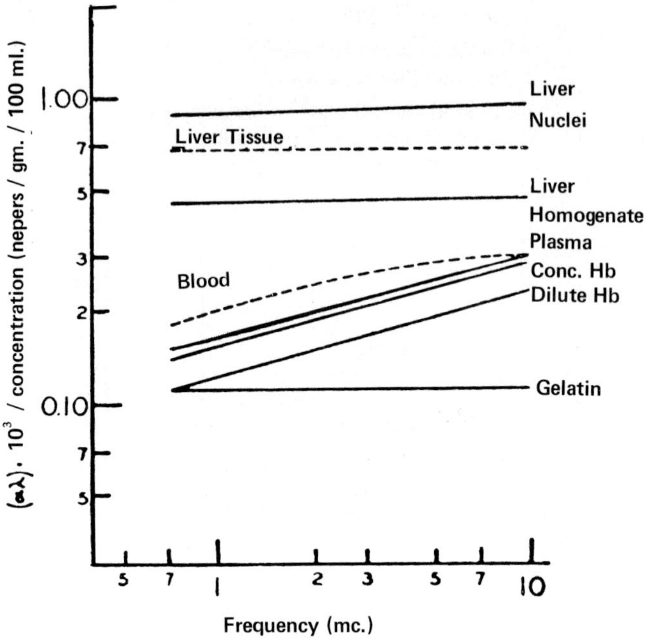

Fig. 4 Specific absorption per wavelength in solutions of biological macromolecules. Absorption per gram dry weight for intact liver tissue and blood are shown by dotted lines for comparison of magnitude and frequency dependence with molecular solutions.[47]

globulin and lysozyme in two different solvents, viz., in 0.2 M NaCl and in 0.2 M NaCl - 8 M urea, at 2.82 and 5.04 MHz and 25°C. In the urea solvent, each of the proteins are known to assume the random coil conformation. The absorption magnitude decreased for bovine serum albumin and β-lactoglobulin but increased for lysozyme as the biopolymers went from their native to random coil conformation. O'Brien and Dunn[16] examined the absorption characteristics of hemoglobin in aqueous solution over the frequency range 1 to 50 MHz, at 10°C, as a function of the strong electrolyte guanidine hydrochloride up to a concentration of 5.4 M. With increasing guanidine hydrochloride concentration to 0.7 M, the hemoglobin molecule dissociates into half-molecules without the loss of tertiary structure, and at concentrations greater than 2.5 M, the hemoglobin molecule dissociates further into its four individual polypeptide chains with complete loss of tertiary structure. The ultrasonic absorption appears to be insensitive to the dissociation process, though a small contribution appears at the lower frequencies, owing to the expansion of the polypeptide molecule.

As in the case of amino acid and polypeptide solutions, the pH dependence of aqueous solutions of proteins has been studied to examine ultrasonic absorption processes. Generally, absorption maxima have been observed within the pH ranges 2 to 4 and 11 to 13. It now appears that such maxima can be described quantitatively assuming that the proton-transfer reaction dominates the absorption process within these pH ranges. However, as Lang, Tondre and Zana[48] indicate, the overall conformation of the protein molecule in solution also affects the ultrasonic absorption.

One of the earliest attempts to investigate the changes in pH of aqueous solutions of proteins was by Carstensen and Schwan[35] who throught that since the solubility of hemoglobin in water could be influenced by changes in pH, so then could the ultrasonic absorption. They measured the absorption of hemoglobin over the pH range from 6 to 9.4 but observed no detectable effect.

Carstensen[47] and Pauly and Schwan[49] examined the pH dependence of ultrasonic absorption in gelatin solutions over the range from 0.4 to 11.8 at 35°C and 5 MHz. Carstensen[47] speculated on whether the perturbation of the equilibrium between bound and free water might provide a mechanism of ultrasonic absorption. Thus, the absorption might be related to the degree of hydration. The pH dependence of gelatin exhibited two absorption maxima ranges 1 to 3 and 9 to 11. These maxima correlated well to the degree of hydration of gelatin.

Perhaps the next attempt to examine the pH dependence in protein solutions was a decade later,[40,50] when the ultrasonic absorption behavior of bovine serum albumin aqueous solutions were examined over the frequency range 0.3 to 163 MHz and over the pH range 2.3 to 11.8, all at 20°C. They observed a sharp increase in the excess ultrasonic absorption beyond the range $4.3 < pH < 10.5$. Within the acidic pH range, the excess absorption peaked around pH 3.7 with the amplitude of the peak greater at the lower frequencies. Above pH 10.5, the excess absorption increased markedly and, as was shown later,[51] the peak occurred around pH 11.8 (see Figure 5). Kessler and Dunn[40] correlated the absorption increase in the acidic pH range with a particular conformational change of the BSA molecule and the expansion of the molecule above pH 10.5, both of which are mechanisms representative of molecular conformational changes.

Additional work dealing with the examination of the mechanism responsible for the pH dependent absorption maxima in BSA solutions has been conducted.[31,48,51-54] Further, aqueous solutions of hemoglobin[43,53,55] and, to a lesser degree, gelatin,[53] β-lactoglobulin and lysozyme[48,51] have been studied as they exhibit the same characteristic pH dependent absorption as does BSA.

Zana and Lang[51] measured the ultrasonic absorption in aqueous solutions of bovine serum albumin, β-lactoglobulin and lysozyme over the extended

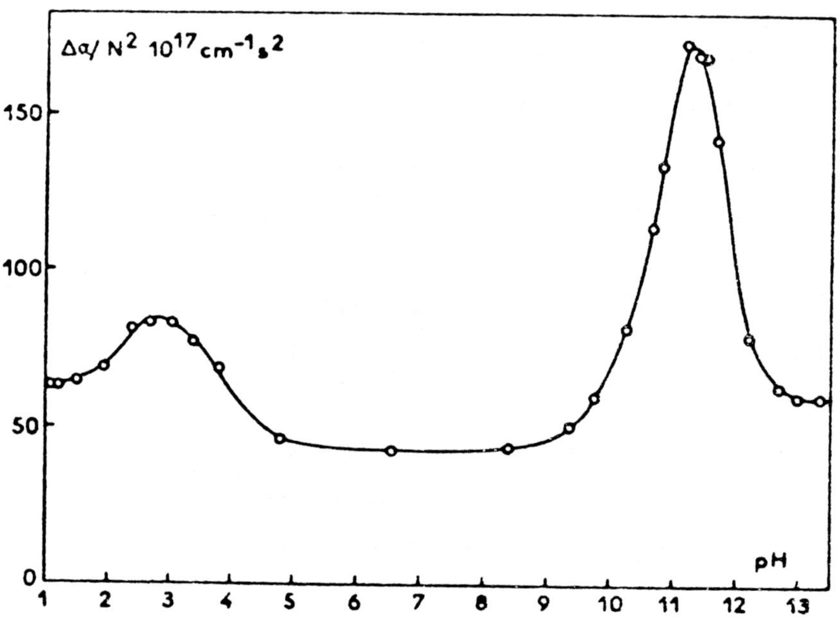

Fig. 5 Bovine serum albumin in H_2O - 0.2 M NaCl. Variation of $\Delta\alpha/f^2$ with pH at 2.82 MHz. T = 25°C. Protein concentration = 0.01 g/cm³.[31]

pH range of 1 to 13.3 at 2.28 MHz and 25°C. All three proteins exhibited maxima in ultrasonic absorption around pH 2 and 3 and around pH 11 to 12. Kessler and Dunn[40] did not observe the peaking of BSA since their upper pH value, 11.8, was the value at which the absorption was maximum. Zana and Lang[51] showed that the absorption maxima could not be explained in terms of conformational changes or dissociation of the biopolymer.

In a more complete and comprehensive report, Lang, Tondre and Zana[48] concluded that the acidic and alkaline absorption maxima are due to the proton-transfer reaction of the carboxyl and amino groups, respectively, and, in addition, the overall ultrasonic absorption is affected by the biopolymer conformation, but not by equilibria between two conformation states. Again they used bovine serum albumin, β-lactoglobulin and lysozyme but under different solvent conditions. The pH dependent properties were examined by comparing the absorption of each of the proteins in a water solvent and in an 8 M aqueous urea solvent. The latter solvent is known to unfold the proteins into coils and thus eliminate any chance of conformational changes as the solution pH is adjusted. In both cases, for all three proteins, the absorption maxima exhibited in both the acidic and alkaline pH regions were similar enough to rule out conformational changes as the primary mechanism. The shift in the pH at which the absorption peaks, pH_m, in both the acidic and alkaline pH regions by approximately +1 pH unit when going from the water solvent to the 8 M aqueous urea solvent could be explained in terms of the proton-transfer theory. Also, the excess absorption at pH_m is less in the 8 M aqueous urea solvent since the standard volume change is smaller therein.

The pH dependent ultrasonic absorption of hemoglobin has been investigated by Hussey and Edmonds,[53] O'Brien and Dunn[43] and Slutsky and White.[55] O'Brien and Dunn[43] measured the ultrasonic absorption in aqueous solutions of bovine (met) hemoglobin (two preparation grades: uncrystallized and 2X crystallized) over the frequency range 2 to 50 MHz and over the pH range 1.5 to 13.3 at 10°C (see Figure 6). They provided a qualitative discussion re-

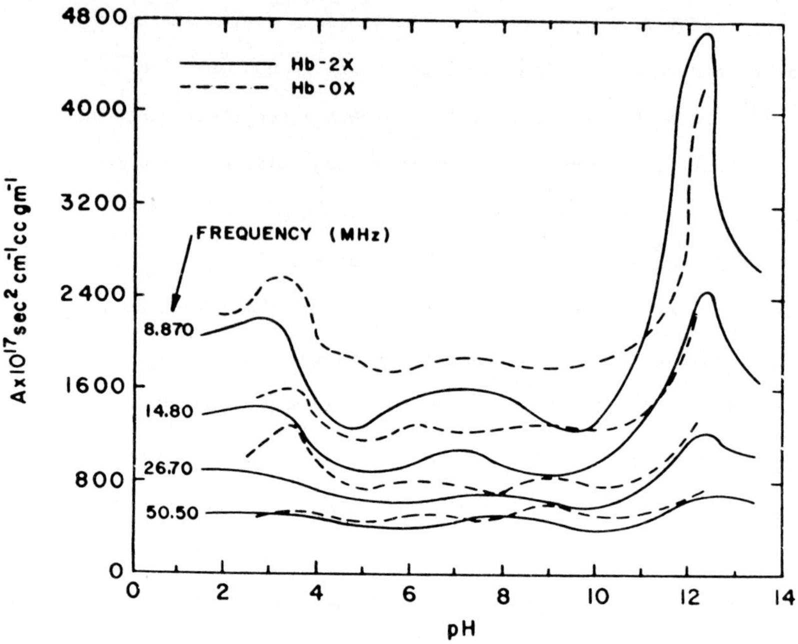

Fig. 6 Composite ultrasonic absorption titration curve in aqueous solutions of bovine hemoglobin ($T = 10^{o}$).[43]

garding the absorption peaks around pH 2 to 4 and 11 to 13 for the hemoglobin solutions and also compared the results to those of bovine serum albumin solutions.

The ultrasonic absorption characteristics of a number of amino acids and polypeptides in aqueous solution have been discussed previously, and the observed maxima in the acidic and alkaline pH ranges were discussed. This absorption peaking has been described quantitatively for the amino acids studies, assuming the proton-transfer reaction dominates the absorption process, and also quantitatively for some of the polypeptides. However, the task of dealing with the similar peaking for proteins, which contain a large number of side chains able to participate in the proton-transfer reactions, is far more difficult since details of reaction coupling among sub-chain groups of amino acids aspartic acid (ASP) and glutamic acid (GLU) is primarily responsible for the absorption peak. At the alkaline pH values, the ε-amino group of lysine (LYS), the phenolic hydroxyl group of tyrosine

(TYR), and the sulfhydryl group of cysteine (CYS) and the guanidinium group of arginine (ARG) contribute to the absorption maximum.

For two proteins O'Brien and Dunn[43] examined concentration of amino acid side chains which can participate in the proton-transfer reaction in the alkaline (LYS, ARG, TYR, CYS) and acidic (ASP, GLU) pH regions and correlated these concentrations to the excess absorption at pH_m in both the alkaline and acidic pH ranges. Specifically, they showed that the concentration ratio of LYS, ARG, TYR and CYS to ASP and GLU was similar for both bovine hemoglobin and bovine serum albumin. They then compared this to the fact that the ratio of excess absorption at the alkaline pH_m to the acidic pH_m was also similar for both proteins. The similarity of these ratios is considered to lend support to the view that the similar ultrasonic absorption magnitude of the maxima of these two globular proteins results from the same mechanism and they suggest, therefore, that the explanation for such maxima may be the proton-transfer reaction. However, they caution that the role of conformation cannot be completely discounted as the evidence for each mechanism (conformation versus proton-transfer) is less convincing than one would like and does not exclude the other from being present, simultaneously, to some degree of effectiveness.

Hussey and Edmonds[53] calculated the excess ultrasonic absorption due to the proton-transfer reaction, based on the theory of Eigen and de Maeyer,[21] for solutions of bovine serum albumin, hemoglobin and gelatin. They then compared the calculated values with experimental data and concluded that the proton-transfer reaction is very likely one of the major mechanisms contributing to the excess ultrasonic absorption in the pH ranges less than 6 and greater than 8 in solutions of bovine serum albumin, hemoglobin and gelatin. This represents one of the first attempts to account quantitatively for the absorption maxima. However, they point out that their methods of calculation need to be refined since the quantitative agreement with the experimental data was not as desired.

Zana and Lang[54] examined the ultrasonic absorption and velocity in *in vitro* amniotic fluid at 25°C as a function of frequency and pH. The ultrasonic velocity at 2.81 and 5.04 MHz was 1510 ± 3 m/s which is approximately 10 m/s greater than water at 25°C. The authors indicate that approximately two-thirds of the excess velocity appears to be due to sodium chloride and one-third to serum albumin. The ultrasonic absorption demonstrated the existence of relaxation processes over the frequency range of 1 to 14 MHz and absorption peaks as a function of pH in the acidic and alkaline pH ranges. The amniotic fluid was divided into low and high molecular weight components by dialysis. Thus, the authors attributed the absorption peak around pH 2.8 as primarily due to low molecular weight component. Two maxima were observed in the alkaline pH region, one at pH 9.5 which was associated with the low molecular weight components, and the other at pH 11.4 due to a high molecular weight component. The latter maximum is probably due to the serum albumin which also is the principal contributor to the absorption within the pH range from 4 to 7.

2.4 Carbohydrates

Hawley, Kessler and Dunn[56,57] have reported absorption data on aqueous solutions of dextran, a linear $\alpha(1 \rightarrow 6)$ anhydroglucose polysaccharide by-product of bacterial metabolism, which is commercially available in crystalline form. The molecular weights of the dextrans were 73,000, 186,000, 370,000 and 2,000,000. The absorption data for these polysaccharides -- in the frequency range 3 to 69 MHz and concentration range 2.2 to 11.4 g/100 ml -- have been compared with those for hemoglobin (molecular weight 68,000) in an endeavor to identify differences of ultrasonic absorption characteristics which might be due to differences in the molecular structure of proteins and polysaccharides, for example, the presence or absence of tertiary structure. The polysaccharide is primarily a linear structure with 5 to 10 percent glucose residues existing as branching elements, and it assumes random coil configurations in aqueous solution. In contrast, hemoglobin is a globular protein displaying a specifically folded structure, which is probably stabilized by hydrogen bonds and van der Walls forces.[58,59]

In view of the uncertainties in interpretation of data, a simplified procedure for comparing the polysaccharides and hemoglobin is adopted. The partial specific absorption coefficient is defined as $\bar{\alpha}_2 = (\partial\alpha/\partial\eta_2)_{\eta_1 T.P.f}$, and the partial specific absorption per wavelength as $\bar{\alpha}_2 \nu/f$. These quantities are defined to be analogous to the partial specific volume $\bar{\nu}_2 = (\partial V/\partial\eta_2)_{1 \cdot T.P.}$ where $\eta_1 \equiv$ concentration of solvent, and $\eta_2 \equiv c' \equiv$ concentration of solute. Therefore $\bar{\alpha}_2$ represents the incremental absorption at frequency f resulting from the addition of a small amount of solute to a solution. At lower concentrations (less than 10 g/100 ml approximately), the measured absorption coefficients are proportional to concentration for both polysaccharides and hemoglobin.

The comparison of partial specific absorption per wavelength is shown on Figure 7. The data for polysaccharides lie below those for hemoglobin in all instances, although not by more than a factor 0.4. Therefore, it was suggested that the tertiary structure might be responsible for the excess incremental absorption of hemoglobin.[56] This suggestion was supported at the time by the absorption data from another protein, gelatin, which does

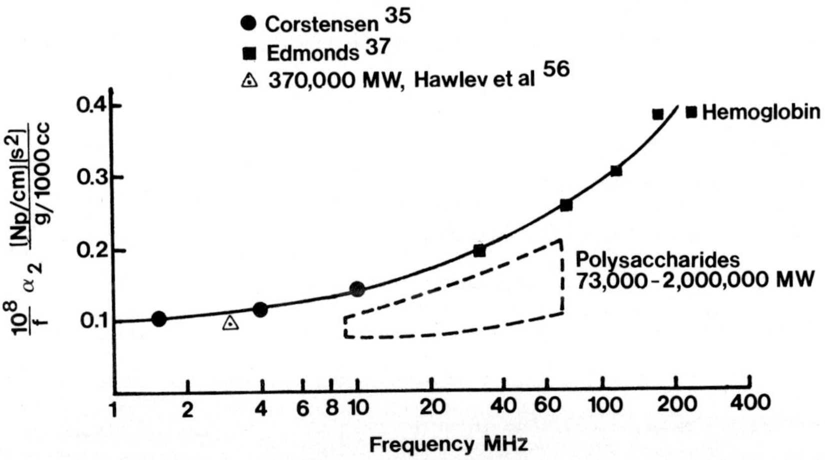

Fig. 7 Partial specific absorption parameter $\bar{\alpha}_2/f$ for aqueous solutions of bovine hemoglobin and polysaccharides.[10]

not exhibit tertiary structure and possesses an absorption coefficient about half as great as that of hemoglobin at pH 6 to 7 at the same concentration. However, as was demonstrated in the previous section, the macromolecular conformation cannot account for this difference.

Hawley and Dunn[60,61] continued the examination of the ultrasonic absorption mechanism responsible in aqueous solutions of dextran and have made comparisons to data on aqueous solutions of polyethylene glycol, another linear polymer. It is shown that the relaxational behavior of dextran can be described by the distribution of relaxation times specified by Rouse[62] for the dynamic behavior of free-draining random coils. However, the actual absorption magnitude is somewhat greater than can be accounted by solution viscoelasticity.

The ultrasonic absorption properties of dextran have been examined as a function of molecular weight and are shown in Figure 8.[60] Similar data from another random coil polymer, polyethylene glycol,[41] a synthesized long-chain flexible macromolecule, is shown in Figure 9 for comparison. It can be observed that beyond a molecular weight corresponding to approximately 100 monomer units, the ultrasonic absorption is independent of molecular weight for both polymers.

The ultrasonic absorption in aqueous solutions of chondroitin sulfate,[63] a mucopolysaccharide, was determined as a function of frequency (2 to 62 MHz), temperature (10 to 37°C) and pH (0.85 to 13.3). A broad distribution of relaxation frequencies is required to describe the absorption spectra at neutral pH. It is suggested that the same mechanism may be responsible for the ultrasonic absorption in the polysaccharides since the absorption magnitude and frequency dependence are similar over the frequency range from 2 to 62 MHz for both chondroitin sulfate and dextran aqueous solutions. A specific mechanism speculated for dextran[60] is that the normal modes of the molecule merely determine the mode of coupling to other relaxation process involving the solvent, since the magnitude of the ultrasonic absorption would require a much greater volume concentration than would be warranted by straight-for-

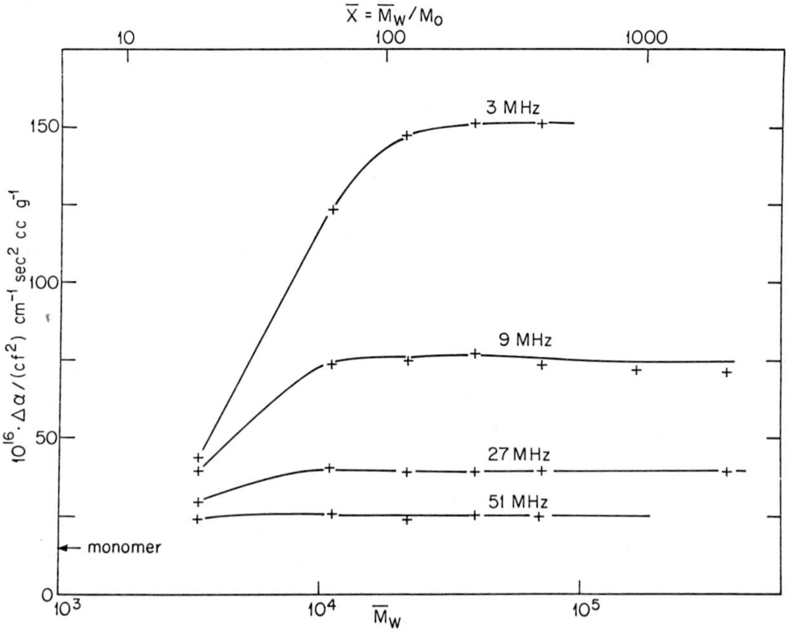

Fig. 8 $\Delta\alpha/(cf^2)$ as a function of weight average molecular weight at four frequencies, 20.0°C. The average degree of polymerization $\bar{X}$ is denoted on the superior abscissa.[60]

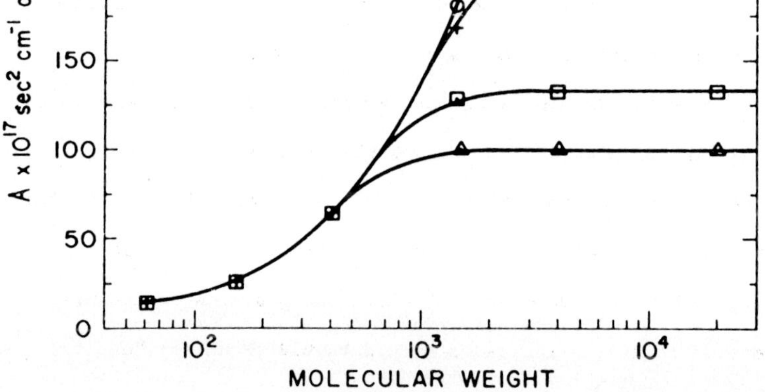

Fig. 9 Dependence of absorption on molecular weight: ⊞ all frequencies; ⊙, 6 MHz; +, 10 MHz; ⊡, 30 MHz; △, 60 MHz.[41]

ward predictions from a random-coil model. As a function of pH, the absorption peaks near pH 2.5 and 11.5 are typical of those observed for the proton-transfer reaction. The absorption peaks in the acidic pH range can be attributed to the carboxyl group, but additional study is required to explain the absorption peak in the alkaline pH region since neither the chondroitin sulfate molecule nor protein impurities can be invoked.

2.5 Bases, Nucleotides and Nucleosides

Measurements of ultrasonic absorption have been reported by Lang, Sturm and Zana[64] on the constituent components of nucleic acids. They measured the absorption at 2.82 MHz and 25°C over the extended pH range 2 to 13 in aqueous solutions of thymine (base), D-ribose 5-phosphate (sugar, phosphate), seven nucleosides (base, sugar) and eleven nucleotides (base, sugar, phosphate). All but three nucleosides exhibited an absorption maximum in the alkaline pH range; all eleven nucleotides exhibited an absorption maximum in the acidic pH region and five of these also exhibited an absorption maximum in the alkaline pH region as is summarized in Table I. Both thymine and the seven nucleosides exhibited no pH dependence in the acidic pH region and D-ribose 5-phosphate showed an absorption peak only in the acidic pH region. The ultrasonic absorption maxima occurred within the pH ranges 4.1 to 5.9 and 9.7 to 11.9.

Within the alkaline pH range, the following points are made. Ultrasonic absorption maxima are not observed in the nucleotides and nucleosides which contain the bases adenine or cytosine since the lactam-lactim tautomerism is not present. Thymine exhibits two absorption maxima and this base contains two lactam groups. As shown in Table I there is agreement between the calculated and experimental values for the pH at which the maxima occur. These calculations assume the absorption mechanism is associated with the ionization of a lactam or lactim group. Also, the position of the maximum for the nucleoside thymidine increases by 0.2 to 0.4 pH units when its concentration is increased from 0.02 to 0.19 M, which is also in agreement with the calculated pH increase. Calculated values for the excess absorption

Table I. Summary of the Results Obtained with 0.02 M Solutions at 2.82 MHz and 25°C.[64]

	Acid range			Alkaline range		
	pH_A		$(\Delta\alpha/f^2)_A \times 10^{17}\ sec^2/cm$	pH_B		$(\Delta\alpha/f^2)_B \times 10^{17}\ sec^2/cm$
	Expt	Calcd[a]		Expt	Calcd[a]	
5'dAMP	5.05 ± 0.1	5.12 (b)	76 ± 4			0
5'AMP	5.05 ± 0.1	5.12 (b)	76 ± 4			0
5'AMPNH$_2$	3.15 ± 0.1	3.1 (a)	13.5 ± 2			
5'XMP	5.9 ± 0.1	5.88 (b)	102 ± 5			0
5'dCMP	5.5 ± 0.1	5.42 (b)	141 ± 7			0
5'CMP	5.35 ± 0.1	5.37 (b)	103 ± 5			0
5'IMP	4.1 ± 0.2	3.97 (c)	9 ± 2	10.55 ± 0.3	10.8 (e)	11 ± 2
5'TMP	4.2 ± 0.2	4.07 (c)	9 ± 2	11.1 ± 0.15	11.22 (e)	24 ± 3
5'TMPNH$_4^+$	4.1 ± 0.2	4.09 (c)	8.5 ± 2	9.7 ± 0.1	9.69 (d)	315 ± 10
5'UMP	4.3 ± 0.3	4.04 (c)	7.5 ± 2.5	10.8 ± 0.3	11.02 (e)	12 ± 3
5'GMP	4.2 ± 0.2	4.1 (c) 4.3 (b)	12.5 ± 2.5	10.9 ± 0.2	11.02 (e)	19 ± 3
Deoxyadenosine			0			0
Xanthosine	Precipitation occurs at pH <4.9					0
Deoxycytidine			0			0
Inosine			0	10.6 ± 0.2	10.6 (e)	11 ± 3
Thymidine			0	11.1 ± 0.2	11.05 (e)	20 ± 2
Uridine			0	10.45 ± 0.3	10.7 (e)	12.5 ± 2.5
Deoxyguanosine			0	10.7 ± 0.15	10.8 (e)	23 ± 2
Thymine			0	11 ± 0.1	11.1 (e)	30.5 ± 3
				11.9 ± 0.1		31.5 ± 3
D-Ribose 5-phosphate	4.15 ± 0.2	3.98 (c)	6 ± 2			

[a] The calculated values of $pH_{A,B}$ have been obtained using (a) $pH_A = \frac{1}{2}(pK_{P1} + pK_N)$; (b) $pH_A = \frac{1}{2}(pK_{P2} + pK_N)$; (c) $pH_A = \frac{1}{2}(pK_{P2} - \log c)$; (d) $pH_B = \frac{1}{2}(pK_N + pK_{NH_4^+})$ with $pK_{NH_4^+} = 9.26$; and (e) $pH_B = \frac{1}{2}(14 + pK_N + \log c)$.

of thymidine and inosine, assuming the lactam-lactim tautomerism mechanism with a standard volume change of 10 ± 2 cm^3/mole, yield (20 ± 7) X 10^{-17} sec^2/cm and (8 ± 3) X 10^{-17} cm^2/cm which is in agreement with the experimental values (Table I). Even though xanthosine and 5' XMP show lactam-lactim tautomerism in the purine xanthine, the hydroxyl concentration is at least 100 times less than that for the other compounds in the alkaline pH region and, therefore, absorption maxima would not be detectable. Thus it is concluded by the authors that the absorption maximum in the alkaline pH region is due to the equilibrium associated with the ionization of either a lactam or lactim group.

The following observations are made from the data associated with the acidic pH region. Only the nucleotides and D-ribose 5-phosphate exhibit absorption maxima. These compounds all consist of a phosphate group as contrasted to the nucleosides and thymine which do not. Also, the absorption amplitude for the five nucleotides which also have a maxima in the alkaline pH region are quite small compared to the other six nucleotides; the small amplitude is in the same range as that of D-ribose 5-phosphate and occur within a very narrow pH range of 4.1 to 4.3. Since the maxima appears to be independent of the base, the authors suggest that it must be attributed to the phosphate group and, specifically, to the protolysis of the secondary acid function in phosphoric acid. Assuming this mechanism, the calculations of the position of the pH maxima for these five nucleotides and D-ribose 5-phosphate are in agreement with the experimental findings. Amplitude calculations are in agreement also except too small for 5' GMP. In fact the protolysis of the secondary phosphoric acid function occurs in all the nucleotides. However, additional reactions tend to mark this reaction. In the case of the five nucleotides 5' dAMP, 5' AMP, 5'XMP, 5'dCMP and 5' CMP, the calculated values of pH position and amplitude agree with the experimental values and thus it is suggested that the proton exchange mechanism is involved between the secondary phosphoric acid function and the protonable nitrogen atom. They indicate that the proton exchange could either be intramolecular or bimolecular.

Three of the nucleosides studied by Lang, Sturm and Zana[64] contain nitrogen atoms which are protonable in the acidic pH range but did not exhibit absorption maxima, viz., deoxyadenosine, deoxycytidine and deoxyguanosine. The principal reason attributed to the lack of a pH dependence is that the standard volume change, less than 5 cm^3/mole, yielded a calculated amplitude which would be too small to detect. The excessively large amplitude of 5' $TMPNH_4^+$ observed in the alkaline pH region is explained as a coupled proton-transfer reaction between NH_4^+ and $5'TMP^-$.

In a follow-up study Lang, Sturm and Zana[65] reported the details of proton-transfer reaction kinetics in aqueous solutions of the nucleotide adenosine 5' monophosphate (5' AMP). They measured the ultrasonic absorption over the frequency range from 1 to 115 MHz at the concentrations 0.016, 0.050, 0.099 and 0.148 M at 25°C around the pH range from 3 to 7. As shown in Table I, the pH at which the ultrasonic absorption peaks is 5.05 ± 0.1.

This value is independent of concentration, ionic strength (0 to 0.15 M KCl) and frequency, within experimental error, over the indicated ranges. A strong argument is put forth to suggest that the mechanism responsible for the amplitude maximum at pH 5.05 is the proton-exchange reaction occurring between two differently ionized forms of 5' AMP coupled with the dimerization of 5'AMP.

2.6 Nucleic Acids

Measurements of the absorption coefficients of DNA have been reported by Lang and Cerf,[66] Sturm, Lang and Zana[67] and O'Brien, Christman and Dunn.[68] Lang and Cerf's[66] measurements were performed in the frequency range of 3 to 58 MHz on samples of calf thymus DNA of molecular weight 4.7 - 7.0 X 10^6 at concentrations of 0.3 and 0.6 g/100 ml in an aqueous solution of 1.18 sodium chloride and 0.02 M sodium citrate. The temperature was maintained at 5° and 25^oC and the pH range was 6 to 12.5. Figure 10 shows the ultrasonic absorption as a function of pH in the alkaline pH region. The data at frequencies less than 27 MHz exhibit incipient relaxation and absorption maxima at pH 11.9 (actually 11.7 as determined later by Sturm, Lang and Zana.[67]) Thermal denaturation, followed by the addition of formaldehyde to prevent renaturation, produced no change in the ultrasonic absorption. This work supersedes previous papers from the same laboratory, reporting a maximum in α/f^2 as a function of frequency for native DNA. A monotonic decrease with increasing frequency is now found under all conditions. The authors suggested that relaxation processes characterized by at least three states may account for this more reliable data. Consequently, it may not be necessary to utilize the theory of "stochastic resonance", developed by Cerf to include both relaxational and resonance behavior.[69-71] It is of interest to note there that Zana and Cerf[72] have calculated the absorption to be assigned to viscosity in a solution of rigid ellipsoidal particles, representing DNA molecules.

In their earlier work Lang and Cerf[66] attributed the absorption amplitude maximum to alkaline denaturation. In a follow-up study Sturm, Lang and Zana[67] measured the ultrasonic absorption in aqueous solutions of native

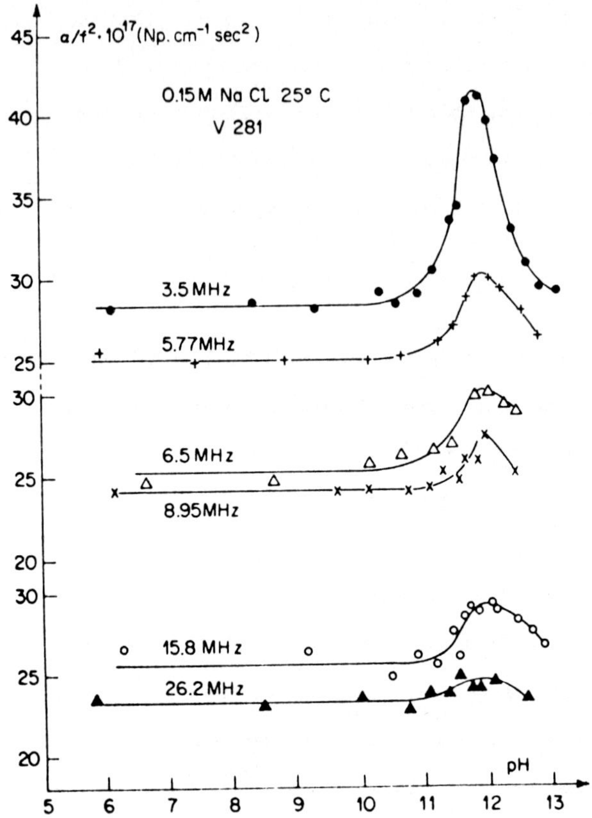

Fig. 10 Dependence of α/f^2 upon pH and frequency during alkaline denaturation of calf-thymus DNA in solution at 25°C and 0.3 g DNA/100 ml.[66]

(double helix) calf thymus DNA over the pH range from 6 to 13 and denatured calf thymus DNA over the pH range from 1.7 to 12.5. They observed absorption maxima in the alkaline pH range for both the native and denatured DNA of comparable amplitudes and pH position and concluded that the conformational equilibrium associated with the double helix breakdown is negligible. On the other hand they suggested the absorption peaks result from the proton-transfer reaction of the thymine and guanine bases. Additionally, the denatured DNA exhibited an ultrasonic absorption increase as the pH was decreased below 5 but did not peak. They suggest that the proton-transfer reaction on the guanine, adenine and cytosine amino groups may be responsible for the increased absorption in the acidic pH range.

O'Brien, Christman and Dunn[68] measured salmon sperm DNA (molecular weight $10^6 - 10^7$) in aqueous solutions of 0.15 M sodium chloride and 0.015 M sodium citrate at a temperature of 25°C at a concentration of 0.4 g/100 ml over the frequency range of 5 to 50 MHz and pH range of 1.7 to 7.5. A maximum in the ultrasonic absorption is observed around pH 2.6 as shown in Figure 11. A maximum at pH 2.6 in salmon sperm DNA was observed at 10°C[42] and in calf thymus DNA of molecular weight around 10^5 at 25°C.[73] They point out that the protonation of the carboxyl group, which occurs between pH 3.5 and 2.5, should produce absorption maxima between pH 1.75 and 1.25. Also, if the amino group is protonated between pH 11 and 12, the absorption maxima would occur between pH 12.5 and 13. Since absorption maxima have occurred around pH 2.6 and 11.7, there is reasonable doubt that the proton-transfer reaction can be invoked to account for such maxima in DNA solutions.

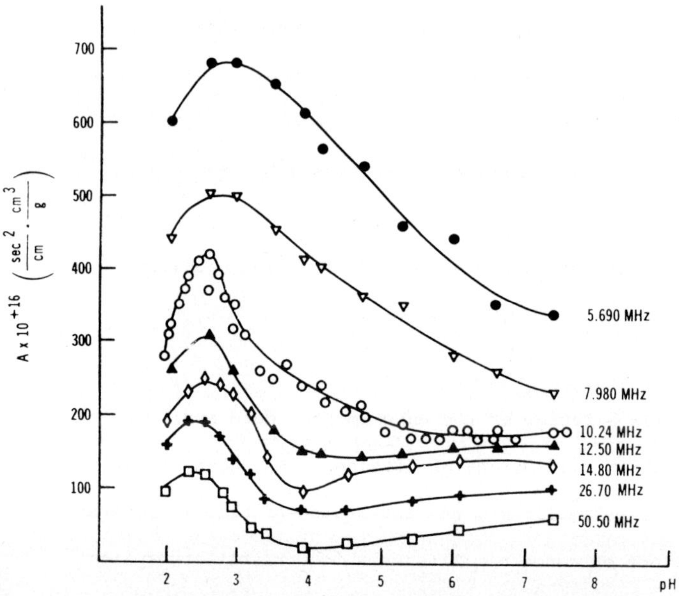

Fig. 11 Ultrasonic absorption acidic titration curve of aqueous solutions of salmon Sperm DNA at 25°C.[68]

As a result of their examination of nucleotides and nucleosides, Lang, Sturm and Zana[64] re-evaluated the DNA data and suggested that the absorption maximum at pH 11.7 appears to be attributable to the proton-transfer reaction involving the lactam or lactim groups of guanine and thymine. In the acidic pH range, it is suggested that the absorption maximum of DNA solutions appears to be due to the proton-transfer reaction between the adenine, guanine and cytidine nitrogen atoms and the phosphoric acid function.

With respect to RNA solutions, Lang, Sturm and Zana[64] suggest that the mechanism is similar to that of DNA solutions except for the alkaline region where the lactam group of uracil and guanine are involved.

2.7 Lipids

Hammes and Roberts,[74] in a study undertaken to examine model membrane, measured the ultrasonic attenuation in suspensions of two phospholipids, phosphatidylserine and phosphatidylcholine, over the frequency range from 10 to 165 MHz at a temperature of 25°C. They postulate four main classes of mechanisms by which ultrasonic energy interacts with phospholipid dispersions: (1) liposomal aggregation, (2) breaking the reforming liposomes, (3) intraliposomal conformational changes and (4) liposome hydration sphere equilibria. A broad distribution of relaxation frequencies are observed with the mean relaxation times approximated to be within the range of 10^{-7} to 10^{-8} seconds. They suggested that the observed relaxation processes are either intraliposome conformational changes or solvation equilibria.

3. TISSUES

This section comprises a review of the experimental data available for the ultrasonic absorption coefficient of tissues. The acoustic field variables, frequency and wave amplitude, and the state variables, temperature and concentration, are of concern. Tissues which exhibit absorption properties greatly different from most soft tissues such as fat, muscle, liver, nerve, etc., viz., refractive media of the eye, lung, the high collagen content supportatial tissues, and mineralized tissues, are discussed separately. While no formal theory is presented for explaining the observed dependencies of the ultrasonic absorption coefficient, allusion is made to several of the dependencies of the absorption coefficient upon specific physical variables where they appear to resemble those of known mechanisms arising from the interaction of ultrasound and nonbiological materials.

3.1 Soft Tissues

3.1.1 *Frequency dependence.* The dependence of the ultrasonic absorption coefficient upon the acoustic frequency has been studied by numerous investigators. Goldman and Hueter[75] have compiled ultrasonic velocity and absorption data in animal soft tissues. Therein it is seen that the velocity, excluding lung, is very nearly that of dilute salt solutions and varies only slightly among those tissues. Fatty tissues are exceptions in having a velocity about 10% less than the others. Figure 12, taken from their paper, is a graphical representation of the acoustic amplitude absorption coefficient per wavelength for several mammalian tissues in the frequency range of approximately 200 kHz to 10 MHz. They attempted to include all measurements available at that time by numerous investigators employing different experimental techniques. The scatter of the data exhibited by the bands or broad shaded regions is thus not wholly surprising since many neglect to give either complete specifications of their experimental procedure and/or a description of the state of the specimen used. For example, it is not possible, in many cases, to determine from the literature the temperatures employed by the investigators reporting the data. (The influence of temperature on the absorption coefficient is discussed below). It is, however, possible to discern several relatively simple relationships. For example, the absorption per cycle α/f, is generally constant over the frequency range considered. For fat, α/f, increases slightly in the frequency range from 1 to 10 MHz. The experimental results for striated muscle and liver appears to exhibit a minimum in the neighborhood of 2 MHz. Kessler[76] has shown that for kidney the linear dependence of the attenuation

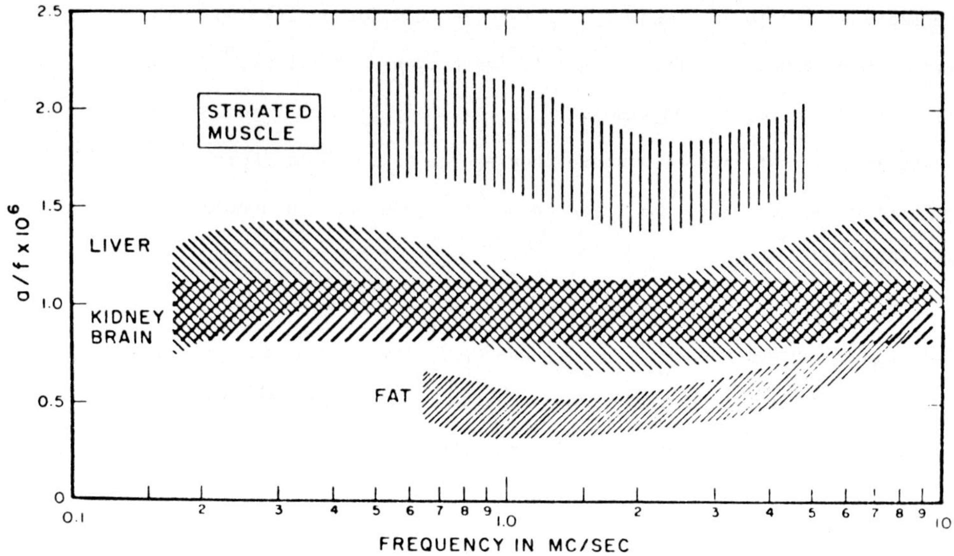

Fig. 12 Acoustic amplitude absorption coefficient (in dB/cm) per wavelength versus frequency for several mammalian tissues.[75]

on frequency exists to about 100 MHz, after which a square law (or greater) dependence exists. Fry[77] has considered a viscous mechanism for the absorption of ultrasound in tissue, in which it is shown that the viscous forces acting between a suitable chosen distribution of suspended particles (or structural elements) and a suspending liquid can account for the experimentally observed linear relationship between acoustic absorption coefficient and ultrasonic frequency. The frequency band over which linearity obtains (in the model) is determined by the limits of the distribution of values for the parameters chosen to describe the structural elements. Below the linear range the theory predicts a quadratic dependence, in agreement with the experiment.

Figure 13 is another way of presenting the data which may be suggestive for determination of mechanisms of absorption.[78] Here, the logarithm of the absorption coefficient is plotted as a function of the logarithm of the ultrasonic frequency, and the slopes of the resulting curves are examined (the slope is the exponent on frequency upon which the magnitude of the ab-

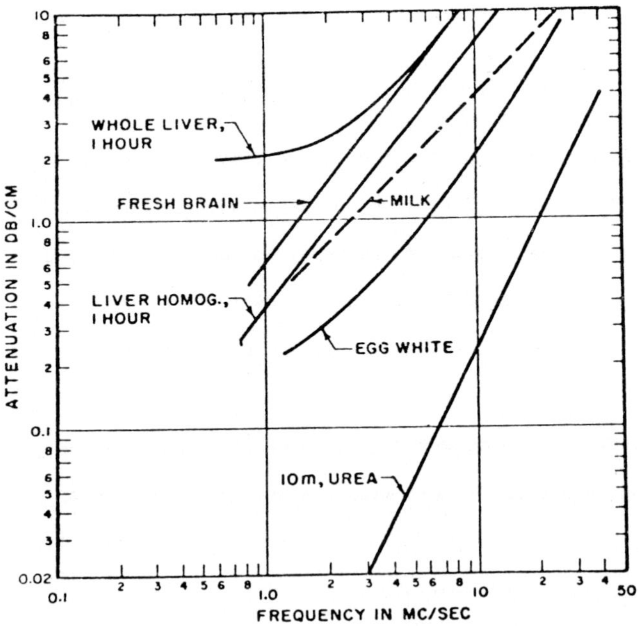

Fig. 13 Acoustic amplitude absorption coefficient versus frequency for materials of different biological complexity.[78]

sorption coefficient depends). Figure 13 shows data for several materials of increasing biological complexity, and they exhibit corresponding graded complexity in absorptive behavior. The urea solution exhibits a slope of 2, indicative of classical viscous absorption for which α/f^2 = constant. Homogenized milk, a suspension of fat particles and hydrated casein complexes, exhibits a slope of nearly unity from approximately 1 to 40 MHz. Behavior of this type cannot be explained in terms of simple viscosity or scattering theories. The curves of the absorption coefficients of egg albumin, brain tissue, liver, and striated muscle (not shown in Figure 13) exhibits slopes between 1 and 2 in the neighborhood of 1 MHz and approach a slope of 2 at higher frequencies. Hueter[78] has suggested that this type of frequency dependence can be described for specific muscle preparations by a double relaxation process in which the bulk (volume) viscosity of the tissue possesses a relaxation frequency near 40 kHz, and the shear viscosity possesses a relaxation near 400 kHz. Although it is conceded that this is an over simplification of a complicated process, it will be shown below that the temperature dependence of the acoustic absorption coefficient lends support to this view.

3.1.2 *Concentration dependence.* The dependence of the ultrasonic absorption coefficient upon the concentration of tissue components in solution is mentioned briefly below, and it is interesting to compare, as Hueter[78] has done, the absorption in brain tissue, reported by numerous investigators, with the absorption of concentrated solutions of red cells, as shown in Figure 14. The curve through the nerve-tissue data is drawn with a slope found for concentrated red cells.[34] The figure suggests that nerve tissue behaves very much like a concentrated aggregate of protein with the absorption increased

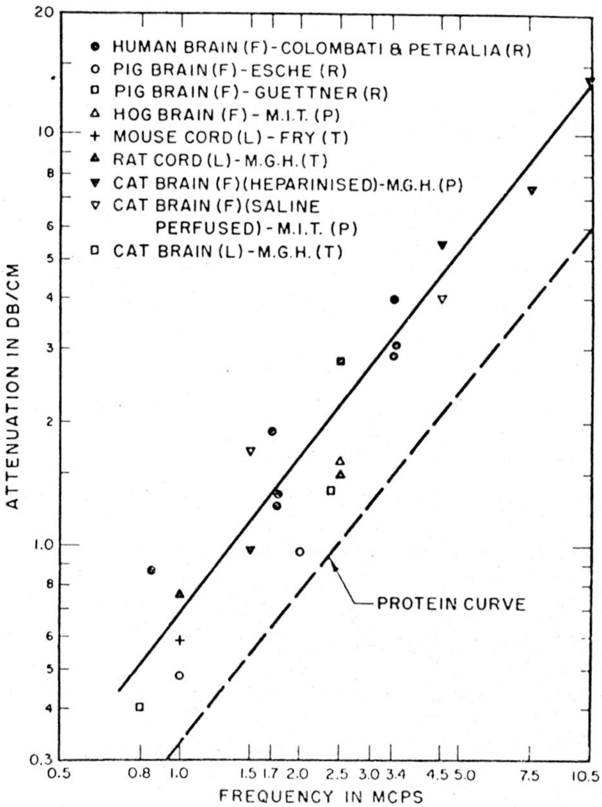

Fig. 14 Ultrasonic absorption versus frequency for nervous tissue and for a solution of 85 percent red cells from human blood.[78]

by approximately a factor of 2, perhaps due to the protein-protein interactions associated with the formation of tissue structure.

Studies by Schwan and coworkers[49,79] on liver and solutions of liver fractions has shown that tissue cell nuclei do not contribute significantly to tissue absorption. The fractions exhibit absorption spectra similar to those of tissues shown in Figure 13. The studies suggested that tissue absorption arises on a molecular level and is largely independent of architectural features. It was found that individual molecular constituents display somewhat different absorption coefficients which depend upon pH and molecular integrity. It is suggested that the tissue proteins account for at least two-thirds of the observed tissue absorption.

3.1.3 *Temperature dependence.* Details of the absorption coefficients as a function of temperature, and frequency, have recently become available. Figure 15 shows observations on mammalian central nervous tissue, the only tissue for which such data are known. The curves for 0.26, 0.5, 0.7 and 1 MHz represent in vivo measurements in the spinal cords of neonatal mice (essential poikilotherms)[1,80,81] and those for 4.2 MHz are in vitro measurements in brains of adult cats (homoeotherms).[82] The relatively complex behavior of the frequency-free absorption coefficient on frequency and temperature suggests a family of curves whose maxima decrease in magnitude and occur at ever higher temperatures as frequency increases. It is not known whether other soft tissues exhibit similar behavior but Kishimoto[83] has observed a positive temperature coefficient for the absorption coefficient of bone in the frequency range 1.4 to 4.5 MHz. However, these data illustrate the necessity for complete specification of the state of specimens when reporting experimental results.

3.1.4 *Amplitude dependence.* The transient thermoelectric method employed in obtaining the value of the absorption coefficient as a function of temperature of the young mouse spinal cord included exposure of each specimen to different acoustic intensity levels of incident sound energy.[1] Thus, the method made data available on the dependence of the acoustic absorption coefficient upon the sound amplitude. For this method, the initial time rate of change of temperature, as a linear function of the acoustic intensity of the heat capacity per unit volume and the absorption coefficient are constant. Figure 16 shows that, in the range of incident acoustic intensities from a few watts per square centimeter to nearly 200 w/cm^2 and at three base temperatures of the animals, such linear relationships indeed are obtained. On the assumption that the heat capacity per unit volume is not dependent on the acoustic intensity, it can be concluded that in the range of intensities employed, the absorption coefficient is independent of the intensity of the radiation for the preparation of this study.

The following comments are of interest relative to the lack of variation of the value of the absorption coefficient of nerve tissue with the

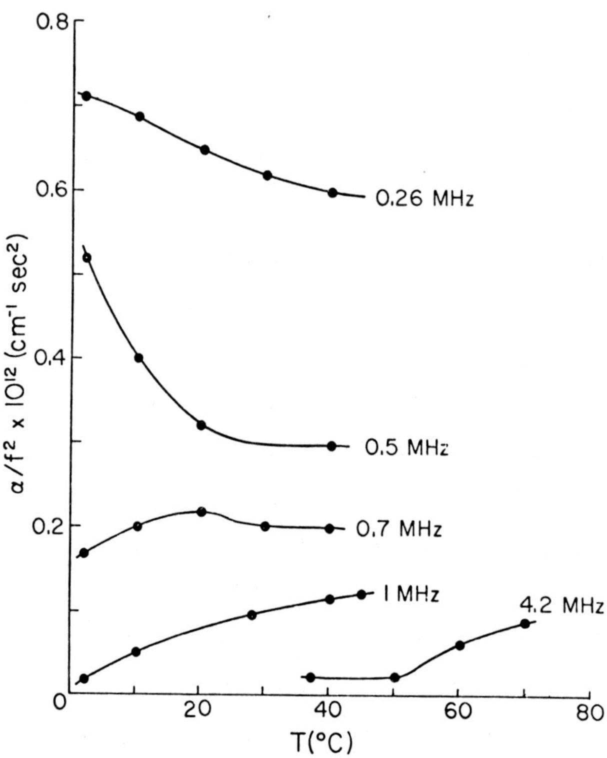

Fig. 15 Frequency and temperature dependence of ultrasonic absorption in mammalian central nervous tissue.[81]

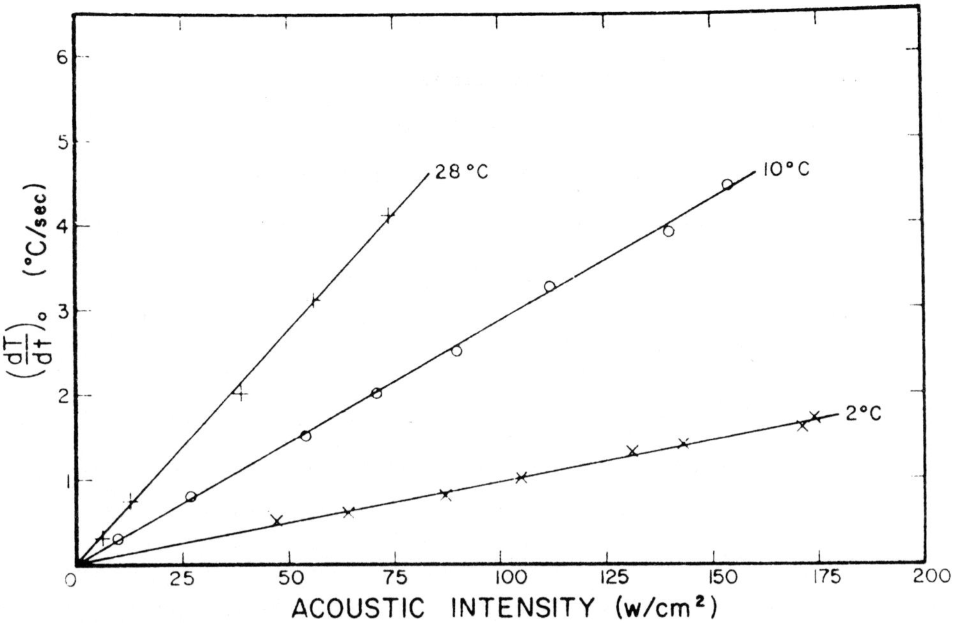

Fig. 16 Time rate of change of temperature in spinal cord of neonatal mice produced by absorption of 1 MHz sound vs. incident acoustic intensity at three temperatures.[1]

sound intensity (for the range of values indicated). The propagation distance from sound source via degassed mammalian saline to the tissue was approximately 5 cm in the experimental work of reference. This distance is sufficiently short in the saline so that initially monochromatic waves produced by the vibrating element (Z-cut quartz plate) remain virtually monochromatic during propagation to the specimen; that is, the energy transferred from the fundamental to the harmonics, which results from the nonlinear equation of state of the liquid medium (see for example, Fry and Dunn),[84] is negligible in this path length at the highest intensities employed. Thus, the wave incident on the tissue is essentially undistorted from the initial single-frequency form. On the assumption that the coefficients in the equation of state of tissue are approximately the same as those of water, and since the propagation distance in the tissue is relatively short (approximately 0.5 mm), the transfer of energy from the incident fundamental wave to its harmonics is negligible within the specimen.

3.2 Absorption in Bone

Bone is a tissue possessing acoustic propagation properties greatly different from those of the soft tissues discussed previously. An early study of specially prepared skull bone, in the frequency range 0.6 to 3.5 MHz (25 to 35°C), exhibited a quadratic dependence of the absorption coefficient upon frequency with a transition to a linear dependence beyond about 2 MHz.[85] An average value found for the acoustic amplitude absorption coefficient per unit path length in skull bone, in the neighborhood of 1 MHz, was of the order of 1 cm^{-1}, approximately an order of magnitude greater than that of soft tissues of the same temperature. However, recent observations have called these values into question and Adler and Cook[86] have obtained absorption measurements of 1.5 cm^{-1} and 2.2 cm^{-1}, in freshly frozen dog tibia at room temperature at, respectively, 2 and 5 MHz.

Reports of measurements of the longitudinal speed of sound in bone are largely in agreement that it is approximately twice that of soft tissues.[83,86-88] Anisotrophy of elastic properties and variations in density of bone present special problems for measurement and for interpretation of results.[86,88]

3.3 Absorption in Tissues of High Structural Protein Content

Dussik and Fritch[89] found that tissues having a high collagen content such as cartilage, tendon, integuement, etc., exhibit speed of sound values approximately 10 to 15% greater than those of soft tissues, but that attenuation values approximated the root-mean-square of organs and bone. These tissues also exhibited anisotrophy of the acoustic parameters.

3.4 Refractive Media of the Eye

Begui[90] has studied the acoustic properties of the refractive media of the eye *in vitro*. He determined the ultrasonic absorption coefficient of the aqueous and vitreous humors at 30 MHz and that of the lens at 3 MHz. The specimens were obtained from excised fresh calf eyes. At 30 MHz and 27.5^{o}C, the aqueous and vitreous humors both exhibit an acoustic amplitude absorption coefficient of 0.35 cm^{-1}. Since this is approximately 50 percent greater than the absorption coefficient of dilute salt solutions, it suggests that the absorption coefficients of the humoral media of the calf eye possess a viscous-type dependence upon frequency; that is the absorption coefficient probably increases as the square of the frequency.

The lens of the calf eye exhibits a value of 0.7 cm^{-1} for the acoustic amplitude absorption coefficient at 3 MHz and 28°C. Since the lens contains a relatively high concentration of protein, it is not unreasonable to assume, in the absence of further information, that the frequency dependence of the absorption coefficient of the lens resembles that of other soft tissue for which the absorption appears to be dominated by the protein content., i.e., it is probable that the absorption coefficient per unit path length of the lens varies approximately with the first power of the frequency.

Some investigators currently using ultrasonic methods for diagnosing disorders of the human eye feel that the lens absorption value given by Begui is larger than that for the human lens (*in vivo*). The possible discrepancy may result because of species differences. Indeed, Begui observed that the viscosity of the intraocular fluid of calf eyes is greater than the values normally stated for the fluid media of human eyes. Further, the specimens used by Begui were first stored (at temperatures in the neighborhood of 0 to 5°C) and were used for measurement purposes within a time interval of 10 days.

Begui obtained for the speed of sound in refractive media of the eye 1497 m/sec for the aqueous humor, 1516 for the vitreous humor, and 1616 for the lens.

3.5 Pulmonary Tissue

Two earlier studies[91,92] showed that ultrasonic attenuation in freshly excised dog lung was unusually high, that the speed of sound was considerably less than that of water, and that both of these quantities had a strong dependence on pulmonary inflation and acoustic frequency. It was also shown that a pathological condition involving an accumulation of liquid-like matter within the pulmonary architecture, had the effect of appreciably reducing both the attenuation and the velocity. Two recent studies have provided details of the frequency and inflation dependencies. Dunn[93] has shown that for excised dog lung of inflation to a fraction of residual air such that the specimen density if 0.4 g/cm^3, the attenuation increases exponentially from 4 cm^{-1} at 1 MHz to 12 cm^{-1} at 5 MHz. In this same frequency range the speed of sound increased linearly from 0.66×10^5 cm/sec to 1.2×10^5 cm/sec. These findings are in general agreement with those of Bauld and Schwan[94] who also showed that, for formalin fixed inflated specimens, the energy reflected at the lung-liquid interface ever increases with gaseous inflation allowing for lesser amounts of energy to enter the lung.

REFERENCES

1. F. Dunn: Temperature and Amplitude Dependence of Acoustic Absorption in Tissue. J. Acoust. Soc. Amer., 34 (1962) p. 1545.

2. J. M. Reid and M. R. Sikov, eds., Interaction of Ultrasound and Biological Tissues -- Workshop Proceedings, DHEW Publication (FDA) 73-8008 BRH/DBE 73-1, U. S. Gov't. Printing Office, Washington, D.C. (1972).

3. G. G. Hammes and C. N. Pace: Ultrasonic Absorption Measurements in Aqueous Solutions of Glycine, Diglycine, and Triglycine. J. Phys. Chem. 72 (1968) p. 2227.

4. R. C. White, L. J. Slutsky and S. Pattison: Kinetics of the Proton-Transfer Reactions of Serine and Threonine. J. Phys. Chem. 75 (1971) p. 161.

5. M. Hussey and P. D. Edmonds: Ultrasonic Examination of Proton-Transfer Reactions in Aqueous Solutions of Glycine. J. Acoust. Soc. Amer. 49 (1971) p. 1309.

6. M. Hussey and P. D. Edmonds: Ultrasonic Examination of Proton-Transfer Reactions at the α-Amino and Side-Chain Groups of Arginine and Lysine in Aqueous Solution. J. Acoust. Soc. Amer. 49 (1971) p. 1907.

7. F. W. Kremkau and E. L. Carstensen: Macromolecular Interaction in Sound Absorption. In Interaction of Ultrasound and Biological Tissues--Workshop Proceedings, ed. by J. M. Reid and M. R. Sikov, p. 37, DHEW Publication (FDA) 73-8008 BRH/DBE 73-1, U.S. Gov't. Printing Office, Washington, D.C. (1972).

8. F. W. Kremkau: Macromolecular Interaction in the Absorption of Ultrasound in Biological Material. Ph.D. Thesis, The University of Rochester (1972).

9. R. E. Barrett, R. T. Beyer and F. D. McNamara: Ultrasonic Absorption in Acetate Solutions. J. Acoust. Soc. Amer. 26 (1954) p. 966.

10. F. Dunn, P. D. Edmonds and W. J. Fry: Absorption and Dispersion of Ultrasound in Biological Media. In Biological Engineering, ed. by H. P. Schwan, McGraw-Hill Book Co., New York (1969)Chapt. 3, pp. 203-332.

11. A Passynsky: Solvation of Non-Electrolytes and the Compressibility of Their Solutions. Acta Physicochim. USSR 22 (1947) p. 137.

12. F. T. Gucker and R. M. Haag: Compressibility of Solutions: II, An Ultrasonic Study of Aqueous Solutions of Some Simple Amino Acids and Their Uncharged Isomers at 25°C. J. Acoust. Soc. Amer. 25 (1953) p. 470.

13. S. Gotto and T. Isemura: Studies of the Hydration and the Structure of Water and Their Roles in Protein Structure: IV, The Hydration of Amino Acids and Oligopeptides. Bull. Chem. Soc. Japan 37 (1964) p. 1697.

14. J. Stuehr and E. Yeager: The Propagation of Ultrasonic Waves in Electrolytic Solutions. In Physical Acoustics, by W. P. Mason, Academic Press, Inc., New York (1956) Vol. 2, Part A. Chapt. 6, p. 351.

15. D. V. Beauregard and R. E. Barrett: Ultrasonics and Water Structure in Urea Solutions. J. Chem. Phys. 49 (1968) p. 5241.

16. W. D. O'Brien, Jr. and F. Dunn: Ultrasonic Examination of the Hemoglobin Dissociation Process in Aqueous Solutions of Guanidine Hydrochloride. J. Acoust. Soc. Amer. 50 (1971) p. 1213.

17. G. Mass: Ph.D. Thesis, University of Gottingen (1962).

18. K. Applegate, L. J. Slutsky and R. C. Parker: Kinetics of Proton-Transfer Reactions of Amino Acids and Simple Polypeptides. J. Am. Chem. Soc. 90 (1968) p. 6909.

19. M. Hussey: Proton Transfer at Side Chain Groups: A Mechanism for the Absorption of Ultrasound in Protein Solutions. Ph.D. Thesis, University of Pennsylvania (1970).

20. M. Hussey: Proton Transfer Processes: One of the Modes of Interaction of Ultrasound with Proteins and DNA. In Interaction of Ultrasound and Biological Tissues, ed. by J. M. Reid and M. R. Sikov, p. 31, DHEW Publication (FDA) 73-8008 BRH/DBE 73-1, U.S. Gov't. Printing Office, Washington, D.C. (1972).

21. M. Eigen and L. deMaeyer: Relaxation Methods. In Technique of Organic Chemistry, ed. by S. L. Friess, E. S. Lewis and A. Weissberger, Interscience Publishers, Inc., New York (1963) Vol. 8, Part 2, p. 895.

22. J. J. Burke, G. G. Hammes and T. B. Lewis: Ultrasonic Attenuation Measurements in Poly-L-Glutamic Acid Solutions. J. Chem. Phys. 42 (1965) p. 3520.

23. G. Schwarz: On the Kinetics of Helix-Coil Transition of Polypeptides in Solution. J. Mol. Biol. 11 (1965) p. 64.

24. P. Doty, A. Wada, J. T. Yang and E. R. Blout: Polypeptides: VIII. Molecular Configurations of Poly-L-Glutamic Acid in Water-Dioxane Solution. J. Polymer Sci. 23 (1957) p. 851.

25. Y. Wada, H. Sasabe and M. Tomono: Viscoelastic Relaxations in Solutions of Poly-(glutamic acid) and Gelatin at Ultrasonic Frequencies. Biopolymers 5 (1967) p. 887.

26. T. K. Saksena, B. Michels and R. Zana: Absorption Ultrasonore et Transition Helice-Polate Statistique dans les Solutions Aqueuses de Polypeptides. J. Chim. Phys. 65 (1968) p. 597.

27. A. D. Barksdale and J. E. Stuehr: The Kinetics of the Helix-Coil Transition in Aqueous Poly-L-Glutamic Acid. J. Am. Chem. Soc. 94 (1972) p. 3334.

28. R. C. Parker, K. R. Applegate and L. J. Slutsky: Ultrasonic Study of the Helix-Coil Transition in Poly-L-Lysine. J. Phys. Chem. 70 (1966) p. 3018.

29. R. C. Parker, L. J. Slutsky and K. R. Applegate: Ultrasonic Absorption and the Kinetics of Conformational Change in Poly-L-Lysine. J. Phys. Chem. 72 (1968) p. 3177.

30. R. Zana and C. Tondre: Ultrasonic Studies in Relation to the Helix-Coil Transition in Aqueous Solutions of Poly-L-Lysine. Biopolymers 10 (1971) p. 2635.

31. R. Zana and C. Tondre: Ultrasonic Studies of Proton Transfer in Solutions of Poly (Lysine) and Poly (Ornithine). Implications for the Kinetics of the Helix-Coil Transition of Polypeptides and for the Ultrasonic Absorption of Proteins. J. Phys. Chem. 76 (1972) p. 1737.

32. G. G. Hammes and P. B. Roberts: Dynamics of the Helix Coil Transition in Poly-L-Ornithine. J. Am. Chem. Soc. 91 (1969) p. 1812.

33. R. Zana: On the Detection of the Helix-Coil Transition of Polypeptides by Ultrasonic Absorption Measurements in the Megahertz Range. Case of Poly-L-Glutamic Acid. J. Am. Chem. Soc. 94 (1972) p. 3646.

34. E. L. Carstensen, Kam Li and H. P. Schwan: Determination of the Acoustic Properties of Blood and its Components. J. Acoust. Soc. Amer. 25 (1953) p. 286.

35. E. L. Carstensen and H. P. Schwan: Acoustic Properties of Hemoglobin Solutions. J. Acoust. Soc. Amer. 31 (1959) p. 305.

36. H. Gramberg: Doctoral Dissertation, Johann-Wolfgang-Goethe Universitat, Frankfurt (1956).

37. P. D. Edmonds: Ultrasonic Absorption of Hemoglobin Solutions. Biochim. Biophys. Acta 63 (1962) p. 216.

38. F. Schneider, F. Muller-Laudau and A. Mayer: Acoustical Properties of Aqueous Solutions of Oxygenated and Deoxygenated Hemoglobin. Biopolymers 8 (1969) p. 537.

39. P. D. Edmonds, T. J. Baulk, III, J. F. Dyro and M. Hussey: Ultrasonic Absorption of Aqueous Hemoglobin Solutions. Biochim. Biophys. Acta 200 (1970) p. 174.

40. L. W. Kessler and F. Dunn: Ultrasonic Investigation of the Conformal Changes of Bovine Serum Albumin in Aqueous Solutions. J. Phys. Chem. 73 (1969) p. 4256.

41. L. W. Kessler, W. D. O'Brien, Jr. and F. Dunn: Ultrasonic Absorption in Aqueous Solutions of Polyethylene Glycol. J. Phys. Chem. 74 (1970) p. 4096.

42. W. D. O'Brien, Jr.: The Absorption of Ultrasound in Aqueous Solutions of Biological Polymers. Ph.D. Thesis, University of Illinois (1970).

43. W. D. O'Brien, Jr. and F. Dunn: Ultrasonic Absorption Mechanisms in Aqueous Solutions of Bovine Hemoglobin. J. Phys. Chem. 76 (1972) p. 528.

44. W. D. O'Brien, Jr. and F. Dunn: Ultrasonic Absorption by Biomacromolecules. In Interaction of Ultrasound and Biological Tissues, ed. by J. M. Reid and and M. R. Sikov, p. 13, DHEW Publication (FDA) 73-8008 BRH/DBE 73-1, U.S. Gov't. Printing Office, Washington,D. C. (1972).

45. S. A. Goss and F. Dunn: Concentration Dependence of Ultrasonic Absorption in Aqueous Solutions of Bovine Serum Albumin. In IEEE Ultrasonics Symposium Proceedings, (1974) p. 65.

46. F. W. Kremkau, E. L. Carstensen and W. G. Aldridge: Macromolecular Interaction in the Absorption of Ultrasound in Fixed Erythrocytes. J. Acoust. Soc. Amer. 53 (1973) p. 1448.

47. E. L. Carstensen: The Mechanism of the Absorption of Ultrasound in Biological Materials. IRE Trans. Med. Elect. ME-7 (1960) p. 158.

48. J. Lang, C. Tondre and R. Zana: Effect of Urea and Other Organic Substances on the Ultrasonic Absorption of Protein Solutions. J. Phys. Chem. 75 (1971) p. 374.

49. H. Pauly and H. P. Schwan: Mechanism of Absorption of Ultrasound in Liver Tissue. J. Acoust. Soc. Amer. 50 (1971) p. 692.

50. L. W. Kessler: The Absorption of Ultrasound in Aqueous Solutions of Bovine Serum Albumin and Polyethylene Glycol. Ph.D. Thesis, University of Illinois (1968).

51. R. Zana and J. Lang: Effect of pH on the Ultrasonic Absorption of Aqueous Solutions of Proteins. J. Phys. Chem. 74 (1970) p. 2734.

52. F. Dunn and L. W. Kessler: Further Remarks on the Ultrasonic Properties of Bovine Serum Albumin Solutions. J. Phys. Chem. 74 (1970) p. 2736.

53. M. Hussey and P. D. Edmonds: Proton-Transfer Reactions. A Mechanism for the Absorption of Ultrasound in Aqueous Solutions of Proteins. J. Phys. Chem. 75 (1971) p. 4012.

54. R. Zana and J. Lang: Interaction of Ultrasound and Amniotic Liquid. Ultrasound in Med. Biol. 1 (1974) p. 253.

55. L. J. Slutsky and R. D. White: Proton Transfer and Acoustic Absorption in Protein Solutions. In Interaction of Ultrasound and Biological Tissues--Workshop Proceedings, ed. by J. M. Reid and M. R. Sikov, p. 27, DHEW Publication (FDA) 73-8008 BRH/DBE 73-1, U.S. Gov't. Printing Office, Washington, D.C. (1972).

56. S. A. Hawley, L. W. Kessler and F. Dunn: Ultrasonic Absorption in Aqueous Solutions of High-Molecular-Weight Polysaccharides. J. Acoust. Soc. Amer. 38 (1965) p. 521.

57. S. A. Hawley, L. W. Kessler and F. Dunn: Absorption in Aqueous Solutions of High-Molecular-Weight Polysaccharides. J. Acoust. Soc. Amer. 38 (1965) p. 1064.

58. J. C. Kendrew: Side-Chain Interactions in Myoglobin. In Enzyme Models and Enzyme Structure, Brookhaven Symp. Biol. 15 (1962) p. 216.

59. G. Braunitzer, K. Hilse, V. Rudloff and N. Hilschmann: The Hemoglobins. Advan. Protein Chem. 19 (1964) p. 1.

60. S. A. Hawley and F. Dunn: Ultrasonic Absorption in Aqueous Solutions of Dextran. J. Chem. Phys. 50 (1969) p. 3523.

61. S. A. Hawley and F. Dunn: Ultrasonic Relaxation Spectra in Aqueous Solutions of Dextran and Polyethylene Glycol. J. Chem. Phys. 52 (1970) p. 5497.

62. P. E. Rouse: A Theory of the Linear Viscoelastic Properties of Dilute Solutions of Coiling Molecules. J. Chem. Phys. 21 (1953) p. 1272.

63. W. D. O'Brien, Jr.: Ultrasonic Absorption in Aqueous Solutions of Chondroitin Sulfate. J. Acoust. Soc. Amer. 57 (1975) p. 956.

64. J. Lang, J. Sturm and R. Zana: Ultrasonic Absorption in Aqueous Solutions of Nucleotides and Nucleosides. I. Effect of pH and Concentration. J. Phys. Chem. 77 (1973) p. 2329.

65. J. Lang, J. Sturm and R. Zana: Ultrasonic Absorption in Aqueous Solutions of Nucleotides and Nucleosides. II. Kinetics of Proton Exchange in Adenosine 5' Monophosphate. J. Phys. Chem. 78 (1974) p. 80.

66. J. Lang and R. Cerf: Absorption Ultrasonore dans des Solutions c/Acide Desoxyribonucleique: Etude de la Denaturation Alcaline. J. Chim. Phys. 66 (1969) p. 81.

67. J. Sturm, J. Lang and R. Zana: Ultrasonic Absorption of DNA Solutions: Influence of pH. Biopolymers 10 (1971) p. 2639.

68. W. D. O'Brien, Jr., C. L. Christman and F. Dunn: Ultrasonic Investigation of Aqueous Solutions of Deoxyribose Nucleic Acid. J. Acoust. Soc. Amer. 52 (1972) p. 1251.

69. R. Cerf, S. Candau and R. Zana: Sur les Anomalies des Spectres d'Absorption Ultrasonores dans les Solutions de Hauts Polymeres: Essai d'Interpretation. Rep. 4th Int. Cong. Acoustics, p. J13, Copenhagen (1962).

70. R. Cerf: Sur la Resonance Stochastique. Rep. 5th Int. Cong. Acoustics, p. C13, Leige (1965).

71. R. Cerf: Compt. Rend. 260 (1965) p. 3910, 5225.

72. R. Zana and R. Cerf: Absorption of Ultrasound by Solutions of Rigid Particles. J. Chim. Phys. 61 (1964) p. 1203.

73. C. L. Christman: Methods for Measuring the Ultrasonic Absorption of Aqueous Solutions of Deoxyribonsenucleic Acid. M. S. Thesis, University of Illinois (1970).

74. G. G. Hammes and P. B. Roberts: Ultrasonic Attenuation Measurements in Phospholipid Dispersions. Biochim. Biophys. Acta 203 (1970) p. 220.

75. D. E. Goldman and T. F. Hueter: Tabular Data of the Velocity and Absorption of High-Frequency Sound in Mammalian Tissues. J. Acoust. Soc. Amer. 28 (1956) p. 35; 29 (1957) p. 655.

76. L. W. Kessler: VHF Ultrasonic Attenuation in Mammalian Tissue. J. Acoust. Soc. Amer. 53 (1973) p. 1759.

77. W. J. Fry: Mechanism of Acoustic Absorption in Tissue. J. Acoust. Soc. Amer. 24 (1952) p. 412.

78. T. F. Hueter: Viscoelastic Losses in Tissues in the Ultrasonic Range. WADC Tech. Rep. 57-706 (1958).

79. A. Smith and H. P. Schwan: Acoustic Properties of Cell Nuclei. J. Acoust. Soc. Amer. 49 (1971) p. 1329.

80. F. Dunn and J. K. Brady: Pogloshchenie Ul'trazvyeka V Biologicheskikh Sredakh. Biofizika 18 (1973) p. 1063; Translation: Ultrasonic Absorption in Biological Materials. Biophysics 18 (1974) p. 1128.

81. F. Dunn and J. K. Brady: Temperature and Frequency Dependence of Ultrasonic Absorption in Tissue. In Proc. 8th Intl. Congress on Acoustics, Goldcrest Press, Trowbridge, Wilts. (1974) Vol. I, p. 366c.

82. T. C. Robinson and P. P. Lele: An Analysis of Lesion Development in the Brain and in Plastics by High Intensity Focused Ultrasound at Low Megahertz Frequencies. J. Acoust. Soc. Amer. 51 (1972) p. 1333.

83. T. Kishimoto: Ultrasonic Absorption in Bones. Acustica 8 (1958) p. 179.

84. W. J. Fry and F. Dunn: Ultrasound: Analysis and Experimental Methods in Biological Research. In Physical Techniques in Biological Research, ed. by W. L. Nostuk, Academic Press, Inc., New York (1962) Vol. 4, Chapt. 6, p.261.

85. T. F. Hueter: Messung der Ultraschallabsorption in Menschlichen Schädelknochen und ihre Abhängigkeit von der Frequenz. Naturwissenschaften 39 (1952) p. 21.

86. L. Adler and K. V. Cook: Ultrasonic Wave Propagation in Animal Bones. J. Acoust. Soc. Amer. 58 (1975) p. 1107.

87. H. Thiesmann and F. Pfander: Uber die Durchlassigkeit des Knochens fur Ultraschall. Strahlenterapie 80 (1949) p. 607.

88. S. B. Lang: Ultrasonic Method for Measuring Elastic Coefficients of Bone and Results on Fresh and Dried Bovine Bones. IEEE Trans. Biomed. Eng. BME-17 (1970) p. 101.

89. K. T. Dussik and D. J. Fritch: Progress Report, NIH Gcant A 545, Boston Dispensary, New England Med. Ctr., Sept. (1956).

90. Z. E. Begui: Acoustic Properties of the Refractive Media of the Eye. J. Acoust. Soc. Amer. 26 (1954) p. 365.

91. F. Dunn and W. J. Fry: Ultrasonic Absorption and Reflection by Lung Tissue. Phys. Med. Biol. 5 (1961) p. 401.

92. J. M. Reid: Ultrasonic Diagnostic Methods in Cardiology. Ph.D. Thesis, University of Pennsylvania (1965).

93. F. Dunn: Attenuation and Speed of Sound in Lung. J. Acoust. Soc. Amer. 56 (1974) p. 1638.

94. T. J. Bauld and H. P. Schwan: Attenuation and Reflection of Ultrasound in Canine Lung Tissue. J. Acoust. Soc. Amer. 56 (1974) p. 1630.

95. P. D. Edmonds and T. J. Bauld: Unpublished data.

96. J. L. Hunter and P. Durdel: Personal communication.

CHAPTER VII

SCATTERING OF ULTRASOUND BY HUMAN TISSUE

C. R. Hill, R. C. Chivers, R. W. Huggins and D. Nicholas

1. INTRODUCTION

"...The induction which is to be available for the discovery and demonstration of sciences and arts, must analyse nature by proper reflections and exclusions...but in order to furnish this induction or demonstration well and duly for its work, very many things are to be provided which no mortal has yet thought of; insomuch that greater labour will have to be spent in it than has hitherto been spent."

Francis Bacon 1561-1626

The processes of interaction between ultrasound and human tissues are fundamentally of two kinds: absorption processes, in which acoustic energy is transformed to thermal or other forms of energy which appear at the site of interaction, and scattering processes, in which energy is re-radiated from the interaction site with some change, relative to the incident wave, in the amplitude and directionality of the wave front and also, possibly, in its phase and frequency. In considering the biological and therapeutic effect of ultrasound it is the absorption processes that are of primary interest, whereas the major proportion of both established and proposed diagnostic techniques are based on the exploitation of scattering processes.

In spite of its fundamental importance over a large area of the field of medical ultrasonics, the scientific basis of acoustic scattering processes by human tissues has hitherto received very little explicit attention. Much of the historical development of pulse echo diagnostic techniques has been guided by the assumption that detectable echoes are the result of a process analogous to specular reflection of light, although it is now becoming plain that specular reflection is only an extreme case of a much more general scattering process. The purpose of this chapter is to bring together the rather small amount of knowledge that at present exists on the acoustic scattering properties of tissue and to indicate some of the very exciting possibilities in the field of medical diagnosis that may result from a fuller and more systematic understanding of the basic scattering process.

It is interesting that, in the pioneering work on pulse-echo visualization of tumors carried out by Wild and Reid,[1] echoes were demonstrated that almost certainly arose as a result of an omni-directional scattering process, rather than by specular reflection. However, this point was apparently lost sight of with the demonstration by Howry of the possibility of producing rather clear anatomical outline diagrams, by use of the technique of compound scanning, together with suppression of low-level echoes, and emphasis moved towards the specular reflection concept, which has dominated the design and application of pulse-echo scanners for much of the past two decades. During this time there has certainly been some awareness of the true situation and of its potentialities: W. J. Fry in 1958[2] commented that "Many problems remain for study in this field (of pulse echo diagnosis), in particular the mechanism of reflection of the sound as related to tissue structure .... this area of investigation contains possibilities for tremendous advances in the near future." In the same year, Donald,[3] who has often commented on the possible significance of the particular "speckled" appearance that he obtains on scans of hydatidiform mole; remarked that "The temptation to try to distinguish benign from malignant tissue by such simple, quick and harmless means is well nigh irresistible.... our experience indicates that pre-operative diagnosis of histological structure is still far off, although such a possibility in the future is an exciting prospect." However, it is only in the past two or three years, helped perhaps by the success of grey-scale echography in visualizing tissue structure detail corresponding to low-level scattered echoes, that the significance of scattering in the diagnostic process has received more general recognition.

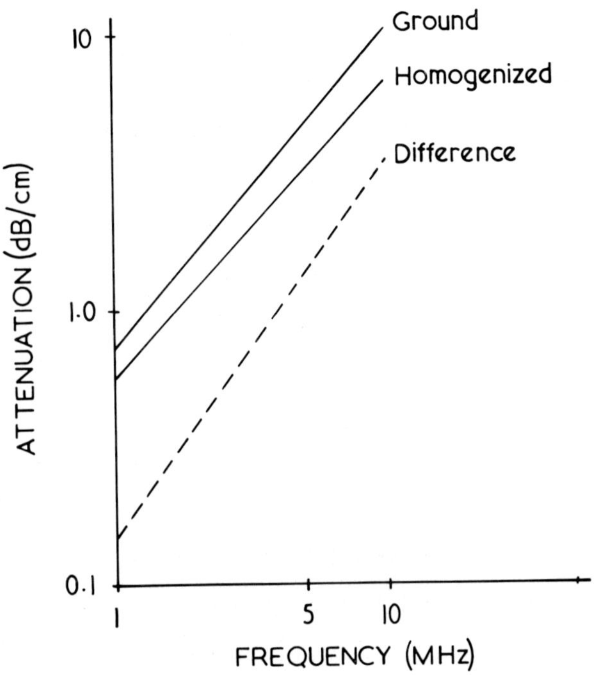

Fig. 1 Attenuation of "ground" and "homogenized" liver as a function of frequency. Data of Pauly and Schwann[4] replotted.

Very little explicit quantitative data yet exists on the magnitude of the contributions of scattering to the total acoustic attenuation coefficient of human tissues, although some indications can be derived from analysis of published measurements. Pauly and Schwan [4] for example published a series of separate measurements on what they referred to as "absorption" (more correctly "attenuation" which includes contributions from true absorption and scattering) of ground and homogenized beef liver at various frequencies. Their data, re-plotted on a single graph, are shown in Figure 1, from which it appears that the decrease in attenuation due to homogenization is about 21% at 1 MHz, rising to 34% at 10 MHz. This difference is presumably due largely to the difference in acoustic inhomogeneity, and thus in scattering coefficient, between the liver preparations.

The general phenomenon of scattering is of importance in many branches of physics and a number of particular results are very well known. Among these are the classical work of Rayleigh on the relationships of the amplitude of a wave scattered by a sphere and the ratio of wave-length to radius of the sphere (a result which "explains" the blueness of the sky), and the work of von Laue and Bragg on the scattering of X-rays by three-dimensional arrays of scatterers in the form of crystals (the basis of X-ray crystallography and all that has stemmed from it). As discussed below, both of these aspects of scattering are of considerable relevance in medical ultrasonics.

Any thoroughly satisfactory treatment of acoustic scattering will include a discussion of the phenomena from a theoretical point of view and an analysis of experimental observations in the light of such theory. Human tissues, however, are well known to be highly complex in structure, and it is to be expected that this complexity will be reflected in the properties of tissues as acoustic scatterers. Exact theoretical descriptions are, therefore, likely to be impracticable and only rather general predictions will be possible. One useful line of approach in this situation, however, will be to use tissue models as intermediaries between theoretical and experimental work on tissues: it may be possible to construct simple scattering objects or matrices which have some reasonable structural relationship with particular human tissues, and which are, at the same time, of sufficient simplicity to be amenable to theoretical treatment. By such means it may be possible to achieve at least a general understanding of the observed acoustic properties of a human tissue in relation to parameters describing its characteristic mechanical structure. In this chapter we shall attempt to outline both theoretical and experimental approaches to the subject and demonstrate the constructive relationship between them, even though, as will become evident, at the present state of development of the subject much of this relationship is far from being clearly established. The practical emphasis of the work of our own research group in this field, and of that of most of the rather limited amount of other relevant work of which we are aware and able to review, is towards improving the accuracy and specificity of the pulse-echo diagnostic process. In particular there is a strong interest in working towards techniques for remote characterization of particular defined regions of human tissue in terms of their structure and pathology.

The organization of this chapter is as follows. First, we consider, from a rather general point of view, the theory of acoustic scattering by acoustically inhomogeneous matrices. Here it is shown that the macroscopic phenomena of diffraction and specular reflection comprise only one aspect of the total scattering process, and particular consideration is also given to predictions that can be made about the frequency dependence of such scattering. As this treatment is one that does not appear to have been given before in this form, it is followed by a brief appraisal of its relevance to some instructive examples of acoustic scattering, taken from outside the medical context. This then leads to a more detailed discussion of the design of experimental techniques and resulting practical observations of the frequency dependence of acoustic scattering by tissues and tissue models. At this point we introduce the specific concept that tissues may comprise spatially ordered, or semi-ordered, arrays of scatterers. Some theoretical consequences of this concept are worked out, and these are compared with further sets of experimental observations on tissue models and real tissues.

The mathematical symbols used in this chapter are listed in Appendix B.

2. ACOUSTIC SCATTERING BY INHOMOGENEOUS STRUCTURES

In each of the many physical guises that scattering occurs, the essential phenomenon is a change in the amplitude, frequency, phase, velocity and direction of propagation of a wave as a result of a spatial or temporal nonuniformity of the medium of propagation. In the particular case of acoustic waves scattered by complex inhomogeneous structures such as human tissues, complete analytical description of this phenomenon will necessarily be highly complex but the rewards for its achievement may be very great, particularly in terms of the diagnostic interpretations that may be derived from measured values of the scattered wave parameters. At present such an analysis is no more than a distant prospect and it is the purpose of the present section to bring together into a consistent context the various existing strands of theoretical groundwork of the subject and to indicate the extent to which these may suggest constructive directions for experimental investigation. No significant attempt to synthesize a theoretical approach to acoustic scattering by human tissues has been published hitherto and we have therefore drawn widely on various relevant sources in fundamental and applied theory.

Provided that continuous waves are considered, and the behavior of the scatterers is not affected by the incident waves, the local behavior of a medium is described by the wave equation:

$$\nabla^2\Psi = \frac{1}{c^2} \cdot \frac{\partial^2\Psi}{\partial t^2} \qquad (2.1)$$

The conditions at infinity control the form of the solutions and the boundary conditions at the scatterers specify their shape, structure and distribution.

The linearity of the wave equation and the consequent validity of the principle of superposition has enabled use to be made of the single body scattering solution in multiple scattering phenomena. Three of the four theoretical approaches to multiple scattering use the single body solution, the fourth considering the medium as an inhomogeneous compound body. The single body solution is also of importance in considering the tests made on transducer performance, as has been discussed by Chivers.[5]

In view of the evidence[6] that there is no measurable biological effect produced by the levels of ultrasound used in diagnosis, it will be assumed that the physical processes of attenuation and scattering are stationary. For studies in vitro, it would be reasonable to assume that the processes are strictly stationary, but the effects of breathing and circulation in vivo suggest that they should be treated as weakly stationary. In either case the ergodic principle may be applied. This immediately makes available the theoretical work done in the field of hydroacoustic transmission. The results are generally produced as mean square fluctuations with time of amplitude, phase and scattering angle, and as such are inapplicable since for any given *in vitro* experimental arrangement, the measured values of any physical quantities will remain unchanged with time. However, by application of the ergodic hypothesis, the mean square fluctuations of a quantity with time are equivalent to the mean square variation expected in different measurements of that quantity. Thus, the results obtained by these workers may be used to give some idea of the variation that might be expected in the measurements on biological systems, provided that meaningful values can be given to some of the parameters involved.

The main reason for the lack of investigation of the physics of sound wave transmission in biological tissue is almost as much due to the absence of a satisfactory acoustic model for the tissue as it is due to the history of ultrasonic diagnostics. Two distinct theoretical approaches are available. A given tissue may be considered as particulate in nature, consisting of discrete inclusions in a homogeneous matrix, or it may be considered as an inhomogeneous continuum wherein the physical properties vary spatially but in a strictly continuous manner. These will be discussed under the heading 'multiple scattering' and 'inhomogeneous continuum scattering', respectively.

2.1 Multiple Scattering

The simplest model of an inhomogeneous medium is one which consists of an array of identical discrete scatterers set in a homogeneous matrix. To a first (Born) approximation, the scattering will simply be the sum of the signals scattered from each of the obstacles, i.e., the sum of the single body scattering solutions for each obstacle considered independently. If the obstacles are arranged regularly, e.g., the rectangular array considered by Rayleigh[7] or the atoms in a crystalline structure,[8] interference effects will be observed which can be used to gain information on the form of the array and the characteristics of the obstacles (see Section 5, below).

A closer approximation than that above will allow for multiple scattering. This can be done by one of three methods: a summation of successive orders of scattering (adding the components from 1,2,3 ... etc. interactions), a procedure in which the total multiple scattering field is postulated and evaluated by applying the criterion of self-consistency, and, thirdly, solution of the boundary value problem for the scatterers as a 'compound body'. The range of problems to which this last method can successfully be applied is very limited, often relying upon analytical forms for the distribution of scatterers.

Twersky[9] has provided a comprehensive review of the above analyses for one, two and three-dimensional distributions of both regular and random arrays, but some discussion of the terminology and basic principles is enlightening. Before the work of Foldy[10] most of the multiple scattering theory had been developed using the Boltzmann integro-differential equation for transport processes. This classical approach is valid for a wide range of physical problems - molecular scattering in gases, scattering of charged particles and γ-rays - but is essentially a 'ray' approach. It holds good only when there is no correlation between the positions of the scatterers and where the distance between scatterers is large compared to the wavelength. When the distance between scatterers becomes comparable to the wavelength, correlation effects between scatterers make an important contribution to the scattering, as in the case of sound waves being scattered by droplets in a fog. In such situations the wave treatment of Foldy and subsequent workers is needed.

Although Foldy only dealt with distribution of identical isotropic scatterers, his distinction between coherent and incoherent scattering has been useful to many other authors. Consider an irradiating wave Ψ_o, and a total scattered wave Ψ_s. Ψ_s will be a function of both the position of the observation and the positions of the individual scatterers. The radiation density associated with it will be the mean square value of the function, $\langle|\Psi_s|^2\rangle$. Allowing Ψ_s to interfere with Ψ_o produces for the total radiation density:

$$\langle|\Psi_o + \Psi_s|^2\rangle = |\Psi_o + \langle\Psi_s\rangle|^2 + \langle|\Psi_s|^2\rangle - |\langle\Psi_s\rangle|^2 \qquad (2.2)$$

The form of this suggests that the mean value of the scattered wave $\langle|\Psi_s|\rangle$ is <u>coherent</u> with the incident wave Ψ_o, i.e., $\langle\Psi_s\rangle$ bears a definite phase relation to Ψ_o . Further analysis shows that the coherent scattering represented by $\langle \Psi_s \rangle$ satisfies the wave equation for a continuous medium. The velocity of sound, c, in this medium is different from that for the original medium without scatterers, c_o, and is given by

$$\frac{1}{c^2} = \frac{1}{c_o^2} + \frac{4\pi G(\underline{r},\omega)}{\omega^2} \tag{2.3}$$

It is a function of the position of the observer, $\underline{r}$, and is in general complex, implying attenuation (see Section 2.3.2 where the scattering function G is also defined (2.28)). This coherent portion of the scattering leads to the macroscopic phenomena of refraction and specular reflexion. It arises from considering only single scattering processes from independent scatterers.

The term $\langle |\Psi_s|^2 \rangle - |\langle \Psi_s \rangle|^2$ in equation 2.2 is a contribution to the total radiation density, caused only by scattering processes, single and multiple. Furthermore, it has to be added as an 'intensity' rather than as an 'amplitude', implying that it is of arbitrary phase relative to Ψ_o, i.e., it is <u>incoherent</u>. This portion of the scattering satisfies an equation similar to the Boltzmann transport equation and has contributions proportional to both $\langle |\Psi_s|^2 \rangle$ and a scattering cross section per unit volume, for every point in the medium.

Lax[11] has extended Foldy's work[10] by considering the wave aspects of particle scattering for situations involving anisotropic and inelastic scattering and scattering by partially ordered arrays. More recently Glotov[12] in the Soviet Union, has used definitions of coherence and incoherence comparable to Foldy's to discuss plane and spherical wave scattering from a layer in the ocean. For high enough concentrations of scatterers the coherent component may be dominant. Kur'yanov[13] has generalized Glotov's work[12] on pulsed emission. Considering a pulse defined only for $0 \leq t \leq T$, and of arbitrary shape, he finds that a medium with a meaningful average scatterer density produces a negligible amount of coherent scattering. However, this component becomes noticeable when there are boundaries present or when there are irregularities in the mean particle distribution. For layerwise inhomogeneous media, he quotes

$$\frac{\text{coherent intensity}}{\text{incoherent intensity}} \leq \frac{\pi}{k^2} n_{max} \frac{cT}{2} \qquad (2.4)$$

where n_{max} is the maximum scatterer density.

2.2 Inhomogeneous Continuum Scattering

The ultrasonic physical properties of the microscopic components of tissue are not well known - indeed the macroscopic properties still require objective evaluation.[14] If discrete scatterers are considered, the only candidates for selection must be organelles, cells and groupings of cells. Without knowledge of the properties of any of these relative to the containing medium, the continuum approach may well be realistic in spite of the fact that some vigorous gradients may be involved. The scattering situation has been discussed in detail elsewhere (Chivers[5]), the results being reproduced here to form the basis of the discussion.

The theory of wave propagation in continuous inhomogeneous media has been lucidly discussed by Chernow,[15] and duplication of his analysis would be purposeless. However, his primary concern was with the problems of hydroacoustics and the analysis requires some modifications to be applicable to the present case. It is intended, therefore, simply to quote his results on the ray approximation and fluctuations of various quantitites, where these are enlightening. His small section on scattering is closely paralleled by that of Chivers,[5] the results of which are quoted in the present text. It will be implicitly assumed that there is no turbulence present and (less accurately for <u>in vivo</u> work) that the positions of the scatterers do not change with time.

For the description of the statistical properties of an inhomogeneous medium, the correlation function of the variation (μ) of the refractive index (n) is a fundamental quantity:

$$N_{12} = \langle \mu(x_1, y_1, z_1, t)\, \mu(x_2, y_2, z_2, t) \rangle \qquad (2.5)$$

where $n = 1+\mu$ and the brackets $\langle \; \rangle$ mean an average over time or over the possible values of μ (using the ergodic principle). (2.6)

If the inhomogeneities are truly (i.e., isotropically) random in distribution, the value of the function will depend only on the distance r between (x_1, y_1, z_1) and (x_2, y_2, z_2). The correlation coefficient N is defined by:

$$N = \frac{N_{12}}{\langle \mu^2 \rangle} \qquad (2.7)$$

The form of the correlation coefficient can only be determined theoretically for homogeneous isotropic turbulence. In other cases, e.g., hydroacoustics, an empirical determination is required. Liebermann[16] measured N for temperature fluctuations in the ocean, finding his data fitted a curve of the form $N(x) = e^{-|x|/a}$, where a is a measure of the linear scale of the inhomogeneities. However, as Chernow[15] points out, this has a finite derivative for x = 0, which corresponds to $\mu(x)$ being a discontinuous function. A better function for theoretical work which also fits Liebermann's data is:

$$N(x) = e^{-x^2/a^2} \qquad (2.8)$$

which for a statistically isotropic medium becomes:

$$N(r) = e^{-r^2/a^2} \qquad (2.9)$$

This last form is the one which will be used in subsequent discussion.

2.2.1 The Ray Approximation

If the scale of the inhomogeneities is much greater than the wavelength (e.g., human organs), the ray approximation may be used for regions of dimension L where

$$\lambda L < a^2 \qquad (2.10)$$

Chernow[15] has developed this theory to obtain expressions for the angular deviation, lateral displacement and phase fluctuation of a wave propagating in such a medium (expressed as r.m.s. values) and for the mean amplitude fluctuations that may be expected. Substitution of arbitrary but reasonable values for the parameters concerned (Chivers,[5]) gives figures in accordance with the limitations on organ visualization discovered by experience or less sophisticated arguments.

2.2.2 Diffraction Theory

If the scale of the inhomogeneities is of the order of the wavelength, or the region for which a theory is required (L) is too large, the ray approximation must cede to diffraction theory. The inhomogeneities which will be discussed from a theoretical point of view in this section will thus correspond to cells or groupings of cells within organs. The theory at present is only tractable for situations where the parameters in the wave equation deviate only slightly from their mean value. There are two ways of using the method of small perturbations, which is the best approach to the problem. The usual form has been used by Ellison[17] and Mintzer[18-20] while Chernow[15] and Obukhov[21] use Rytov's method. Chernow obtains:

$$\langle |p_1|^2 \rangle = \frac{VA_o^2 \langle (\Delta T)^2 \rangle k^4}{4\pi^2 R^2} \cdot \left[\frac{1}{c_o} \cdot \left(\frac{\partial c}{\partial T}\right)_{p_o} + \frac{1}{\rho_o} \left(\frac{\partial \rho}{\partial T}\right)_{p_o} \sin^2 \frac{\theta}{2} \right]^2 \cdot \int_V N(\underline{r}) e^{i\underline{k}_o \cdot \underline{r}} dv \tag{2.11}$$

for the average scattered intensity $\langle |p_1|^2 \rangle$ caused by variations in density and sound velocity. He thus obtains agreement with Pekeris,[22] Staras,[23] Booker and Gordon[24] and Fejer.[25] His expression (equation 2.11) describes variations due to changes in the density and velocity of propagation of the waves, and although these are the two easiest parameters to measure, they are not independent. To a good approximation, the density and adiabatic bulk modulus may be regarded as independent and one may use[78,79] these parameters in the discussion of the scattering theory. This has the advantage of showing the remarkable similarity between this theory for an inhomogeneous continuum, and Rayleigh's theory for discrete inhomogeneities. He obtains:

$$\frac{\langle |p_1|^2 \rangle}{A_o^2} = \frac{k^4 V}{16\pi^2 R^2} \left[\frac{\langle (\Delta\rho)^2 \rangle}{\rho_o^2} \cdot \cos^2 \theta + \frac{\langle (\Delta m)^2 \rangle}{m_o^2} \right] a^3 \cdot \pi^{3/2} \cdot \exp(-k^2 a^2 \sin^2 \frac{\theta}{2}) \tag{2.12}$$

which can be compared with Rayleigh's[26] expression:

$$\left(\frac{\Delta s}{s_o^2}\right)^2 = \frac{\pi^2 V}{\lambda^4 R^2}\left[\frac{\langle(\Delta\rho)^2\rangle}{\rho_o^2}\cdot\cos^2\theta + \frac{\langle(\Delta m)^2\rangle}{m_o^2}\right]\cdot V \qquad (2.13)$$

The limits of applicability are defined to be $a \ll L \ll R$ and $a \ll \lambda \ll R$. If, in addition the condition that $a \ll \lambda$ be imposed upon equation 2.12, the exponential disappears and the form is strikingly like Rayleigh's.

The angular dependence of the first term is of interest. For $\theta<1$, we find

$$\cos^2\theta \simeq 1 - \theta^2 + \frac{\theta^4}{3} - \frac{2\theta^6}{45} \qquad (2.14)$$

and

for large inhomogeneities ($ka \gg 1$) the scattering is strongly directional, most of the energy being contained within the angle:

$$\theta = \frac{1}{ka}\ . \qquad (2.15)$$

Scattering cross sections for various cases have been evaluated in Section 2.3.4 and Appendix A. It is important, however, to stress the implicit assumption underlying the work in this section that the inhomogeneities differ only slightly in their physical properties from the mean physical properties of the medium in which they occur. Thus, whereas the analysis may be applicable to a suspension of drops of one liquid in another, it need not be valid for suspensions of solids in liquids or liquid-like media where the multiple scattering discussion of Section 2.1 may be more appropriate.

2.3 Scattering Cross-Sections

2.3.1 Cross-Sections of Individual Scatterers

A single discrete scatterer in a homogeneous medium irradiated with a wave Ψ_o will give rise to a scattered wave, Ψ_s, such that

$$\Psi_s = g(a,\omega)\ \Psi_o \qquad (2.16)$$

The scattered wave is proportional to the incident wave, the constant of proportionality being a function of frequency, ω, and of some parameter of the scatterer, a. Typically, a would be the radius of a spherical scatterer.

The total wave at $\underline{r}$ due to a scatterer at $\underline{r}_1$ will be

$$\Psi(\underline{r}) = \Psi_o(\underline{r}) + E(\underline{r}, \underline{r}_1) \cdot g(a,\omega) \cdot \Psi_o(\underline{r}_1), \tag{2.17}$$

where $$E(\underline{r}, \underline{r}_1) = \frac{\exp(-ik|\underline{r} - \underline{r}_1|)}{|\underline{r} - \underline{r}_1|} \tag{2.18}$$

The function E describes the propagation of the wave from $\underline{r}$ to $\underline{r}_1$, and in this instance (equation 2.18) assumes a scattered field that is spherically symmetrical.

The total flux* of the wave flowing into a sphere centered on the scatterer is found by integrating Ψ as defined by equation 2.17 over the sphere. This is equal to

$$4\pi(g^*_1 - g_1)\, A^2 - 4\pi\, ik|g_1|^2 A^2 \tag{2.19}$$

for an incident plane wave $\Psi_o(\underline{r}) = A\exp(-i\underline{k}\cdot\underline{r})$. (2.20)

The flux arising from scattering is $(\Psi_s^*\nabla\Psi_s - \Psi_s\nabla\Psi_s^*)$ where Ψ_s is given by the second term in equation 2.17. This flux is

$$+\, 4\pi\, ik|g_1|^2 A^2 \tag{2.21}$$

out of the sphere. Equation 2.19 thus consists of the flux of the incident wave entering the sphere (the first term) minus the flux egressing from the sphere as a result of scattering. It therefore constitutes the net flux incident on the scatterer that is absorbed or captured. Dividing it by the flux per unit area in the incident wave ($ik\, A^2$) we obtain the capture, or absorption <u>cross-section</u>:

$$\sigma_a = -4\pi|g_1|^2 - \frac{8\pi\, \mathrm{Im}(g_1)}{k}, \tag{2.22}$$

representing an area which, when multiplied by the incident flux per unit area, gives the flux that is absorbed and seen no more.

We can similarly define a <u>scattering cross-section</u>, σ_s, by dividing

* The flux per unit area of a function Ψ is defined as $(\Psi^*\nabla\Psi - \Psi\nabla\Psi^*)$ where the asterisk indicates a complex conjugate.

the total scattered flux (equation 2.21) by the incident flux per unit area:

$$\sigma_s = 4\pi |g_1|^2 \tag{2.23}$$

The sum of the absorption and scattering cross-sections is called the extinction or <u>attenuation cross-section</u>, σ_T, where

$$\sigma_T = \sigma_a + \sigma_s \tag{2.24}$$

This is a measure of the loss in flux of the ongoing incident wave due to the presence of the scatterer. Using equations 2.22 and 2.23, we see that

$$\sigma_T = -\frac{8\pi}{k} \operatorname{Im}(g_1) \tag{2.25}$$

Rearrangement of equations 2.23 and 2.25 gives:

$$g_1 = \frac{\sigma_s}{4\pi} - \frac{k^2\sigma_T^2}{64\pi^2} - \frac{ik\,\sigma_T}{8\pi} \tag{2.26}$$

2.3.2 Cross-section for volume distributions of discrete scatterers

If at the point $\underline{r}$, there are $n(a,\underline{r})da$ scatterers per unit volume having values of a between a and a+da, we can define an attenuation cross-section per unit volume:

$$\begin{aligned} S_T(\underline{r},\omega) &= \int \sigma_r(a,\omega)\, n(\underline{r},a)\, da \\ &= -\frac{4\pi}{k} \int \operatorname{Im}[g(a,\omega)]\, n(\underline{r},a)\, da \\ &= -\frac{4\pi}{k} \operatorname{Im}[G(\underline{r},\omega)] \ , \end{aligned} \tag{2.27}$$

$$\text{where } G(\underline{r},\omega) = \int g(a,\omega)\, n(\underline{r},a)\, da \ . \tag{2.28}$$

The scattering cross-section per unit volume is

$$\begin{aligned} S_s(\underline{r},\omega) &= \int \sigma_s(a,\omega)\, n(\underline{r},a)\, da \\ &= 4\pi \int |g(a,\omega)|^2 \cdot n(r,a)\, da, \end{aligned} \tag{2.29}$$

provided that the scatterers are similar, i.e., they have the same scattering function $g(a,\omega)$.

2.3.3 Attenuation of waves by discrete scatterers

The average value of the wave function satisfies the linear wave equation, as discussed in Section 2.1. Its solution is given by:

$$\langle \Psi(r) \rangle = A \exp(-i\, \underline{k}' \cdot \underline{r}) \tag{2.30}$$

where $|\underline{k}'| = k$, and $k'^2 = k^2 + 4\pi G$ (2.31)

from equation 2.30, assuming G is a constant, independent of position. If the number of scatterers per unit volume is small enough to allow expansion of the square root in equation 2.31, we have:

$$k' \simeq k + \frac{2\pi}{k} \mathrm{Re}(G) + \frac{2\pi i}{k} \mathrm{Im}(G) \tag{2.32}$$

and

$$\langle \Psi(\underline{r}) \rangle \simeq A \exp -i[r \cdot \cos\theta (k + \frac{2\pi}{k} \mathrm{Re}(G)] \cdot \exp [r \cdot \cos\theta \cdot \frac{2\pi}{k} \mathrm{Im}(G)] \tag{2.33}$$

The second exponential in equation 2.33 is an attenuation term, showing that the flux of $\Psi(r)$ in equation 2.30 is reduced by a factor

$$\exp [\frac{2\pi}{k} \mathrm{Im}(G)] = \exp (-S_T) = \exp (-S_a - S_s) \tag{2.34}$$

per unit distance along the wave. Writing the incident flux as I_o, and the net flux at a point x, I_x we obtain the familiar expression

$$I_x = I_o \exp (-x S_T) \tag{2.35}$$

for a plane wave travelling in the x-direction.

2.3.4 Attenuation of waves by scattering in an inhomogeneous continuum

Equation 2.35 will be valid for this case also. Here, however, we will assume a lossless medium, i.e., $S_a = 0$, thus

$$\frac{I_x}{I_o} = \exp(-xS_T) = \exp(-xS_a - xS_s) = \exp(-xS_s) ; \tag{2.36}$$

and evaluate S_s for the inhomogeneous continuum discussed in Section 2.2.2.

The detailed analysis is given in Appendix A, resulting in the expression:

$$S_s \simeq \frac{k^2 a \pi^{1/2}}{4} \left[\frac{\langle (\Delta m)^2 \rangle}{m_o^2} \left((1 - \exp(-k^2 a^2) \right) - \frac{\langle (\Delta\rho)^2 \rangle}{\rho_o^2} \exp(-k^2 a^2) \right] \tag{2.37}$$

From this we can see that if the acoustic radius ka<<1, S_s has two components: that arising from density variations being proportional to k^2 and that from elastic modulus variations being proportional to k^4. If, however, ka>>1, both components are proportional to k^2.

2.4 Discussion

In this section, consideration has been given to both the multiple scattering and continuum approaches to the analysis of scattering by tissues. Both may be valid and they may provide complementary sources of illumination on the nature of the scattering process.

The discussion of multiple scattering suggests that it may be useful to distinguish between coherent and incoherent components of the scattered energy. Of these the former apparently corresponds to the echo signals which arise from refraction and specular reflexion and which are exploited in conventional organ imaging techniques, while the latter may comprise the volume scattering that provides the basis for grey scale imaging of tissues.

Some development of the continuum approach has been attempted and its predictions in relation to the deviations to be expected from unperturbed ray acoustics, as a result of introduction of a scattering medium, are consistent with practical experience in imaging systems. At a more fundamentally interesting level, however, the formulation of the diffraction theory given here brings out a remarkable similarity between this analysis and Rayleigh's analysis of discrete scatterers.

Evaluation of scattering cross-sections gives a fourth power frequency dependence for the multiple scattering situation, while the continuum approach gives a fourth power or a square law dependence or a combination of both, depending upon the physical nature of the inhomogeneities. Numerical estimation of the magnitude of the cross-sections rests entirely upon the correct choice of the parameters involved. These are *a priori* unknown but may be found from comparison of measurements from tissues with those from models, and extension of experimental work in this direction would seem to be very worthwhile.

It is not at all clear that the results of continuous wave scattering theory for single scatterers, let alone scattering arrays, are applicable in situations where short pulses are used for interrogation although Chivers[5] argues in favor of this premise for the former case at least. The interference phenomena, which are a feature of continuous wave scattering and which to a large extent define the structure of the scattered spectrum, do not necessarily occur. Although a little work has been reported on the scattering of time-limited functions, the scattering has been defined in terms of a time function - a sequence of pulses - rather than the frequency spectrum of the scattered sequence of pulses which is of interest in the present discussion.

An answer to this problem may be achieved along the following lines. At the instant before the pulse first hits the scatterer, it may be considered to be transformed into a distribution of infinitely long sine waves which irradiate the target, are scattered and interfere independently. The distribution of the sine waves scattered in a particular direction will be the spectrum scattered in that direction. If this spectral conception of the scattering process and the temporal conception described in the literature are, as they should be, equivalent descriptions of the same physical process, the transform of the scattered spectrum should give the scattered sequence of pulses.

The application of any theory to the problem of characterizing tissue structure is made harder by the fact that the problem to be solved is an <u>inverse</u> scattering problem. That is to say that, given a particular form of scattering behavior, the first task is to describe the acoustical composition of the scatterer. Having achieved this, the next task is to systematize the information in a sufficiently simple form for it to be used by physically-minded doctors. The first task is complicated by the fact that there is no way of knowing that the inverse scattering solution is unique, and also that there is no way, bar intuition, of deciding which the parameters are that control the scattering process, nor the range of values they may be expected to take.

3. ACOUSTIC SCATTERING OUTSIDE THE MEDICAL CONTEXT

The theoretical discussion given in Section 2 provides a useful skeleton upon which to hang the loose-fitting clothes of experimental results. The degree of success of the tailoring, at this stage in the development of the art, can only be judged by the closeness with which clothes made by other people fit the same frame. Although scattering situations abound in nearly all branches of physics, this discussion of experimental achievement will be restricted to acoustic scattering contexts. These include in particular the realms of underwater acoustics and non-destructive testing.

3.1 Marine Scattering

The use of sound for echo-ranging at sea was begun by L.F. Richardson after the Titanic disaster of 1912. During the last twenty years, increasing use has been made of SONAR as a means of studying the location and behavior of various marine organisms, especially those that humans find palatable. The physical principle - that of recognizing an object or group of objects by the manner in which they scatter sound - is the theme of this chapter. The relevance of work in this field of marine scattering is slightly enhanced by the fact that the scatterers are often composed of biological tissue and, although the frequencies used are typically two orders of magnitude smaller than those commonly used in ultrasonic diagnostics, the successful use of scale models confirms the intuition that the ratio $a:\lambda$ is more significant than the absolute frequencies used.

3.1.1 Boundary Scattering

Marine scattering may be caused by bubbles, fish, vegetation, discontinuities in temperature, turbulence and irregularities of the sea surface and bed. The surface and sea floor scattering are nearly always much greater than the volume scattering (Vigoureux and Hersey,[27]); and although they are not directly relevant in the present context they will be important in considering reflection of ultrasound at tissue interfaces. An understanding of the mechanism of this process may well be a prerequisite for the efficient application of any successful outcome of the scattering approach adopted in the present discussion.

Fortuin[28] has reviewed scattering from the ocean surface in detail, but the fundamental fact that the surface approximates a perfectly reflecting, pressure release boundary diverts attention to results from the sea-bed as being a closer analogy to the biological situation. Breslau[29] has shown a correlation between the reflectivity and the particulate structure of the sea-floor, as could have been reasonably expected on the basis of the theory of Foldy[10] and Kur'yanov[13] discussed in Section 2.4. The calm induced by this agreement is disturbed by the report of McKinney and Anderson.[30] Working between 12 and 290 kHz they find a backscattering strength per unit area proportional to $f^{1.6}$, suggesting some sort of aggregation of the sand which fails to produce the Rayleigh f^4 dependence measured by Jitkovskiy and Volovova[31] between 5 and 20 kHz. Below 5 kHz, the latter workers report an absence of frequency dependence, indicating that irregularities of dimensions large compared with the wavelength have become the principle scatterers (Kur'yanov[32]). McKinney and Anderson[30] suggest that even if the frequency, grazing angle, pulse length, beam width, sediment type and sea-bed relief were all specified, it would be dangerous to try to describe the backscattering from the ocean floor by a single number. This is a caution that requires little modification for it to be applicable to the tissue interface situation.

3.1.2 Scattering by Marine Organisms

Scattering by marine organisms has largely been studied as a means of locating shoals of a particular type of fish, most workers using frequencies around 12 kHz. Two types of scattering behavior are observed on a depth-range recorder: hyperbolic echoes which are assumed to come from small groups of strong scatterers, and diffuse echoes corresponding to a large population of small scatterers.[33-35] On the basis of measurements of the quantity of zooplankton present, Bary[36] concludes that the diffuse scattering layer must be composed of fish. The increase with layer depth of the frequency of sound most efficiently reflected,[35,37,38] fits in with the evidence that the air-filled swimbladder of the fish is a major contributor to the scattering. Cushing[39] estimates that an air-bladder of 20% of the volume of a fish produces between 30% and 75% of the scattered signal amplitude due to the fish itself.

Laboratory measurements on isolated marine organisms tend to be of more interest than those made at sea. This is largely due to the measurements being made as a function of frequency for optimization of the frequency used at sea, or for target recognition by multiple frequency techniques.[39] Cushing's comment that the maximum scattered signal was obtained for a fish of length $L \simeq \lambda$ must have been published at almost the same time as Smith[40] published his work on the variation with frequency of the scattering cross-sections of fish, squid and shrimp. Working between 8 and 31 kHz, Smith found a general decrease in cross-section as the frequency increased. However, as expected from the behavior of regular bodies with $a \simeq \lambda$, and as Haslett[41,42] subsequently confirmed with fish, there are a good many maxima and minima present. Haslett's use of a scale model[41] enabled him to detect an increase of scattering cross-section with frequency at low frequencies (Rayleigh region) which Smith was not able to do.

One of the most interesting comments was made by Andreeva[43] on the frequency dependence of the scattering strength of a layer of marine organisms. At frequencies between 5 and 10 kHz, there is a sharp peak which he attributes to the resonant swimbladders of fish 5-10 cm long. Above 10 kHz there is a sharp increase with frequency which Andreeva attributes to the "scattering effect of animal tissue". He does not indicate whether he considers this to arise from the nature of the tissue itself or from the configurational aspects of a large number of pieces of fish flesh suspended in the sea.

3.2 Non-Destructive Testing

Medical diagnostic ultrasonics has developed from and alongside commercial ultrasonic non-destructive testing techniques. The testing of bodies - fleshy or inorganic - by a process that does not cause irreversible harmful changes in their structure, and where other techniques (e.g., γ-rays) cannot be used, is the feature that is responsible for the increased use of ultrasonic testing in both industry and medicine. The two main areas of interest in the metallurgical field have been the recognition and classification of discontinuities (localized defects), and the determination of the grain size and structure of a specimen by scattering measurements.

3.2.1 Grain Scattering in Metals

Mason and McSkimin[44] seem to have planted the seed of this particular tree of knowledge with their measurements of the attenuation of ultrasound by metals and glasses. Working with a pulse technique between 2 and 15 MHz, they found the attenuation coefficient of aluminum to be of the form $\alpha = B_1 f + F_2 f^4$. The first term describes absorption due to hysteresis effects and is negligible for $\bar{a} \leq \frac{\lambda}{6}$ (where $\bar{a}$ is the mean grain radius). The second term describes Rayleigh scattering. The coefficient B_1 and B_2 depend on the grain structure, and Mason and McSkimin[44] suggest an $(\bar{a})^3$ dependence for B_2. However, as Roth[45] points out, this was based on only two experimental points. Using purer samples and more measurements he finds B_2 to be proportional to $^1/\bar{a}$, between 5 and 100 MHz. Mason and McSkimin[46] clarified their evidence in the light of Roth's comment and find that for $\bar{a} \leq \frac{\lambda}{6}$, is proportional to $\bar{a}^3$; but for $\bar{a} \geq 1.5\ \lambda$, B_2 is proportional to $^1/\bar{a}$, agreeing with Roth. Between these two limits they can say nothing.

This intermediate region (for $\lambda \simeq \bar{a}$) has been considered by Huntingdon[47] who finds the attenuation due to scattering to be proportional to f^2. His measurements in the region 10 to 30 MHz provide reasonable agreement with the theory. Closer agreement with theory has largely been the result of Papadakis' determination[48,49] of the grain size distribution. The considerable contribution Papadakis has made in this field in the interpretation of anomalous results in terms of the microstructure and phase transition of the sample used, are discussed in Papadakis.[50]

The statistics of scattering in this context have been discussed by Beecham,[51] who shows that the transducer requirements for optimization of defect recognition and measurement of scatter are not identical. Taking Bhatia's[52] analysis of the scattering theory, with its angular dependence of the form $\left(\frac{1-3\cos 2\theta)}{4}\right)^2$, he derives an expression for the scattering cross-section per unit volume of:

$$\sigma = \frac{4\pi^3 \bar{a}^6 f^4}{c^4} P$$

where P is a measure of the grain anisotropy. This shows the impossibility of attempting to separate the effects of grain size and anisotropy in the Rayleigh region by a measurement of the level of ultrasonic scatter alone.

3.2.2 Defect Recognition

Krautkramer[53] seems to have started the most recent phase of development with his interest in determining the size of a flaw. He suggested using a calibrated gain control on a standard flaw detector so that the magnitude of the echo produced by the defect could be compared to that produced by a standard target. In spite of the excellent experimental results that he obtained, using a circular plate at a fixed distance as a standard, he was only able to find the minimum size of the defect.

Gericke[54] suggests that white noise is a better tool for interrogation than the Gaussian pulse of Krautkramer. Using model defects - holes classified according to number, size and angulation - he finds that, provided the wavelength is smaller than the size of the defect, a rectilinear scan produces a good estimate of the size of the hole. However, as the wavelength becomes comparable to the size of the hole, little can be said. In 1964, Gericke[55] described the use of an electrical-response equalization technique to produce a wide interrogating spectrum. Taking measurements at 2.25 MHz and 6.75 MHz, he was able to discriminate quite well between different holes in an aluminum block, provided they were not too close to the transducer. This successful dual frequency approach led almost inevitably to complete spectral analysis as a tool for defect description.

3.2.3 Spectroscopy

There is still some difference of opinion between workers in the field as to whether a spectroscopic approach to flaw recognition has any possibility of being successful. The early work of Gericke[56] showed that different defects certainly do produce different echo spectra, an observation that Clipson[57] has confirmed. The main problem that arises is that of obtaining a method of analyzing and displaying the information that is meaningful (e.g., Lloyd[58]). The aim at the moment is not for information that is reproducible by other workers, but for information that is consistent and useful in the context of the application and the calibration procedures used. The comparison of results is further hindered by the singular reluctance of workers in the field (Legge,1971; Whaley and Adler, 1971; Adler and Whaley, 1972) to recognize and specify all the vital parameters of their measurements.

3.3 Conclusion

Although the potential usefulness of the results of marine acoustics in ultrasonic diagnostics is great, since both fields span the frequency range from ray propagation to the Rayleigh region, the practical value of the results is limited by the coarse end-point dictated by economic factors in the fishing industry. All that can be gleaned is: a qualitative confirmation of scattering behavior in the Rayleigh and diffractive regions, a warning on the ease with which air-bubbles may be mistaken for more substantial objects (McCartney and Bary[59]), and an ambiguous comment on the scattering properties of fish flesh.

In the field of non-destructive testing the experimental determination of the frequency dependence of grain scattering follows the pattern that might be expected from the theory of Section 2, namely f, f^2, f^4 for the ray, diffraction and Rayleigh regions, respectively. The most surprising is probably the last. Rayleigh's original theory assumed that the scatterers were to have physical properties only slightly different from the mean, and that they were to be independent scatterers. Although the first of these assumptions is valid in this context, we have seen in Section 2 that the law holds good even when this condition is not met. It is somewhat surprising, however, to find that the second assumption is also not crucial to the frequency dependence: we could hardly describe a solid matrix of particles, each of which has its shape and motion directly determined by the particles contiguous to it, as a system of independent scatterers.

The lack of a completely successful application of spectroscopy is not really surprising in the light of the many parameters that are involved both in the description of the flaw and the analysis of its spectrum. Nevertheless, reports of recent developments (Brown[60]) are encouraging. The measurements reported in Section 5 are concerned with volume scattering and its spectroscopy, rather than the spectroscopy of single features, and demand a slightly different treatment.

4. MEASUREMENT OF VOLUME SCATTERING BY TISSUES

The theoretical treatment propounded in Section 2 and the experimental observations on acoustic scattering in other contexts, described in Section 3, both lead to the expectation that acoustic volume scattering by human tissues may show appreciable structure-related dependence on acoustic frequency. The actual experimental arrangements used to observe and record such data, either in the laboratory or in practical examination of patients, may take a variety of detailed forms but many features will necessarily be common, as will a number of practical constraints. In this section we outline the problems and limitations that are commonly to be encountered in this type of measurement, with particular reference to a specific measuring system, and we review the rather limited amount of experimental data that exists in this field.

4.1 Problems and Limitations

The basic requirement for a system of acoustic scattering measurement is for a source of acoustic energy, coupled to the scattering volume of interest, and an appropriate detector capable of receiving energy scattered from that volume in any specific direction. Both the energy source and the detector will generally be directional, having a defined beam profile, and may be either broad band or monochromatic (narrow band) with variable frequency. In the special case of 180° backscattering measurement, the same device will be used for both transmission and reception. In all cases it will be necessary to define the actual volume that is to be interrogated and for this purpose the transmitted beam must be pulsed, so that it is switched on for a finite period T, and the detector time gated so that it is active only for a finite period τ, whose commencement is delayed by a certain period t following the start of the transmitted pulse.

One problem that is immediately apparent is that, in any attempt to measure the variation of scattering amplitude with scattering angle, the geometry of the situation will dictate that the actual scattering volume interrogated cannot remain constant and will, in fact, vary in a rather complex manner (Chivers and Hill,[61] Chivers[5]). In the absence,therefore, of theoretical indications that such measurements would be particularly informative it will generally be best to avoid this particular complication and confine measurements to those of true (i.e., 180°) backscattering.

In order that meaningful measurements can be made of the variation of backscattering amplitude with frequency it will be necessary to recognize that the actual beam profiles of the transducers employed will themselves be somewhat frequency dependent. In particular the frequency spectrum of the echo signal returned by a given target placed in the beam will be found to vary with respect to movement both along and normal to the beam axis. In practice, however, it will be possible to limit the extent of corresponding uncertainty to an acceptable degree by restricting the use of any one transducer to a particular spectral window: in one particular experimental system, a set of five broad band transducers with center frequencies between 1 and 6 MHz were found to have usable band widths, over which the variations in the spectrum of a plane interface echo varied by not more than $\pm$ 1.5 dB, of between 27 and 43%.[5] In this way, a broad frequency range (0.6 - 7.5 MHz in the above case) can reliably be covered with a limited number of different transducers.

Frequency spectrum analysis can be implemented by either digital or analogue methods. Although digital techniques offer advantages in versatility, the application calls for equipment of high sampling rate and large capacity and most work reported so far has consequently been done using commercially available swept filter spectrum analyzers. An inexpensive alternative, however, is to drive the interrogating beam from a time gated, swept frequency sine wave and to record the received echo signal as a function of transmitted frequency.

The backscattering cross-section 'σ_s' for a particular volume is defined by the expression:

$$I_s(f) = \sigma_s(f) \cdot I_o(f) \quad (4.1)$$

where $I_o(f)$ is the intensity of sound per unit area at a frequency 'f'. In practice, the calculation of 'σ_s' is complicated by the use of a time-gate to select the signal for analysis. The measured backscattered spectrum $I_m(f)$ is given by:

$$I_m(f) = I_s(f) * \frac{\mathrm{Sin}(\pi f\tau)}{2\pi f} \quad (4.2)$$

where 'τ' is the gate length (Chivers[5]).

In other words, before 'I_s' can be used to calculate 'σ_s' the function $\mathrm{Sin}(\pi f\tau)/2\pi f$, (which represents the Fourier Transform of the gate) must be deconvoluted from the measured spectrum. Good computational facilities would enable 'σ_s' to be calculated, with a view to solving the inverse scattering problem and obtaining some idea of the morphology of the tissue volume that produced a given spectrum. The complication introduced by the gate length does not, however, preclude comparative studies aimed at tissue differentiation, provided that the gate length is kept fixed.

One further important consideration is that, in any practical measurement of backscattering cross-section, the actual received signal will be attenuated by an amount corresponding to the double passage of the ultrasonic beam through the medium overlying the scattering volume of interest, and a corresponding correction must be made to derive a true value of 'σ_s'. This is not a fundamental problem, but implies in practice that it will be necessary to carry out appropriate measurements of tissue attenuation coefficients in parallel with actual backscattering measurements if any attempt is to be made to derive absolute values for the latter. If, however, measurements are carried out on a comparative basis at a fixed depth of overlying tissue, it may be possible to avert this problem.

4.2 Experimental Approach to Backscattering Measurement

A practical system for acoustic backscattering measurements on human tissues has recently been developed in the authors' laboratory and it will be useful to describe this briefly in order to illustrate some solutions to the problems discussed above and to indicate the derivation of some of the experimental data that is reviewed in Sections 4.3 and 5.

The system is based on the pulse-echo principle and thus operates by generating an interrogating pulse and subsequently detecting and analyzing the echoes that return from a tissue volume of interest. Referring to the schematic diagram in Figure 2, the echoes returning from the tissue under examination are amplified and then time gated, thus defining the interrogated

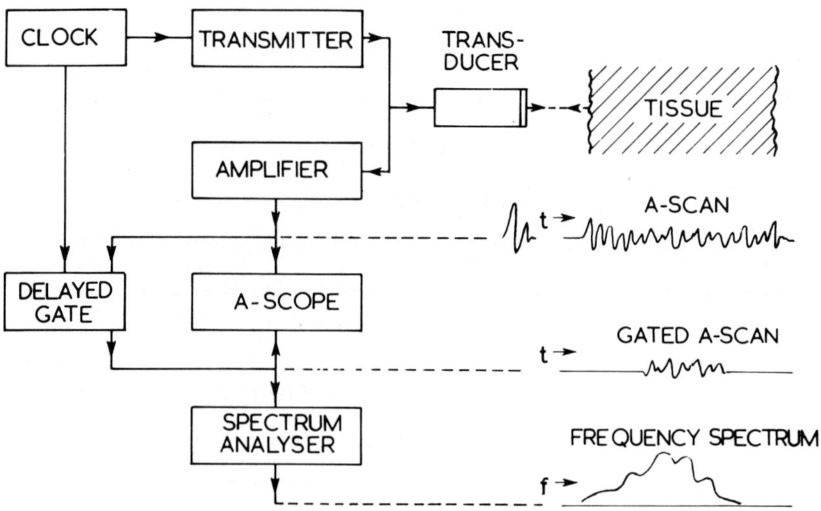

Fig. 2 Experimental arrangement for investigation of the frequency dependence of acoustic scattering by tissues.

volume. The gated section of the echo train is then frequency analyzed on a Hewlett-Packard 8552/8553 Spectrum Analyzer. The amplifiers are designed to be linear, with an essentially flat frequency response over the range 0.5-9 MHz, and provide a dynamic range of 35 dB.

The time gate selects the volume of tissue to be analyzed but high spatial resolution (thinness of the tissue slice examined) can only be obtained at the sacrifice of frequency spectral information: the resolution in frequency that can be associated with a given gate length τ is found to be given by $\Delta f . \tau = 0.5$, provided that the lowest frequency of interest, f_o, is such that $f_o . \tau > 1$ (Chivers and Hill[61]). The other factor that defines the volume of tissue that is interrogated is the effective beam width and, here again, fine spatial resolution can only be achieved at the expense of frequency spectral information, and a working compromise must be found.

The mounting of tissue specimens in vitro, in such a way that the acoustic characteristics of the specimens are not significantly modified and that extraneous scattering or attenuation effects are not introduced, poses considerable problems, for which there may be no perfect solutions. A satisfactory arrangement, however, is to fit the tissue specimen into a fluid-filled polythene bag, which can be suspended in an open-work frame within the water tank that is used to couple the beam and the specimen. The advantages of this arrangement are the ease with which tissue specimens of different size and shape can be accommodated (using different holder frames) and obviation of the need to replace the water in the sound tank for each new specimen. The thin polythene sheet is found to have no significant effect on either the amplitude or spectral distribution of the transmitted echoes.

4.3 Quantitative Measurements of Scattering Amplitude

Credit for making the first systematic attempt to record, and exploit for diagnostic purposes, the amplitude of scattering from human tissues is due to Wild and Reid[1], who claimed to be able to make significant differential diagnoses of various benign and malignant conditions of the human breast on the basis of the echo amplitudes that they recorded as A and B scans. Although, in one sense, the subsequent more general development of pulse-echo visualization techniques has been based on the recording of scattered echo amplitude, this early pioneering work has not until recently been seriously followed up and the general approach to pulse echo diagnosis has been towards qualitative techniques of visualization. Some groups have adopted a somewhat quantitative approach: Donald[62], for example, has used the gain setting of his machine at which detectable echoes occur as a criterion for recognizing the presence of hydatidiform moles. Buschmann[63] has established a systematic approach to the differential diagnosis of orbital tumors based on standardized measurements of A-scan amplitude, Kikuchi[64] has suggested the extension to medical diagnosis of quantitative ideas employed in ultrasonic non-destructive testing, and Fry et al[65] have investigated the echo amplitudes returned by liver tissue in various species.

The recent development of grey-scale visualization techniques has greatly facilitated the adoption of this kind of approach and Taylor et al[66], for example, are able both to identify malignant lesions in the liver and to monitor their response to radio- or chemo-therapy on the basis of observation of scattered echo amplitude and of progressive changes in such amplitudes.

A more quantitative approach to this kind of assay has recently been described by Mountford and Wells,[67] who made a detailed numerical analysis of the distribution of echo amplitudes recorded in liver A-scans and were able to show in this way that significant differentiation could be made between normal and cirrhotic livers. A method for carrying out this type of analysis by on-line digital techniques has subsequently been described by McCready and Milan.[68]

Apart from a few rather general comments on the increased level of the returned echoes that appears to result from increase in the ultrasonic working frequency employed diagnostically, there has been little interest until recently in the specific use of the frequency dependence of scattering as a diagnostic criterion, and very little in the way of experimental data has been reported. Some possible approaches to the subject have been discussed in papers by Hill and Chivers,[69] Sigelman and Reid[70] and Waag and Lerner.[71] Some preliminary measurements of the frequency dependence of the surface scattering that occurs at a fluid/tissue interface have been reported by Lele and Namery[72] who showed that marked changes occur in the frequency spectrum of echoes from the surface of exposed amphibian muscle as a result of lesions induced by vascular occlusion and designed to simulate a cardiac infarct. This work was complemented by studies (Senapati et al[73]) both on frog tissue surfaces and on metal surfaces of various measured degrees of roughness, which confirmed the expected decrease with increasing frequency in the rate of fall off of reflected echo amplitude with increasing departure from normal incidence.

The only published data at present available on the frequency dependence of volume scattering of human tissues are those of Chivers et al;[74] Chivers[5] who carried out measurements on the lines described in Section 4.2. Some typical volume backscattering frequency spectra obtained in this study from different human tissues are shown in Figure 3, in which comparison is made with the spectrum of the interrogating pulse. In these there is little evidence of a systematic increase in scattering amplitude with increasing frequency, such as would be expected for scattering occurring predominantly in the Rayleigh region ($\lambda >> a$), but there is considerable spectral structure such as would be expected from diffraction. Comparison of the appearances of the sets of spectra obtained from the three different tissues does suggest that there are distinguishable differences between the physical structures responsible for this diffraction in the three different cases but more precise and quantitative assessment of these differences has not so far been achieved. One feature of these results is that the detailed appearance of each spectrum may change considerably with changes in the positioning of the tissue in the ultrasonic beam: the pairs of tissue spectra presented in Figure 3 were obtained at two different positionings of each tissue. This behavior presents a practical problem in terms of the achievement of re-

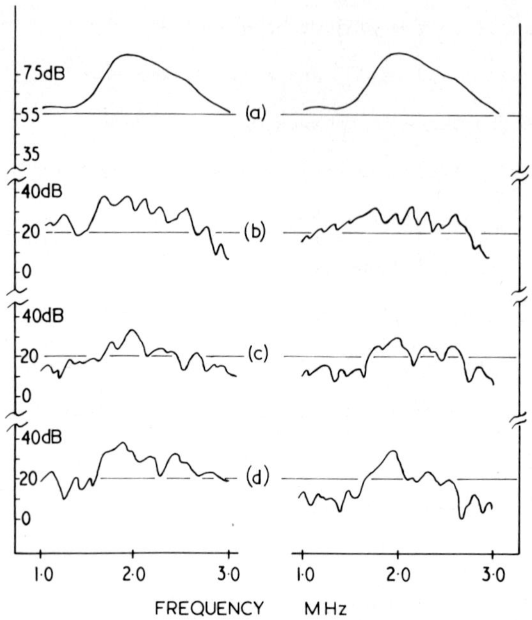

Fig. 3 Frequency spectra of ultrasonic backscattering from three types of human soft tissue (two samples of each type) compared with the interrogating spectrum. (a) transducer reference spectrum, (b) fat, (c) liver, (d) spleen. (If filter width 30 kHz; gate length 8 μs).

peatable spectral measurements, unless resort is taken to statistical analysis of the data, but it also suggests, as an alternative practical approach to the underlying problem of characterization of tissue structure, the application of the Bragg diffraction technique. A discussion of this approach is given in the next section.

5. BRAGG SCATTERING

In the foregoing Sections of this chapter the conventional assumption has generally been followed that human tissues are, acoustically, randomly ordered structures. That this is not strictly the case, however, is readily apparent from the evidence of microscopic histology: muscular tissue, for example, generally has the form of rather regularly arranged bundles of fibers, while some glandular tissues, and particularly liver, appear to consist of three-dimensional matrices of packed, pseudo-spherical objects. Taken together with the evidence, described above, for diffractive scattering of ultrasound by human tissues, this suggests that a useful approach to understanding and exploiting such scattering processes may be possible along the lines of Bragg X-ray diffraction crystallography.

A suitable experimental arrangement for such a study can be achieved as a minor modification of the arrangement for frequency spectroscopy, as illustrated in Figure 4. Mechanically, it is only necessary for the tissue volume of interest to be rotated steadily about its center-point, which will be positioned on the beam axis and at the midpoint of the time gate. In the laboratory the tissue is rotated relative to the beam, but it is clearly entirely equivalent, and more appropriate for _in vivo_ examination of a patient, to rotate the beam relative to the tissue. Electrically, a narrow frequency band width arrangement is required, in order to define a particular ultrasonic wavelength, and this can readily be achieved by selecting a restricted frequency band from the full returned broad band spectrum. The amplitude of the returned echo at this selected frequency is then recorded

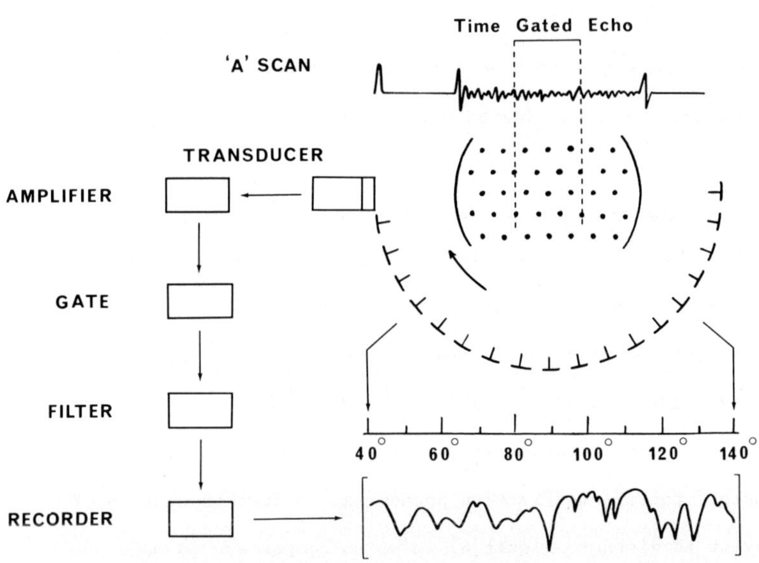

Fig. 4 Experimental arrangement for investigation of the backscattered Bragg diffraction characteristics of human tissues.

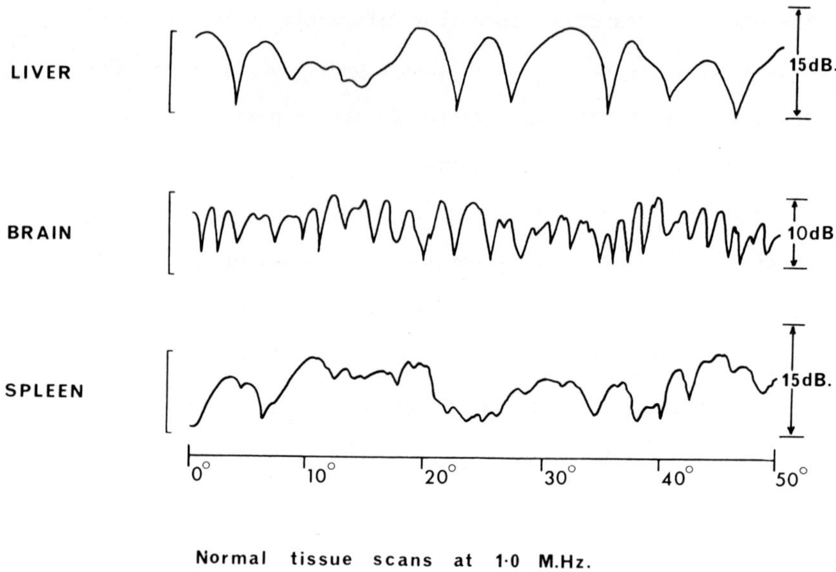

Fig. 5 Observed backscattered Bragg diffraction patterns from three types of human soft tissue at 1 MHz. Top trace: Liver; center: brain; bottom: spleen.

as a function of the angle of rotation of the tissue relative to the beam axis. It will be seen that this is not an exact analogue to the X-ray diffraction arrangement, which employs near-forward scattering in contrast to the backscattering arrangement considered here, although the principle is essentially the same.

For illustrative purposes a typical orientation dependence (Bragg diffraction pattern) has been included in Figure 4, and comparative measurements for three different human tissues are shown in Figure 5 (see also Hill[75]). Visual examination of these records strongly suggests the existence of significant differences between them and this, again, seems to imply the existence of significant and characteristic differences between the physical structures responsible for the diffraction phenomena. A further example of this type of behavior is given to Figure 6, in which comparison is made between records obtained from normal and diffusely metastatically involved (i.e., cancerous) human liver.

Without further analysis of the data these results cannot be claimed as evidence for the existence of any particular degree of orderliness in the diffracting structure. It is possible, however, to make some assessment of the extent to which a particular diffraction pattern might be consistent with a certain hypothetical arrangement of scatterers in the tissue. It is appropriate to consider here two such model arrangements: a system of parallel filaments as a model for skeletal muscle and a system of close packed spheres as a model for liver parenchyma.

In the case of the muscle model, consisting of an orthogonal array of evenly spaced, parallel threads lying along the y axis of the interrogated volume indicated in Figure 7, it is possible to calculate the amplitude of the total backscattered wave with respect to the interrogating pulse.

If we assume a matrix of scatterers such that a scatterer exists, at the center of the interrogated volume, which scatters the incident pulse with an amplitude:

$$A_s = A\exp\,(i\omega t) \text{ (assuming inelastic scattering)} \qquad (5.1)$$

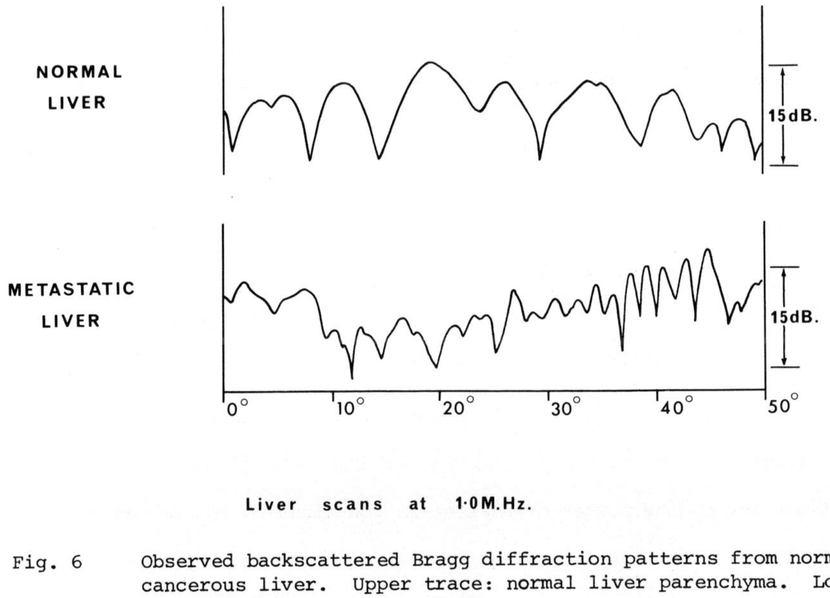

Fig. 6 Observed backscattered Bragg diffraction patterns from normal and cancerous liver. Upper trace: normal liver parenchyma. Lower trace: secondary liver tumour. 1 MHz.

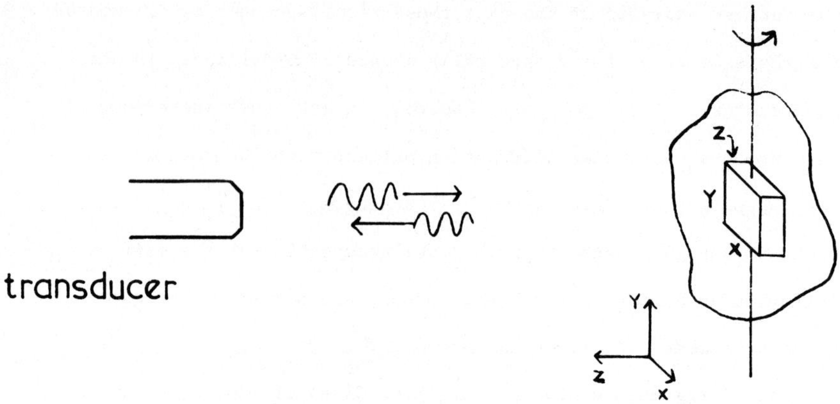

Fig. 7 Computer simulation model for experimental scattering geometry.

then, by summation, the total backscattered amplitude from M equally spaced scatterers in one dimension is:

$$A_s = A\exp(i\omega t) \cdot [\cos(nd) - \cos(n+1)d] \,/\, [1-\cos(d)] \quad (5.2)$$

where $n = (M-1)/2$ (5.3)

(Nicholas)[81]

here d is the phase difference between adjacent scatterers and can be expressed as:

$$d = 2S\ \mathrm{Sin}(\theta) \quad (5.4)$$

where θ is the angle of rotation and S the spacing between adjacent fibers in the muscle model.

Expanding to an orthogonal three dimensional array of scatterers simply involves expanding 5.2 to:

$$A_s = A\ \exp(i\omega t) \cdot \Psi_1 \cdot \Psi_2 \cdot \Psi_3 \quad (5.5)$$

where

$$\Psi_i = [\cos(n_i d_i) - \cos(n_i+1)d_i\ ] / [1-\cos(d_i)\ ] \quad (5.6)$$

for i = 1,2,3 - corresponding to the three cartesian coordinates of the three dimensional frame.

The limitation on the number of scatterers contributing to the backscattered amplitude is solely determined by their spacing and the dimensions of the volume under investigation. Equation 5.5 may be modified for matrices

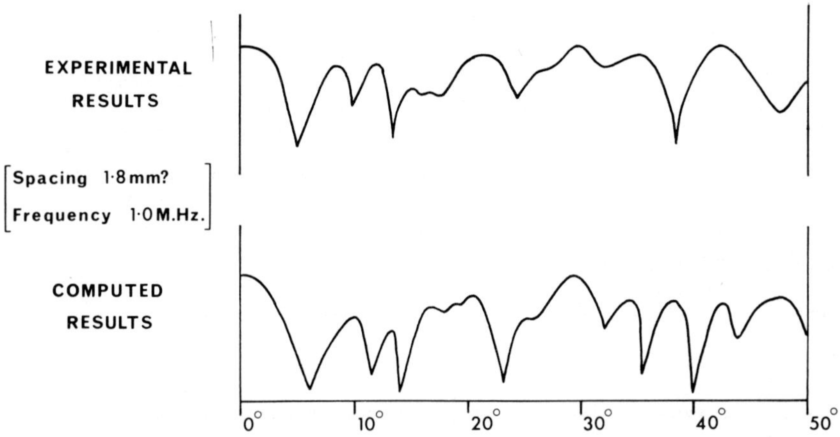

Variation of amplitude with orientation of the returned echoes from vertically orientated muscle tissue.

Fig. 8 Acoustic Bragg diffraction records for skeletal muscle at 1 MHz (log scale). Upper trace: measured. Lower trace: computed from parallel filament array model with 1.8 mm spacing.

other than orthogonal by re-evaluating the values of d_i with respect to the angle of orientation.

In Figure 8 the Bragg diffraction behavior predicted from this simple model is compared with measurements obtained from a specimen of human skeletal muscle rotated in the beam about an axis parallel to the muscle fibers. Evidently this measured behavior is not inconsistent with a model of the general form suggested although the only definite conclusion that can be drawn is that a regular two-dimensional array of spacing 1.8 ± 0.1 mm yields results more comparable to the measured diffraction pattern for muscle than those yielded by arrays of different spacings. Anatomically it was determined that the muscle macroscopic fiber spacing was 1.5 ± 0.3 mm, in good agreement with the predicted results. However, before any special credence can be given towards characterizations of muscle structure by this method, a more refined approach to both the data collection and theoretical predictions must be evolved.

A somewhat similar acoustic modeling approach can be attempted in the case of liver. Under a low-power microscope liver parenchyma, particularly in certain species such as the pig, is seen to be separated by trabeculae into well defined lobules. These lobules are polyhedral, usually showing five to seven sides in a cross-section. Running through the middle of each lobule is the central vein (intralobular vein) which drains via a radial network of sinusoids into a sub-lobular tributary of the hepatic vein. At the periphery of the lobule, in the connective tissue, are found the interlobular vein (which is a tributary of the portal vein), the branches of the hepatic artery, the interlobular bile ducts and the lymphatic vessels: the whole is termed a portal canal. Thus we can visualize a system of pseudo-spherical scatterers of approximately 1-2 mm spacing in a somewhat regular three-dimensional matrix, although whether it is the network of portal canals or the actual lobular structure which acts predominantly as a scattering matrix has yet to be determined. However, the structure is in any case one that macroscopically exhibits an appreciable degree of regularity when considered over a finite number of interlobular spacings.

For modeling purposes, the backscattered intensity from a regular three-dimensional randomly orientated array of point scatterers has been computed for a geometry which simulates that of the experiment as closely as possible. It seems reasonable to assume a close packing array, and two arrays that have been investigated are the close packed hexagonal (Figure 9a),

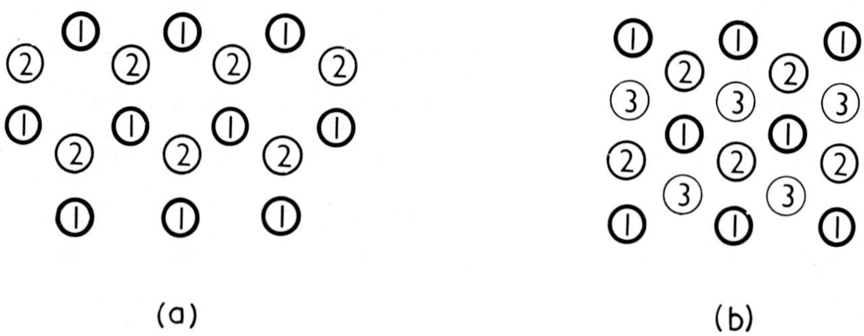

Fig. 9 Possible close-packing arrays for spheres: (a) close-packed hexagonal; (b) close-packed cubic.

and the close packed cubic (Figure 9b). The arrangement of points (the sphere centers) in individual planes of these arrays is identical, and the arrays differ only in the stacking of the planes. In the close packed hexagonal array, points in planes 3 and 4 lie below points in planes 1 and 2 and so on, and in the close packed cubic array, points in planes 4, 5, and 6 lie below points in planes 1, 2 and 3.

The experimental geometry is simulated for the purposes of computation in the manner shown in Figure 7. The interrogated volume is assumed to be a square slab inside the array and centered at the origin of the array. The slab is normal to the beam and has dimensions x = y = the transducer beam width, and $z = c\tau/2$ where c is the ultrasound propagation velocity (taken equal to the velocity in water), and τ = electronic gate width. The array is rotated about the Y axis with the slab remaining fixed, and only points whose coordinates lie within the slab at any particular angle are assumed to contribute to the backscattered signal.

The random orientation of the arrays was obtained as follows. The coordinates of the center of each sphere comprising the array were defined in a coordinate system X', Y', Z', with the distance between two points in a plane (=a where a/2 is the sphere radius) taken as the scale length. The coordinates were then transformed into the laboratory coordinate system X, Y,Z, by assuming nine random but self-consistent direction cosines (l_1,m_1,n_1) between X' and X, etc. The laboratory coordinate system was then rotated about the Y axis so that the coordinates of each point in the array are now defined in the laboratory system by the random direction cosines (l_1, m_1, n_1), (l_2, m_2, n_2), and (l_3, m_3, n_3) and the angle of rotation. Thus, a point in the array having coordinates x'y'z' will have coordinates

$$x = l_1x' + m_1Y; + n_1z'$$

$$y = l_2x' + m_2Y' + n_2z'$$

$$z = l_3x' + m_3Y' + n_3z'$$

in the laboratory system, and upon rotation through an angle θ about the Y axis

$$x'' = x\cos\theta - z\sin\theta$$

$$y'' = y$$

$$z'' = x\sin\theta + z\cos\theta$$

Only z" is required for the computation of the backscattered intensity; however, all three coordinates are required in order to ascertain whether the scattering point lies within the interrogated slab.

The general expression for the scattered intensity is[76]

$$I = \frac{|f(\theta)|^2}{r^2} \sum_{i,j=1}^{N} \exp(i\underline{k}\cdot\underline{R}_{ij})$$

where

$\underline{k} = k(\underline{n} - \underline{m})$,

$|k| = |2k\sin(\theta/2)|$,

$\underline{n}$ and $\underline{m}$ are unit vectors for the incident and scattered waves respectively,

$\underline{R}_{ij} = \underline{r}_i - \underline{r}_j$ is the vector from the ith to the jth scatterer,

and

N is the total number of scatterers.

In the present case, r - z" and is assumed to be much greater than the dimensions of the interrogated volume, and θ = 180 degrees, giving

$$I \propto \left[N + \sum_{i \neq j} \exp 2ki\,(z_1 - z_j)\right]$$

where $k = 2\pi/\lambda$.

The first term is the intensity that would result if all N scatterers sent out their waves independently of each other and is the incoherent part of the scattering. The second term is the coherent part which takes into account the various phase relations of the scattered waves. If the origins of the coordinate systems are taken to be coincident with the center of the interrogated volume, N is determined by the criteria:

$$|x''| < x_{lim}, \quad |Y''| < Y_{lim}, \quad \text{and} \quad |z''| < z_{lim},$$

where

$2x_{lim} = 2y_{lim}$ and is the length of the side of the slab, and $2z_{lim}$ is the

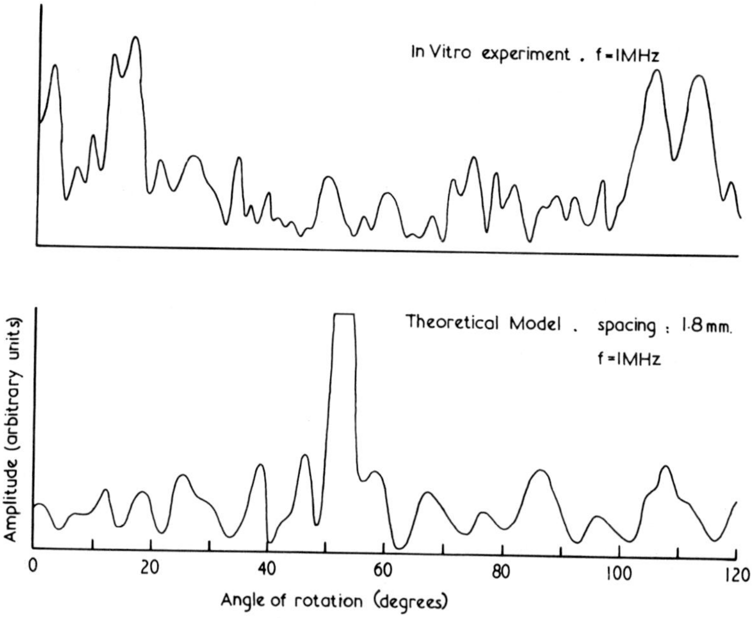

Fig. 10 Acoustic Bragg diffraction records for liver at 1 MHz (linear scale). Upper trace: measured. Lower trace: computed from close-packed hexagonal array model with 1.8 mm spacing.

thickness of the slab. All points lying inside the slab are included in the summation and, as the array is rotated, the number of points included in the summation varies by a small fraction.

Examination of histological sections of pig liver suggests a typical lobular size of 1.5 to 2.0 mm.[77] The computed amplitude ($I^{1/2}$) for a spacing of 1.8 mm is shown in Figure 10 along with experimental data taken at 1 MHz. Slab dimensions of 6 mm x 6 mm x 4 mm were assumed, which compares with an experimental 6 dB beam width of 6 mm and an equivalent slab thickness (set by the gate) of 4 mm. The angle between the Y axis in the laboratory coordinate system and the Y axis in the array coordinate system (Y') was approximately 25 degrees. This theoretical result is for the close packed hexagonal array: the result for the close packed cubic array was found to exhibit a similar variation of amplitude with θ.

The comparison between the computed result based on the structure of liver from pig, where the lobular structure is well defined, and the experimental result from excised human liver, where the lobular structure is poorly defined, is remarkably good if the rapid variation (∿ 0.25 cycles per degree) in the experimental data is smoothed out. The model does not purport to be an accurate description of the scattering mechanism in human liver: however, the coarse agreement with the experimental result suggests that scattering from a quasi-regular array of scatterers is responsible for part of the observed backscatter amplitude. The rapid variation with angle is not described by the model but is nevertheless remarkable on account of its phase coherence as shown by autocorrelation of the data. The source of this component of the backscattered signal is not clear at this time.

The Bragg diffraction record in the form presented here appears to be a qualitatively useful means for characterizing tissue structure but does not readily provide a quantitative basis for this process. For this purpose it seems useful to obtain the Fourier Transforms of the original records (i.e., their power spectra with respect to angular frequency). In this form,

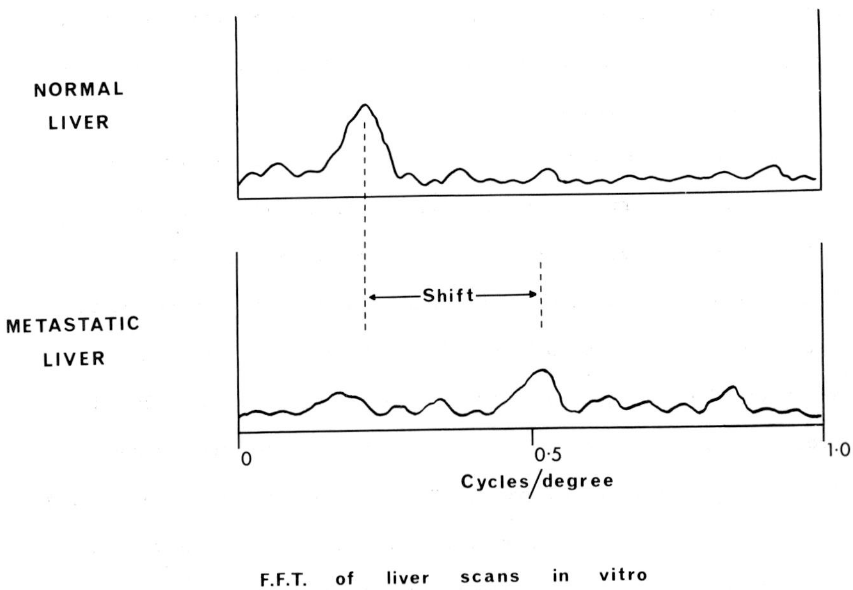

Fig. 11 Power spectra derived from the Bragg diffraction records of Figure 6. Upper trace: normal liver parenchyma. Lower trace: secondary liver tumour.

structural similarities between different scattering structures may be expected to result in numerically significant similarities between the corresponding power spectra (Hill and Nicholas[78]). Power spectra derived from the two Bragg diffraction records of Figure 6 are shown in Figure 11.

6. CONCLUSIONS

This chapter constitutes the first comprehensive review of the subject of acoustic scattering by human tissues. It is a topic that has received very little serious attention hitherto: a situation that is somewhat surprising in view of its fundamental importance as the basis of almost all ultrasonic diagnostic methods.

Useful advances in science and technology commonly arise from constructive interactions between theoretical ideas and experimental measurement. The theoretical analysis of acoustic scattering tends to be fairly complex, even where relatively simple mechanical structures are involved, and any attempt at a successful treatment for the complex structures exhibited by human tissues must necessarily entail extensive simplifying departures from the real situation. Nonetheless, theoretical ideas have an important part to play in this field and a number of useful advances have already been made: the generalization, to the case of an inhomogeneous continuum, of Rayleigh's analysis of the behavior of discrete scatterers is one example, and another is the introduction of the concept of a tissue as a semi-ordered scattering array, which suggests an experimental analogy to X-ray diffraction phenomena.

Pure theory has little to say about the relative importance of Rayleigh region scattering and diffraction processes in the total scattering situation but the small amount of experimental data that now exists on volume scattering points strongly to diffraction as the predominant contributor in the Megahertz frequency region and to structural features on the order of scale of a millimeter as being responsible. From histological considerations this would seem to be consistent with the suggestion of Fields and Dunn[79] that collagen, because of its relatively high bulk modulus, is the tissue component that shows the highest acoustic mismatch with its environment and is thus effective in providing scattering interfaces.

The general measurement of angular scattering cross-sections is likely to be impracticable but the particular use of the 180° backscattering cross-section is tractable and is of considerable practical importance. Frequency analysis of the signal backscattered from a specific volume appears to provide a possible basis for tissue characterization although a possibly more informative approach is to analyze the dependence of backscattered amplitude, at one or more frequencies, on the relative orientation of the tissue structure and the interrogating beam. Analyses of the results of such measurements appear to be consistent with semi-ordered scattering array models for some tissues and offer an exciting prospect for the achievement of practical quantitative methods for remote, in vivo characterization of selected volumes of human soft tissues on the basis of their structure, and thus their anatomy and pathology.

In the achievement of a fuller fundamental understanding of the mechanisms by which sound is scattered by tissue structures, the employment of tissue models is likely to play an important part in bridging the gap between theoretical analysis and the observed response of tissues. It should be possible to construct artificial scattering matrices, in the form of random suspensions or ordered arrays of simple scattering objects, in forms that are both amenable to somewhat exact theoretical analysis and, at the same time, constitute reasonable approximations to real tissues in respect to their observable acoustic scattering properties. Advances on these lines may be expected to indicate relationships between the behavior of tissues as acoustic scatterers and the microscopically observable features of their structure.

The practical relevance of improvements in the understanding of ultrasonic scattering processes in human tissues has several aspects. Of greatest immediate importance may be the implications in the field of ultrasonic imaging, where a fuller appreciation of the nature of the echo formation process is likely to lead to the development of more efficient and systematic techniques for the collection of the basic information on which the imaging process relies. An extension of the imaging process, which will rely even more heavily on knowledge of scattering behavior, is likely to develop in the direction of remote quantitative characterization of tissue structure, perhaps based on some of the experimental approaches outlined in the present chapter. Apart from its direct implication in techniques for imaging and tissue characterization, acoustic scattering plays a role in the complex of processes that attenuate ultrasonic energy in tissues, and improvements in its understanding will be needed as a part of the process of establishing good and reliable data on this important but currently poorly documented subject.

Preparation of this chapter was completed late in 1974 and, in the ensuing year, a number of new publications emphasizing certain aspects of scattering by human tissues have appeared. Amongst these the following should be mentioned.

At a specialist meeting on ultrasonic characterization of human tissues (Maryland, U.S.A., 1975) several papers relevant to scattering processes were presented. Reid[82] and Shung and Reid[83] reported measurements of the magnitude of frequency dependence and concentration dependence of scattering by suspensions of erythrocytes. In relation to organized soft tissues Waag[84] and Hill[85] discussed methods by which the scattering from such tissues may be related to their macroscopic structures. This later topic has met with some success of late, in particular Nicholas and Hill[78,86] have been able to show a statistically significant degree of differentiation between normal, freshly excised, human liver, spleen and brain (white matter), and have also indicated that back-scattering analysis is capable of assessing the macroscopic structure of muscle tissue. Hill[87] briefly summarized the more pertinent aspects of scattering from human tissues, indicating lines for future research.

Finally, the aspects of ultrasound scattering by soft human tissues presented in this chapter have been considerably extended by Nicholas,[81] where results on both frequency and orientation dependence indicate the potential usefulness of this partially neglected field.

APPENDIX A

Evaluation of the scattering cross-section per unit volume for a lossless inhomogeneous continuum.

The energy flow of a plane wave of amplitude A into a cube of side L is

$$I_o = \beta A_o^2 L \tag{A.1}$$

where β is a constant of proportionality. Assuming that the correlation functions of density and compressibility variations are identical and isotropic we have for the scattered wave:

$$\langle |P_1|^2 \rangle = \frac{VA_o^2 k}{16\pi^2 R^2} \left[\frac{\langle (\Delta\rho)^2 \rangle}{p_o^2} \cdot \cos^2\theta + \frac{\langle (\Delta m)^2 \rangle}{m_o^2} \right] \frac{4\pi}{k_o} \int_0^\infty N(r) \sin k_o r \cdot rdr \tag{A.2}$$

Nicholas[89]. The total intensity scattered out of the cube is

$$I_s = \beta \oint \langle |P_1|^2 \rangle \, ds \tag{A.3}$$

where the integration is taken over a sphere of radius R. Combining equations A.1, A.2 and A.3, we obtain:

$$\frac{I_s}{I_o} = \frac{k^4 L}{4\pi} \left[\frac{\langle (\Delta\rho)^2 \rangle}{p_o^2} \cdot \cos^2\theta + \frac{\langle (\Delta m)^2 \rangle}{m_o^2} \right] \cdot \int d\Omega \cdot \int N(r) \cdot \frac{\sin k_o r}{k_o} \cdot rdr \tag{A.4}$$

where $d\Omega$ is an element of solid angle.

If $I_s \ll I_o$, as is true for biological tissue in the medical frequency range, we may write equation A.4 in the form

$$\frac{\Delta I}{I} = - S_s L \text{ (c.f. equation 2.36)} \tag{A.5}$$

The scattering cross-section per unit volume is given by

$$S_s = \frac{k^4}{4\pi} (A + B) \tag{A.6}$$

where

$$A = \frac{\langle(\Delta\rho)^2\rangle}{p_o^2} \int d\Omega \int_o^\infty \cos^2\theta \exp(-r^2/a^2) \frac{\sin k_o r}{k_o} \cdot rdr \qquad (A.7)$$

and

$$B = \frac{\langle(\Delta m)^2\rangle}{m_o^2} \int d\Omega \int_o^\infty \exp(-r^2/a^2) \frac{\sin k_o r}{k_o} \cdot rdr$$

$$= \frac{\langle(\Delta m)^2\rangle}{m_o^2} \cdot \frac{\pi^{3/2}a}{k^2} \cdot [1 - \exp(-k^2a^2)], \qquad (A.8)$$

Thus:

$$S_s = \frac{k^4}{4\pi}\left[\frac{\langle(\Delta m)^2\rangle}{m_o^2} \frac{\pi^{3/2}a}{k^2}[1 - \exp(-k^2a^2)] + \frac{\langle(\Delta\rho)^2\rangle}{\rho_o} \cdot \frac{2\pi}{k} \cdot Z\right] \qquad (A.9)$$

where Z is the integral in equation A.7.

Neglecting density variations, we obtain

$$S_s = \frac{k^2 a\pi^{1/2}}{4} \frac{\langle(\Delta m)^2\rangle}{m_o^2} \cdot [1 - \exp(-k^2a^2)] \qquad (A.10)$$

in agreement with Chernow.[15] For small scale inhomogeneities, ka << 1 and we obtain a 'Rayleigh' k^4 dependence for S_s in agreement with Bihari and Szilard;[80] but for large scale variations, ka >> 1 and we obtain a square law.

APPENDIX B

List of Symbols employed

a = Linear dimension of inhomogeneity

c = velocity of sound at a point in space and time

c_o = velocity of sound in undisturbed medium

f = frequency

g = scattering function for individual body (defined in 2.16)

k = wave number

k^1 = wave number for coherent scattering (equation 2.31)

m = bulk modulus of elasticity at a point in space and time

m_o = bulk modulus of undisturbed medium

p = pressure at point in space and time

p_1 = pressure of scattered wave

s = condensation of medium at a point in space and time

s_o = condensation of undisturbed medium

A, A_o = wave amplitude

G = volume scattering function (equation 2.28)

I = intensity of wave

L = linear dimension of volume under consideration

N = correlation function

R = distance of observation

S = cross-section per unit volume

S_a = absorption cross-section per unit volume

S_s = scattering cross-section per unit volume

S_r = attenuation cross-section per unit volume

T = pulse length or absolute temperature

V = volume of integration

Z = integral (A.9)

Θ = scattering angle

λ = wavelength

μ = deviation of refractive index from unity

σ_a = single body absorption cross-section

σ_s = single body scattering cross-section

σ_r = single body attenuation cross-section

τ = duration of a time interval (gate)

ψ = wave function

ψ_o = interrogating wave

ψ_s = scattered wave

ω = angular frequency

$\langle\,\rangle$ = time average

∇ = Laplacian operator

Δ = small change

REFERENCES

1. J. J. Wild and J. M. Reid: Further Pilot Echographic Studies on the Histologic Structure of Tumours of the Intact Living Breast. Amer. J. Path. 28 (1952) pp. 839-861.

2. W. J. Fry: Biological and Medical Acoustics. J. Acoust. Soc. Amer. 30 (1958) pp. 387-393.

3. I. Donald, J. MacVicar and T. G. Brown: Investigation of Abdominal Masses by Pulsed Ultrasound. Lancet 1 (1958) p. 1188.

4. H. Pauly and H. P. Schwann: Mechanism of Absorption of Ultrasound in Liver Tissue. J. Acoust. Soc. Amer., 50 (1971) pp. 692-699.

5. R. C. Chivers: The Scattering of Ultrasound by Human Tissues. Ph.D. Thesis (1973) University of London.

6. C. R. Hill: The Possibility of Hazard in Medical and Industrial Applications of Ultrasound. Brit. J. Radiol. 41 (1968) pp. 561-570.

7. Lord Rayleigh: On the Influence of Obstacles Arranged in Rectangular Order on the Properties of a Medium. Phil. Mag. 34 (1892) p. 481.

8. W. H. Bragg: X-Rays and Crystal Structure. Phil. Trans. Roy. Soc. (London) A.215 (1915) p. 253.

9. V. Twersky: On Multiple Scattering of Waves. J. Res. Nat. Bur. Standards 64D, No. 6 (1960) p. 715.

10. L. L. Foldy: Multiple Scattering of Waves: General Theory of Isotropic Scattering by Randomly Distributed Scatterers. Phys. Rev. 67 (1945) p. 197.

11. M. Lax: Multiple Scattering of Waves. Rev. Mod. Phys. 23 (1951) pp. 287-310.

12. V. P. Glotov: Coherent Scattering of Plane and Spherical Waves in Deep Water Layers Containing Discrete Inhomogeneities. Sov. Phys. Doklady 1 (1962) p. 211.

13. B. F. Kur'yanov: Coherent and Incoherent Scattering of Waves by a Set of Point Scatterers Distributed Randomly in Space. Sov. Phys. Acoustics 10 (1964) p. 160.

14. R. C. Chivers and C. R. Hill: Ultrasonic Attenuation in Human Tissue. Ultrasound in Med. and Biol. 2(1) (1975) pp. 25-29.

15. L. A. Chernow: Wave Propagation in a Random Medium. Dover (1960) New York.

16. L. J. Liebermann: The Effect of Temperature Inhomogeneities in the Ocean on the Propagation of Sound. J. Acoust. Soc. Amer. 23 (1951) p. 563.

17. T. H. Ellison: The Propagation of Sound Waves Through a Medium With Very Small Random Variations in Refractive Index. J. Atmos. Terrest. Phys., 2 (1951) p. 14.

18. D. Mintzer: Wave Propagation in a Randomly Inhomogeneous Medium. I. J. Acoust. Soc. Amer. 25 (1953) p. 922.

19. D. Mintzer: Wave Propagation in a Randomly Inhomogeneous Medium. II. J. Acoust. Soc. Amer. 25 (1953) p. 1107.

20. D. Mintzer: Wave Propagation in a Randomly Inhomogeneous Medium. III. J. Acoust. Soc. Amer. 26 (1954) p. 186.

21. A. M. Obukhov: On the Influence of Weak Atmospheric Inhomogeneities on the Propagation of Sound and Light. (In Russian) Izv. Akad. Nauk. SSSR Ser. Geograf. Geofiz. No. 2 (1953) pp. 155-165.

22. C. L. Pekeris: A Note on the Scattering of Radiation in an Inhomogeneous Medium. Phys. Rev. 71 (1947) p. 268L.

23. H. Staras: Scattering of Electromagnetic Energy in a Randomly Inhomogeneous Atmosphere. J. App. Phys. 23 (1952) p. 1152-1156.

24. H. G. Booker and W. E. Gordon: A Theory of Radio Scattering in the Troposphere. Proc. IRE 38 (1950) p. 401.

25. J. A. Fejer: The Diffraction of Waves Passing through an Irregular Refracting Medium. Proc. Roy. Soc. A. 220 (1953) pp. 455-471.

26. Lord Rayleigh: The Theory of Sound. Dover (1945) London.

27. P. Vigoureux and J. B. Hersey: In The Sea, Ideas and Observations on its Study, ed. M. N. Hill, Wiley, New York (1962) pp. 475-497.

28. L. Fortuin: Survey of Literature on the Reflexion and Scattering of Sound Waves at the Sea Surface. J. Acoust. Soc. Amer. 47 (1970) p. 1209.

29. L. R. Breslau: The Normally Incident Reflectivity of the Sea Floor at 12 kc/s and its Correlation with Physical and Geological Properties of Naturally Occurring Sediments. AD655 245 (WHO IR 6716), Woods Hole Oceanographic Institute, U.S.A. (1967).

30. C. M. McKinney and C. D. Anderson: Measurements of the Backscattering of Sound from the Ocean Bottom. J. Acoust. Soc. Amer. 36 (1964) pp. 158-163.

31. Y. Y. Jitkovskiy and L. A. Volovova: Sound Scattering from the Ocean Bottom. Proc. 5th Intl. Cong. Acoust., Liege (1965) paper E67.

32. B. F. Kur'yanov: The Scattering of Sound at a Rough Surface with Two Types of Irregularity. Sov. Phys. Acoustics 8 (1963) p. 252.

33. J. B. Hersey and R. H. Backus: In The Sea, Ideas and Observations on its Study, ed. M. N. Hill, Wiley, New York (1962) pp. 498-539.

34. B. McK. Bary: Qualitative Observations of Scattering of 12 kc/s Sound in Saanich Inlet, B. C. Deep Sea Res. 13 (1966) p. 667.

35. R. H. Backus, J. E. Craddock, R. L. Haedrich, D. L. Shores, J. M. Teal, A. S. Sing, G. W. Mead and W. D. Clarke: Ceratoscopelus Mederensis: Peculiar Sound Scattering Layer Associated with this Fish. Science 160 (1968) p. 991-993.

36. B. McK. Bary: Backscattering at 12 kc/s in Relation to Biomass and Numbers of Zooplanktonic Organisms in Saanich Inlet, B. C. Deep Sea Res. 13 (1966) p. 655.

37. J. R. Marshall and R. P. Chapman: Reverberation from the Deep Scattering Layer Measured with Explosive Sound Sources. J. Acoust. Soc. Amer. 36 (1964) p. 164.

38. R. L. Capen: Swimbladder Morphology of Some Mesopelagic Fishes in Relation to Sound Scattering. AD657 443 (NEL 1447) Navy Electronics Lab., San Diego, CA U.S.A. (1967).

39. D. H. Cushing: Some Echo Sounding Experiments on Fish. J. de Conseil 20 (1954) p. 266.

40. P. F. Smith: Further Measurements of the Sound Scattering Properties of Several Marine Organisms. Deep Sea Res. 2 (1954) pp. 71-79.

41. R. W. G. Haslett: Physics Applied to Echo Sounding for Fish. Ultrasonics 2 (1964) pp. 11-22.

42. R. W. G. Haslett: The Target Strengths of Fish. J. Sound-Vibration 9 (1969) p. 181.

43. I. Andreeva: Acoustical Characteristics of Sonic Scattering Layers in the Ocean. Proc. 5th Intl. Cong. Acoustics, Liege (1965) paper E68.

44. W. P. Mason and H. J. McSkimin: Attenuation and Scattering of High Frequency Sound Waves in Metals and Glasses. J. Acoust. Soc. Amer. 19 (1947) p. 464.

45. W. Roth: Scattering of Ultrasonic Radiation in Polycrystalline Metals. J. App. Phys. 19 (1948) p. 901.

46. W. P. Mason and H. J. McSkimin: Energy Loss in Metals Due to Scattering and Diffusion. J. App. Phys. 19 (1948) p. 940.

47. H. B. Huntingdon: On Ultrasonic Scattering by Polycrystals. J. Acoust. Soc. Amer. 22 (1950) pp. 362-364.

48. E. P. Papadakis: Grain Size Distribution in Metals and its Influence on Ultrasonic Attenuation Measurements. J. Acoust. Soc. Amer. 33 (1961) p.1616.

49. E. P. Papadakis: From Micrograph to Grain Size Distribution with Ultrasonic Applications. J. App. Phys. 36 (1964) pp. 1586-1594.

50. E. P. Papadakis: Ultrasonic Attenuation Caused by Scattering in Polycrystalline Media. In Physical Acoustics, ed. by W. P. Mason, Academic Press, New York (1968) Vol. IVB, pp. 296 ff.

51. D. Beecham: Ultrasonic Scatter in Metals, Its Properties and Its Application to Grain Size Determination. Ultrasonics 4 (1966) pp. 67-76.

52. A. B. Bhatia: Scattering of High Frequency Sound Waves in Polycrystalline Materials. J. Acoust. Soc. Amer. 31 (1959) p. 16.

53. J. Krautkrämer: Determination of the Size of Defects by the Ultrasonic Impulse-Echo Method. Brit. J. App. Phys. 10 (1959) pp. 240-245.

54. O. R. Gericke: Determination of the Geometry of Hidden Defects by Ultrasonic Pulse Analysis Testing. J. Acoust. Soc. Amer. 35 (1963) pp. 364-368.

55. O. R. Gericke: Dual Frequency Ultrasonic Pulse-Echo Testing. J. Acoust. Soc. Amer. 36 (1964) p. 313.

56. O. R. Gericke: Experimental Determination of Ultrasonic Transducer Frequency Response. Mat. Eval. 24 (1966) pp. 409-411.

57. W. R. Clipson: Ultrasonic Spectroscopy Development for Inclusion Cloud

Assessment. Proc., Symp. on Future of Ultrasonic Spectroscopy, Brit. Non-Farrous Metals Res. Assoc., London (1970) paper #4.

58. E. A. Lloyd: Proc. Symp. on Future of Ultrasonic Spectroscopy, Brit. Non-Farrous Metals Res. Assoc., London (1970) paper #2.

59. B. S. McCartney and B. McK. Bary: Echo Sounding on Probable Gas Bubbles From the Bottom of Saanich Inlet, B. C. Deep Sea Res. 12 (1965) p. 285.

60. A. F. Brown: Materials Testing by Ultrasonic Spectroscopy. Ultrasonics 11 (1973) pp. 202-210.

61. R. C. Chivers and C. R. Hill: A Spectral Approach to Ultrasonic Scattering from Human Tissue: Methods, Objectives and Backscattering Measurements. Phys. Med. Biol. 20 (1975) p. 799.

62. I. Donald: Sonar Examination of the Abdomen. Ultrasonics 4 (1966) pp. 119-124.

63. W. H. Buschmann: Reproducible Calibrations: The Basis of Ultrasonic Differential Diagnosis in A-Mode and B-Mode Examination of Eye and Orbit. Intl. Ophthal. Clinics 9 (1969) pp. 761-792.

64. Y. Kikuchi: Way to Quantitative Examination in Ultrasonic Diagnosis. Med. Ultrasonics (Japan) 6(1) (1968) pp. 1-8.

65. E. Kelly Fry, D. Okuyama and F. J. Fry: The Influence of Biological and Instrumentation Variables on the Characteristics of Echosonograms. In Ultrasono Graphia Medica, Vienna Acad. of Med. (1970) Vol. 3, pp. 387-393.

66. K. J. W. Taylor, D. A. Carpenter and V. R. McCready: Grey Scale Echography in the Diagnosis of Intrahepatic Disease. J. Clin. Ultrasound 1 (1973) pp. 284-287.

67. R. A. Mountford and P. N. T. Wells: Ultrasonic Liver Scanning: The A-Scan in the Normal and Cirrhosis. Phys. Med. Biol. 17 (1972) pp. 261-269.

68. R. McCready and J. Milan: The Use of Computers in Ultrasound. In Ultrasonics in Medicine, Excerpta Medica, Amsterdam (1974) pp. 318-324.

69. C. R. Hill and R. C. Chivers: Investigations of Backscattering in Relation to Ultrasonic Diagnosis. In Conf. Proc. "UBIOMED-70", Polish Acad. of Sciences, Warsaw (1972).

70. R. A. Sigelman and J. W. Reid: In Interaction of Ultrasound and Biological Tissues, US DHEW Pub. (FDA) 73-8008 (1972) pp. 245-249.

71. R. C. Waag and R. M. Lerner: Tissue Macrostructure Determination with Swept-Frequency Ultrasound. In 1973 Ultrasonics Symposium Proceedings, IEEE, New York (1973).

72. P. P. Lele and J. Namery: Detection of Myocardial Infarction by Ultrasound. In Proceedings of 25th ACEMB (1972) p. 135.

73. N. Senapati, P. P. Lele and A. Woodin: A Study of the Scattering of Sub-Millimeter Ultrasound from Tissues and Organs. In 1972 Ultrasonics Symposium Proceedings, IEEE New York (1972) pp. 59-63.

74. R. C. Chivers, C. R. Hill and D. Nicholas: Frequency Dependence of Ultrasonic Backscattering Cross-Sections: An Indicator of Tissue Structure Characteristics. In Ultrasonics in Medicine, Excerpta Medica, Amsterdam (1974) pp. 300-303.

75. C. R. Hill: Interactions of Ultrasound with Tissues. In Ultrasonics in Medicine, Excerpta Medica, Amsterdam (1974) pp. 14-20.

76. M. P. Morse and H. Feshbach: Methods of Theoretical Physics, McGraw-Hill (1953) p. 1496.

77. S. Bradbury: Histology for Medical Students, 9th Ed., Heinemann (1969) p. 309.

78. D. Nicholas and C. R. Hill: Acoustic Bragg Diffraction from Human Tissues. Nature 257 (1975) pp. 305-306.

79. S. Fields and F. Dunn: Correlation of Echographic Visualizability of Tissue with Biological Composition and Physiological State. J. Acoust. Soc. Amer. 54 (1973) pp. 809-812.

80. I. Bihari and J. Szilard: Ultrasonic Testing of Inhomogeneities in Heterogeneous Media. Periodica Polytech. Elect. Eng. Elektrotech. 11 (1967) pp. 261-288.

81. D. Nicholas: Ultrasonic Backscattering and the Structure of Human Tissues. Ph.D. Thesis (1976) University of London.

82. J. M. Reid: Scattering of Ultrasound by Tissues. In Ultrasonic Tissue Characterization (NBS Publication 453) U.S. Gov't. Printing Office, (1976) pp. 21-28.

83. K. K. Shung, R. A. Sigelman and J. M. Reid: The Scattering of Ultrasound by Red Blood Cells. In Ultrasonic Tissue Characterization, (NBS Publication 453) U. S. Gov't. Printing Office (1976) pp. 207-212.

84. R. C. Waag, R. M. Lerner and R. Gramiak: Swept-Frequency Ultrasonic Determination of Tissue Macrostructure. In Ultrasonic Tissue Characterization, (NBS Publication 453) U. S. Gov't. Printing Office, (1976) pp. 213-228.

85. C. R. Hill: Frequency and Angular Dependence of Ultrasonic Scattering from Tissue. In Ultrasonic Tissue Characterization, (NBS Publication 453) U.S. Gov't. Printing Office (1976) pp. 197-206.

86. D. Nicholas and C. R. Hill: Tissue Characterization by an Acoustic Bragg Scattering Process. In Ultrasonics Intl. 1975 (Cont. Proc.) London, IPC Science and Technology Press (1975) pp. 269-272.

87. C. R. Hill: Echoes from Human Tissues. In Ultrasonics Intl. 1975 (Cont. Proc.) London, IPC Science and Technology Press (1975) pp. 20-22.

88. R. C. Chivers: The Scattering of Ultrasound by Human Tissues - Some Theoretical Models. Ultrasound in Med. and Biol. 3 (1977) in press.

89. D. Nicholas: An Introduction to the Theory of Acoustic Scattering by Biological Tissues. In Recent Advances in Ultrasound in Biomedicine, Vol. 1, ed. D. N. White, Research Studies Press (in press).

CHAPTER VIII

ACOUSTIC IMPEDANCE PROFILING

An Analytical and Physical Model Study with Reference to Soft Tissue Characterization

A. C. Kak, F. J. Fry and J. P. Jones

1. INTRODUCTION

Accurate classification of tissue types in the live patient by non-invasive atraumatic means is a continuing quest in diagnostic medicine. Ultrasound offers the possibility of obtaining quantitative information on acoustic parameters in living tissue while satisfying the stated criteria. The degree of success achieved by ultrasound tissue classification depends on the uniqueness of a set of acoustic properties for specific tissue types and the accuracy with which the acoustic parameters can be derived. It is probable that no single acoustic parameter is likely to qualify as a unique determinant in tissue characterization. For this reason it is important to assess a number of acoustic properties in order to identify indices which may be definitive. Laboratory measurements on physical models and excised tissue are an important prelude to the more difficult problems encountered when quantitative measurements of acoustic properties are to be made in the clinical setting. Additionally, any attempt to measure only one specific acoustic parameter in the tissue complex presents considerable difficulty, unless the other parameters are assumed to be known *a priori*. Acoustic impedance profiling typifies this difficulty.

A basic premise of the theoretical developments presented here is that the reflected waveform is a convolution of the incident waveform and the impulse response associated with the medium for the path of propagation of the acoustic pulse. This is certainly true in the linear regime of operation. Clearly then the impulse response, which can be calculated given the incident and the received waveforms, characterizes the path in the medium taken by the acoustic pulse in going from the illuminating transducer to the receiving transducer (the two can be the same element).

From the impulse response one can, in principle, determine the impedance profile even in the presence of loss and multiple reflections. The methods that have been suggested to date for this are recursive in nature.[1,2] By this we mean a medium is considered to be composed of a large number of layers each of constant acoustic impedance and one calculates the impedance for one layer at a time starting from the first. The accuracy of calculation at each step depends upon how accurately the previous impedance values have been calculated. In practice these techniques cannot often be used because their recursive nature is a source of instability. In other words, the impulse response has to be known highly accurately for the calculations to be stable. It is well known that the calculation of the impulse response is mathematically an ill-conditioned problem and one can at best calculate it only to a good approximation. It is because of these reasons we have attempted to derive non-recursive relationships between the impulse response and the impedance profile. For the sake of completeness we have also included a brief discussion on the previously mentioned recursive methods.

In any tissue or organ system, the medium is always lossy. An average value of attenuation loss of soft tissues of the human body interpreted as an acoustic pressure absorption coefficient is 1 db per centimeter of path length at 1 MHz frequency. Soft tissues also show an approximately linear absorption loss as a function of frequency. Since there is a characteristic frequency dependence of attenuation in tissue, it is apparent that the temporal character of the acoustic interrogating pulse will be modified depending on its bandwidth as it passes through the tissue.

Attenuation loss is composed generally of a number of components (viz., intrinsic absorption, scattering). In order to arrive at an appropriate acoustic impedance profile for medically relevant systems, the intimate relationship between attenuation and impedance determinations is apparent. It is anticipated that practical methods will evolve permitting the simultaneous determination of the absorption profile and the impedance profile. Substantially no more will be said here about the more general problem of characterizing the broad spectrum of acoustic parameters and their interrelationships. In this work the attenuation losses have been ignored in the theoretical developments and inserted in *a priori* fashion for the calculation of impedances for experimental models. Individual layer thickness variations leading to variable amounts of insertion loss are computed from determination of the transit time between interfaces after an *a priori* insertion of sound velocity for individual layers.

The rationale for the physical model approach used is based on a number of factors including the fact that, to the present, substantially all medical diagnostic procedures using ultrasound have relied heavily on a qualitative presentation and interpretation of acoustic impedance differences at tissue interfaces, and that conceptually the individual layer methodology can be extended to include any number of reflective targets in a tissue or organ system. Extension of this physical model study to the living tissue situation encountered in the diagnostic clinic will obviously require adaptations including acoustic beam orientation to maximize return signals from specific targets, appropriate signal temporal synchronization to keep tissue targets in geometric registration and knowledge of insertion losses and velocities.

Methods of signal processing and computer technology have been applied to acoustic interrogating pulses in this study so that ultimately real time, or near real time quantitative descriptions of acoustic impedance profiles can be obtained. Detailed descriptions of the analytic approach to the layered model and specific algorithm considerations and computer simulation studies are discussed later. The algorithm implemented for impulse response determination for the experimental studies uses the matched filtering technique which is applicable to a rather generalized acoustic waveform (broadband or narrowband having a damped oscillatory wave character).

The first major subdivision presents the analytical descriptions, from a physical acoustics viewpoint, signal processing methods and computer simulations. Experimental results are presented in the second major subdivision.

2. RELATIONSHIP BETWEEN THE IMPEDANCE PROFILE AND THE IMPULSE RESPONSE

Let us consider a target interrogated by a plane acoustic wave. In the linear regime of operation, the reflected wave can be considered to be a convolution of the input signal and an impulse response associated with the medium. (This is also true for the transmitted wave if the transmission impulse response is used). This impulse response is strictly a property of the medium in the direction of propagation of the incident wave. Relationships between this impulse response and the impedance profile of the medium have been a matter of investigation in the past.[1,2] The relationships that have been developed in these investigations are recursive in nature, an example of which is presented in Section 2.4. In recursive relationships a medium is considered to be composed of a large number of layers each of constant acoustic impedance, and one calculates the impedance of one layer at a time starting from the first. The accuracy of the calculation at each step depends upon how accurately the previous impedance values have been calculated. In practice these techniques cannot often be used because their recursive nature is a source of instability.

Non-recursive relationships are a topic of current research. In a recent paper by Jones[3] one such relationship was presented for the restricted case of monotonically increasing or decreasing impedance profiles in the absence of multiple reflections. We will show that the same relationship holds for any arbitrary impedance profile.

2.1 Monotonically Increasing or Decreasing Profiles

We will first derive the noniterative relationship for the monotonically increasing profile by a method different from that given by Jones.[3] The derivation given here provides insight and is mathematically essential to the derivation for the general case of arbitrary impedance profiles, which will be presented later.

Let a monotonically increasing or decreasing profile be approximated by discrete layers of constant impedance. Evidently as the number of layers in such an approximation approaches infinity for an acoustic target of finite extent the impulse response calculated from the discrete approximation should approach that from the continuous profile. The discrete approximation may be constructed in such a way that the reflection coefficient at each impedance boundary is the same; let this value be denoted by r. It is readily seen that the impulse response of such a discrete profile is a train of impulses. Let the time instances at which the reflected impulses are received be denoted by $t_1, t_2, t_3, \ldots$. If $h(t_m)$ denotes the reflected impulse at t_m, then it can be shown

$$h(t_m) = r(1 - r^2)^{m-1} \, \delta(t - t_m) \tag{1}$$

Let us integrate the impulse response from 0 to a time instant t with $t_m \leq t < t_{m+1}$. Then

$$S(t) = \int_0^t h(t)\, dt = \sum_{k=1}^{m} r(1 - r^2)^{k-1} \tag{2}$$

This can be expressed as

$$S(t) = \frac{1 - (1 - r^2)^m}{r} \tag{3}$$

For a given time instant t as $m \to \infty$ and $r \to 0$, the discrete impedance profile will approach the continuous profile. As $m \to \infty$, the time intervals between successive reflections from the impedance discontinuities approach zero, so that t_m approaches t. Also, as $m \to \infty$, r must approach zero in such a way that the following is always satisfied:

$$\frac{Z_m}{Z_0} = \left(\frac{1+r}{1-r}\right)^m \tag{4}$$

where Z_m is the impedance of the mth layer in the discrete profile, and Z_0 the impedance of the water bath in which the target it immersed. From (4) we get the following relationship between m and r:

$$m = \frac{\log (Z_m/Z_0)}{\log \frac{1+r}{1-r}} \tag{5}$$

Substituting (5) in (3) and taking the limit $r \to 0$, we get

$$\int_0^t h(t')dt' = \lim_{r\to 0} \frac{1 - (1-r^2)^{\frac{\log (Z_m/Z_0)}{\log\{(1+r)/(1-r)\}}}}{r} \tag{6}$$

It is seen that if r=0 is directly substituted in Eq. (6) the resulting expression becomes indeterminate. The limit on the right hand side may therefore be evaluated by taking the partial derivative of both the numerator and the denominator, separately, with respect to r and r=0 may then be substituted in the resulting expression. The result is

$$\int_0^t h(t')dt' = \lim_{r\to 0}\{-\frac{\partial}{\partial r} K(r)\} \tag{7}$$

$$= - \log \frac{Z_t}{Z_0} \lim_{r\to 0} K(r)\frac{\partial}{\partial r} \frac{\log (1-r^2)}{\log \frac{1+r}{1-r}} \tag{8}$$

where we have made use of the fact that $r \to 0$ and $m \to \infty$, $t_m \to t$ and hence $Z_m \to Z_t$; and where

$$K(r) = (1-r^2)^{\frac{\log (Z_m/Z_0)}{\log\{(1+r)/(1-r)\}}} \tag{9}$$

By taking the logarithm of both sides in (9) and then taking the limit $r \to 0$, we get

$$\lim_{r\to 0} \log K(r) = \log \frac{Z_t}{Z_0} \lim_{r\to 0} \frac{\log(1-r^2)}{\log(1+r)/(1-r)} \tag{10}$$

$$= \log \frac{Z_t}{Z_0} \lim_{r\to 0} \frac{\frac{\partial}{\partial r} \log(1-r^2)}{\frac{\partial}{\partial r} \log \frac{1+r}{1-r}} \tag{11}$$

$$= 0.$$

Therefore

$$\lim_{r\to 0} K(r) = 1 \tag{12}$$

Substituting this in (8), we get

$$\int_0^t h(t')dt' = - \log \frac{Z_t}{Z_0} \lim_{r\to 0} \frac{\partial}{\partial r} \frac{\log(1-r^2)}{\log \frac{1+r}{1-r}} \tag{13}$$

Taking the derivative on the right hand side and noting that as $r \to 0$, $\log(1-r^2) \to -r^2$ and $\log \frac{1+r}{1-r} \to 2r$, we get

$$\int_0^t h(t')dt' = \frac{1}{2} \log \frac{Z_t}{Z_0} \tag{14}$$

Clearly $\frac{Z_t}{Z_0}$ is the normalized acoustic impedance at that point in the medium where the acoustic impulse arrives in $\frac{t}{2}$ seconds. (We assume the medium to be nondispersive).

2.2 Arbitrary Impedance Profiles

In deriving the impulse response-impedance profile relationship given by (5), we assumed in the discrete approximation that all reflection coefficients are equal. This assumption constrains the impedance profile to be either monotonically increasing or decreasing. We will now show that Equation (5) is valid even when the impedance profile does not satisfy this condition. We will continue to assume the absence of multiple reflections. In the discrete approximation of a continuous medium, let r_k be the reflection coefficient at the kth step, the kth step being a transition from impedance Z_{k-1} to Z_k. By definition

$$r_k = \frac{Z_k - Z_{k-1}}{Z_k + Z_{k-1}} \tag{15}$$

Clearly,

$$\frac{Z_m}{Z_0} = \prod_{k=1}^{m} \frac{1+r_k}{1-r_k} \tag{16}$$

As the continuous medium is approximated by increasingly finer steps each $r_k \to 0$. In such a limit each $\log \frac{1+r_k}{1-r_k} \to 2r_k$ and, therefore, from Eq. (16), we get

$$\log \frac{Z_m}{Z_0} \underset{r_k \to 0}{=} 2 \sum_{k=1}^{m} r_k \tag{17}$$

Therefore,

$$\frac{1}{2} (\log Z_m - \log Z_{m-1}) \underset{r_k \to 0}{=} r_m \tag{18}$$

Again, if the multiple reflections can be assumed to be negligible,

$$h(t_k) = r_k \prod_{\ell=1}^{k-1} (1-r_\ell^{\,2}) \tag{19}$$

As each $r_\ell \to 0$, we get from (18) and (19)

$$\log h(t_k) - \log h(t_{k-1}) = \log[\tfrac{1}{2} (\log Z_k - \log Z_{k-1})]$$

$$- \log [\tfrac{1}{2} (\log Z_{k-1} - \log Z_{k-2})]$$

$$- \tfrac{1}{4} [\log Z_k - \log Z_{k-1}]^2$$

The impedance Z is a function of distance. It can also be considered to be a function of time, in the sense that Z(t) is the impedance of the medium at a position which to arrive at the incident wave must travel $\frac{t}{2}$ seconds. In fact, in the absence of any knowledge of acoustic velocities in the medium, the best one can do is to find the impedance profile as a function of time. Let $\Delta t = t_k - t_{k-1} = t_{k-1} - t_{k-2} = \ldots.$; then

$$\Delta t \frac{\log h(t_k) - \log h(t_{k-1})}{\Delta t} = \log [\frac{\log Z_k - \log Z_{k-1}}{\Delta t}]$$

$$- \log [\frac{\log Z_{k-1} - \log Z_{k-2}}{\Delta t}] \tag{20}$$

$$- \frac{(\Delta t)^2}{4} [\frac{\log Z_k - \log Z_{k-1}}{\Delta t}]^2$$

As $\Delta t \to 0$, we get from the above equation

$$h(t) = C \frac{d}{dt} \log Z(t) \tag{21}$$

where C is a constant of integration. To evaluate this constant we make use of the boundary condition that for some non-zero distance near the first change in impedance, the impedance variation is increasing or decreasing, making Eq. (14) valid for times t that correspond to reflections from such regions. This implies $C = \frac{1}{2}$. Substituting this in Equation (21) and integrating, we get

$$\int_0^t h(t')dt' = \frac{1}{2} \int_{t'=0}^{t'=t} d(\log Z(t')) = \frac{1}{2} \log \frac{Z(t)}{Z_0} \tag{22}$$

which completes the proof.

2.3 Multiple Reflections Included

In the derivations in the preceding subsections we assumed that multiple reflections made negligible contributions to the reflected waveform at the receiver. There are experimental situations where such an assumption would be justified. For illustration, consider a medium in which the impedance changes in discrete steps, and let all these changes be of equal magnitude. The reflection coefficients at each impedance change will then be the same and let r denote its value. It is easily seen that the ratio of third order reflections to first order reflections is proportional to r^2, the ratio of fifth order to first order reflection is r^4, and so on. Evidently, if r is equal to, let us say, 0.1, all the reflections higher than the first order can be neglected. By extending this argument one concludes that so long as the total impedance change is small, and so long as its rate of change is small, neglecting higher order reflections would introduce negligible error. In biological experiments, soft tissue with tumors would often satisfy this condition. But if the interrogation is made through bone, this assumption would be grossly violated.

We will now include third order reflections in deriving a relationship between the impulse response and the impedance profile. Because of mathematical complexity, our treatment is limited to the case of discrete approximation to a continuous medium where all the reflection coefficients can be considered to be equal. In other words, we are again dealing with only the monotonically increasing or decreasing impedance profiles.

To find the effect of third order reflection on $\int_0^t h(\tau)d\tau$, consider the pulse bounce diagram in Fig. 1. Each $l_{\tau,j}$ represents the amplitude of a positive traveling impulse through section $Z_{j-\tau-1}$ between times t_i and t_{i+1} where i is the largest integer less than or equal to $\frac{j+\tau-1}{2}$. From the pulse bounce diagram it can be shown that $l_{j-k,j}$ has k-1 second order reflections. Therefore, each $h(t_j)$ has $L(j) = 1 + 2 + \ldots + (j-3) + (j-2) = \frac{(j-2)(j-1)}{2}$ third order reflection components (each component having an am-

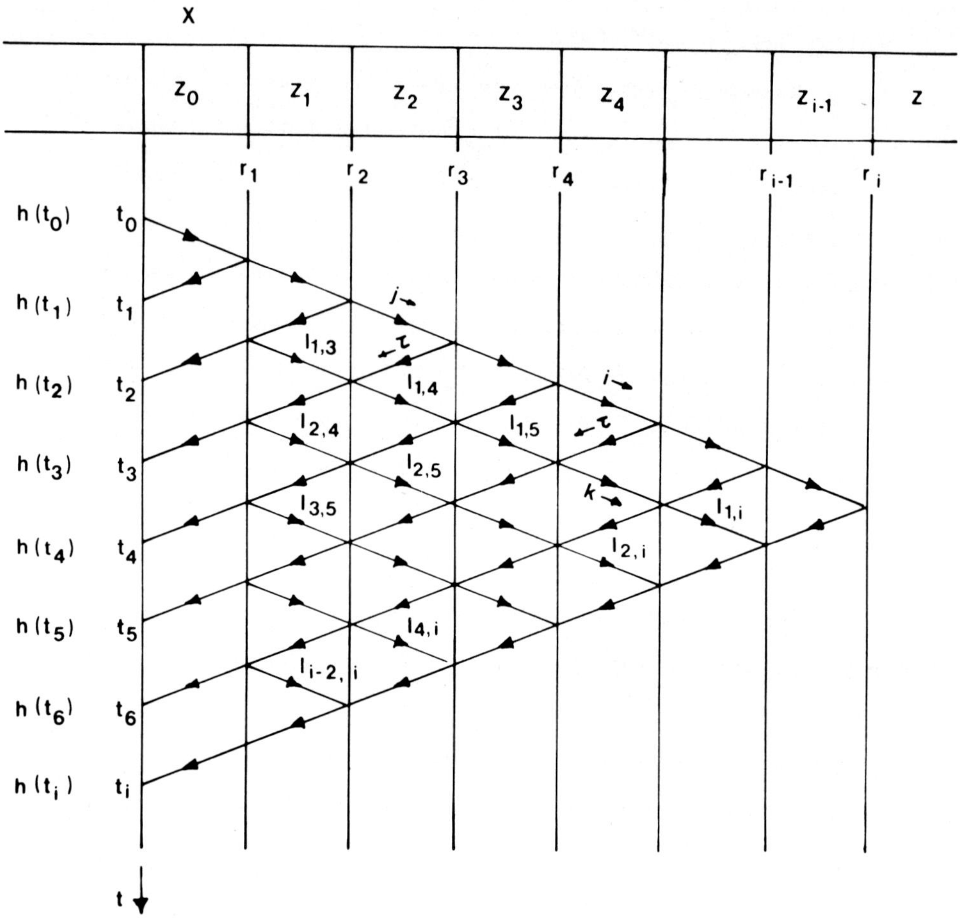

Fig. 1 Pulse bounce diagram for multilayer medium showing multiple internal reflections. X = distance; t = time; Z = impedance; r = reflection coefficient; and $h(t_i)$ = reflected impulse at time t_i.

plitude proportional to $-r^3$). It is easily shown that each $h(t_m)$ is now given by

$$h(t_m) = [r(1-r^2)^{m-1} - L(m)\ r^3(1-r^2)^{m-2}]\ \delta(t-t_m) \tag{23}$$

where $\delta(t-t_m)$ is the impulse at time $t = t_m$. Where

$$L(m) = \frac{(m-2)(m-1)}{2} \tag{24}$$

Therefore,

$$\int_0^t h(\tau)d\tau = \sum_{k=1}^{m} r(1-r^2)^{k-1} - \sum_{k=1}^{m} L(k)\ r^3(1-r^2)^{k-2} \tag{25}$$

where $t_m \leq t < t_{m+1}$. In the limit as the discrete profile becomes continuous $r \to 0$ and $m \to \infty$ in such a way that Eq. (5) is satisfied. The first term on the right in (25) is the same as the expression in Eq. (2). The problem of finding the limit of this expression was rigorously solved earlier, the result being the Eq. (14). The result for this expression could also be obtained by a less rigorous method by recognizing that as $\{{r \to 0 \atop m \to \infty} (1-r^2)^{m-1} \to 1\}$. This makes the first term in (25) equal to mr. Now as $r \to 0$, we get from (5) $m \to \frac{1}{2r} \log \frac{Z_m}{Z_0}$. This implies

$$\lim_{\substack{m\to\infty \\ r\to 0}} mr = \frac{1}{2} \log \frac{Z_m}{Z_0} \tag{26}$$

when substituted in $\int_0^t h(\tau)d\tau = mr$ leads to the same result as in Eq. (14). We will use this line of reasoning to find the limit of the second term. We can write in the limit $m \to \infty$ and $r \to 0$

$$\begin{aligned} \lim_{\substack{m\to\infty \\ r\to 0}} \int_0^t h(t')dt' &= mr - r^3 \sum_{k=1}^{m} L(m) \\ &= mr - r^3 \frac{m(m-1)(m-2)}{6} \end{aligned} \tag{27}$$

where we have made use of (24). Since $m \to \infty$, the above expression becomes

$$\lim_{\substack{m\to\infty \\ r\to 0}} \int_0^t h(t')dt' = mr - \frac{(mr)^3}{6} \tag{28}$$

Substituting (26) in (28) and recognizing that in the limit $t_m \to t$, we get

$$\int_0^t h(t')dt' = \frac{1}{2} \log \frac{Z(t)}{Z_0} - \frac{1}{6} (\log \frac{Z(t)}{Z_0})^3 \tag{29}$$

It is immediately clear that in the presence of third order reflections the impedance profile is not uniquely related to the integral of the impulse response. We refer to uniqueness in the sense that Eq. (29) is a cubic equation in log (Z(t), therefore, in general, it has three roots for every t.

One can similarly derive non-recursive relationships of the type (29) which would include reflections of, say, nth order, the right hand side in such a case being an nth order polynomial in log Z(t). The futility of such an effort is clear from the fact that for any value of t, $\int_0^t h(t')dt'$

would not yield a unique value for Z(t) because of reasons mentioned above.

Limitations such as above should not reduce the importance of Eq. (22) which holds in the absence of multiple reflections. As was mentioned before, experimental situations of importance do exist which satisfy this condition.

2.4 Recursive Determination of Impedance Profile Following Ware and Aki

Now for the sake of completion we will briefly present a recursive method of determining the acoustic impedance profile of a stratified elastic medium from the reflection of plane waves normally incident from a homogeneous medium of known impedance. The medium is modeled as a series of discrete layers, each layer being internally uniform. The thickness of the layers is such that the wave propagation time through each layer is T/2 seconds. It is obvious that the reflected impulse response of such a medium will be a train of impulses spaced T seconds apart. The material that follows draws heavily on Section V of the paper by Ware and Aki[1] and, to a lesser extent, upon the paper by Boakye and Wing.[2]

In this subsection we will take as the dependent variable pressure (r.m.s.) divided by the square root of impedance (this is essentially the square root of intensity for traveling waves). The reflection coefficient for this variable is the same as that for pressure waves as given by (15), since the incident and reflected waves are measured at the same characteristic impedance. The advantage of working in terms of $p/Z^{1/2}$ is that conservation of energy across the interface is expressed as

$$r_k^2 + \tau_k^2 = 1 \qquad (30)$$

where τ_k is the transmission coefficient; τ_k is thus the same for both forward and backward traveling waves, a fact that imposes a useful symmetry on the equations we need relating waves in two adjacent layers.

Let F(k,z) and B(k,z) be the z transforms of the forward and backward traveling waves in the kth layer (see Fig. 2). Let the phase of the waves be determined at the left edge of the layer. We recall that division by z

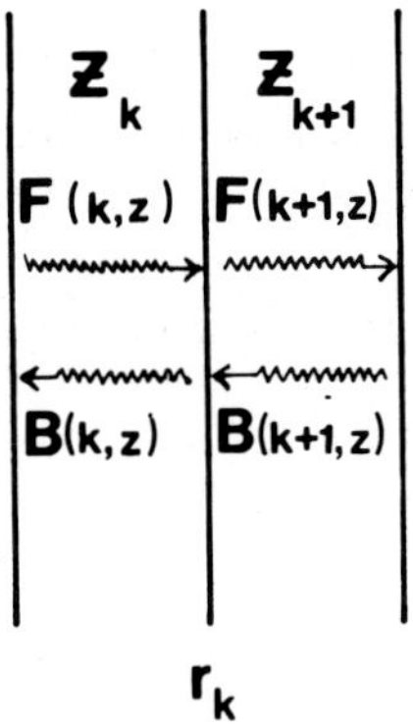

$$r_k = \frac{Z_{k+1} - Z_k}{Z_{k+1} + Z_k}$$

$$\begin{bmatrix} B(k,z) \\ F(k,z) \end{bmatrix} = \frac{1}{w\tau_k} \begin{bmatrix} 1 & r_k \\ z r_k & z \end{bmatrix} \begin{bmatrix} B(k+1,z) \\ F(k+1,z) \end{bmatrix}$$

Fig. 2 Schematic representation of the Z transforms in wave propagation at an interface.

is equivalent to a delay of the round-trip propagation time through a layer and define a variable $W = z^{1/2}$ corresponding to one-way propagation through a layer. The forward traveling wave in layer k+1, or F(k+1,z), will be the sum of that part of F(k,z) transmitted through the interface and that part of B(k+1,z) reflected from the same interface. Thus we can write

$$F(k+1,z) = (\tau_k/W)F(k,z) - r_k B(k+1,z) \tag{31}$$

It takes T/2 seconds for F(k,z) to propagate to layer k+1; hence, the division of F(k,z) by W. The kth interface is at the left edge of layer k+1; therefore, there is no time delay or advance between the reflection of B(k+1,z) and F(k+1,z).

Similarly, we can write B(k,z), the backward traveling wave in layer k, as the sum of the reflected component of F(k,z) and the transmitted component of B(k+1,z).

$$B(k,z) = (r_k/z)\ F(k,z) + (\tau_k/W)\ B(k+1,z) \tag{32}$$

Because the kth interface is at the right edge of layer k, the reflected part of F(k,z) in B(k,z) is delayed by the round-trip time T; thus, F(k,z) is divided by z.

We can combine Eqs. (31) and (32) into a single matrix equation

$$\begin{bmatrix} B(k,z) \\ F(k,z) \end{bmatrix} = \frac{1}{W\tau_k} \begin{bmatrix} 1 & r_k \\ r_k z & z \end{bmatrix} \begin{bmatrix} B(k+1,z) \\ F(k+1,z) \end{bmatrix} \quad (33)$$

This corresponds to Eq. (36) in Ware and Aki.[1] Note that the determinant of the layer matrix is $z(1-r_k^2)/(W\tau_k)^2 = 1$. We can easily invert Eq. (33) to obtain the waves in layer k+1 in terms of those in layer k.

$$\begin{bmatrix} B(k+1,z) \\ F(k+1,z) \end{bmatrix} = \frac{1}{W\tau_k} \begin{bmatrix} z & -r_k \\ -r_k z & 1 \end{bmatrix} \begin{bmatrix} B(k,z) \\ F(k,z) \end{bmatrix} \quad (34)$$

Now suppose that we have n layers separating two semi-infinite regions of uniform impedance. We hypothesize that F(0,z) the wave incident from the left, is a single impulse so that the reflected wave B(0,z) is the impulse response of the system. There will be no backward wave B(n,z) in the region beyond the last interface since there is no further impedance discontinuity to cause reflections. We can thus write

$$\begin{bmatrix} H_1(z) \\ 1 \end{bmatrix} = \begin{bmatrix} A_1 \end{bmatrix} \begin{bmatrix} 0 \\ T_1(z) \end{bmatrix} \quad (35)$$

where H_1 is the z-transform of the reflected impulse response for a wave impinging from the left on the first interface. T_1 is the transmitted impulse response, and A_1 is the overall transfer matrix, the product of all the individual matrices like that of Eq. (33).

If we measure the impulse response one "layer" to the left of the leftmost interface, as shown in Fig. 1, then it is clear that the z transform of the impulse response will be of the form

$$H_o(z) = \frac{h(t_1)}{z} + \frac{h(t_2)}{z} + \frac{h(t_3)}{z} + \ldots . \quad (36)$$

The leading term, $h(t_1)$, is delayed by T seconds because the measurement is made to the left of the r_1 interface by one layer thickness. It is also clear that since the leading term in the impulse response is just the reflection from the leftmost interface,

$$h(t_1) = r_1 \tag{37}$$

Thus, given the impulse response, we immediately know r_1; we can then construct the inverse layer matrix of Eq. (34) to determine B(1,z) and F(1,z). We premultiply both sides of eq. (35) by the matrix of eq. (34), obtaining

$$\begin{bmatrix} B(1,z) \\ \\ F(1,z) \end{bmatrix} = \frac{1}{W\tau_1} \begin{bmatrix} zH_1(z)-r_1 \\ \\ 1-zr_1H_1(z) \end{bmatrix} = \begin{bmatrix} \\ A_2 \\ \\ \end{bmatrix} \begin{bmatrix} 0 \\ \\ T_1(z) \end{bmatrix} \tag{38}$$

where A_1 is the transfer matrix of all interfaces save the leftmost or r_1 interface which has effectively been "peeled off" by the premultiplication.

We see that the pulse train $1-zr_1H_1(z)$ incident from the left on the r_2 interface will produce the reflected pulse train $zH_1(z)-r_1$. Thus this impulse response looking at the r_2 interface from the left is

$$H_2(z) = B(1,z)/F(1,z) = (zH_1(z)-r_1)/(1-zr_1H_1(z)) \tag{39}$$

The leading term of $H_2(z)$ will be r_2/z. Knowing r_2, we can "peel off" the r_2 interface exactly as we did the r_1 interface. Layer by layer, we can thus recover the reflection coefficients by repeatedly premultiplying by the inverse layer matrix. Knowing the source impedance Z_o, we can then construct the impedance profile with repeated application of eq. (39).

Let us look at the physical interpretation of this. From eq. (38)

$$\tau_o B(1,z) = (zH_1(z)-r_1)/W = \left(\frac{h(t_2)}{z^2} + \frac{h(t_3)}{z^3} + \frac{h(t_4)}{z^4} + \ldots\right) W \tag{40}$$

In other words, that part of the backwards traveling wave B(1,z) transmitted into the Z_o region is identical to the reflected impulse response $H_1(z)$ except that the first term of H_1 (the reflection from the r_1 interface) is missing and everything happens t/2 seconds sooner (multiplication by W). This agrees with our physical conception as shown in Fig. 1.

Similarly, from eq. (38)

$$F(1,z) = (\tau_1/W) - \tau_1 B(1,z) \qquad (41)$$

Referring again to Fig. 1, we can identify τ_1/W with the transmitted (and delayed) portion of the original incident impulse and $r_1B(1,z)$ with the reflected component of the backward traveling wave in the first layer. (The minus sign is a result of the reflection coefficient being $-r_1$ for waves traveling leftwards toward the Z_o region from the Z_1 layer).

Actually beginning the division of eq. (39) gives as the first term $h_2/((1-r_1^2)z$; thus

$$r_2 = h_2/(1-r_1^2) \qquad (42)$$

as can be seen from Fig. 1. This agrees with Ware and Aki eqs. (49) and (50). The second reflected pulse, h_2, results from a round trip transmission through the r_1 interface (hence the factor $1-r_1^2$) and a reflection from the r_2 interface (hence the factor r_2). Clearly, in order to find r_2 we must have $|r_1| < 1$; otherwise, eq. (42) becomes indeterminate. We cannot expect to resolve anything on the far side of a perfectly reflecting interface.

It is not necessary to carry out a full deconvolution at each layer, since the next reflection coefficient is given by the leading term of the impulse response. All we need to compute is F and B or two other sequences proportional to these. The following computational scheme will require a total computational labor not much in excess of one complete deconvolution. First, we are given (presumably from experimental data) two sequences, one incident, one reflected, from which we could compute the impulse response of the entire system. In our present notation these would be B(0,z) and F(0,z). We define these as the first of two sets of sequences R and S.

$$\begin{aligned} R(0,z) &= B(0,z) \\ S(0,z) &= F(0,z) \end{aligned} \qquad (43)$$

Now we compute, not $H_1(z)$, but only its leading term in order to obtain r_1. We then generate the next two sequences through

$$\begin{aligned} R(k+1,z) &= zR(k,z) - r_k\, S(k,z) \\ S(k+1,z) &= S(k,z) - zr_k\, R(k,z) \end{aligned} \tag{44}$$

The sequences R(k,z) and S(k,z) are identical to B(k,z) and F(k,z) except for a shift in the time origin and a scaling in magnitude. Carrying two time sequences through the calculation instead of reducing to one at each layer (the impulse response) considerably reduces the total labor. At each stage the reflection coefficient is obtained from the quotient of the two leading terms of R(k,z) and S(k,z).

3. SIGNAL PROCESSING FOR THE DETERMINATION OF THE IMPULSE RESPONSE

In section 2 we considered the theoretical problem of noniteratively relating in a continuous medium the acoustic impedance profile in a given direction and the impulse response calculated from the incident and the reflected waveforms for that direction. We assumed the medium to be lossless and dispersionless and we assumed that the contours of constant impedance were locally perpendicular to the direction of propagation. Evidently, in order to use these relationships one must first calculate the impulse response. In this section we will deal with the problem of determining the impulse response from the incident and the reflected wave-forms.

Let x(t) and y(t) denote the incident and the reflected waveforms, respectively. Let h(t) be the impulse response of the medium along the path of propagation. Then, in the linear regime of operation, the theory of linear systems requires that

$$y(t) = \int h(t')x(t-t')dt' \tag{45}$$

In the above equation given y(t) and x(t), the aim is to solve for h(t). This problem is also known as signal deconvolution. We will now briefly discuss some of the important signal deconvolution techniques that are available today. A much more detailed survey of these techniques can be found in Reference 21.

3.1 Various Signal Deconvolution Techniques

It is interesting to examine the equation (45) in the light of what is actually displayed in A-mode or B-mode echograms. If the incident pulse $x(t)$ is sufficiently narrow in duration, then for all practical purposes, the displayed waveform $y(t)$ is given by

$$y(t) \simeq h(t) \tag{46}$$

that is one actually sees the reflectivity function of the object for that direction. The significance of the reflectivity function is clear from the fact that in the absence of multiple reflections and refraction effects for a plane parallel target, each spike in the reflectivity function corresponds to an impedance discontinuity in the medium; the strength of the spike being proportional to the magnitude of the discontinuity.

If the input signal $x(t)$ is not small in duration, equation (45) must then be solved numerically for $h(t)$. We will come to the discussion of these numerical techniques later. First let us see what happens if $y(t)$ itself is used as an approximation to the reflectivity function even in this case. Since most types of incident pulses have most of the energy at lower frequencies, it follows from the transform theory that $y(t)$ is a smoothed version of $h(t)$. This implies that if $y(t)$ is displayed on a scope without any processing, as is usually done, the resolution capabilities of the system would be limited to primarily the duration of the incident waveforms.

Clearly, for superior spatial resolution and tissue discrimination capability the received $y(t)$ must be processed for a more precise determination of the reflectivity function $h(t)$.

In the hypothetical case of no noise, the determination of the reflectivity function of the target from $y(t)$ and $x(t)$ is simple indeed. For example, taking the Fourier transforms of two sides of (45), we get

$$Y(\omega) = H(\omega)\ X(\omega)$$

or (47)

$$H(\omega) = \frac{Y(\omega)}{X(\omega)}$$

where $X(\omega)$, $Y(\omega)$, and $H(\omega)$ are Fourier transforms of $x(t)$, $y(t)$, and $h(t)$, respectively. A number of problems arise when one attempts to make practical use of eq. (47). There may be points on the frequency axis where $X(\omega) = 0$. In the absence of noise, the transform $Y(\omega)$ would also be zero at these frequencies leading to indeterminate ratios. So we see that even in the absence of noise it is, in general, impossible to determine $h(t)$ exactly if $X(\omega)$ has zeroes. One notes, however, that if $X(\omega)$ has at most a countably infinite number of zeroes, then $h(t)$ is, in principle, perfectly recoverable. That is because the inverse Fourier transform of $\frac{Y(\omega)}{X(\omega)}$ is merely an integration process, to which any single frequency component contributes no area.

In practice the reflected signal $y(t)$ is always corrupted by noise. Various phenomena contribute to this noise. Some of these are the general acoustic back scatter from the target, the noise in the receiver electronics and the quantization noise; the quantization noise being the dominant component for signals digitized for computer processing. In the presence of noise the equation for the output waveform $y(t)$ becoming

$$y(t) = \int_{-\infty}^{\infty} h(t-t')x(t')dt' + n(t) \tag{48}$$

Clearly now

$$\frac{Y(\omega)}{X(\omega)} = H(\omega) + \frac{N(\omega)}{X(\omega)} \tag{49}$$

In the neighborhood of zeroes of $X(\omega)$, it may take values much smaller than those of $N(\omega)$. Thus the term $\frac{N(\omega)}{X(\omega)}$ may have a much larger magnitude than $H(\omega)$. The inverse transform of $\frac{Y(\omega)}{X(\omega)}$ will then be strongly influenced by this large term, and will no longer resemble $h(t)$. Instead it will contain many rapid noise-like variations.

Given the waveforms x(t) and y(t) and statistical properties of the noise n(t), the determination of h(t) from eq. (48) has been of interest in many areas of science, engineering, and mathematics. The techniques can be roughly divided into the following categories.

(1) Inverse Filtering:[4] Statistical properties of the signals and noise are not required. Basically the inverse filter is the digital implementation of (47) with protection against division by $X(\omega)$ at those frequencies where the signal to noise ratio is low. Brigham, et al.[4] have developed an iterative time domain implementation of the filter.

(2) Algebraic Methods:[5] One of the more recent developments in deconvolution theory has been the application of the projection method for solving a set of linear equations to the general problem of extracting a signal which is convolved with a position variant operator. This is one of the techniques we hope to try for the determination of reflectivity function in dispersive and attenuating media.

(3) Matched Filtering:[6-8] We have successfully applied the concepts of matched filtering to detect extremely small ultrasonic impedance differences (about 1%) behind the skull bone. The matched filter, which involves correlating the incident ultrasonic signal with the received waveform, is optimum in the presence of white Gaussian noise. If the noise spectrum cannot be considered to be white the optimality of the deconvolution can be preserved by using a prewhitening filter.[26]

(4) Least Squares or Wiener Filtering Techniques:[9-13] If the correlation properties of the signal and noise ensembles are known and if the ensembles can be assumed to be stationary, then this is the optimum (in the least-squares sense) procedure for determining the reflectivity function. Often, however, gross approximations are made for the signal and noise power spectral densities required in this technique. For example, the ratio of signal to noise power spectral densities, which is a function of frequency, may simply be replaced by a constant equal to

the signal to noise power ratio. With this approximation the optimality of the technique is open to question.

(5) Constrained Deconvolution:[14-17] Note that in the presence of noise n(t) the function h(t) in eq. (48) can never be determined exactly. The method of least squares mentioned above assumes the signal and noise waveforms and the impulse response to belong to ensembles, and then gives a best estimate in an average sense for all the members of the ensemble. If an optimum estimation is required for the one reflectivity function under consideration, and if the signal and the noise ensembles cannot be assumed to be stationary and, also, if all that is known about the noise is its average power, then the method of constrained deconvolution can be used.

(6) State-Space Techniques:[18-21] In the literature on seismic signal processing attempts have been made to represent the reflectivity function h(t) by a first order differential equation called the state equation. If this is possible, one can use the Kalman filtering technique to estimate the function h(t). One advantage of this technique is that it does not require the signals to be stationary in the statistical sense. Also, because of the recursive nature of the formulation the technique can be implemented in real time. To apply this method considerable research yet needs to be done in determining the parameters of the first-order differential equation representation of the reflectivity function.

(7) Homomorphic Deconvolution:[22] Homomorphic systems are a class of nonlinear systems which satisfy a generalized principle of superposition. Such systems are particularly useful in separating signals which have been combined through convolution by suitable operations on a time domain function called the complex cepstrum. This method is deterministic in nature and does not require explicit assumptions regarding the character of the reflectivity function. However, the method's success

depends upon the separability of the incident pulse x(t) and the reflectivity function h(t) in the complex cepstrum.

(8) The Method of Backus and Gilbert:[23] Backus and Gilbert have recently proposed a method which allows a direct **and** simple control over the tradeoff between noise and resolution in deconvolution. In this approach it is assumed that every point in the reflected waveform is an integral of the product of a weighting function and the reflectivity function. Suppose we desire a certain average of the reflectivity function. This averaging function can be approximated by a linear combination of the original weighting functions. The coefficients of this expansion are important because the desired estimate of the reflectivity function is a linear combination of the observed data in the reflected waveform, the combination being formed by the same set of coefficients. These coefficients also determine the variance of the noise in the estimate. Therefore, for a given desired spatial resolution, one knows a priori the uncertainty of the estimate. A method similar and in some respects superior to that of Backus and Gilbert was developed independently by McGillem and Riemer.[23]

3.2 Matched Filtering for Media with Discrete Layers

In the beginning stages of a research effort dealing with the determination of acoustic impedance profiles one is more likely to use a target made of a small number of parallel layers. For a discrete target like this one does not need to use the theoretical expression derived in sections 2.1 - 2.3 to obtain the impedance profile from the impulse response. The impedance profile can now be obtained by the direct approach on a layer by layer basis as discussed in section 2.4.

For example, consider a target consisting of N lossless layers of impedances $Z_1, Z_2, \ldots, Z_N$ immersed in a medium of characteristic impedance Z_0. The impulse response of such a medium can be expressed as

$$h(t) = \sum_{i=1}^{N+1} A_i \, \delta(t-t_i) \tag{50}$$

where $\delta(t-t_i)$ represents an impulse located at $t=t_i$, and where, in the absence of multiple reflections, t_i is the time at which the reflection from the ith discontinuity arrives at the receiver, the strength of this reflected impulse being A_i. It is clear from the definition of the reflection coefficient that

$$A_1 = \frac{Z_1-Z_0}{Z_1+Z_2}$$

$$A_2 = \frac{Z_2-Z_1}{Z_2+Z_1} \left[1 - (\frac{Z_1-Z_0}{Z_1+Z_0})\right]^2 \qquad (51)$$

$$\begin{matrix} \cdot & \cdot \\ \cdot & \cdot \\ \cdot & \cdot \\ \cdot & \cdot \end{matrix}$$

Therefore, if one knows A_1 one can use the first equation in the above set to find Z_1. Knowing A_2 and Z_1 one can use the second equation to find Z_2, etc. It can be shown that even in the presence of multiple reflections, given the impulse response coefficients A_1, A_2, ..., one can uniquely obtain the impedances Z_1, Z_2, ...[6] These iterative calculations for impedances are stable only for small N.

Therefore, for the case of a discrete medium, the basic question that remains to be answered is that given the incident and the received waveforms how does one optimally calculate the coefficients A_i's. In the rest of this section we will address this problem. Substituting (50) in (49) we get for the received waveform y(t):

$$y(t) = \sum_{i=1}^{N+1} A_i \, x(t-t_i) \qquad (52)$$

Given the received waveform the aim is to estimate the values of the coefficients A_i. The waveform y(t) is in practice recorded in the presence of noise. Also, in the digital calculation of the impedance profiles, further noise is introduced by the sampling and quantization of the waveforms. If the noise can be considered to be white Gaussian the following correlation technique yields the maximum likelihood estimate for the coefficients

A_i. In the presence of colored noise the optimality of the estimate can be preserved by using a prewhitening filter.[26] However, this case will not be treated here.

Multiplying both sides of (52) by $x(t+t')$ and integrating over the entire axis, we get

$$R_{xy}(t') = \sum_{i=1}^{N+1} A_i R_{xx}(t'-t_i) \tag{53}$$

where $R_{xy}(t')$ and $R_{xx}(t')$ are the cross-correlation and the autocorrelation functions, respectively. If the cross-correlation function is sampled at $t = t_1', t_2', t_3', \ldots$, we can then generate a set of simultaneous equations given by

$$\begin{aligned} R_{xy}(t_1') &= A_i R_{xx}(t_1' - t_i) \\ R_{xy}(t_2') &= A_i R_{xx}(t_2' - t_i) \\ &\cdots \\ &\cdots \end{aligned} \tag{54}$$

Let us for a moment assume that it is known a priori that the impulse response consists of N discrete impulses which in the presence of sufficiently weak multiple reflection would mean an N layer target. Now, if one samples the cross-correlation function at N+1 instants of time $t_1', t_2', \ldots, t_{N+1}'$, one would end up with N+1 equations in N+1 unknowns $A_1, A_2, \ldots, A_{N+1}$. The coefficients A_i can thus be determined provided $t_1, t_2, \ldots, t_{N+1}$ are known. An analysis of the reflected waveforms for the case of oscillatory inputs shows that the time instants $t_1, t_2, \ldots, t_{N+1}$ correspond to the maximas of the magnitude of the cross-correlation function with an uncertainty of half a cycle. The errors caused by using the maximas of the magnitude of the cross-correlation function for t_i's depend upon how rapidly the autocorrelation function of the incident waveform drops off in the vicinity of the origin. Experimental results indicate that for the types of waveforms and targets used in our experiments these errors could easily be ignored.

For the case of discrete targets the above technique works well for both the oscillatory and the non-oscillatory incident waveforms we have used. For non-oscillatory (highly damped oscillatory) waveforms we have developed another deconvolution technique that appears very promising. In this technique one essentially models the initial rise of the incident waveform by an exponential. More details on this can be found in Reference 6.

4. EXPERIMENTS

Essentially all experimental data collected has been on physical models involving layered media having plane parallel faces, which can be considered of essentially infinite extent in the direction perpendicular to the incident longitudinal acoustic wave propagation. This wave is at normal incidence to the interfaces of the model. The exception to the plane parallel face structures involves acoustic wave transmission through excised sections of human skull bone which can considerably modify the amplitude and temporal character of the acoustic pulse.

Acoustic interrogation of the models has involved both damped oscillatory and unipolar pulse type waveforms, as will be expanded on subsequently.

4.1 Instrumentation and Technique

A block diagram of the system components used for this study is shown in Fig. 3 and a photograph of the equipment is shown in Fig. 4. For the damped oscillatory wave interrogation, a focused transceiver is used. Unipolar type pulses are generated by an unfocused plane face one inch diameter disc, and input and output waves from the model are received by a small probe. Each of these schemes is presented in Fig. 5.

When the unipolar system is used for interrogation, the system configuration involves the unipolar transducer source driven by an appropriate ramp voltage. The small piezoceramic probe records the input wave before entering the target and the received reflected wave from the target. An example of these waveforms for a single

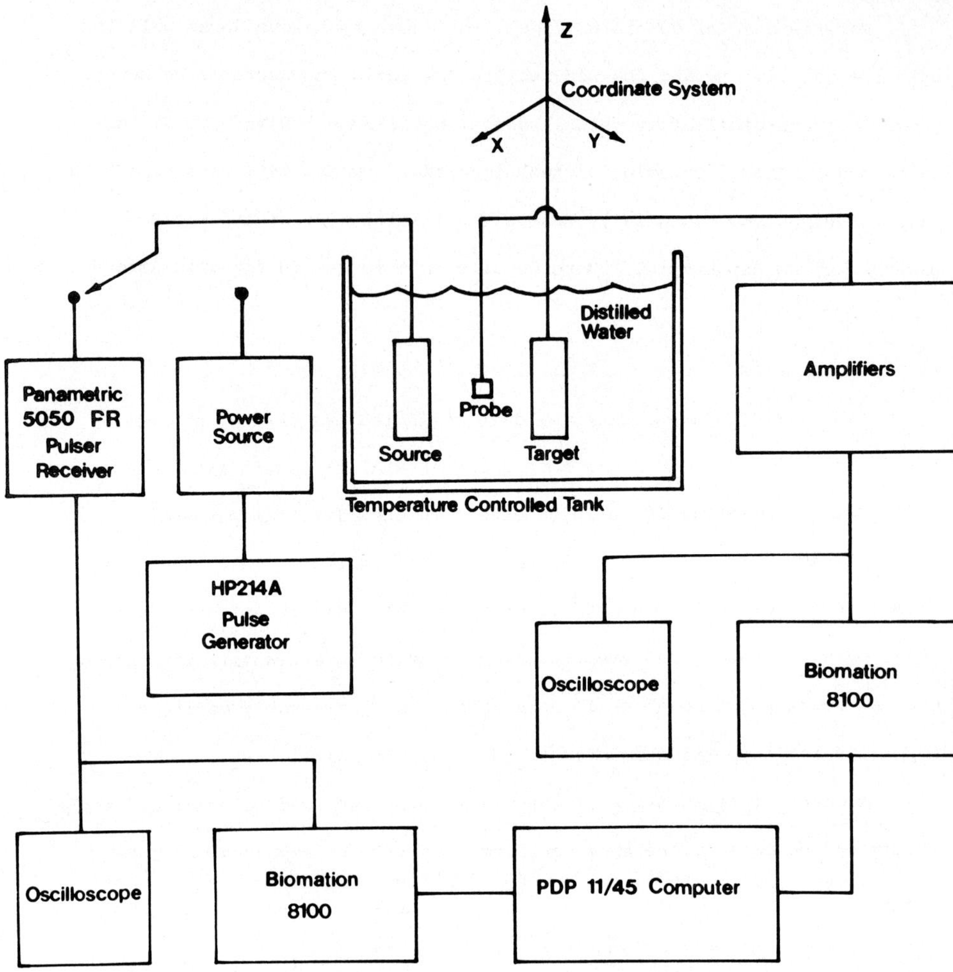

Schematic Diagram-System for Acoustic Parameter Determination

Fig. 3 Schematic diagram of the system used for acoustic parameter determinati

---

Fig. 4 Photograph of the system used for acoustic parameter determination. A, the Pulser Amplifier; B, the Oscilloscope; C, the Digitizer; D, Pulse Generator; E, Power Supply; F, Coordinate System; G, Amplifiers, and H is the Sound Tank.

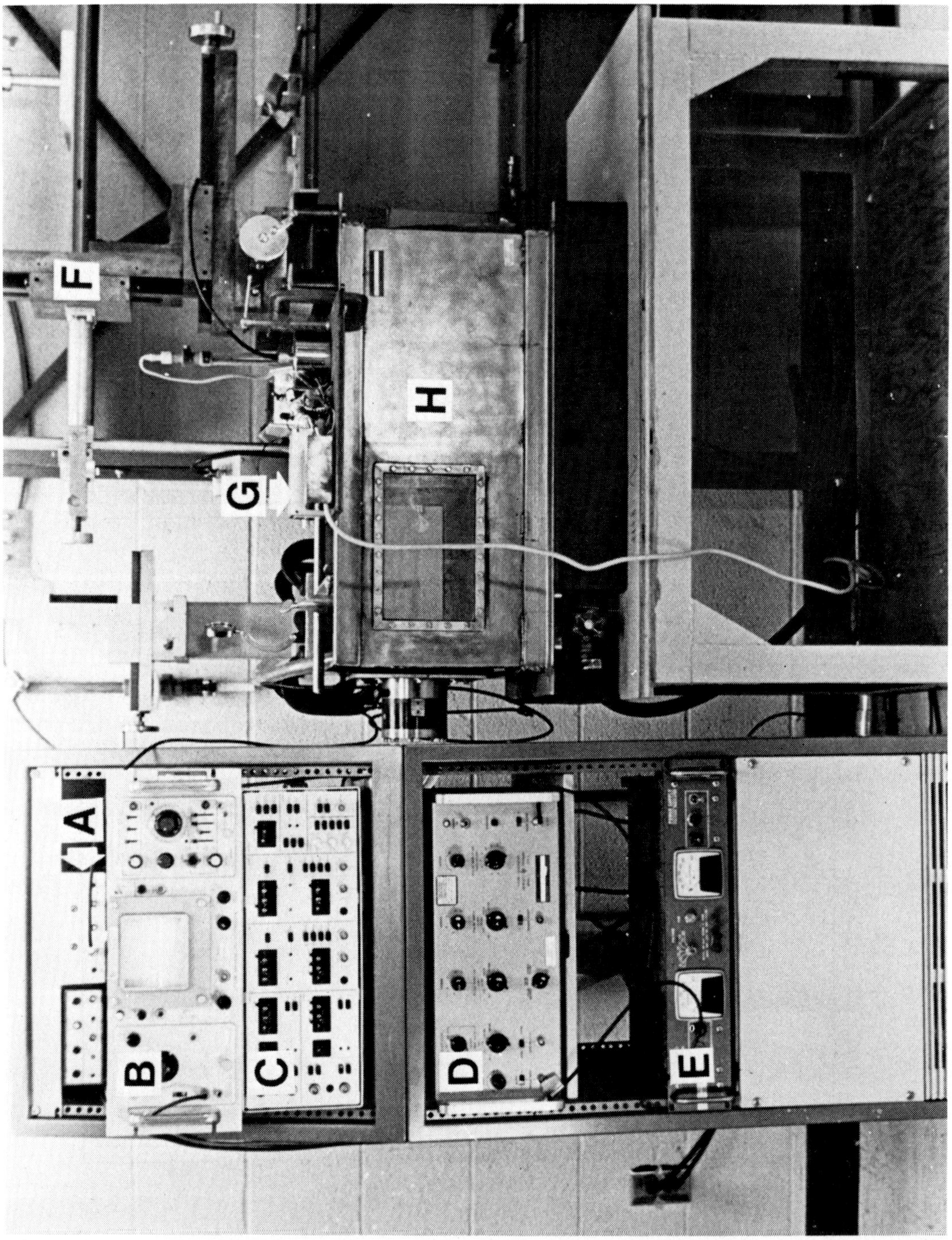
A
B
C
D
E
F
G
H

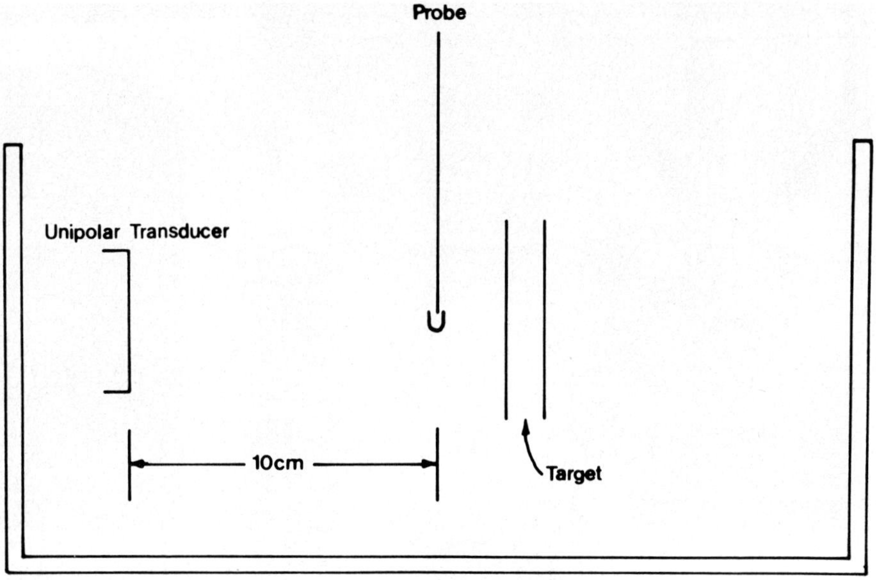

A. Unipolar Pulse Interrogation

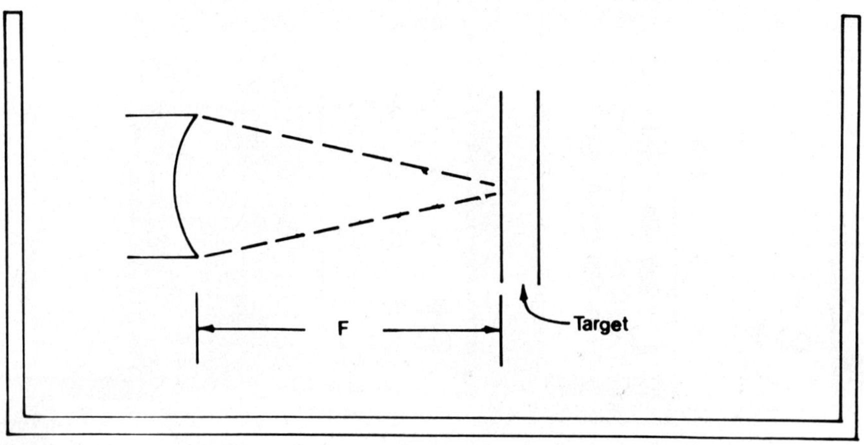

B. Focused Transceiver Interrogation

Fig. 5 Schematic diagram of transducer-target arrangements used for target interrogation.

layer lucite target is shown in Fig. 6. Transceiver operation involves a transducer having an on-axis longitudinal beam plot (taken by recording reflections from a 1.0 mm diameter steel ball target) shown in Fig. 7, and a lateral beam plot recorded with the same ball in the focal plane is also shown. The resonant frequency of this particular transceiver is 1.7 MHz, and the focal plane is 20 cm from the transceiver face. This transducer has a total aperture angle of 14 degrees. No programs have been implemented correcting for amplitude drop in the longitudinal direction, angulation variations across the beam profile for planar targets and out-of-phasing occurring as the targets move out of the focal zone for the focusing transceiver. Of necessity, interrogated thin layered targets will yield data with the greatest absolute accuracy under these circumstances.

Digitized waveforms are stored in the digitizer (Biomation 8100 ---8-bit with 10 nsec. minimum sample interval) in 2,000 word lengths. This word sequence is transferred to DEC tape on the dedi-

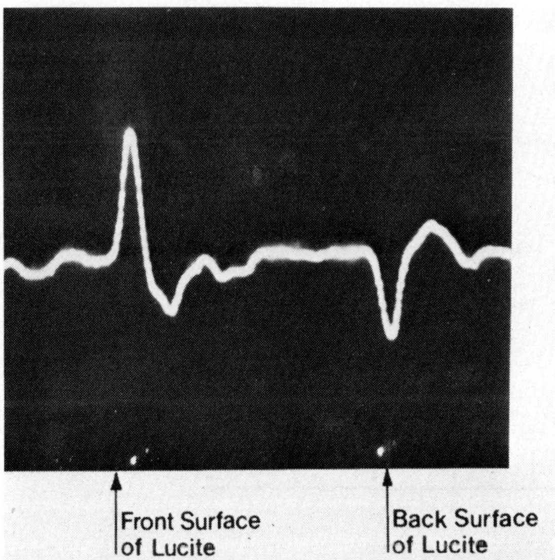

Fig. 6 Probe output for signals from a single layer lucite target in water with the unipolar acoustic interrogating pulse.

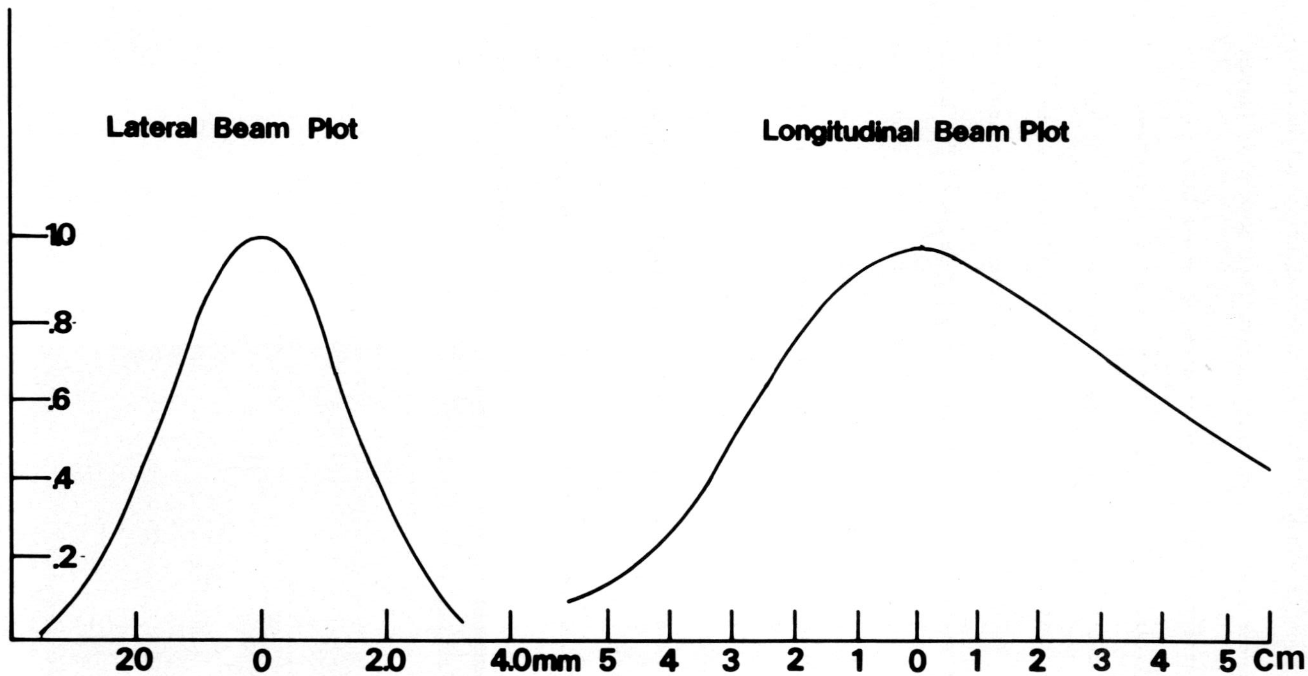

Fig. 7 Field plots of the interrogating beam from the focused transceiver.

cated PDP-11/45 computer. The input of incident waveform and the received or reflected waveforms are separately recorded. For the unipolar pulse the input waveform is recorded on the probe with no target in place, while for the transceiver arrangement, the input waveform is recorded from an essentially perfectly reflecting target placed at the focal plane. The target is then placed in position for the received waveform recording in either case. This recording method facilitates the program methodology which performs a cross-correlation as discussed previously. From this correlation function the impulse response of the system is computed. Subsequent processing as discussed previously yields the impedance profile.

In order to project from the present status of these analytical and experimental considerations to the realm of human tissue parameter characterization, questions of spatial resolution and sensitivity require discussion. Focused transceivers of the type used for this experimental program have a lateral resolution which is substantially equal to one wavelength at the transceiver resonant frequency operation. This resolution capability is demonstrated in Fig. 8. Two converging monofilament nylon threads (0.010" diameter) are scanned by the transceiver and the position of coalescence of two separate target delineations into one single reflection is taken as a meaningful indication of lateral resolution. Depth resolution is determined by using a similar target. An indication of depth resolution capability can be obtained from Fig. 9. With the electric output wave shown it is evident that depending on the type of target, spacing distances of one or two wavelengths can be resolved.

Some soft tissue interface acoustic impedance differences are listed in Table II.[25] It is apparent that acoustic interrogating overall system sensitivity should be such that they should be capable

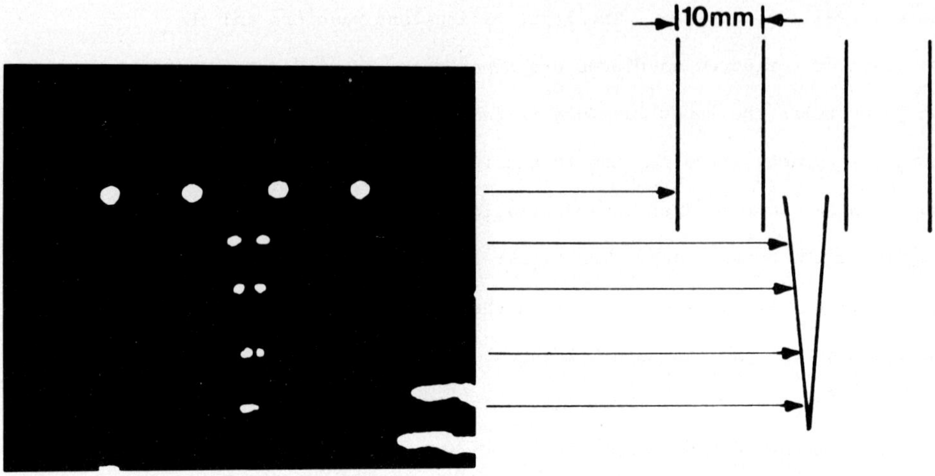

Fig. 8 Target representation of lateral resolution capability of the focused transceiver. The four (4) horizontally aligned spots in the photograph represent echoes received from four monofilament nylon strings (0.010" dia.) spaced 10 mm on centers. The lower V configuration of spots represents the scanning of a V-shaped figure of the same nylon string. The horizontal arrows represent the direction and position of transceiver scanning of the target.

of detecting reflection differences at interfaces down to the order of one part per thousand. Sensitivity data taken with a typical transceiver used in these experiments is shown in Fig. 10. A linear scan of the transceiver is made across a specially constructed target which has step gradiations in acoustic impedance ranging from 0.2% to 1.0%. By appropriate settings of the gain and reject controls in the electronic system, it is possible to detect the 0.2% acoustic impedance mismatch in the target. Questions involving the accuracy with which acoustic impedances can be determined at deep body sites remain to be answered. The subsequently presented experimental data on physical models will show the present status of accuracy and repeatability as the models are interrogated and the digitized data transmitted to the computer for processing.

TABLE I

ACOUSTIC REFLECTION COEFFICIENTS - ABSOLUTE VALUES (25)

	Water	Fat	Muscle	Skin	Brain	Liver	Blood	Skull Bone	Lucite
Water	0.0	.047	.02	.029	.007	.035	.007	.57	.35
Fat			.067	.076	.054	.049	.047	.61	.39
Muscle				.009	.013	.015	.02	.56	.33
Skin					.022	.0061	.029	.56	.32
Brain						.028	.00*	.57	.34
Liver							.028	.55	.32
Blood								.57	.35
Skull Bone									.29

* Clotted and freshly injected blood can be detected ultrasonically in the live and formalin fixed brain.

TABLE II

DAMPED OSCILLATORY WAVE (1.7 MHz) INTERROGATION OF SINGLE LUCITE LAYER IN DISTILLED WATER

		Derived Acoustic Impedance $gr.cm^{-2}\ sec^{-1} \times 10^{-5}$	
Lucite Thickness (mm)	Lucite Thickness (computed mm)	Lucite Z_1	Water Z_2
4.89	4.85	3.25	1.46
6.64	6.59	3.23	1.52
9.33	9.19	3.25	1.58
12.28	12.17	3.17	1.59

Used velocity of 2.68 mm/μs for lucite

Used α of 0.35 cm^{-1} for lucite

Quoted values for lucite: $\alpha = 0.2\ cm^{-1}$ at 1 MHz

$$Z = 3.2 \times 10^5\ gr.cm^{-2}\ sec^{-1}$$

UNIPOLAR WAVEFORM

Probe pickup of input and received wave

6.64	6.51	3.10	1.52

Used α of 0.30 cm^{-1} for lucite

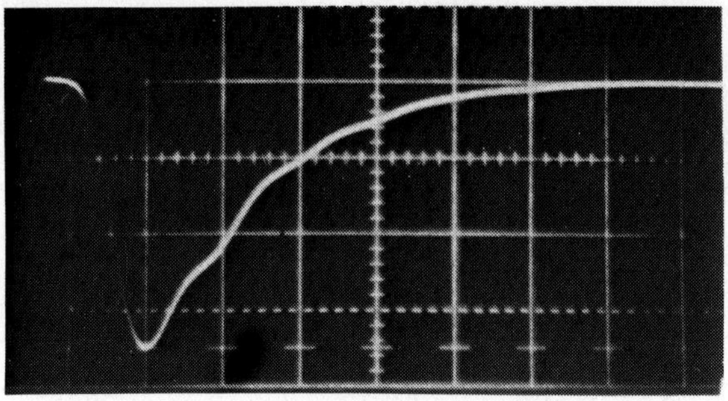

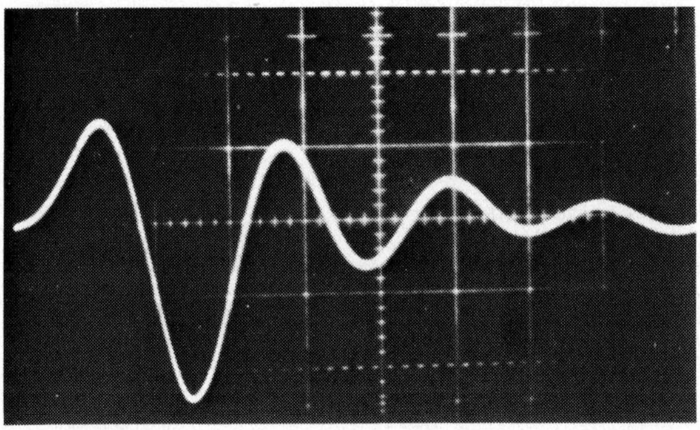

Fig. 9 Electrical waveforms associated with focused transceiver operation.

Fig. 10 Visualization of interfaces with a graded set of acoustic impedance differences. The horizontal echo lines shown between the 6 vertical bands (5 zones) show reflections from regions characterized by reflections from interfaces with acoustic impedance differences across the interface ranging from 0.2 to 1.0%. The dB values shown in the upper right hand corner represent the effective relative gain of the system. At 40 dB (least sensitive) the 0.2% acoustic impedance mismatch does not show, while it begins to show at 30 dB (i.e., 10 dB less attenuation in the system).

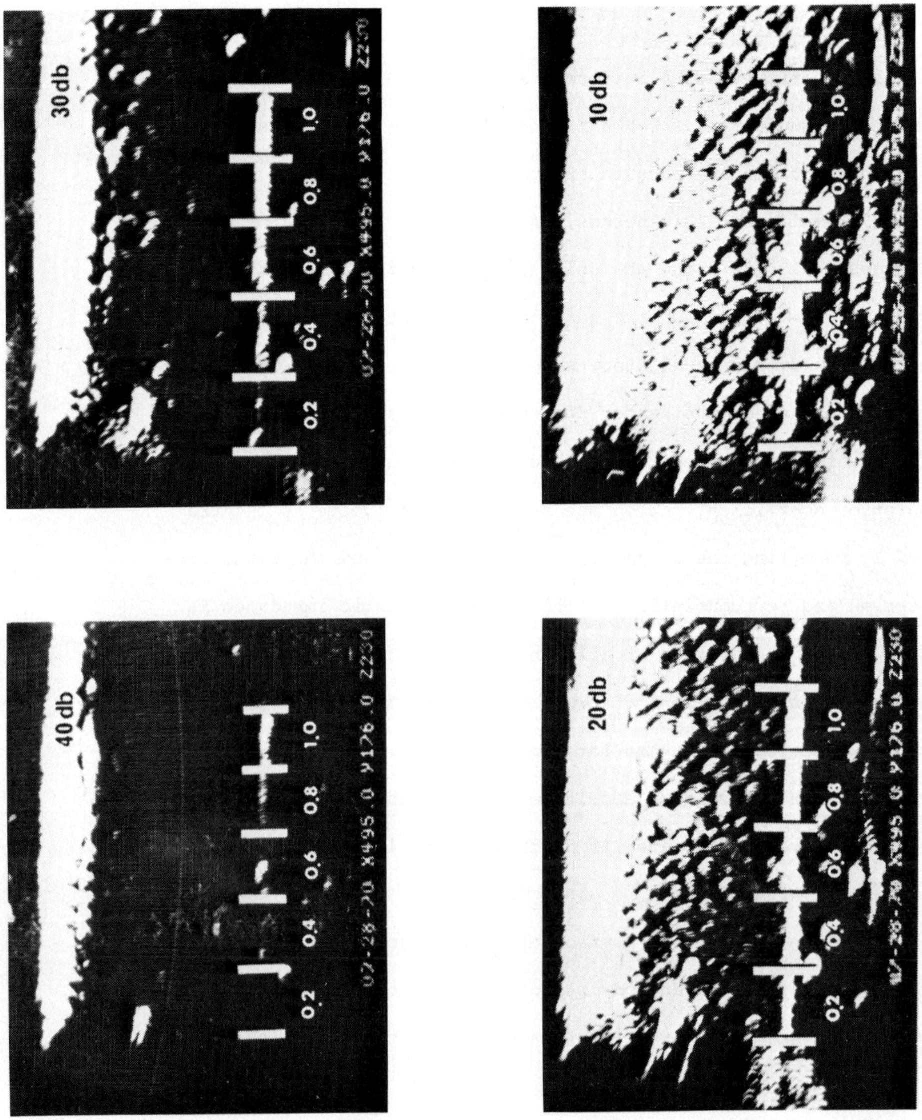
40 db
30 db
20 db
10 db

4.2 Experimental Data

All experiments were performed in distilled water maintained at 37°C in the sound tank. Table II shows the results of runs involving damped oscillatory wave interrogation for single layer lucite targets of different thicknesses. A single run is shown for one thickness of lucite using the unipolar pulse system. A different $\alpha(0.30\ cm^{-1})$ was used for the unipolar pulse interrogation than was used for the damped oscillatory wave base $\alpha(0.35\ cm^{-1})$. These values are different since the center frequency and frequency bandwidth is different for each case. Values are inserted for the pressure absorption coefficient. When the same piece of lucite is interrogated by resetting the target in the sound field and the signal reprocessed and reproducibility of the lucite acoustic impedance is within ±1% of the average of all runs. Table III shows the results of separate interrogations of two different multilayer targets (lucite-water-lucite) in the interrogating tank. Since there is no correction for focusing or amplitude drop off in the interrogating field the Z_4 value for water is deviating from its true value by approximately 20%.

Lucite in water is of course not a very appropriate model for soft tissues of the human body because of its large reflection coefficient (approximately 30%). More appropriate targets which have

TABLE III
Damped Oscillatory Wave (1.7 MHz) Interrogation of Lucite-Water-Lucite Target

Model Thickness mm			Computed Thickness mm			Derived Acoustic Impedance $gr.cm^{-2}sec^{-1}x10^{-5}$			
Lucite	Water	Lucite	Lucite	Water	Lucite	Lucite Z_1	Water Z_2	Lucite Z_3	Water Z_4
5.98	9.07	6.12	5.98	9.06	6.21	3.14	1.56	2.99	1.93
5.98	9.07	6.12	5.98	9.05	6.22	3.05	1.47	3.12	1.81
4.86	9.07	4.86	4.88	8.90	4.77	3.22	1.47	3.10	1.77

been interrogated include ρc rubber in water and silastic rubber multilayer sandwiches in water. The silastic sandwiches have been constructed to have impedance differences across the interfaces of approximately 1%. These targets have yielded values which are approximately correct and the results of the ongoing studies will demonstrate what the yield will be with this approach.

This model study has additionally involved interrogation of layered media behind human skull bone sections. Human skull itself presents a multilayered media which considerably modifies the amplitudes and temporal character of a transmitted ultrasonic wave. An appreciation of this acoustic wave modification can be obtained by observing the waveforms of Figs. 11, 12, and 13. In Fig. 11 the damped oscillatory wave is that received on a large aperture focused transceiver when the wave is reflected from a multilayer target in water. Figs. 12 and 13 show the result of inserting a human skull bone section at two different skull regions between the transceiver and the target. Fig. 14 shows the received signal from an approximately 1% acoustic impedance difference silastic-silastic (Dow Corning RTV) interface with an intervening human skull section.

A set of data taken on a tissue simulation target involving a silastic sandwich (approximately 6% difference in acoustic impedance between the sandwich elements) immersed in water is shown in Table IV. For the case of no intervening skull the target acoustic impedance values are normalized by taking $Z_1 = 1.5 \times 10^{-5}$. When skull bone is inserted between the transceiver and the target it is necessary to assume a value of Z_2 which in this case was set to be substantially that derived from the non-skull case. With the exception of the Z_4 values, all computed values of common elements for the various conditions are within a few percentage points of each other.

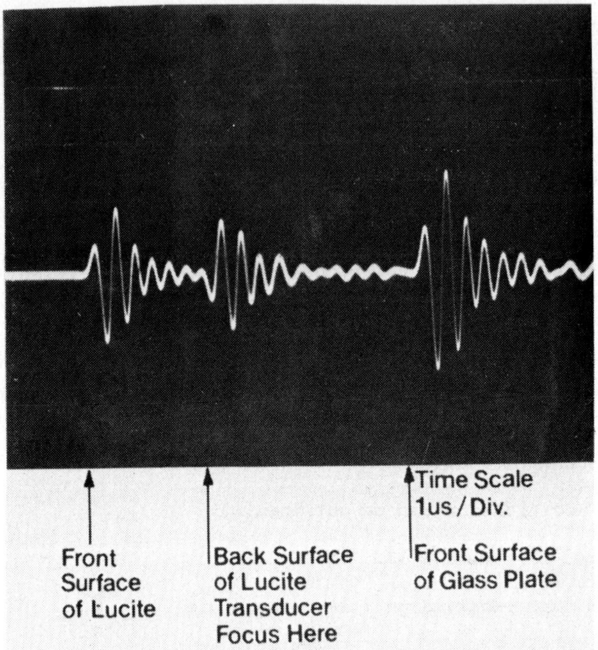

Fig. 11 Focused transceiver output from a water-lucite-water-glass series of interfaces without intervening skull bone.

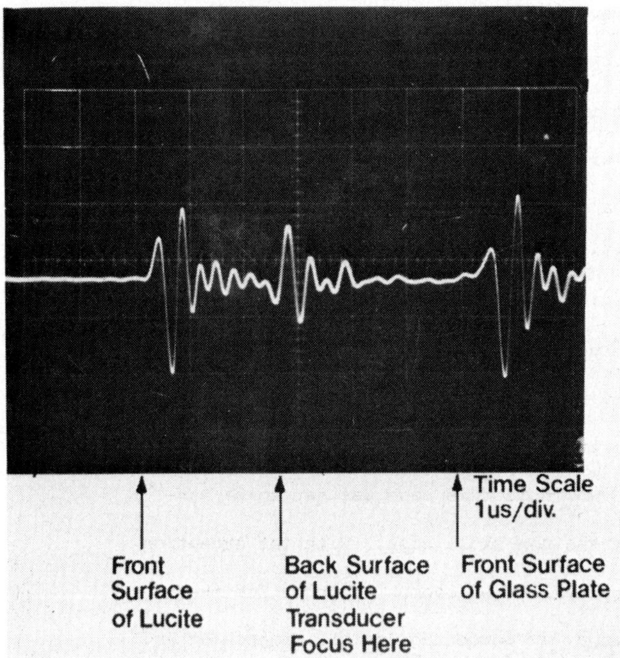

Fig. 12 Focused transceiver output from a water-lucite-water-glass series of interfaces with intervening skull bone between the transceiver and the first water-lucite interface.

TABLE IV

ACOUSTIC IMPEDANCE VALUES DETERMINED FOR MULTILAYER TARGERS

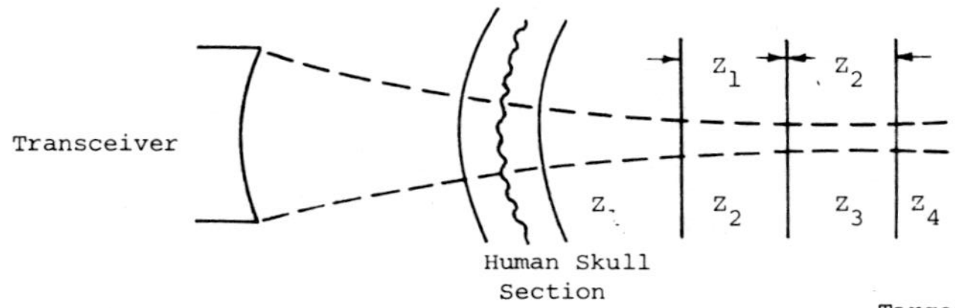

Assumed velocity in silastic = 1.1 mm/μs

Assumed α = 0.06 mm^{-1}

Run	Z_1	Z_2	Z_3	Z_4		Human Skull	Target Thickness Computed in mm Z_1	Z_2	Acoustic Impedance $gr.cm^{-2}\ sec^{-1} \times 10^{-5}$ Assumed Z_1	Computed Z_2	Z_3	Z_4
	Water	Silastic	Silastic	Water								
1						No	6.21	5.54	1.50	1.05	.980	1.54
2						No	6.23	5.56	1.50	1.07	.995	1.53
3						No	6.21	5.56	1.50	1.07	.998	1.52
					Average		6.217	5.553	1.50	1.063	.991	1.53
									Z_2			
4						Yes	6.14	5.49	1.065		1.014	1.43
5						Yes	6.17	5.49	1.065		1.002	1.50
6						Yes	6.05	5.65	1.065	–	1.007	1.53
					Average		6.12	5.543	1.065		1.007	1.487
						REVERSED SANDWICH						
	Water	Silastic	Silastic	Water								
7						Yes	5.51	6.14		1.005	1.083	1.64
	SANDWICH BACK TO ORIGINAL ORIENTATION - THICKER SKULL SECTION REGION THAN FOR RUNS 4-7											
	Water	Silastic	Silastic	Water								
8						Yes	6.01	5.67	1.065		1.037	1.39
9						Yes	6.07	5.69	1.065		.999	1.65
10						Yes	6.01	5.65	1.065		.992	1.66
					Average		6.01	5.67	1.065		1.009	1.567

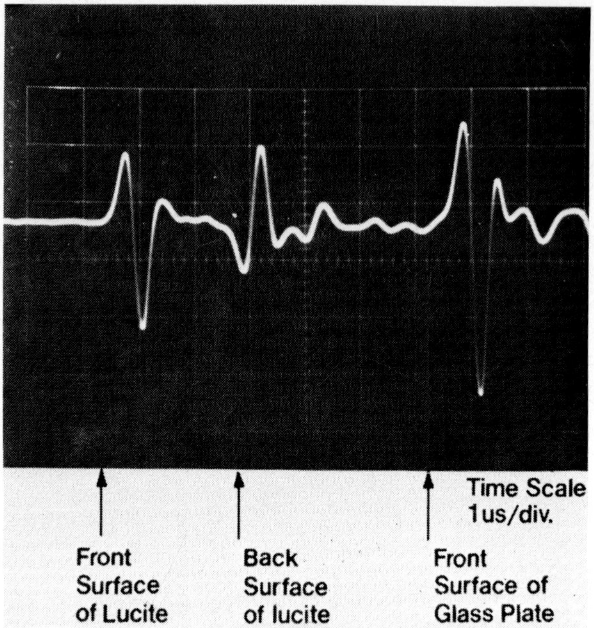

Fig. 13 Focused transceiver output from a water-lucite-water-glass series of interfaces with intervening skull bone at a thicker skull bone region than that used for the information in Fig. 11.

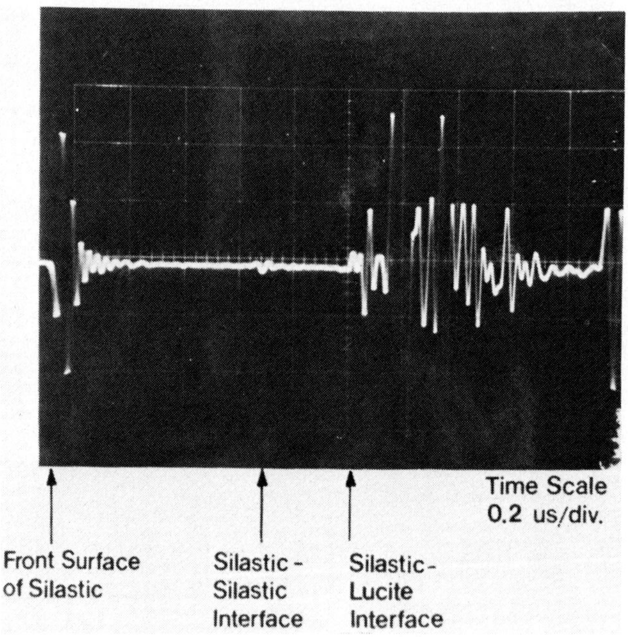

Fig. 14 The received signal from an approximately 1% acoustic silastic-silastic interface with an intervening human skull section.

TABLE V

ACOUSTIC IMPEDANCE VALUE DETERMINED FOR BRAIN TISSUE

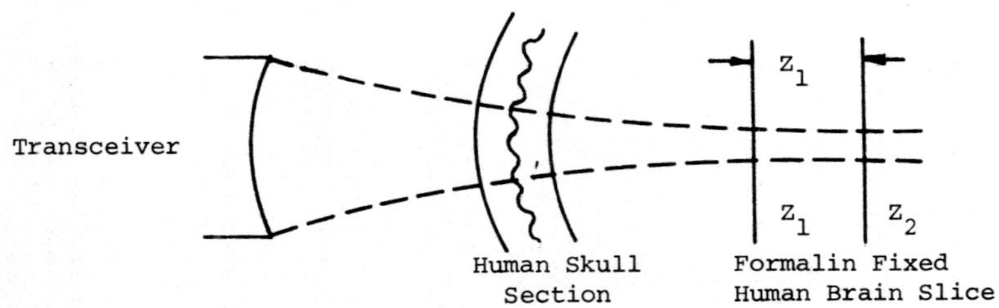

Run	Target Description		Human Skull	Target Thickness Computed in mm	Acoustic Impedance gr. cm^{-1} sec^{-1} x 10^5		
	Z_1	Z_2			Assumed Z_1	Computed Z_1	Computed Z_2
1	Brain	Water	No	10.86		1.012	.999
2	Brain	Water	Yes	11.01	1.020		1.001

Interrogation of a formalin fixed human brain slice in water with and without intervening skull yielded the values shown in Table V. The Z_1 value (1.530×10^{-5}) was inserted as known for the skull intervening case, and the point of significance is the computed Z_2 value of 1.001×10^{-5}.

It is our feeling that this research in the quantitative determination of acoustic impedance as it might be applied to soft tissue is in its early stages of development. Much work needs to be done to relate this particular parameter to the other interacting acoustic parameters mentioned previously. Progress towards the goal of tissue typing through acoustic parameter characterization will be aided by accurate assessment of this parameter.

5. ACKNOWLEDGEMENTS

The authors would like to thank Leland R. Beaumont, John Wolfley, and Kris A. Dines for their contributions to the development of the theoretical ideas presented here, and Narendra T. Sanghvi for help and assistance in implementing the experimental program.

REFERENCES

1. J. A. Ware and K. Aki: Continual and Discrete Inverse Scattering Problems in a Stratified Elastic Medium. J. Acoust. Soc. Amer., 45 (1969) pp. 911-921.

2. K. A. Boakye and O. Wing: On the Analysis of Realization of Cascaded Trahsmission Line Networks in the Time Domain. IEEE Trans. on Circuit Theory, Vol. CT-20 (1973) pp. 301-307.

3. J. P. Jones: A New Broadband Ultrasonic Technique With Biomedical Implications. 1: Background and Theoretical Discussion. Presented at the 83rd Meeting of the Acoustical Society of America, Buffalo, New York.

4. E. O. Birgham, H. W. Smith, F. X. Bostik, Jr., and W. C. Duesterhoeft, Jr.: An Iterative Technique for Determining Inverse Filters. IEEE Trans. on Geoscience Electronics, Vol GE-6 (May 1968) pp. 86-96.

5. T. S. Huang, D. A. Barker and S. P. Berger: Iterative Image Restoration. Applied Optics, Vol. 14 (May 1975) pp. 1165-1170.

6. A. C. Kak, L. R. Beaumont and J. Wolfley: Signal Processing in the Determination of Acoustic Impedance Profiles. Research Report TR-EE-75-5, (Dec. 1974) School of Electrical Engineering, Purdue University.

7. A. C. Kak and F. J. Fry: Acoustic Impedance Profiling: An Analytical and Physical Model Study. in Ultrasonic Tissue Characterization, edited by M. Linzer, NBS Publication (To be submitted for publication).

8. A. C. Kak, F. J. Fry, K. A. Dines and N. Sanghvi: Determination of the Acoustic Impedance and the Attenuation Coefficient Profiles. To be submitted for publication to the IEEE Trans. on Sonics and Ultrasonics.

9. G. K. C. Clark: Time-Varying Deconvolution Filters. Geophysics, Vol. 33 (1968) pp. 936-944.

10. C. W. Helstrom: Image Restoration by the Method of Least Squares. JOSA, Vol. 57 (March 1967) pp. 297-303.

11. M. J. Levin: Optimum Estimation of the Impulse Response in the Presence of Noise. IRE Trans. on Circuit Theory (March 1960) pp. 50-60.

12. K. L. Peacock and S. Trietel: Predictive Deconvolution: Theory and Practice. Geophysics, Vol. 27 (1962) pp. 4-18.

13. R. B. Rice: Inverse Convolution Filters. Geophysics, Vol. 27 (Feb. 1962) pp. 4-18.

14. B. R. Hunt: Deconvolution of Linear Systems by Constrained Regression and Its Relationship to the Wiener Theory. IEEE Trans. on Automatic Control, Oct. 1972.

15. D. L. Phillips: A Technique for the Numerical Solution of Certain Integral Equations of the First Kind. JACM, Vol. 9 (1962) pp. 84-97.

16. S. Twomey: On the Numerical Solution of the Fredholm Equations of the First Kind by the Inversion of the Linear System Produced by Quadrature. JACM, Vol. 10 (1963) pp. 97-101.

17. S. Twomey: The Application of Numerical Filtering to the Solution of Integral Equations Encountered in Indirect Sensing Measurements. J. Franklin Institute, Vol. 279 (Feb. 1965) pp. 95-109.

18. J. W. Bayless and E. D. Brigham: Application of the Kalman Filter to Continuous Signal Restoration. Geophysics, Vol. 35 (1970) pp. 2-23.

19. N. D. Crump: A Kalman Filter Approach to the Deconvolution of Seismic Signals. Geophysics, Vol. 27 (Feb. 1962) pp. 4-18.

20. N. E. Nahi: Estimation Theory and Applications, New York, Wiley (1969).

21. A. Rosenfeld and A. C. Kak: Digital Picture Processing, New York, Academic Press (1975).

22. T. J. Ulrych: Application of Homomorphic Deconvolution to Seismology. Geophysics, Vol. 36 (Aug., 1971) pp. 650-660.

23. G. Backus and F. Gilbert: Uniqueness in the Inversion of Inaccurate Gross Earth Data. Trans. of the Phil. Soc., Vol. 266 (March, 1970) pp. 123-192.

24. T. E. Reimer and C. D. McGillem: Constrained Optimization of Image Restoration Filters. Applied Optics, Vol. 12 (Sept. 1973) pp. 27-29.

25. K. R. Erikson, F. J. Fry and J. P. Jones: Ultrasound in Medicine: A Review. IEEE Trans. on Sonics and Ultrasonics, Vol. SU-21 (July, 1974).

26. H. L. Van Trees: Detection, Estimation and Modulation Theory, New York, John Wiley (1968).

INDEX TO PART I